METHODS IN MOLECULAR BIOLOGY™

Series Editor
John M. Walker
School of Life Sciences
University of Hertfordshire
Hatfield, Hertfordshire, AL10 9AB, UK

For other titles published in this series, go to
www.springer.com/series/7651

Antiviral RNAi

Concepts, Methods, and Applications

Edited by

Ronald P. van Rij

Department of Medical Microbiology, Radboud University Nijmegen Medical Centre, Nijmegen Centre for Molecular Life Sciences, Nijmegen Institute for Infection, Inflammation and Immunity, Nijmegen, The Netherlands

Humana Press

Editor
Dr. Ronald P. van Rij
Department of Medical Microbiology
Radboud University Nijmegen Medical Centre
Nijmegen Centre for Molecular Life Sciences
Nijmegen Institute for Infection, Inflammation and Immunity
Nijmegen
The Netherlands
r.vanrij@ncmls.ru.nl

ISSN 1064-3745 e-ISSN 1940-6029
ISBN 978-1-4939-5825-2 ISBN 978-1-61779-037-9 (eBook)
DOI 10.1007/978-1-61779-037-9
Springer New York Dordrecht Heidelberg London

Printed on acid-free paper

Humana Press is part of Springer Science+Business Media (www.springer.com)

Preface

Little over a decade ago, Andrew Fire, Craig Mello, and colleagues demonstrated that double-stranded (ds)RNA induces sequence-specific gene silencing in the nematode *Caenorhabditis elegans* (RNA interference, RNAi). This work converged with research in plants, in which related RNA-based silencing processes were known to exist. Ever since, research in the field has progressed at an astonishing rate, resulting in our appreciation of small silencing RNAs as central regulators of gene expression, as guards of genome integrity, and as essential mediators of antiviral defense. The discovery that synthetic small interfering RNA (siRNA) induces gene silencing in mammals, by Thomas Tuschl and colleagues in 2001, has further boosted the development of novel therapeutics and experimental tools based on RNAi technology.

Viruses and RNAi share an intricate relationship at many levels. Early work in plants indicated that viruses can be both inducers and targets of RNA-based post-transcriptional gene silencing (which we now know as RNAi or RNA silencing). The concept of RNAi as an antiviral defense mechanism is now well-established in plants and other organisms, including insects. In vertebrates, viruses also interact with a related RNA silencing mechanism, the microRNA (miRNA) pathway. Many nuclear DNA viruses encode their own set of miRNAs, by which they regulate viral or host gene expression and modify, for example, the transition from latent to lytic infection and the recognition of infected cells by the host immune system. Furthermore, cellular miRNAs likely regulate expression of many genes that are important for virus biology, but they have been suggested to directly target viral RNA as well.

The therapeutic potential of RNAi-based antiviral drugs was recognized early on. It is now clear that replication of many, if not all, mammalian viruses can be suppressed by RNAi in cell culture. While these results have raised considerable optimism about the potential of RNAi-based drugs, important hurdles remain, including issues related to the delivery and stability of siRNAs and the risk of viral escape.

From this brief overview it will be apparent that a great — and increasing — number of tools and techniques are available for those interested in the interface of viruses and RNAi. *Antiviral RNAi: concepts, methods, and applications* provides a collection of protocols for the analysis of natural antiviral RNAi responses and viral miRNAs, as well as for the development and optimization of RNAi-based antiviral drugs. As RNAi is a central regulatory mechanism in the cell, the methods in this volume can also be applied out of the context of a virus infection. In the established tradition of the *Methods in Molecular Biology* series, *Antiviral RNAi: concepts, methods, and applications* provides detailed step-by-step protocols and extra tools and tricks that should be useful to those new to the field and experienced scientists alike.

This volume consists of five parts. Part 1 reviews important basic concepts in the field of antiviral RNAi. Part 2 provides experimental and bio-informatic tools for the analysis of small silencing RNAs. Part 3 covers methods to biochemically dissect RNAi-based antiviral defense and viral counter-defense mechanisms. Part 4 describes methods for the design, expression, and delivery of therapeutic antiviral siRNAs. Part 5 presents genome-wide

RNAi approaches for the identification of factors involved in virus replication, which may represent novel targets for antiviral therapy.

I am grateful to all authors for providing their outstanding contributions and to John Walker for guidance while editing this volume. I thank members of my lab, especially Walter Bronkhorst, Koen van Cleef, Marius van den Beek, and Joël van Mierlo, for discussions. I am thankful to Raul Andino for having been a great mentor and for introducing me to this exciting field of research. Finally, I would like to apologize for doing little justice to the seminal work in plants; space limitations forced me to focus this volume on the animal system.

Nijmegen, The Netherlands ***Ronald P. van Rij***

Contents

Preface . . . . *v*
Contributors . . . . *ix*

PART I GENERAL INTRODUCTION

1 Defense and Counterdefense in the RNAi-Based Antiviral Immune System in Insects 3
Joël T. van Mierlo, Koen W.R. van Cleef, and Ronald P. van Rij

2 RNAi and Cellular miRNAs in Infections by Mammalian Viruses 23
Joost Haasnoot and Ben Berkhout

3 Viral miRNAs 43
Karlie Plaisance-Bonstaff and Rolf Renne

4 Progress in RNAi-Based Antiviral Therapeutics 67
Jiehua Zhou and John J. Rossi

5 Chemical Modification of Small Interfering RNA 77
Jesper B. Bramsen and Jørgen Kjems

PART II VIRAL SMALL RNAS

6 Viral Small RNA Cloning and Sequencing 107
Valérie Gausson and Maria-Carla Saleh

7 Visitor, An Informatic Pipeline for Analysis of Viral siRNA Sequencing Datasets 123
Christophe Antoniewski

8 Computational Prediction of Viral miRNAs 143
Adam Grundhoff

9 Detection of Viral microRNAs by Northern Blot Analysis 153
Lydia V. McClure, Yao-Tang Lin, and Christopher S. Sullivan

10 Detection of Viral microRNA with S1 Nuclease Protection Assay 173
Matthias John and Sébastien Pfeffer

11 Characterization of RISC-Associated Adenoviral Small RNAs 183
Ning Xu and Göran Akusjärvi

PART III RNAI-BASED ANTIVIRAL DEFENSE

12 Identification of Viral Suppressors of RNAi by a Reporter Assay in *Drosophila* S2 Cell Culture 201
Koen W.R. van Cleef, Joël T. van Mierlo, Marius van den Beek, and Ronald P. van Rij

13 Dicer Assay in Drosophila S2 Cell Extract 215
Baojun Yang and Hongwei Li

14 Slicer Activity in *Drosophila melanogaster* S2 Extract 231
Arabinda Nayak and Raul Andino

15 Gel Mobility Shift Assays for RNA Binding Viral RNAi Suppressors 245
Tibor Csorba and József Burgyán

16 dsRNA Uptake in Adult *Drosophila* 253
Benjamin Obadia and Maria-Carla Saleh

PART IV ANTIVIRAL RNAI THERAPY

17 Design of Small Interfering RNAs for Antiviral Applications 267
Diana Rothe, Erik J. Wade, and Jens Kurreck

18 RNAi-Inducing Lentiviral Vectors for Anti-HIV-1 Gene Therapy 293
Ying Poi Liu, Jan-Tinus Westerink, Olivier ter Brake, and Ben Berkhout

19 Production of Multicopy shRNA Lentiviral Vectors for Antiviral Therapy 313
Scot D. Henry, Quiwei Pan, and Luc J.W. van der Laan

20 Intranasal Delivery of Antiviral siRNA 333
Sailen Barik

21 Antibody-Mediated Delivery of siRNAs for Anti-HIV Therapy 339
Sang-Soo Kim, Sandesh Subramanya, Dan Peer, Motomu Shimaoka, and Premlata Shankar

22 Aptamer-Targeted RNAi for HIV-1 Therapy 355
Jiehua Zhou and John J. Rossi

PART V RNAI SCREENS TO STUDY VIRUS–HOST INTERACTIONS

23 RNAi Screening for Host Factors Involved in Viral Infection Using *Drosophila* Cells 375
Sara Cherry

24 Genome-Wide RNAi Screen for Viral Replication in Mammalian Cell Culture 383
Bhupesh K. Prusty, Alexander Karlas, Thomas F. Meyer, and Thomas Rudel

25 RNAi Screening in Mammalian Cells to Identify Novel Host Cell Molecules Involved in the Regulation of Viral Infections 397
Carolyn B. Coyne and Sara Cherry

Erratum E1

Index *407*

Contributors

GÖRAN AKUSJÄRVI • *Department of Medical Biochemistry and Microbiology, Uppsala Biomedical Center (BMC), Uppsala, Sweden*
RAUL ANDINO • *Department of Microbiology and Immunology, University of California, San Francisco, CA, USA*
CHRISTOPHE ANTONIEWSKI • *Drosophila Genetics and Epigenetics, Institut Pasteur, CNRS URA 2578, Paris, France*
SAILEN BARIK • *Center for Gene Regulation in Health and Disease, Cleveland State University, Cleveland, OH, USA*
BEN BERKHOUT • *Laboratory of Experimental Virology, Department of Medical Microbiology, Center for Infection and Immunity Amsterdam (CINIMA), Academic Medical Center, University of Amsterdam, Amsterdam, The Netherlands*
JESPER B. BRAMSEN • *Department of Molecular Biology, Interdisciplinary Nanoscience Center (iNANO), University of Aarhus, Aarhus, Denmark*
JÓZSEF BURGYÁN • *Istituto di Virologia Vegetale, Consiglio Nazionale dell Ricerche, Torino, Italy*
SARA CHERRY • *Department of Microbiology, Penn Genome Frontiers Institute, University of Pennsylvania, Philadelphia, PA, USA*
CAROLYN B. COYNE • *Department of Microbiology and Molecular Genetics, University of Pittsburgh, Pittsburgh, PA, USA*
TIBOR CSORBA • *Agricultural Biotechnology Center, Gödöllö, Hungary*
VALÉRIE GAUSSON • *Institut Pasteur, Viruses and RNAi group, CNRS URA 3015, Paris, France*
ADAM GRUNDHOFF • *Heinrich-Pette-Institute for Experimental Virology and Immunology, Hamburg, Germany*
JOOST HAASNOOT • *Laboratory of Experimental Virology, Department of Medical Microbiology, Center for Infection and Immunity Amsterdam (CINIMA), Academic Medical Center, University of Amsterdam, Amsterdam, The Netherlands*
SCOT D. HENRY • *Department of Surgery, Erasmus MC – University Medical Center, Rotterdam, The Netherlands*
MATTHIAS JOHN • *Roche Kulmbach GmbH, Kulmbach, Germany*
ALEXANDER KARLAS • *Department of Molecular Biology, Max Planck Institute for Infection Biology, Berlin, Germany*
SANG-SOO KIM • *Department of Biomedical Sciences, Center of Excellence for Infectious Diseases, Paul L. Foster School of Medicine, Texas Tech University Health Sciences Center, El Paso, TX, USA*
JØRGEN KJEMS • *Department of Molecular Biology, Interdisciplinary Nanoscience Center (iNANO), University of Aarhus, Aarhus, Denmark*

JENS KURRECK • *Institute of Biotechnology, University of Technology Berlin, Berlin, Germany*
HONGWEI LI • *Department of Microbiology, University of Hawaii at Manoa, Honolulu, HI, USA*
YAO-TANG LIN • *Department of Molecular Genetics & Microbiology, The University of Texas at Austin, Austin, TX, USA*
YING POI LIU • *Laboratory of Experimental Virology, Department of Medical Microbiology, Center for Infection and Immunity Amsterdam (CINIMA), Academic Medical Center, University of Amsterdam, Amsterdam, The Netherlands*
LYDIA V. MCCLURE • *Department of Molecular Genetics & Microbiology, The University of Texas at Austin, Austin, TX, USA*
THOMAS F. MEYER • *Department of Molecular Biology, Max Planck Institute for Infection Biology, Berlin, Germany*
ARABINDA NAYAK • *Department of Microbiology and Immunology, University of California, San Francisco, CA, USA*
BENJAMIN OBADIA • *Institut Pasteur, Viruses and RNAi group, CNRS URA 3015, Paris, France*
QUIWEI PAN • *Department of Gastroenterology & Hepatology, Erasmus MC – University Medical Center, Rotterdam, The Netherlands*
DAN PEER • *Laboratory of Nanomedicine, Department of Cell Research and Immunology, George S. Wise Faculty of Life Sciences, Center for Nanoscience and Nanotechnology, Tel Aviv University, Tel Aviv, Israel*
SÉBASTIEN PFEFFER • *Architecture et Réactivité de l'ARN, Institut de Biologie Moléculaire et Cellulaire du CNRS, Université de Strasbourg, Strasbourg, France*
KARLIE PLAISANCE-BONSTAFF • *Department of Molecular Genetics and Microbiology, University of Florida Shands Cancer Center, University of Florida, Gainesville, FL, USA*
BHUPESH K. PRUSTY • *Department of Microbiology, Biocenter, University of Würzburg, Würzburg, Germany*
ROLF RENNE • *Department of Molecular Genetics and Microbiology, University of Florida Shands Cancer Center, University of Florida, Gainesville, FL, USA*
JOHN J. ROSSI • *Division of Molecular and Cellular Biology, Irell and Manella Graduate School of Biological Sciences, Beckman Research Institute of City of Hope, City of Hope, Duarte, CA, USA*
DIANA ROTHE • *Institute of Biotechnology, University of Technology Berlin, Berlin, Germany*
THOMAS RUDEL • *Department of Microbiology, Biocenter, University of Würzburg, Würzburg, Germany*
MARIA-CARLA SALEH • *Institut Pasteur, Viruses and RNAi group, CNRS URA 3015, Paris, France*
PREMLATA SHANKAR • *Department of Biomedical Sciences, Center of Excellence for Infectious Diseases, Paul L. Foster School of Medicine, Texas Tech University Health Sciences Center, El Paso, TX, USA*

Motomu Shimaoka • *Program in Cellular and Molecular Medicine, Children's Hospital Boston, Immune Disease Institute, Harvard Medical School, Boston, MA, USA*
Sandesh Subramanya • *Department of Biomedical Sciences, Center of Excellence for Infectious Diseases, Paul L. Foster School of Medicine, Texas Tech University Health Sciences Center, El Paso, TX, USA*
Christopher S. Sullivan • *Department of Molecular Genetics & Microbiology, The University of Texas at Austin, Austin, TX, USA*
Olivier ter Brake • *Laboratory of Experimental Virology, Department of Medical Microbiology, Center for Infection and Immunity Amsterdam (CINIMA), Academic Medical Center, University of Amsterdam, Amsterdam, The Netherlands*
Koen W.R. van Cleef • *Department of Medical Microbiology, Radboud University Nijmegen Medical Centre, Nijmegen Centre for Molecular Life Sciences, Nijmegen Institute for Infection, Inflammation and Immunity, Nijmegen, The Netherlands*
Marius van den Beek • *Department of Medical Microbiology, Radboud University Nijmegen Medical Centre, Nijmegen Centre for Molecular Life Sciences, Nijmegen Institute for Infection, Inflammation and Immunity, Nijmegen, The Netherlands*
Luc J.W. van der Laan • *Department of Surgery, Erasmus MC – University Medical Center, Rotterdam, The Netherlands*
Joël T. van Mierlo • *Department of Medical Microbiology, Radboud University Nijmegen Medical Centre, Nijmegen Centre for Molecular Life Sciences, Nijmegen Institute for Infection, Inflammation and Immunity, Nijmegen, The Netherlands*
Ronald P. van Rij • *Department of Medical Microbiology, Radboud University Nijmegen Medical Centre, Nijmegen Centre for Molecular Life Sciences, Nijmegen Institute for Infection, Inflammation and Immunity, Nijmegen, The Netherlands*
Erik J. Wade • *Sealed BioBusiness Consulting, Weil am Rhein, Germany*
Jan-Tinus Westerink • *Laboratory of Experimental Virology, Department of Medical Microbiology, Center for Infection and Immunity Amsterdam (CINIMA), Academic Medical Center, University of Amsterdam, Amsterdam, The Netherlands*
Ning Xu • *Department of Medical Biochemistry and Microbiology, Uppsala Biomedical Center (BMC), Uppsala, Sweden*
Baojun Yang • *Department of Microbiology, University of Hawaii at Manoa, Honolulu, HI, USA*
Jiehua Zhou • *Division of Molecular and Cellular Biology, Beckman Research Institute of City of Hope, City of Hope, Duarte, CA, USA*

Part I

General Introduction

Chapter 1

Defense and Counterdefense in the RNAi-Based Antiviral Immune System in Insects

Joël T. van Mierlo, Koen W.R. van Cleef, and Ronald P. van Rij

Abstract

RNA interference (RNAi) is an important pathway to combat virus infections in insects and plants. Hallmarks of antiviral RNAi in these organisms are: (1) an increase in virus replication after inactivation of major actors in the RNAi pathway, (2) production of virus-derived small interfering RNAs (v-siRNAs), and (3) suppression of RNAi by dedicated viral proteins. In this chapter, we will review the mechanism of RNAi in insects, its function as an antiviral immune system, viral small RNA profiles, and viral counterdefense strategies. We will also consider alternative, inducible antiviral immune responses.

Key words: Viral suppressor of RNAi (VSR), RNA silencing, Insect antiviral immunity, Small RNA profiles, Insect virus

1. Introduction

Invertebrates lack the innate and adaptive immune responses that mediate antiviral defense in vertebrate animals. Nevertheless, they are able to effectively fight viral infections, suggesting that they rely on alternative mechanisms for antiviral defense. Indeed, in both plants and invertebrates, the RNA interference (RNAi) pathway serves as an antiviral defense system (1, 2). The presence of double-stranded RNA (dsRNA), which is absent from uninfected cells (3), is sensed as a danger signal that triggers an antiviral RNAi response. Viral dsRNA is processed by an RNase of the Dicer family into 21-nt small interfering RNAs (siRNAs). These viral siRNAs (v-siRNAs) are incorporated in an Argonaute (Ago)-containing RNA-induced silencing complex (RISC), where they guide the recognition and cleavage of viral target RNAs and thereby restrict viral replication. In this chapter, we review the

Ronald P. van Rij (ed.), *Antiviral RNAi: Concepts, Methods, and Applications*, Methods in Molecular Biology, vol. 721,
DOI 10.1007/978-1-61779-037-9_1,

RNAi-based antiviral immune system and viral counterdefense strategies in insects. We will start with an overview of the mechanism of RNAi in *Drosophila melanogaster* (for more details, see refs. (4, 5)).

2. Mechanism of RNAi

The RNAi pathway starts with processing of long dsRNA into siRNAs by the ribonuclease Dicer-2 (Dcr-2) (6) (Fig. 1). Dicer proteins are type III members of the RNaseIII family that contain a DExD/H ATPase domain, a DUF283 domain, a Piwi/Argonaute/Zwille (PAZ) domain, two tandem RNaseIII domains, and a dsRNA-binding domain. The PAZ domain interacts with the terminus of long dsRNA, which is then positioned along the surface of the protein towards the processing center of Dicer (7). An intramolecular dimer of the two RNaseIII domains forms the processing center of the enzyme. Each RNaseIII domain cleaves one strand of the long dsRNA molecule, thereby generating 21-nt siRNAs with 2-nt 3′ overhangs and bearing characteristic 5′ monophosphate and 3′ hydroxyl moieties (8–10). The distance between the PAZ domain and the RNaseIII active sites determines the characteristic 21-nt size of siRNAs (10, 11). The DExD/H ATPase domain may convert ATP to provide energy required for dsRNA cleavage. Dcr-2 activity in *Drosophila* extracts is indeed enhanced by addition of ATP; the activity of human Dicer, however, is not (12, 13). Remarkably, the DExD/H ATPase domain of Dcr-2 is also implicated as a sensor for viral dsRNA in an alternative antiviral defense pathway (see Subheading 7). No functions have thus far been assigned to the other Dcr domains. Efficient processing of dsRNA by *Drosophila* Dcr-2 requires Loquacious isoform PD (Loqs-PD). Loqs-PD probably acts as an adaptor molecule that enhances the affinity of Dcr-2 for long dsRNA (14). A similar activity has been proposed for Arsenic resistance protein 2 (Ars2), which promotes the efficiency and fidelity of Dcr-2-mediated cleavage (15).

Following dsRNA cleavage, the resulting siRNA is bound by Dcr-2 and its dsRNA-binding protein partner R2D2, generating a RISC loading complex (RLC). The Dcr-2/R2D2 heterodimer will load the siRNA duplex into an Ago-2-containing RISC. Binding of R2D2 to the siRNA is enhanced by the phosphate group at the 5′ terminus of the siRNA, thereby ensuring that only authentic siRNAs are efficiently bound by the complex (12, 14, 16). Within RISC, the PAZ domain of Ago-2 binds the 3′ terminus of the strand of the siRNA duplex that will be retained in RISC (the guide strand), probably at the 2-nt overhang (17–19).

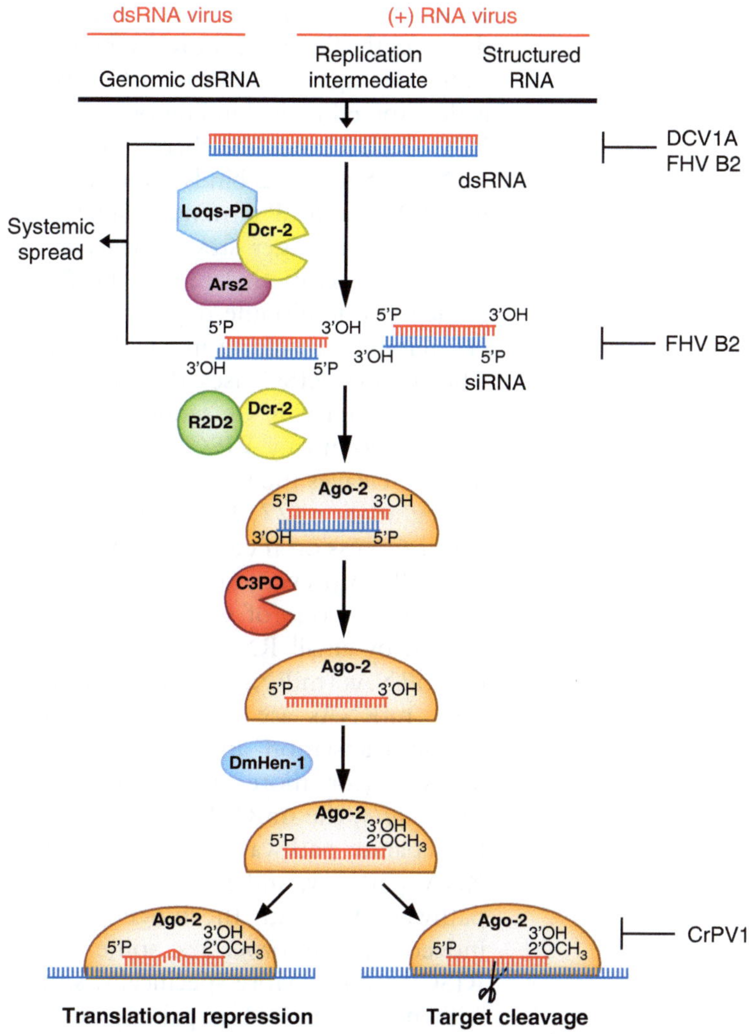

Fig. 1. Schematic overview of the RNAi pathway in insects. Viral dsRNA is processed by Dicer-2 (Dcr-2) into 21-nt small interfering RNAs (siRNAs), which are incorporated in an Argonaute-2 (Ago-2) containing RNA-induced silencing complex (RISC). Within RISC, these siRNAs guide the recognition and cleavage of viral target RNAs and thereby restrict viral replication (for details, see Subheading 2). Suppressors of RNAi from Cricket paralysis virus (CrPV 1A), Drosophila C virus (DCV 1A), and Flock House virus (FHV B2) interfere at different stages of the antiviral RNAi pathway (indicated on the *right*). Systemic spread of RNAi is thought to be essential for effective antiviral defense (52). The nature of the sequence specific silencing signal (dsRNA or siRNA) is still unknown. Dicer-2 was shown to interact both with Loquacious isoform PD (Loqs-PD) (14) and with Arsenic resistance protein-2 (Ars-2) (15). Please note, a trimeric complex (as indicated in this figure) has not been demonstrated.

The 5′ phosphate of the guide strand is bound in a pocket in the Ago-2 middle (Mid) domain (20, 21). Upon loading of the siRNA into RISC, the endonucleolytic activity of the Piwi domain of Ago-2 cleaves the phosphodiester bond between nucleotides 9 and 10 of the strand that will be excluded from

RISC (the passenger strand) (22–24). The adaptor molecule C3PO degrades the 9- and 12-nt RNA fragments that result from passenger strand cleavage (25). The fate of the two strands of an siRNA duplex is determined by its binding orientation to R2D2 and Dcr-2 in the RLC. R2D2 binds the thermodynamically most stable end of the siRNA, while Dcr-2 binds the other end. The strand that interacts with R2D2 at its 3′ terminus will become the guide strand (16).

After passenger strand cleavage and elimination, the guide strand is 2′-*O*-methylated at the 3′ terminal nucleotide by the *S*-adenosylmethionine-dependent methyltransferase Hen-1 (DmHen-1), resulting in the formation of a mature RISC (26). The mature RISC uses the incorporated guide strand to bind complementary RNA sequences. When a fully complementary RNA is bound by RISC, the RNase activity of the Ago-2 Piwi domain cleaves the target RNA (Slicer activity), thereby inducing its degradation (27, 28). Alternatively, Ago-2 may induce translational repression if central mismatches between guide strand and target RNA prevent Slicer activity (29).

RNAi is one of several gene silencing mechanisms that are guided by small RNAs. Another class of small silencing RNAs, microRNAs (miRNAs), is genomically encoded by all eukaryotic cells. miRNAs are central regulators of gene expression that inhibit translation and/or induce degradation of their target mRNAs (for more detail see ref. (30) and Chapter 3). In *Drosophila*, the miRNA and siRNA pathways were considered to be separate pathways, with Dcr-2 and Ago-2 dedicated to the RNAi pathway, and Dcr-1 and Ago-1 dedicated to the miRNA pathway. Although the biogenesis of siRNAs and miRNAs are indeed separate processes, some cross-talk occurs at the level of RISC loading. More specifically, some miRNAs, especially those with more extensive base pairing, can be loaded into Ago-2 (31, 32). The asymmetric miRNA duplex consists of the miRNA (miR) and miRNA star (miR*) strands. In analogy to the guide and passenger strands in the RNAi pathway, the mature miR strand is retained in Ago-1, while the miR* strand is released from the mature RISC complex. Recently, it was shown that both the miR and miR* strands are frequently loaded into Ago-2 by a *Dcr-2* and *R2D2*-dependent mechanism (31, 33). A third class of small silencing RNAs, Piwi-interacting RNAs (piRNAs), suppresses the activity of transposons in the germline (for more details, see Subheading 6 and ref. (34)).

3. RNAi as an Antiviral Defense Mechanism

Viral dsRNA forms a danger signal that is sensed and processed by Dcr-2, resulting in the production of viral siRNAs (v-siRNAs) and their incorporation into Ago-2/RISC. Since the Ago-2-associated guide strand is of viral origin, viral RNA will be the target of the RISC complex, enabling the RNAi pathway to exert antiviral activity.

Three major lines of evidence support a crucial role for the RNAi pathway in antiviral defense in insects. First, RNAi-deficient flies (*Dcr-2*, *R2D2*, or *Ago-2* null mutants) are more sensitive than wild-type flies to infection with several (+) strand RNA viruses, such as Drosophila C virus (DCV) (*Dicistroviridae*), Cricket paralysis virus (CrPV) (*Dicistroviridae*), and Flock House virus (FHV) (*Nodaviridae*), resulting in higher viral RNA copy numbers, higher viral titers, and increased mortality (35–38). *Ago-2* and *R2D2* mutants are also hypersensitive to the dsRNA virus Drosophila X virus (DXV) (*Birnaviridae*), but, surprisingly, *Dcr-2* mutants are not (38). The observation that *Ago-2* mutants (in which Dcr-2 is fully functional) are also hypersensitive to virus infection indicates that cleavage of viral dsRNAs by itself is not sufficient to control virus infection. RNAi also proved to be essential for controlling replication of several arthropod-borne (arbo-) viruses in mosquitoes. Knockdown of *Dcr-2* or *Ago-2* expression in *Aedes aegypti* results in higher viral titers after infection with either Dengue virus (DENV) (*Flaviviridae*) or Sindbis virus (SINV) (*Togaviridae*) (39, 40). Comparable results were obtained upon infections of *Anopheles gambiae* mosquitoes with O'nyong-nyong virus (ONNV) (*Togaviridae)* after *Ago-2* depletion (41). Second, the accumulation of v-siRNAs during virus infections in both *Drosophila* and mosquitoes provides direct evidence for processing of viral RNAs by Dcr-2 (42–48) (see Subheading 4). Third, insect viruses encode viral suppressors of RNAi (VSRs) as a counterdefense to the RNAi-based immune response (35, 48, 49) (see Subheading 5). VSRs allow viruses to replicate in the presence of a functional antiviral RNAi response, but they are dispensable for replication in RNAi mutants (36).

In addition to cell-autonomous silencing, activation of the RNAi pathway can lead to systemic silencing in plants. In the setting of a viral infection, noncell-autonomous RNAi may thus generate systemic protective immunity in noninfected tissues. Systemic RNAi in plants is a composite of short-range and long-range movement of a silencing signal. Movement of a silencing signal to distal sites in the plant requires amplification of the signal by a cellular RNA-dependent RNA polymerase (RdRP) (50). In contrast, local spread of RNAi to 10–15 neighboring cells is independent of RdRP activity. The mobile silencing molecules in plants were recently identified as siRNAs; it remains to be established whether they are protein-bound or not (51).

Virus infections in *Drosophila* also trigger a systemic RNAi response, which depends on a mechanism for active uptake of dsRNA (52). Accordingly, fly mutants with defects in dsRNA uptake are hypersensitive to virus infection. These observations suggest that, in addition to the cell-autonomous response, a systemic

RNAi response is essential for effective antiviral immunity in flies. The nature of the mobile silencing signal remains unknown. It was long thought that flies lack RdRP activity. Recently, however, RdRP activity was attributed to the Elongator complex protein 1 (Elp1), a subunit of the polymerase II core elongator complex (53). Whether this RdRP activity is required for systemic antiviral RNAi remains to be established.

4. Small RNA Profiles During Virus Infection

The antiviral RNAi pathway in plants and insects is triggered by viral dsRNA molecules which are processed by Dcr-2 into v-siRNAs. In theory, there are several potential sources of viral dsRNA that can serve as a substrate for Dcr-2. These sources include genomic dsRNA, structural RNA elements in the viral genome or in viral transcripts, dsRNA replication intermediates and convergent transcripts of viral genes. Recently, the introduction of massive parallel sequencing has given more insight into the origin of the v-siRNAs for several insect viruses (Table 1).

Table 1
Overview of insect viruses of which the v-siRNA profiles have been determined by massive parallel sequencing

Virus	Family	Genome	Host[1]	References
Drosophila A virus[2]	*Tetraviridae*	(+) RNA	*Drosophila*	(47)
Drosophila C virus	*Dicistroviridae*	(+) RNA	*Drosophila*	(47)
Noravirus	Unassigned	(+) RNA	*Drosophila*	(47)
Sindbis virus	*Alphaviridae*	(+) RNA	*Aedes aegypti*	(46)
Mosquito nodavirus	*Nodaviridae*	(+) RNA	*Aedes aegypti*	(47)
West Nile virus	*Flaviviridae*	(+) RNA	*Culex pipiens quinquefasciatus*	(43)
Flock House virus[3]	*Nodaviridae*	Bipartite (+) RNA	*Drosophila*	(2, 44, 45)
American nodavirus[3]	*Nodaviridae*	Bipartite (+) RNA	*Drosophila*	(47)
Drosophila totivirus	*Totiviridae*	dsRNA	*Drosophila*	(47)
Drosophila X virus	*Birnaviridae*	Bipartite dsRNA	*Drosophila*	(47)
Drosophila birnavirus	*Birnaviridae*	Bipartite dsRNA	*Drosophila*	(47)

[1]The v-siRNA profiles were derived from *Drosophila* cell lines and adult *Aedes aegypti* and *Culex pipiens quinquefasciatus* mosquitoes

[2]Drosophila A virus is described as Drosophila tetravirus by Wu et al. (47)

[3]Given the close similarity between Flock House Virus and American nodavirus at the nucleotide level, these viruses likely represent two variants of the same virus species within the *Nodaviridae* family

The first insect v-siRNA profiles came from a study designed to profile endogenous small RNAs from a *Drosophila* S2 cell line (54). In addition to endogenous small RNAs, a large subset of the Ago-2-associated siRNAs was found to match the FHV genome. These v-siRNAs most likely arose due to a persistent infection of the S2 cell line. Alignment of the available FHV-derived v-siRNAs to the viral genome indicated that they mapped in roughly equal proportions to both the positive (+) and negative (−) strand of the genome (2). Since FHV contains a (+) strand RNA genome, a bias towards the (+) strand of the genome would be expected if structural elements within the viral genome were the predominant source of v-siRNAs. This notion is inferred from the replication cycle of FHV. The FHV genome consists of RNA1 and RNA2, which encode the RNA-dependent RNA polymerase (A) and the capsid protein, respectively. A subgenomic RNA (RNA-3) encodes the VSR of FHV (B2). During FHV replication, complementary (−) strands are synthesized to generate dsRNA replication intermediates, which serve as templates for the production of new (+) RNA genomes. In infected cells, the (+) and (−) strands accumulate asymmetrically, with the genomic (+) strands being approximately 50–100-fold more prominent than the (−) strands (55). The equal distribution of the v-siRNAs over both the (+) and (−) strands of the FHV genome, therefore, suggests that dsRNA replication intermediates are the major substrates for v-siRNA biogenesis. It can, however, not be excluded that structural elements within the viral genome are processed by Dcr-2 and thereby contribute to the production of v-siRNAs. Interestingly, although the v-siRNAs mapped across the entire viral genome, there were a few specific hotspots from which the majority of the v-siRNAs were derived. These observations suggest that certain regions within the viral genome are more accessible to Dcr-2 than others. The occurrence of hotspots may be explained by different levels of the three viral RNAs as well as by the formation of stalled replication complexes, the production of defective interfering RNAs, and structural elements within the viral genome, which may all be potential substrates for Dcr-2 (2).

Similar observations were done in a second study in which S2 cells were abortively infected with a B2-deficient mutant of FHV (44). The B2-deficient FHV mutant lacks its RNAi suppressor, leading to a high abundance of v-siRNAs (see Subheading 5). Again, the v-siRNAs mapped in roughly equal proportions to both the (+) and (−) strands of the viral genome. These results underscore the notion that the FHV dsRNA replication intermediates are the main substrates for Dcr-2. Hotspots of v-siRNAs were also observed, but the distribution of these hotspots over the viral genome was surprisingly different from that observed by van Rij and Berezikov (2). This discrepancy might be attributed to the severe replication defects of the B2-deficient FHV mutant or, alternatively, to the different types of infection (acute vs. persistent) from which the v-siRNAs were sequenced. In a third study, the FHV small RNA profiles were determined in two persistently infected S2 cell lines (45). Once more, the replication intermediates were identified as the main substrates for Dcr-2. However, abundant FHV-derived v-siRNAs were not effective in silencing reporters that contain their target sites. Furthermore, although the FHV v-siRNAs were preferentially loaded into Ago-2, bulk v-siRNAs were unmethylated and did not associate with either Ago. These results suggest that Dcr-2 not only has a role in the biogenesis of v-siRNAs that exert their antiviral effect through Ago-2, but that dicing of the dsRNA replication intermediates itself also directly contributes to control virus replication, at least in these persistently infected cell lines.

Recently, Wu et al. showed that v-siRNAs are often overlapping in sequence and can be assembled into long continuous fragments (47). Based on this observation, they developed an approach for viral genome assembly and virus discovery, which enabled them to identify seven distinct RNA viruses in two *Drosophila* cell lines. Drosophila A virus (DAV) (likely *Tetraviridae*), American nodavirus (ANV) (*Nodaviridae*), Drosophila totivirus (DTV) (*Totiviridae*), DXV, and Drosophila birnavirus (DBV) (*Birnaviridae*) were identified in an S2 cell line, whereas DAV, DCV, Noravirus (unassigned), ANV, DXV, and DBV were found in an ovarian somatic sheet cell line (Table 1). Among these viruses, of which some had not been identified before, are four viruses with a (+) strand RNA genome (DAV, DCV, Noravirus and ANV) and three viruses with a dsRNA genome (DTV, DXV and DBV). The v-siRNAs mapped in similar proportions to both strands of the viral genome for all of the analyzed (+) strand RNA viruses in both cell lines, indicating that dsRNA replication intermediates are the major Dcr-2 substrates for (+) strand RNA viruses. Interestingly, for all of the analyzed dsRNA viruses in both cell lines, the v-siRNAs also mapped in roughly equal ratios to both strands of the viral genome. This result indicates that genomic dsRNA is the predominant substrate for Dcr-2 for dsRNA viruses.

The dsRNA replication intermediates of (+) strand RNA viruses seem to be a common target for Dcr-2 in insects, since their identification as Dcr-2 substrates is not restricted to viruses in *Drosophila*. More specifically, sequencing of v-siRNAs from SINV-infected *A. aegypti* as well as from West Nile virus (WNV, *Flaviviridae*)-infected *Culex pipiens quinquefasciatus* mosquitoes showed that the dsRNA replication intermediates of these arthropod-borne (+) strand RNA viruses are also a major target for Dcr-2 (43, 46). For both viruses, dicing of structural elements within the viral genomes may also contribute to the production of v-siRNAs, but these elements do not appear to be a predominant substrate for Dcr-2. Interestingly, in contrast to insect (+) strand RNA viruses, structural RNA elements within the viral genome have been identified as the main source for v-siRNA biogenesis for some plant (+) strand RNA viruses, such as Cymbidium ringspot virus (CymRSV) (*Tombusviridae*) (56). This discrepancy might reflect differences in substrate-specificity between insect Dcr-2 and the plant Dcr-like enzymes.

Taken together, v-siRNA profiling strongly suggests that the dsRNA replication intermediates of (+) strand RNA viruses and the dsRNA genomes of dsRNA viruses serve as important triggers for an antiviral RNAi response in insects. To prevent activation of the innate immune defenses of their hosts, viruses protect their dsRNAs from immune sensors. For example, (+) strand RNA viruses shield their dsRNA replication intermediates in virus-induced membrane vesicles, whereas the dsRNA genomes of dsRNA viruses are protected in viral cores (57). Nevertheless, the v-siRNA profiles clearly indicate that the viral dsRNAs are available for Dcr-2-mediated cleavage. This suggests that Dcr-2 is capable of protruding the compartments in which viral dsRNAs are shielded or that, at certain stages of the viral replication cycle, viral dsRNA is released into the cytoplasm where it is exposed to Dcr-2. Which of these scenarios holds true will be an interesting subject for future studies. Furthermore, determination of v-siRNA profiles of other viruses, such as those with (−) strand RNA and DNA genomes, will be of special interest as it will shed more light onto recognition of these viruses by the RNAi machinery.

5. Viral Suppression of RNAi

Despite the potent antiviral activity of the RNAi pathway in plants and insects, many viruses manage to persist in these organisms. Thus, viruses seem to be able to avoid recognition by the RNAi pathway or to counteract its antiviral activity. Indeed, plant and insect viruses encode VSRs that allow them to replicate in the presence of a potent RNAi-based antiviral immune response. Since dsRNA is recognized as a "nonself" immune activator, it bears little surprise that many viruses prevent detection of dsRNA by the immune system. Many VSRs bind dsRNA in a sequence-independent manner, thereby shielding it from Dicer (35, 58–60). Other VSRs, such as P19 from CymRSV, are able to specifically bind siRNAs (61). In biochemical assays, these VSRs sequester siRNAs and prevent their incorporation into RISC. The mechanism of RNAi suppression in viral infection, however, seems to be more complex. P19 is dispensable for virus accumulation in primary infected cells, but prevents cell-to-cell movement of the virus-induced silencing signal. Accordingly, in the absence of P19, the virus accumulates normally within vascular bundles, but is unable to establish systemic infection of the leaves (60, 62).

Binding of dsRNA or siRNA is a feature that is shared by many VSRs. Nevertheless, in analogy to the multitude of mechanisms by which mammalian viruses suppress innate and adaptive immunity, some plant and insect viruses suppress RNAi by a mechanism that is independent of dsRNA or siRNA binding. We will discuss these mechanisms in the next sections.

5.1. RNAi Suppressors Encoded by Plant Viruses

Although dsRNA binding appears to be a common mechanism to suppress RNAi, some VSRs employ other mechanisms to counteract the RNAi pathway (Table 2). These VSRs rely on protein–protein interaction with key components of the RNAi pathway. One example of such a VSR is the P0 protein of *Poleroviruses* (*Luteoviridae)* (63–65). Beet western yellows virus (BWYV) P0, for example, interacts with and induces degradation of AGO1, the main antiviral RISC component in plants. P0 contains an F-box motif, which is commonly found in proteins within the E3 ubiquitin ligase complex. This suggests that P0 induces ubiquitination and subsequent proteosomal degradation of AGO1 (66). Indeed, the VSR activity of P0 depends on an interaction with a protein from the E3 ubiquitin ligase complex. However, blocking proteosomal degradation did not prevent AGO1 degradation. These data suggest an ubiquitin-dependent, proteosomal-independent mechanism for P0 VSR activity.

Through an interaction with part of the PAZ and Piwi domains of AGO1, the Cucumber mosaic virus (CMV) (*Bromoviridae*) 2b protein is able to suppress Slicer activity of a preassembled RISC in vitro (67). In addition, CMV 2b binds

Table 2
Viral suppressors of RNAi (VSRs) of selected plant and insect viruses

Host	Virus	VSR	Mechanism of suppression	Reference
Plant	Cymbidium ringspot virus	P19	siRNA binding	(61)
	Pothos latent virus	P14	siRNA and dsRNA binding	(92)
	Beet western yellows virus	P0	Interaction with and degradation of AGO1	(63–65)
	Turnip crinckle virus	P38	GW motif-based interaction with AGO1	(71)
	Cucumber mosaic virus	2b	siRNA binding and AGO1 interaction, inhibition of Slicer activity	(67–69)
Insect	Flock House virus	B2	siRNA and dsRNA binding	(48, 72)
	Drosophila C virus	1A	dsRNA binding	(35)
	Cricket paralysis virus	1A	Interaction with Ago-2, inhibition of Slicer activity	(49)

small RNAs in vitro, which is suggested to contribute to VSR activity (68, 69). Which activity predominates during an authentic virus infection remains to be established.

The P38 capsid protein of Turnip crinkle virus (TCV) (*Tombusviridae*) employs yet another mechanism of RNAi suppression. P38 contains two glycine-tryptophane (GW) motifs. Different cellular proteins use linear GW or WG motifs as an "Ago hook" for functional interactions with Ago proteins (70). P38 mimics this cellular GW motif-based interaction; its GW motifs allow P38 to bind AGO1 in vitro and to suppress RNAi in vivo (71).

5.2. RNAi Suppressors Encoded by Insect Viruses

Plant viruses are extensively studied and most, if not all, seem to encode a VSR. In contrast, few VSRs have thus far been identified and characterized in insect viruses. The B2 protein of FHV was the first VSR identified in an invertebrate virus (48). Homodimers of B2 bind to dsRNA independent of sequence and length. Indeed, viral dsRNA replication intermediates coimmunoprecipitate with B2 from FHV-infected S2 cells, confirming an interaction of B2 with dsRNA in vivo (44). B2 seems to exert a dual mode of RNAi suppression. Binding of long dsRNA inhibits Dcr-2 cleavage; binding of siRNAs prevents their incorporation into Ago-2 (48, 72). The VSR activity of the DCV 1A protein also depends on dsRNA binding. As a member of the *Dicistroviridae*, DCV encodes two polyproteins from two distinct open reading frames. The first open reading frame, ORF1, encodes the nonstructural proteins, whereas ORF2 encodes the viral capsid proteins (73). DCV 1A maps to the N-terminal part of ORF1. In contrast to FHV B2, DCV 1A binds long dsRNA, but not siRNAs, with high affinity in vitro, thereby inhibiting Dcr-2 cleavage of dsRNA (35). Therefore, DCV 1A likely binds the viral replication intermediate to prevent its degradation by Dcr-2. Protection of the viral dsRNA from Dcr-2 cleavage by VSRs is, however, not complete, as indicated by the detection of v-siRNAs in FHV and DCV infection (see Subheading 4).

CrPV is the closest relative of DCV within the *Dicistroviridae* family. The VSR of CrPV, 1A, maps to the same genomic location as DCV 1A. Interestingly, whereas CrPV and DCV share a high degree of protein sequence identity within ORF1 (~55%), the ORF1 N-terminal region that contains the VSRs is not well conserved. It is, therefore, no surprise that CrPV 1A suppresses RNAi through a different mechanism as DCV 1A. CrPV 1A directly interacts with Ago-2, without affecting RISC assembly or stability. Since CrPV 1A is able to inhibit the activity of a preassembled RISC, it most likely interferes with the Slicer activity of Ago-2 (49).

The different mechanisms and potencies of RNAi suppression might explain the difference in pathogenicity between DCV

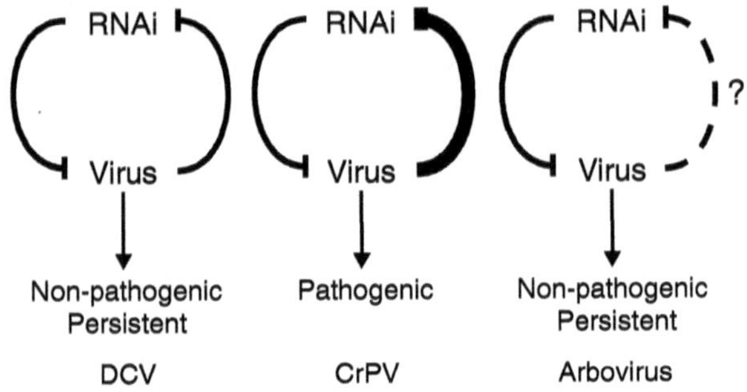

Fig. 2. Defense and counterdefense in virus infection of insects. An antiviral RNAi response restricts virus replication, whereas viruses suppress the antiviral RNAi response via dedicated RNAi suppressor proteins (VSRs). VSR activity may be an important determinant of viral pathogenicity. A potent VSR may render a virus pathogenic to its host (for example, Cricket paralysis virus, CrPV). Absent or mild VSR activity may result in non-pathogenic persistent infections, such as observed in natural Drosophila C virus (DCV) infections or in arbovirus infections in their invertebrate vector.

and CrPV. DCV establishes a nonlethal, persistent infection, whereas CrPV causes high mortality (73, 74). This hypothesis was supported by expression of DCV 1A or CrPV 1A from a recombinant SINV, which is thought not to suppress RNAi by itself (75, 76). In the adult fly, SINV-CrPV 1A replicates to higher titers and causes higher mortality than SINV-DCV 1A, which correlates with a more potent RNAi suppression by CrPV 1A. Thus, VSRs may be important determinants of viral pathogenicity (Fig. 2).

5.3. Do Arboviruses Encode RNAi Suppressors?

Arboviruses are maintained in a cycle that requires transmission by hematophagous arthropod vectors (mainly mosquitoes and ticks) to vertebrate hosts. After ingestion of a virus-infected blood meal, arboviruses replicate in the midgut epithelium of their vector. The virus then spreads through the hemolymph into the salivary glands for further amplification and transmission to a naive vertebrate host. Several arboviruses, including DENV, SINV, and ONNV, were shown to be suppressed by RNAi in their mosquito vectors (see Subheading 3 and refs. (39–41)). Nevertheless, while VSR activity seems to be a common feature in (non-arboviral) insect viruses, suppressors of RNAi have, thus far, not been identified in arboviruses (75, 77, 78).

Arbovirus infection in the insect vector is typically persistent and nonpathogenic. The mosquito's RNAi pathway thus restricts arbovirus replication, but is unable to fully clear these viruses. Since arboviruses rely on survival of their mosquito vector for transmission from host to host, the nonpathogenic phenotype is of key importance for efficient virus spread. Hence, an arbovirus that effectively kills its vector will be selected against in nature, since it would decrease the chance of spread to vertebrate hosts.

Encoding a potent VSR might therefore be deleterious for the survival of arboviruses (Fig. 2). In support of this hypothesis, recombinant SINV expressing the FHV B2 VSR replicates to higher midgut titers, eventually killing the infected mosquito (46, 76).

6. Other Viral Small Silencing RNAs

While analyzing v-siRNAs in a cell line derived from the ovarian somatic sheet (OSS), Wu et al. (47) made a remarkable observation. In addition to v-siRNAs, high levels of viral piRNAs were identified for two viruses: DCV and ANV (47). Although less abundant, viral piRNAs were also identified for other viruses that persist in this cell line. These observations suggest that viruses may be targeted by both the siRNA and piRNA pathways.

piRNAs are small RNAs of ~24–32 nucleotides that interact with members of the Piwi subfamily of the Ago family (34, 79). The Piwi subfamily comprises Piwi, Aubergine (Aub), and Ago-3, which are predominantly expressed in germline tissues. In these tissues, the piRNA pathway serves to prevent the activation of transposons. Their biogenesis is still obscure, but in contrast to siRNAs and miRNAs, piRNAs are generated in a Dicer-independent manner. They most likely arise from long single-stranded precursor RNAs which are derived from genomic regions of defective transposons that are transcribed in an antisense direction. How these precursors are processed into the mature primary piRNAs is not clear. In the presence of a target transposon RNA, these primary piRNAs engage in an amplification loop, known as the ping-pong mechanism. According to the ping-pong model, precursor RNAs are cleaved into antisense primary piRNAs that are loaded into Aub and Piwi. The antisense piRNAs then target the transposon mRNAs which are subsequently cleaved into sense piRNAs that are loaded into Ago-3. These sense piRNAs direct the cleavage of more precursor RNAs, leading to a feedback loop that results in amplification of the initial pool of piRNAs.

The ping-pong model was deduced from the characteristic features of the piRNAs that are associated with the different Piwi proteins. First, piRNAs bound to Piwi and Aub are in general antisense to the transposon RNAs, whereas piRNAs bound to Ago-3 are most often of sense orientation. Second, Piwi- and Aub-associated piRNAs have a strong bias for a uridine at position one, whereas Ago-3-associated piRNAs have a strong preference for an adenine at position ten. Third, the first ten nucleotides of Piwi- and Aub-associated piRNAs are frequently complementary to Ago-3-associated piRNAs.

The viral piRNAs detected in the OSS cell line were 24–30 nucleotides in length with peaks at 27 and 28 nucleotides (47). The OSS cell line expresses *Piwi*, but not *Aub* and *Ago-3*. Primary piRNAs are therefore produced, but the lack of Aub and Ago-3 precludes ping-pong amplification (80). In agreement, viral piRNAs resemble primary piRNAs with a strong bias for a uridine at position one, but no enrichment for an adenine at position ten (80). Finally, the viral piRNAs were almost exclusively of (+) polarity for both (+) strand RNA and dsRNA viruses. These data suggest that the genomic RNAs of the (+) strand RNA viruses and the sense mRNAs of the dsRNA viruses are processed by the piRNA pathway.

Maintenance of germline integrity is of crucial importance to ensure the proper development of offspring. The germline-specificity of the piRNA pathway and the detection of viral piRNAs led to the hypothesis that the piRNA pathway may protect the germline from invasion by viruses (81). Additional studies are required to elucidate whether the piRNA pathway indeed mediates antiviral defense in insects. In this regard, it will also be interesting to explore whether there is a role for another pathway of small silencing RNAs, the miRNA pathway, during virus infections in insects (82, 83). Such a role for the miRNA pathway has been demonstrated in mammals, where viral replication can be modulated by a complex regulatory network in which both viral and cellular miRNAs can control gene expression of both virus and host (reviewed in Chapter 2).

7. Inducible Antiviral Responses

Although RNAi is a major antiviral mechanism in insects, other immune signaling pathways are also implicated in antiviral defense. The antimicrobial Toll pathway, for example, is suggested to play a role during infection with DXV (84). DXV infection induces the expression of antimicrobial peptides (AMPs) that are known to depend on the Toll and IMD pathways. Even though both pathways are activated during DXV infection, only Toll mutants are more sensitive to DXV. The Toll pathway is also involved in controlling DENV infection in *A. aegypti* mosquitoes (85). An antiviral role for the IMD pathway is suggested during infection with CrPV (86). Flies defective in the IMD pathway are more sensitive to virus infection and display higher viral loads. The differential antiviral function of Toll and IMD in these studies might be due to differences in viral genome organization and replication strategy, host species, and cell tropism in the host. More studies are required to elucidate the exact role of IMD and Toll in antiviral defense, as well as the viral activators and downstream antiviral effectors of these pathways.

A study of the global transcriptional response during DCV infection identified a third inducible antiviral pathway in *Drosophila*. Approximately 90 genes, which are not induced during microbial infections, are upregulated during DCV infection. Induction of one of these genes, *vir-1*, is dependent on the Jak-STAT pathway (87). In accordance, Jak-STAT-deficient flies are more sensitive to viral infection. Knockdown or overexpression of *vir-1*, however, did not affect viral infection. These results indicate that vir-1 is not an antiviral effector, and that other downstream effector molecules mediate the antiviral effect of the Jak-STAT pathway.

Another response to virus infection is the induction of a gene called *Vago* (88). Upon DCV and SINV infection, *Vago* is specifically induced in the fat body, a tissue in the fly abdomen with some functional equivalence to the mammalian liver. In accordance with a role for Vago in antiviral defense, DCV RNA accumulated to higher levels in the fat body of *Vago* mutants. Induction of *Vago* is independent of the Toll, IMD, and Jak-STAT pathways. Strikingly, *Vago* induction is abolished in fly mutants with a missense mutation in the DExD/H-box helicase domain of *Dcr-2*. Other components of the RNAi pathway are not required for induction of *Vago*. The requirement for *Dcr-2* implies that viral dsRNA is the inducer of *Vago* expression. Regulation of *Vago* expression, however, seems to be more complex; expression of a dsRNA hairpin, which is processed by Dcr-2 into functional siRNAs, was not sufficient to induce expression of *Vago*.

Though DCV and FHV are both able to infect and replicate in the fat body, only DCV induced *Vago* expression. An intriguing hypothesis to explain this discrepancy is that the VSR of FHV, B2, suppresses the induction of *Vago*. While both viruses encode a VSR that binds long dsRNA, the FHV B2 protein is able to interact with the PAZ domain of Dcr-2 (89). It would be of interest to investigate whether this interaction inhibits the induction of *Vago* expression. Taken together, these results imply that Dcr-2 plays a dual role in virus infection: it generates the specificity determinants (siRNAs) of the antiviral RNAi response, and it acts as a pattern recognition receptor that senses dsRNA and induces an antiviral program that includes *Vago* expression.

8. Summary and Concluding Remarks

Viruses and RNAi share an intricate relationship, in which the cellular RNAi machinery restricts virus replication and viruses suppress the antiviral RNAi response. In the most straightforward model, viruses produce dsRNAs that, after processing by Dicer into v-siRNAs, guide RISC to cleave viral target RNAs in a sequence specific manner. In recent years, it has become clear that the role of RNAi extends beyond its function in restricting virus replication in a cell-autonomous manner. A systemic RNAi response seems to be essential for mounting an effective antiviral response in adult flies. Furthermore, the RNAi machinery seems to play a critical role in regulating, and perhaps orchestrating, an inducible antiviral response in *Drosophila*. Thus, Dcr-2 not only generates the specificity determinants of the antiviral RNAi response, but also acts as a *bona fide* pattern recognition receptor for dsRNA that alerts the immune system to virus infection. We are also witnessing the detailed characterization of novel mechanisms of action of VSRs. Whereas many VSRs were thought to bind dsRNA or siRNA, thereby preventing production of siRNAs or their incorporation into RISC, it is now clear that several viruses are able to suppress other steps of the RNAi pathway. The advent of deep-sequencing technology has led to unanticipated insights into the biogenesis or v-siRNAs. Viral replication intermediates and genomic dsRNA seem to be the main target for Dcr-2 in infections with (+) strand RNA and dsRNA viruses, respectively. Deep-sequencing also led to the detection of viral piRNAs, implying a role for RNA silencing processes in protecting the germline from virus invasion. It has been more than a decade since the seminal observations, both in plants, that RNA silencing and antiviral defense share conserved features (90), and that small RNAs are produced in virus infection (91). Ever since, we continue to appreciate the complex and central role of RNAi in antiviral defense. We anticipate more surprises.

Acknowledgments

This work was financially supported by a fellowship from the Nijmegen Center for Molecular Life Sciences, by a VIDI fellowship from the Netherlands Organization for Scientific Research (project number 864.08.003), and by Horizon Breakthrough fellowships from the Netherlands Genomics Initiative (project numbers 93519018 and 93518020).

References

1. Voinnet, O. (2005) Induction and suppression of RNA silencing: insights from viral infections, *Nat Rev Genet* **6**, 206–20.
2. Van Rij, R. P. and Berezikov, E. (2009) Small RNAs and the control of transposons and viruses in *Drosophila*, *Trends Microbiol* **17**, 139–78.
3. Weber, F., Wagner, V., Rasmussen, S. B., Hartmann, R., and Paludan, S. R. (2006) Double-stranded RNA is produced by positive-strand RNA viruses and DNA viruses but not in detectable amounts by negative-strand RNA viruses, *J Virol* **80**, 5059–64.
4. Kawamata, T. and Tomari, Y. (2010) Making RISC, *Trends Biochem Sci.* **35**, 368–76.
5. Carthew, R. W. and Sontheimer, E. J. (2009) Origins and mechanisms of miRNAs and siRNAs, *Cell* **136**, 642–55.

6. Bernstein, E., Caudy, A. A., Hammond, S. M., and Hannon, G. J. (2001) Role for a bidentate ribonuclease in the initiation step of RNA interference, *Nature* **409**, 363–6.
7. Macrae, I. J., Zhou, K., and Doudna, J. A. (2007) Structural determinants of RNA recognition and cleavage by Dicer, *Nat Struct Mol Biol* **14**, 934–40.
8. Elbashir, S. M., Martinez, J., Patkaniowska, A., Lendeckel, W., and Tuschl, T. (2001) Functional anatomy of siRNAs for mediating efficient RNAi in *Drosophila melanogaster* embryo lysate, *EMBO J* **20**, 6877–88.
9. Elbashir, S. M., Lendeckel, W., and Tuschl, T. (2001) RNA interference is mediated by 21- and 22-nucleotide RNAs, *Genes Dev* **15**, 188–200.
10. Zhang, H., Kolb, F. A., Jaskiewicz, L., Westhof, E., and Filipowicz, W. (2004) Single processing center models for human Dicer and bacterial RNase III, *Cell* **118**, 57–68.
11. Macrae, I. J., Zhou, K., Li, F., Repic, A., Brooks, A. N., Cande, W. Z., Adams, P. D., and Doudna, J. A. (2006) Structural basis for double-stranded RNA processing by Dicer, *Science* **311**, 195–8.
12. Liu, Q., Rand, T. A., Kalidas, S., Du, F., Kim, H. E., Smith, D. P., and Wang, X. (2003) R2D2, a bridge between the initiation and effector steps of the *Drosophila* RNAi pathway, *Science* **301**, 1921–5.
13. Zhang, H., Kolb, F. A., Brondani, V., Billy, E., and Filipowicz, W. (2002) Human Dicer preferentially cleaves dsRNAs at their termini without a requirement for ATP, *EMBO J* **21**, 5875–85.
14. Marques, J. T., Kim, K., Wu, P. H., Alleyne, T. M., Jafari, N., and Carthew, R. W. (2010) Loqs and R2D2 act sequentially in the siRNA pathway in *Drosophila*, *Nat Struct Mol Biol* **17**, 24–30
15. Sabin, L. R., Zhou, R., Gruber, J. J., Lukinova, N., Bambina, S., Berman, A., Lau, C. K., Thompson, C. B., and Cherry, S. (2009) Ars2 regulates both miRNA- and siRNA- dependent silencing and suppresses RNA virus infection in *Drosophila*, *Cell* **138**, 340–51.
16. Tomari, Y., Matranga, C., Haley, B., Martinez, N., and Zamore, P. D. (2004) A protein sensor for siRNA asymmetry, *Science* **306**, 1377–80.
17. Lingel, A., Simon, B., Izaurralde, E., and Sattler, M. (2003) Structure and nucleic-acid binding of the *Drosophila* Argonaute 2 PAZ domain, *Nature* **426**, 465–9.
18. Lingel, A., Simon, B., Izaurralde, E., and Sattler, M. (2004) Nucleic acid 3′-end recognition by the Argonaute2 PAZ domain, *Nat Struct Mol Biol* **11**, 576–7.
19. Ma, J. B., Ye, K., and Patel, D. J. (2004) Structural basis for overhang-specific small interfering RNA recognition by the PAZ domain, *Nature* **429**, 318–322.
20. Boland, A., Tritschler, F., Heimstadt, S., Izaurralde, E., and Weichenrieder, O. (2010) Crystal structure and ligand binding of the MID domain of a eukaryotic Argonaute protein, *EMBO Rep* **11**, 522–7.
21. Wang, Y., Sheng, G., Juranek, S., Tuschl, T., and Patel, D. J. (2008) Structure of the guide-strand-containing argonaute silencing complex, *Nature* **456**, 209–13.
22. Matranga, C., Tomari, Y., Shin, C., Bartel, D. P., and Zamore, P. D. (2005) Passenger-strand cleavage facilitates assembly of siRNA into Ago2-containing RNAi enzyme complexes, *Cell* **123**, 607–20.
23. Rand, T. A., Petersen, S., Du, F., and Wang, X. (2005) Argonaute2 cleaves the anti-guide strand of siRNA during RISC activation, *Cell* **123**, 621–9.
24. Miyoshi, K., Tsukumo, H., Nagami, T., Siomi, H., and Siomi, M. C. (2005) Slicer function of *Drosophila* Argonautes and its involvement in RISC formation, *Genes Dev* **19**, 2837–48.
25. Liu, Y., Ye, X., Jiang, F., Liang, C., Chen, D., Peng, J., Kinch, L. N., Grishin, N. V., and Liu, Q. (2009) C3PO, an endoribonuclease that promotes RNAi by facilitating RISC activation, *Science* **325**, 750–3.
26. Horwich, M. D., Li, C., Matranga, C., Vagin, V., Farley, G., Wang, P., and Zamore, P. D. (2007) The *Drosophila* RNA methyltransferase, DmHen1, modifies germline piRNAs and single-stranded siRNAs in RISC, *Curr Biol* **17**, 1265–72.
27. Okamura, K., Ishizuka, A., Siomi, H., and Siomi, M. C. (2004) Distinct roles for Argonaute proteins in small RNA-directed RNA cleavage pathways, *Genes Dev* **18**, 1655–66.
28. Rand, T. A., Ginalski, K., Grishin, N. V., and Wang, X. (2004) Biochemical identification of Argonaute 2 as the sole protein required for RNA-induced silencing complex activity, *Proc Natl Acad Sci USA* **101**, 14385–9.
29. Iwasaki, S., Kawamata, T., and Tomari, Y. (2009) *Drosophila* argonaute1 and argonaute2 employ distinct mechanisms for translational repression, *Mol Cell* **34**, 58–67.
30. Fabian, M. R., Sonenberg, N., and Filipowicz, W. (2010) Regulation of mRNA translation and stability by microRNAs, *Annu Rev Biochem* **79**, 351–79.
31. Ghildiyal, M., Xu, J., Seitz, H., Weng, Z., and Zamore, P. D. (2010) Sorting of *Drosophila*

small silencing RNAs partitions microRNA* strands into the RNA interference pathway, *RNA* **16**, 43–56.

32. Okamura, K., Liu, N., and Lai, E. C. (2009) Distinct mechanisms for microRNA strand selection by *Drosophila* Argonautes, *Mol Cell* **36**, 431–44.
33. Czech, B., Zhou, R., Erlich, Y., Brennecke, J., Binari, R., Villalta, C., Gordon, A., Perrimon, N., and Hannon, G. J. (2009) Hierarchical rules for Argonaute loading in *Drosophila*, *Mol Cell* **36**, 445–56.
34. Thomson, T. and Lin, H. (2009) The biogenesis and function of PIWI proteins and piRNAs: progress and prospect, *Annu Rev Cell Dev Biol* **25**, 355–76.
35. Van Rij, R. P., Saleh, M. C., Berry, B., Foo, C., Houk, A., Antoniewski, C., and Andino, R. (2006) The RNA silencing endonuclease Argonaute 2 mediates specific antiviral immunity in *Drosophila melanogaster*, *Genes Dev* **20**, 2985–95.
36. Wang, X. H., Aliyari, R., Li, W. X., Li, H. W., Kim, K., Carthew, R., Atkinson, P., and Ding, S. W. (2006) RNA interference directs innate immunity against viruses in adult *Drosophila*, *Science* **312**, 452–4.
37. Galiana-Arnoux, D., Dostert, C., Schneemann, A., Hoffmann, J. A., and Imler, J. L. (2006) Essential function in vivo for Dicer-2 in host defense against RNA viruses in *Drosophila*, *Nat Immunol* 7, 590–7.
38. Zambon, R. A., Vakharia, V. N., and Wu, L. P. (2006) RNAi is an antiviral immune response against a dsRNA virus in *Drosophila melanogaster*, *Cell Microbiol* **8**, 880–9.
39. Campbell, C. L., Keene, K. M., Brackney, D. E., Olson, K. E., Blair, C. D., Wilusz, J., and Foy, B. D. (2008) *Aedes aegypti* uses RNA interference in defense against Sindbis virus infection, *BMC Microbiol* **8**, 47.
40. Sanchez-Vargas, I., Scott, J. C., Poole-Smith, B. K., Franz, A. W., Barbosa-Solomieu, V., Wilusz, J., Olson, K. E., and Blair, C. D. (2009) Dengue virus type 2 infections of *Aedes aegypti* are modulated by the mosquito's RNA interference pathway, *PLoS Pathog* **5**, e1000299.
41. Keene, K. M., Foy, B. D., Sanchez-Vargas, I., Beaty, B. J., Blair, C. D., and Olson, K. E. (2004) From the cover: RNA interference acts as a natural antiviral response to O'nyong-nyong virus (Alphavirus; Togaviridae) infection of Anopheles gambiae, *Proc Natl Acad Sci U S A* **101**, 17240–5.
42. Sanchez-Vargas, I., Travanty, E. A., Keene, K. M., Franz, A. W., Beaty, B. J., Blair, C. D., and Olson, K. E. (2004) RNA interference, arthropod-borne viruses, and mosquitoes, *Virus Res* **102**, 65–74.
43. Brackney, D. E., Beane, J. E., and Ebel, G. D. (2009) RNAi targeting of West Nile virus in mosquito midguts promotes virus diversification, *PLoS Pathog* **5**, e1000502.
44. Aliyari, R., Wu, Q., Li, H. W., Wang, X. H., Li, F., Green, L. D., Han, C. S., Li, W. X., and Ding, S. W. (2008) Mechanism of induction and suppression of antiviral immunity directed by virus-derived small RNAs in *Drosophila*, *Cell Host Microbe* **4**, 387–97.
45. Flynt, A., Liu, N., Martin, R., and Lai, E. C. (2009) Dicing of viral replication intermediates during silencing of latent *Drosophila* viruses, *Proc Natl Acad Sci U S A* **106**, 5270–5.
46. Myles, K. M., Wiley, M. R., Morazzani, E. M., and Adelman, Z. N. (2008) Alphavirus-derived small RNAs modulate pathogenesis in disease vector mosquitoes, *Proc Natl Acad Sci U S A* **105**, 19938–43.
47. Wu, Q., Luo, Y., Lu, R., Lau, N., Lai, E. C., Li, W. X., and Ding, S. W. (2010) Virus discovery by deep sequencing and assembly of virus-derived small silencing RNAs, *Proc Natl Acad Sci U S A* **107**, 1606–11.
48. Li, H. W., Li, W. X., and Ding, S. W. (2002) Induction and suppression of RNA silencing by an animal virus, *Science* **296**, 1319–21.
49. Nayak, A., Berry, B., Tassetto, M., Kunitomi, M., Acevedo, A., Deng, C., Krutchinsky, A., Gross, J., Antoniewski, C., and Andino, R. (2010) Cricket paralysis virus antagonizes Argonaute 2 to modulate antiviral defense in *Drosophila*, *Nat Struct Mol Biol* **17**, 547–54.
50. Voinnet, O. (2005) Non-cell autonomous RNA silencing, *FEBS Lett* **579**, 5858–71.
51. Dunoyer, P., Schott, G., Himber, C., Meyer, D., Takeda, A., Carrington, J. C., and Voinnet, O. (2010) Small RNA duplexes function as mobile silencing signals between plant cells, *Science* **328**, 912–16.
52. Saleh, M. C., Tassetto, M., Van Rij, R. P., Goic, B., Gausson, V., Berry, B., Jacquier, C., Antoniewski, C., and Andino, R. (2009) Antiviral immunity in *Drosophila* requires systemic RNA interference spread, *Nature* **458**, 346–50.
53. Lipardi, C. and Paterson, B. M. (2009) Identification of an RNA-dependent RNA polymerase in *Drosophila* involved in RNAi and transposon suppression, *Proc Natl Acad Sci U S A* **106**, 15645–50.
54. Czech, B., Malone, C. D., Zhou, R., Stark, A., Schlingeheyde, C., Dus, M., Perrimon, N., Kellis, M., Wohlschlegel, J. A., Sachidanandam, R., Hannon, G. J., and

Brennecke, J. (2008) An endogenous small interfering RNA pathway in *Drosophila*, *Nature* **453**, 798–802.

55. Kopek, B. G., Perkins, G., Miller, D. J., Ellisman, M. H., and Ahlquist, P. (2007) Three-dimensional analysis of a viral RNA replication complex reveals a virus-induced mini-organelle, *PLoS Biol* **5**, e220.

56. Szittya, G., Moxon, S., Pantaleo, V., Toth, G., Rusholme Pilcher, R. L., Moulton, V., Burgyan, J., and Dalmay, T. (2010) Structural and functional analysis of viral siRNAs, *PLoS. Pathog.* **6**, e1000838.

57. Ahlquist, P. (2006) Parallels among positive-strand RNA viruses, reverse-transcribing viruses and double-stranded RNA viruses, *Nat Rev Microbiol* **4**, 371–82.

58. Hemmes, H., Lakatos, L., Goldbach, R., Burgyan, J., and Prins, M. (2007) The NS3 protein of Rice hoja blanca tenuivirus suppresses RNA silencing in plant and insect hosts by efficiently binding both siRNAs and miRNAs, *RNA* **13**, 1079–89.

59. Merai, Z., Kerenyi, Z., Kertesz, S., Magna, M., Lakatos, L., and Silhavy, D. (2006) Double-stranded RNA binding may be a general plant RNA viral strategy to suppress RNA silencing, *J Virol* **80**, 5747–56.

60. Silhavy, D., Molnar, A., Lucioli, A., Szittya, G., Hornyik, C., Tavazza, M., and Burgyan, J. (2002) A viral protein suppresses RNA silencing and binds silencing-generated, 21- to 25-nucleotide double-stranded RNAs, *EMBO J* **21**, 3070–80.

61. Lakatos, L., Szittya, G., Silhavy, D., and Burgyan, J. (2004) Molecular mechanism of RNA silencing suppression mediated by p19 protein of tombusviruses, *EMBO J* **23**, 876–84.

62. Havelda, Z., Hornyik, C., Crescenzi, A., and Burgyan, J. (2003) In situ characterization of Cymbidium Ringspot Tombusvirus infection-induced posttranscriptional gene silencing in Nicotiana benthamiana, *J Virol* 77, 6082–6.

63. Baumberger, N., Tsai, C. H., Lie, M., Havecker, E., and Baulcombe, D. C. (2007) The polerovirus silencing suppressor p0 targets argonaute proteins for degradation, *Curr Biol* **17**, 1609–14.

64. Bortolamiol, D., Pazhouhandeh, M., Marrocco, K., Genschik, P., and Ziegler-Graff, V. (2007) The Polerovirus F box protein P0 targets ARGONAUTE1 to suppress RNA silencing, *Curr Biol* **17**, 1615–21.

65. Csorba, T., Lozsa, R., Hutvagner, G., and Burgyan, J. (2010) Polerovirus protein P0 prevents the assembly of small RNA-containing RISC complexes and leads to degradation of ARGONAUTE1, *Plant J* **62**, 463–72.

66. Ho, M. S., Tsai, P. I., and Chien, C. T. (2006) F-box proteins: the key to protein degradation, *J Biomed Sci* **13**, 181–91.

67. Zhang, X., Yuan, Y. R., Pei, Y., Lin, S. S., Tuschl, T., Patel, D. J., and Chua, N. H. (2006) Cucumber mosaic virus-encoded 2b suppressor inhibits Arabidopsis Argonaute1 cleavage activity to counter plant defense, *Genes Dev* **20**, 3255–68.

68. Gonzalez, I., Martinez, L., Rakitina, D. V., Lewsey, M. G., Atencio, F. A., Llave, C., Kalinina, N. O., Carr, J. P., Palukaitis, P., and Canto, T. (2010) Cucumber mosaic virus 2b protein subcellular targets and interactions: their significance to RNA silencing suppressor activity, *Mol Plant Microbe Interact* **23**, 294–303.

69. Goto, K., Kobori, T., Kosaka, Y., Natsuaki, T., and Masuta, C. (2007) Characterization of silencing suppressor 2b of cucumber mosaic virus based on examination of its small RNA-binding abilities, *Plant Cell Physiol* **48**, 1050–60.

70. El-Shami, M., Pontier, D., Lahmy, S., Braun, L., Picart, C., Vega, D., Hakimi, M. A., Jacobsen, S. E., Cooke, R., and Lagrange, T. (2007) Reiterated WG/GW motifs form functionally and evolutionarily conserved ARGONAUTE-binding platforms in RNAi-related components, *Genes Dev* **21**, 2539–44.

71. Azevedo, J., Garcia, D., Pontier, D., Ohnesorge, S., Yu, A., Garcia, S., Braun, L., Bergdoll, M., Hakimi, M. A., Lagrange, T., and Voinnet, O. (2010) Argonaute quenching and global changes in Dicer homeostasis caused by a pathogen-encoded GW repeat protein, *Genes Dev* **24**, 904–15.

72. Chao, J. A., Lee, J. H., Chapados, B. R., Debler, E. W., Schneemann, A., and Williamson, J. R. (2005) Dual modes of RNA-silencing suppression by Flock House virus protein B2, *Nat Struct Mol Biol* **12**, 952–7.

73. Bonning, B. C. and Miller, W. A. (2010) Dicistroviruses, *Annu Rev Entomol* **55**, 129–150.

74. Manousis, T. and Moore, N. F. (1987) Cricket paralysis virus, a potential control agent for the Olive Fruit Fly, Dacus oleae Gmel, *Appl Environ Microbiol* **53**, 142–8.

75. Attarzadeh-Yazdi, G., Fragkoudis, R., Chi, Y., Siu, R. W., Ulper, L., Barry, G., Rodriguez-Andres, J., Nash, A. A., Bouloy, M., Merits, A., Fazakerley, J. K., and Kohl, A. (2009) Cell-to-cell spread of the RNA interference

response suppresses Semliki Forest virus (SFV) infection of mosquito cell cultures and cannot be antagonized by SFV, *J Virol* **83**, 5735–48.
76. Cirimotich, C. M., Scott, J. C., Phillips, A. T., Geiss, B. J., and Olson, K. E. (2009) Suppression of RNA interference increases alphavirus replication and virus-associated mortality in *Aedes aegypti* mosquitoes, *BMC Microbiol* **9**, 49.
77. Blakqori, G., Delhaye, S., Habjan, M., Blair, C. D., Sanchez-Vargas, I., Olson, K. E., Attarzadeh-Yazdi, G., Fragkoudis, R., Kohl, A., Kalinke, U., Weiss, S., Michiels, T., Staeheli, P., and Weber, F. (2007) La Crosse bunyavirus nonstructural protein NSs serves to suppress the type I interferon system of mammalian hosts, *J Virol* **81**, 4991–9.
78. Li, H. W. and Ding, S. W. (2005) Antiviral silencing in animals, *FEBS Lett* **579**, 5965–73.
79. Ghildiyal, M. and Zamore, P. D. (2009) Small silencing RNAs: an expanding universe, *Nat Rev Genet* **10**, 94–108.
80. Lau, N. C., Robine, N., Martin, R., Chung, W. J., Niki, Y., Berezikov, E., and Lai, E. C. (2009) Abundant primary piRNAs, endo-siRNAs, and microRNAs in a *Drosophila* ovary cell line, *Genome Res* **19**, 1776–85.
81. van Mierlo, J. T., van Cleef, K. W. R., and Van Rij, R. P. (2010) Small silencing RNAs: piecing together a viral genome, *Cell Host Microbe* **7**, 87–9.
82. Hussain, M., Taft, R. J., and Asgari, S. (2008) An insect virus-encoded microRNA regulates viral replication, *J Virol* **82**, 9164–70.
83. Hussain, M. and Asgari, S. (2010) Functional analysis of a cellular microRNA in insect host-ascovirus interaction, *J Virol* **84**, 612–20.
84. Zambon, R. A., Nandakumar, M., Vakharia, V. N., and Wu, L. P. (2005) The Toll pathway is important for an antiviral response in *Drosophila*, *Proc Natl Acad Sci U S A* **102**, 7257–62.
85. Xi, Z., Ramirez, J. L., and Dimopoulos, G. (2008) The *Aedes aegypti* toll pathway controls dengue virus infection, *PLoS Pathog* **4**, e1000098.
86. Costa, A., Jan, E., Sarnow, P., and Schneider, D. (2009) The Imd pathway is involved in antiviral immune responses in *Drosophila*, *PLoS One* **4**, e7436.
87. Dostert, C., Jouanguy, E., Irving, P., Troxler, L., Galiana-Arnoux, D., Hetru, C., Hoffmann, J. A., and Imler, J. L. (2005) The Jak-STAT signaling pathway is required but not sufficient for the antiviral response of *Drosophila*, *Nat Immunol* **6**, 946–53.
88. Deddouche, S., Matt, N., Budd, A., Mueller, S., Kemp, C., Galiana-Arnoux, D., Dostert, C., Antoniewski, C., Hoffmann, J. A., and Imler, J. L. (2008) The DExD/H-box helicase Dicer-2 mediates the induction of antiviral activity in *Drosophila*, *Nat Immunol* **9**, 1425–32.
89. Singh, G., Popli, S., Hari, Y., Malhotra, P., Mukherjee, S., and Bhatnagar, R. K. (2009) Suppression of RNA silencing by Flock house virus B2 protein is mediated through its interaction with the PAZ domain of Dicer, *FASEB J* **23**, 1845–57.
90. Ratcliff, F., Harrison, B. D., and Baulcombe, D. C. (1997) A similarity between viral defense and gene silencing in plants, *Science* **276**, 1558–60.
91. Hamilton, A. J. and Baulcombe, D. C. (1999) A species of small antisense RNA in posttranscriptional gene silencing in plants, *Science* **286**, 950–2.
92. Merai, Z., Kerenyi, Z., Molnar, A., Barta, E., Valoczi, A., Bisztray, G., Havelda, Z., Burgyan, J., and Silhavy, D. (2005) Aureusvirus P14 is an efficient RNA silencing suppressor that binds double-stranded RNAs without size specificity, *J Virol* **79**, 7217–26.

Chapter 2

RNAi and Cellular miRNAs in Infections by Mammalian Viruses

Joost Haasnoot and Ben Berkhout

Abstract

MicroRNAs (miRNAs) play an essential role in the regulation of eukaryotic gene expression. Recent studies demonstrate that miRNAs can also strongly affect the replication of pathogenic viruses. For example, cellular miRNAs can target and repress the expression of viral mRNAs, but there is also at least one example of a cellular miRNA that stimulates virus replication. Furthermore, viruses can encode their own miRNAs, trigger changes in cellular miRNA expression or encode RNA silencing suppressor factors that inhibit cellular miRNAs. These interactions together form a complex regulatory network that controls both viral and host gene expression, which ultimately determines the outcome of viral infection at the cellular level and disease progression in the host. Here, we summarize the literature data on such virus–cell interactions in mammals and discuss how miRNAs can be used as research tools or targets in the development of novel antiviral therapeutics.

Key words: Virus, microRNA, Regulation, RNA interference, RNA silencing suppressor, HIV-1, HCV, Adenovirus

1. Introduction

Early studies in plants showed that small RNA-induced gene silencing is a potent antiviral mechanism that plants need to survive viral infection (1). Since then, it has become clear that there are many more aspects to the interaction between the RNA silencing mechanism – in plants, insects and mammals – and viral infection. Since RNA silencing is a central regulatory mechanism in eukaryotic cell biology, it is involved in many different cellular processes. These processes not only include cell development, differentiation and proliferation, but also cell death, metabolism, transposon silencing and antiviral defences (2). Viruses, being strictly dependent on cellular resources for their

Ronald P. van Rij (ed.), *Antiviral RNAi: Concepts, Methods, and Applications*, Methods in Molecular Biology, vol. 721,
DOI 10.1007/978-1-61779-037-9_2, © Springer Science+Business Media, LLC 2011

replication, therefore interact with the RNA silencing machinery at multiple levels.

RNA silencing represents a general cellular phenomenon in which small RNA molecules of 20–30 nucleotides associate with Argonaute or Piwi proteins to trigger sequence-specific inhibition of gene expression (3). Currently, three classes of small RNAs that are involved in silencing have been identified in mammals, namely microRNAs (miRNAs), endogenous small interfering RNAs (endo-siRNAs) and piwi-associated RNAs (piRNAs). Endo-siRNAs and piRNAs are primarily involved in the repression of transposons, whereas miRNAs regulate cellular gene expression (4). In recent years, miRNA-mediated gene regulation has received much attention. So far, over 700 human miRNAs have been cloned, which are estimated to regulate the expression of at least 30% of human genes (5). This regulation involves miRNA-guided targeting of the multi-protein RNA-induced silencing complex (RISC) to partially complementary sequences in the 3′ untranslated region (3′UTR) of target mRNAs. Once RISC is bound to the target site, it triggers mRNA translational inhibition or destabilization (6).

Besides mediating regulation of gene expression, RNA silencing in plants, insects, nematodes and fungi plays an essential role in the antiviral "immune" response via virus-specific siRNAs (1, 7–9). Despite the fact that the RNA silencing or RNA interference (RNAi) mechanism and machinery is highly conserved in eukaryotes, researchers failed to detect virus-specific siRNAs in virus-infected mammalian cells (10). This initially suggested that RNAi does not have an antiviral role in mammals (11). Interestingly, recent studies used novel and very sensitive deep sequencing technology to show that small virus-derived RNAs do accumulate in virus-infected mammalian cells (12, 13). It currently remains unclear to what extent these RNA molecules contribute to the antiviral defence response.

An increasing number of studies have addressed novel aspects of the interaction between mammalian viruses and small RNA-induced silencing mechanisms. It has become clear that cellular miRNAs and other components of the miRNA pathway can interact with viruses at multiple levels to influence viral replication (Fig. 1). In this chapter we will focus on this multitude of possible interactions. For a good understanding of the way in which viruses interact with cellular miRNAs in mammals, we will first give a brief overview of different aspects of the miRNA and small interfering RNA (siRNA) pathways. Subsequently, we will discuss the effect of cellular miRNAs on viral gene expression, cellular miRNAs as determinants of the viral tropism for certain cell types, virus-encoded RNA silencing suppressors (RSS), and virus-induced changes in the cellular miRNA expression profile. Finally, we will discuss the use of miRNAs or miRNA-inactivating compounds in novel antiviral therapeutic strategies. Virus-encoded miRNAs, as

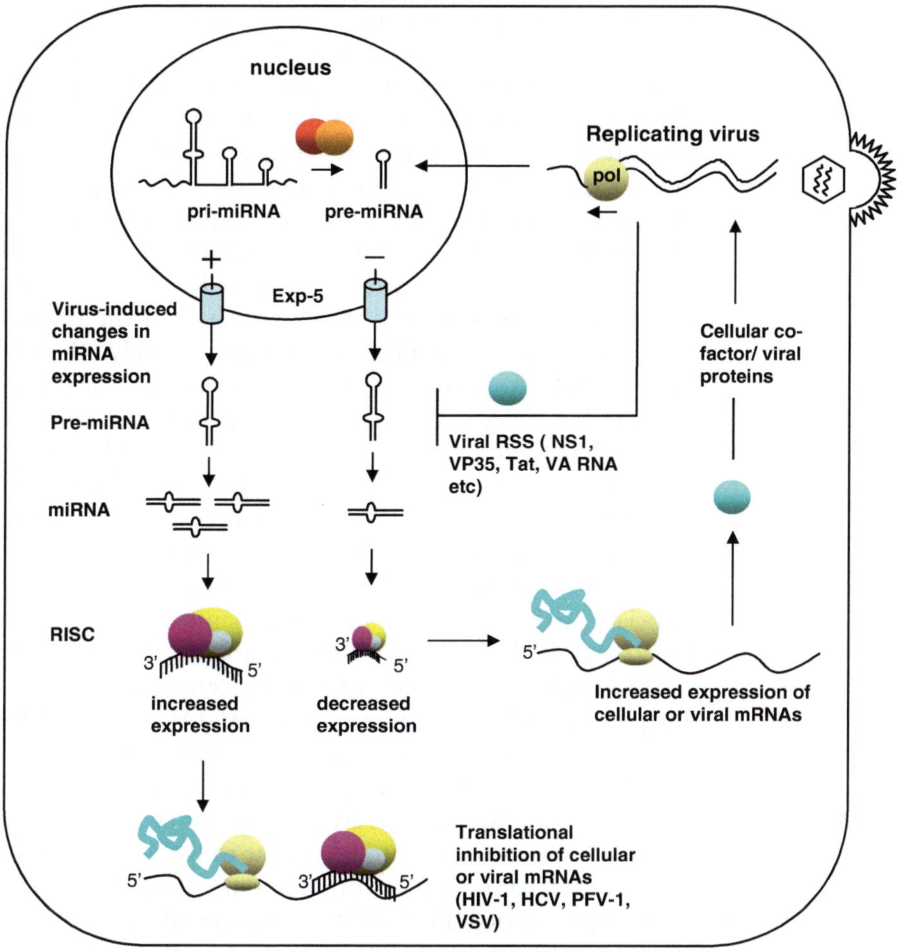

Fig. 1. Overview of the interactions between the cellular RNAi mechanism and an invading virus. After virus entry and start of replication the virus can induce (+) or inhibit (–) the expression of certain cellular miRNAs. These may influence signalling pathways and stress responses. Virus-encoded RSS factors may block specific RNAi actions. RSS factors like influenza virus NS1, Ebola virus VP35, and HIV-1 Tat can bind and sequester small RNAs/pre-miRNAs, whereas the adenovirus VA RNAs inhibit miRNA processing by saturating the miRNA pathway. Some viruses may depend on constitutively expressed miRNAs (HCV-miR-122). Increased expression of cellular miRNAs that target the viral RNA will result in inhibition of viral gene expression and decreased miRNA expression may increase virus replication. Changes in the miRNA profile may also affect cellular gene expression that may either stimulate or inhibit viral gene expression.

expressed by the herpes virus family and other DNA viruses (14), will be discussed in Chap. 3.

2. miRNA-Mediated Silencing in Mammals

Mature miRNAs are single-stranded RNAs of approximately 21–23 nucleotides that are processed from endogenously expressed primary transcripts (pri-miRNAs) (15). These transcripts are produced by polymerase II and contain several imperfectly base paired hairpin structures. In the nucleus the RNA hairpin structure is excised by the RNAse III-like enzyme Drosha and its co-factor DGCR8 to form the precursor miRNA (pre-miRNA), which represents an imperfect hairpin structure (16). There is a separate miRNA pathway that is independent of Drosha processing. In this pathway, introns containing the hairpin are termed mirtrons, which are spliced into pre-miRNA hairpins that access the miRNA pathway. First described in Drosophila and *Caenorhabditis elegans*, these mirtrons have recently also been reported in mammals (17). Both Drosha and the mirtron pathway generate an excised hairpin of approximately 60 nucleotides, the pre-miRNA. This pre-miRNA is exported to the cytoplasm by the nuclear RNA export factor Exportin 5 (Exp5), where it is further processed by the RNAse III-like enzyme Dicer in association with its co-factors TAR binding protein (TRBP) and PACT (18, 19). The Dicer-TRBP-PACT complex cleaves off the terminal loop resulting in the miRNA duplex with a two-nucleotide 3′ overhang on either side. Depending on the structure, miRNAs associate with one of the four Argonaute proteins to form the functional miRNA-loaded RISC complex. In *D. melanogaster* miRNAs with mismatches within the central part of the duplex are preferentially incorporated into AGO1. The miRNAs with more perfect duplexes associate with AGO2, which is the only member of the Argonaute protein family with slicing activity. Similar sorting mechanisms are conserved in C. elegans and mammals (20). Once loaded into RISC, the miRNA duplex is unwound into a guide strand and a complementary passenger strand that is subsequently degraded. RISC programmed by the guide miRNA subsequently targets mRNAs with complementary sequences within the 3′UTR. Pairing of the 5′ 7–8 nucleotides of the miRNA (seed region) to multiple sites within the 3′UTR of an mRNA is important for target recognition and translational inhibition (21). The targeted RNAs are translocated to cellular processing (P)-bodies where they are stored or de-adenylated, de-capped, and subsequently degraded (22). It was initially thought that miRNA targeting could result in cleavage of the mRNA target, but only in rare cases. However, at the recent 2010 Keystone meeting it was reported that, in contrast to general belief, miRNAs do frequently induce RNA destabilization instead of translational inhibition.

3. Virus-Derived Small Interfering RNAs

Unlike miRNAs, siRNAs are perfectly complementary duplexes of 21 nucleotides that originate from dsRNA precursors upon processing by Dicer. The siRNAs are loaded into AGO2-containing

RISC and, in case of sufficient complementarity to the target mRNA, guide RISC towards cleavage of the target. In plants, worms and insects, the dsRNA precursors can be long virus-derived dsRNAs, e.g. replication intermediates. Once incorporated in RISC, the processed virus-specific siRNAs trigger the sequence-specific cleavage of the homologous viral RNAs. Plant and insect viruses counter this RNA-based antiviral response through the production of specific RSS factors (23, 24). In mammals, perfectly base paired long dsRNAs trigger the interferon (IFN) response that signals cells to induce the innate antiviral IFN response, which involves the induction of many different genes (25). Despite serious attempts, virus-specific siRNAs have not been detected in virus-infected mammalian cells via conventional cloning and sequencing techniques (10, 26). Because mammals have the IFN response and virus-specific siRNAs could not be detected, there was little support for an antiviral function of RNAi in mammals. Therefore, studies on siRNAs in mammals mostly focused on the use of artificially introduced synthetic siRNAs or short hairpin RNA vectors as tools to induce sequence-specific knock-down of a gene of interest (27).

Recently however, new highly sensitive sequencing technology has become available that is being used to analyze low abundant species of regulatory RNAs including virus-derived siRNAs in mammalian cells. Watanabe was the first to use 454 sequencing to study small RNAs in mouse oocytes, where transposon-specific RNAs were found to accumulate (28). These so-called endo-siRNAs actively repress the expression of the corresponding transposable elements and certain protein-coding genes (28–30). The endo-siRNAs are processed from naturally occurring dsRNAs formed by hybridization of perfectly complementary transcripts that are made by bidirectional transcription of the genome segments.

Two other recent studies used deep sequencing technology to show that low amounts of small virus-specific RNAs accumulate during virus infection in mammalian cells. Parameswaran et al. reported small virus-derived RNAs in cells infected with dengue virus, vesicular stomatitis virus (VSV), poliovirus, hepatitis C virus (HCV) and West Nile virus (12). In addition, Yeung et al. reported the accumulation of small virus-derived RNAs in HIV-1 infected cells (13). In this study, several discrete small virus-specific RNA species were detected. One of the HIV-specific RNAs detected was a molecule of 18 nucleotides that was complementary to the HIV-1 primer binding site (PBS). This is the sequence element in the HIV-1 RNA genome to which $tRNA^{Lys3}$ binds to prime reversed transcription. This process is essential for the generation of the proviral DNA that integrates in the host genome and from which the viral proteins are subsequently expressed. Expression of the PBS-specific small RNA was found to depend on Dicer. Furthermore, this small RNA was shown to associate with Ago2, suggesting a role in antiviral responses. However, recent studies indicate that cellular tRNA molecules can also be processed by Dicer into 18-nucleotide long fragments in uninfected cells (31). This raises the possibility that the PBS-specific small RNA is of cellular origin.

Thus, recent studies demonstrated that small RNAs do accumulate in virus-infected mammalian cells, but it is currently not clear whether these RNAs represent functional siRNAs that negatively affect virus replication. In any case, the highly sensitive deep sequencing technology will certainly help to further describe and classify this new class of regulatory RNAs.

4. Cellular miRNAs Targeting Viral Sequences

Although the role of antiviral siRNAs in mammalian cells is still unclear, there is evidence that cellular miRNAs are able to target and inhibit viral gene expression. At least four viruses have been reported to be subject to miRNA-mediated repression of gene expression. These include primate foamy virus type 1 (PFV-1), HIV-1, HCV and VSV (Table 1) (32–35). The first reported example of a cellular miRNA that targets a viral RNA genome is miR-32. This miRNA targets the retrovirus PFV-1 in a genome region that is present in the 3′UTR of all viral transcripts and open reading frame 2 that encodes both the viral Bet and EnvBet proteins. Targeting of the viral RNA by miR-32 results in reduced virus replication, and PFV-1 requires the RSS activity of its Tas protein to overcome this inhibition (33).

Another example of a virus that is restricted by cellular miRNAs is VSV. Ostsuka et al. showed that VSV transcripts encoding the viral large protein (L protein) and the phosphoprotein (P protein) are targeted by miR-24 and miR-93, respectively (34). Reduced expression of miR-24 and miR-93 in Dicer-deficient mice caused a strong increase of VSV replication leading to virus-induced lethality. Besides VSV, the authors also tested the susceptibility of Dicer-deficient macrophages for infection by other viruses such as encephalo-myocarditis virus, lymphocytic choriomeningitis virus, Coxsackievirus group B serotype 3, influenza A virus, herpes simplex virus type 1 (HSV-1) and vaccinia virus. Interestingly, only HSV-1 displayed increased virus production, suggesting that Dicer-mediated antiviral RNAi responses do not represent a general antiviral mechanism in mammals. As miRNAs are differentially expressed among different cell types and viruses likewise infect specific subsets of cells, one should thus be extremely careful in drawing general conclusions. The lack of an effect on virus replication in Dicer-deficient macrophages does

Table 1
Cellular miRNAs targeting viral genes

Virus	miRNA	Target	Effect	References
HIV-1	miR-28, miR-125b, miR-382, miR-150, miR-223	TAR, Env, NF-κb, Env, Nef	Induces viral latency	(32, 37)
	miR-29a	Nef	Inhibits translation/ replication	(39, 40)
PFV-1	miR-32	Bet and EnvBet proteins, 3′UTR	Inhibits translation/ replication (Tas protein counters this effect)	(33)
VSV	miR-24, miR-93	L and P protein	Inhibits translation/ replication	(34)
HCV	miR-122	5′UTR (IRES)	Stimulates translation/ replication (liver specific)	(42)
	miR-1, miR-30, miR-128, miR-196, miR-296, miR-351, miR-431, miR-448	C, NS5A (others not specified)	Inhibits translation/ replication (induced by IFN signalling)	(35)
	miR-199a	5′UTR (IRES)	Inhibits translation/ replication	(96)

not exclude the possibility that these viruses are affected by Dicer deficiency in the biologically more relevant cell types.

Besides inhibiting viral gene expression in an acute manner, the set of miRNAs expressed in a certain cell type may also determine the level of evolutionary freedom of the virus on a longer time scale. Certain nucleotide changes in the viral genome sequence will not be tolerated because they create more optimal miRNA target sequences that trigger repression by the miRNA machinery. This evolutionary pressure is illustrated by the observation that the artificial insertion of fully complementary miRNA target sites in viral genes results in severe attenuation of virus replication (36). This fine-tuning of the viral RNA to the miRNA composition of the host cell may also restrict the viral ability to adjust to a change in the environment, such as the invasion of a new host organism upon zoonotic transmission or the intra-host pressure when the immune system is induced. In addition to PFV-1 and VSV, cellular miRNAs have also been shown to target the RNA genome of HIV-1 and HCV. Because of the complex interaction of these viruses with miRNAs, these cases will be discussed separately.

5. Cellular miRNAs and HIV-1 Latency

Huang et al. were the first to show that cellular miRNAs can target and potently repress HIV-1 gene expression (32). HIV-1 is a member of the lentivirus genus within the family of retroviruses. Its genome is encoded by a positive-stranded RNA molecule of 9.8 kb that is reverse transcribed into double-stranded DNA that integrates into the host cell genome. HIV-1 replicates in CD4-positive T cells and thus causes a direct attack on the immune system, which when left untreated will eventually result in the development of AIDS. It was shown that miR-28, miR-125b, miR-150, miR-223, and miR-382 are able to target sequences near the 3′ end of all HIV-1 transcripts and suppress the expression of viral mRNAs in resting $CD4^+$ T cells. The expression of this set of miRNAs is reduced in activated $CD4^+$ T cells, thus allowing active virus replication. This finding suggests that miRNAs play a role in the establishment of viral latency. HIV-1 latency occurs in resting $CD4^+$ T cells when the HIV-1 provirus is stably integrated into the host genome without producing new viral transcripts and proteins. HIV-1 latency may allow the virus to escape from the immune system and latency is a major problem in attempts to eradicate the virus in a patient by therapeutic intervention.

The latency-inducing miRNAs also appear to determine the susceptibility of peripheral blood monocytes for HIV-1 infection (37). These cells express all the essential receptors for HIV-1 entry, but become infrequently infected. New findings suggest that miRNA-mediated suppression protects the cells against productive HIV-1 infection. In other words, the cellular miRNAs form one of the determinants of the viral tropism for certain cell types. Several miRNAs including miR-29a were also predicted to target the HIV-1 RNA genome (38). The miR-29a recognizes a target within the viral *nef* gene to restrict expression of the Nef protein and overall virus replication (39). Nathans et al. subsequently showed that HIV-1 mRNAs that are targeted and inhibited by miR-29a are loaded into RISCs that associate with P bodies (40).

6. Cellular miRNAs Targeting HCV

Not all cellular miRNAs that target viral mRNAs have a negative effect on gene expression and viral replication. It was convincingly demonstrated that miR-122 is required for HCV replication (41, 42). This miRNA is highly expressed in human liver cells. The increased replication is at least in part caused by enhanced translation of the viral RNA via direct interaction of miR-122 with two target sites in the 5′UTR (43). This region in the HCV genome harbours the internal ribosomal entry site (IRES) that is instrumental for efficient HCV mRNA translation.

Besides the positive role of miR-122, a specific set of cellular miRNAs is involved in the restriction of HCV gene expression. These miRNAs (miR-1, miR-30, miR-128, miR-196, miR-296, miR-351, miR-431, and miR-448) were found to be up-regulated in response to IFN signalling that is triggered during HCV replication (35). The transfection of synthetic miRNA-mimics reproduced the antiviral effect of IFN-β on HCV replication. In addition, IFN treatment leads to a significant reduction in expression of the positive miR-122 co-factor. Therefore, the antiviral activity of IFN appears to be caused by up and down-regulation of specific cellular miRNAs. A recent follow-up study did however not reveal a correlation between the miR-122 expression level in HCV-infected individuals and the viral load (44).

7. Virus-Induced Changes in Cellular miRNA Expression

Virus infection triggers cells to mount an antiviral response that involves the expression of many antiviral and stress related genes. Similarly, virus-induced changes in the cellular miRNA expression profile may have a profound effect on the outcome of infection. In an early study, Yeung et al. determined the miRNA expression profile in HeLa cells transfected with HIV-1 DNA (45). The authors reported significant down-regulation of miR-93, miR-148b, miR-221, and miR-16. The miRNA profile in peripheral blood mononuclear cells (PBMCs) from HIV-1 infected patients showed even more dramatic changes compared to cells from uninfected controls (46). Depending on the disease stage of the patient, the T cell specific miR-223, miR-150, miR-146, miR-16, and miR-191 were down-regulated 3–9-fold. Changes in the cellular miRNA expression profile can affect the expression of protein co-factors that are required for viral replication. For example, Triloubet et al. reported increased expression of eleven miRNAs in HIV-1 infected cells, whereas expression of the polycistronic miRNA cluster miR17/92 was strongly decreased (47). This miR17/92 cluster comprises miR-17-(5p/3p), miR-18, miR-19a, miR-20a, miR-19b-1, and miR-92-1 and has been implicated in various types of cancer (48). Computer-assisted target prediction showed that the mRNA encoding histone acetylase PCAF has four target sequences in its 3′UTR for miR-17-5p and miR-20a. PCAF is an important co-factor of HIV-1 transcription in interaction with the viral Tat protein. Thus, HIV-induced down-regulation of miR-17/92 increases PCAF expression, which in turn results in further enhancement of HIV-1 replication.

The miR-198 inhibits the expression of Cyclin T1 in monocytes, which is required for HIV-1 transcription elongation (37). Once monocytes are activated to differentiate into macrophages, miR-198 expression is reduced, resulting in increased Cyclin T1 expression that facilitates HIV-1 replication. Although this is not a virus-induced change, it does indicate that miRNAs can be important determinants of viral replication by affecting the expression of essential host factors. In this manner, miRNAs can have an important role in determining the cell tropism of certain viruses.

A recent study analyzed the miRNA expression profile in bronchoalveolar stem cells upon infection by the SARS coronavirus (SARS-CoV) and the miRNA targets were identified. Intriguingly, the miRNAs that were up-regulated during infection could suppress virus replication. The authors describe this as "co-opting" by means of the viral miRNAs in order to minimize the expression of viral antigens to evade the immune system (49).

In an attempt to analyze the role of miRNAs in influenza virus pathogenicity, the miRNA expression profile was analyzed in mice infected with the 1918 pandemic influenza virus (50). A specific group of miRNAs was differentially expressed in mice infected with the pandemic 1918 virus compared to mice infected with a less pathogenic seasonal influenza strain. Interestingly, several of the corresponding target mRNAs encode proteins that are involved in the immune response. These results suggest that miRNAs and virus-induced changes in the miRNA expression profile may correlate with virus-induced pathogenicity (30). Wang et al. determined the differentially expressed miRNAs in chicken lung and trachea infected with the low pathogenic H5N3 avian influenza virus (51). It was reported that 73 and 36 miRNAs are differentially expressed in lungs and trachea, respectively, upon virus infection. This list includes miR-146a, which has been proposed to be involved in immune-related signalling pathways (52).

Finally, murine cytomegalovirus (MCMV) infection was shown to down-regulate the expression of miR-27a and b (53). This down-regulation does surprisingly not affect the other miRNAs within the same gene cluster. Therefore, the down-regulation was proposed to occur post-transcriptionally. Both miR-27a and b exhibit an antiviral effect, although it is as yet unclear how this works mechanistically.

8. Oncogenic Viruses and miRNAs

Changes in the cellular miRNA expression profile have been implicated in oncogenesis (54). Interestingly, oncolytic viruses also affect miRNA expression and this may contribute to the multi-step oncogenic process. It is currently unclear whether these changes are part of an immune response, viral signalling, or other thus far unidentified mechanisms. Tomita et al. showed that the human T lymphotropic virus type I (HTLV-I) induces mir-146a expression, which in turn increases the growth characteristics of infected cells (55). In an earlier study miR-146a was shown to be increased in tumours (56). The T strain of the reticuloendotheliosis virus induces miR-155, which promotes cell survival (57). Marek's disease virus (MDV), which is a highly oncogenic herpes virus in poultry, triggers the up-regulation of miR-221 and miR-222 (58). These miRNAs have both been implicated in tumorigenesis and were found to be up-regulated in a number of cancers.

9. Virus-Encoded Suppressors of RNAi

Viral RSS factors were first described for plant viruses as multifunctional virulence or pathogenicity factors (24, 59). These RSSs were shown to inhibit RNA silencing via sequestration of antiviral siRNAs, protection of long virus-specific dsRNA from processing into siRNAs, or via direct blocking of specific components of the RNA silencing pathway. The identification of mammalian virus-encoded RSS functions provided one of the first indications that the RNAi mechanism is involved in antiviral responses in mammalian cells. So far, a total of 11 mammalian virus-encoded RSS factors have been described (60). These include the following proteins: influenza A virus NS1 (61, 62), vaccinia virus E3L (62), Ebola virus VP35 (63), HIV-1 Tat (63–65), PFV-1 Tas (33), and HCV coat and envelop protein (66–68). Most of these proteins display RSS activity when co-expressed together with a silenced reporter gene. Other RSS proteins prevent Dicer cleavage in vitro. An interesting way to determine RSS activity is by means of so-called trans-complementation assays in which putative RSS factor functionally replaces a known RSS in the viral context. RSS factors from plant or insect virus suppressors can sometimes be replaced by RSSs encoded by mammalian viruses. Another commonly used assay is to measure mammalian virus-encoded RSS activity in plants.

The first reported mammalian RSS factors were E3L and NS1. Both proteins are able to rescue the replication in insects of a flock house virus (FHV) variant in which the RSS B2 gene is replaced by GFP. The FHV B2 protein inhibits RNAi by sequestration of long viral dsRNAs and siRNAs. These results indicate that E3L and NS1 are the functional equivalents of the B2 protein (62). In addition, the influenza virus NS1 protein was shown to suppress RNA silencing in plants and to increase the virulence of potato virus X in *Nicotiana Benthamiana* (61).

There are a number of reports on RSS activity of the HIV-1 Tat protein. The Tat protein stimulates transcription from the HIV-1 long terminal repeat (LTR) promoter by interacting with the transactivation response (TAR) element, a stable RNA stem-loop structure present at the 5′ terminus of each viral transcript (69). Bennasser et al. were the first to show Tat-mediated RSS suppression in luciferase-based reporter assay (64). They showed that Tat-mediated RSS activity is independent of the transcriptional transactivation activity of the protein. We could confirm these results using an HIV-1 variant that is independent of Tat-mediated transcription activation (63). Only Tat mutants or heterologous factors that exhibit RSS activity could restore virus production of a mutant virus lacking Tat RSS activity. Another study showed that the HIV-1 Tat RSS function can be functionally replaced by the RSS protein P19 of tomato bushy stunt virus and that the Tat protein exhibits RSS activity in *Nicotiana Benthamiana* protoplasts (65). We also showed that the RSS protein NS3 of rice hoja blanca virus (RHBV), a plant virus, is able to complement the RSS function of HIV-1 Tat (70). The NS3 protein of RHBV is an RSS that exclusively binds small dsRNA molecules such as miRNAs and siRNAs. An NS3 mutant that is deficient in dsRNA binding and RSS activity is unable to rescue production of the Tat-negative HIV-1 variant. This result suggests that HIV-1 replication is inhibited either by siRNAs or miRNAs and that Tat RSS activity is required to counter this inhibition. Similarly, the Tas transcriptional transactivator protein of PFV-1 was shown to exhibit RSS activity (33). The Tas protein RSS function represses miR-32 accumulation, thus allowing efficient virus replication. PFV-1 Tas also shows considerable RSS activity in Arabidopsis plants (33).

Most of the mammalian virus-encoded RSS factors are proteins. However, the human adenovirus encodes an RSS function that follows a different strategy. During the late phase of infection, the virus expresses high levels of two non-coding but highly structured virus-associated (VA) RNAs, RNAI and RNAII. Previously, these RNAs were shown to block antiviral PKR activity (71). The structured nature of the VA molecules shows similarity with pre-miRNAs, raising the possibility that the VA RNAs are Dicer substrates. It was demonstrated that the VA RNAs are recognized and processed by Dicer into virus-derived si/miRNAs that are functionally incorporated into RISC (72–76). Because of their extremely high expression level of up to 1×10^8 copies per cell, the VA RNAs cause saturation of the RNAi pathway. Thus, the VA RNAs act as suppressor by saturation of the RNAi machinery. It remains possible that the VA RNAs also represent a virus-encoded miRNA, but their extremely high copy number perhaps argues against this possibility (76).

The majority of mammalian virus-encoded RSS factors described above have also been reported to have IFN-antagonistic properties. For example, the Ebola virus VP35, influenza virus NS1, and vaccinia virus E3L proteins also act as antagonists of the cellular dsRNA sensors RNA helicases retinoic acid-inducible gene I (RIG-I), dsRNA-dependent protein kinase (PKR), or 2′–5′ oligo adenylate synthetase enzymes (OAS) (77–85). In mammals the IFN pathway is activated upon virus infection when viral nucleic acids are sensed by pattern-recognition receptors on the host cell. This activation leads to inhibition of viral replication and cell proliferation, apoptosis, and destruction of virus-infected cells by natural killer cells. The double function of many viral RSS/IFN-antagonistic proteins is perhaps not that surprising considering that both the RNAi and IFN pathways are induced by the same virus-derived molecule, dsRNA. Possibly, the RNAi and IFN pathways co-operate in the innate defence response against invading viruses. The antiviral responses against HCV provide a good example on how the IFN and RNAi mechanisms may co-operate to optimally combat virus infections (35).

10. miRNAs in Novel Antiviral Approaches

Therapeutic RNAi via delivery of synthetic siRNA or intracellular shRNA expression is currently being developed as a novel approach to treat various genetic diseases as well as pathogenic virus infections (27). Indeed, direct targeting of the RNAi mechanism towards viral mRNAs can potently inhibit virus replication. RNAi-mediated inhibition of host factors that are required for virus replication is also being considered in RNAi-based antiviral strategies. The newly discovered interactions between cellular miRNAs and pathogenic viruses likewise provide new possibilities for antiviral drug design. In fact, the therapeutic use of such miRNA co-factors is similar in concept to antiviral silencing of protein co-factors that are required for virus replication. Targeting of the set of cellular miRNAs that have been implicated in HIV-1 latency has been suggested as a strategy to block or reduce HIV-1 latency (32, 86). HIV-1 latency poses a serious problem for the treatment of HIV-1 infected individuals with antiviral drugs. Antiviral drugs can strongly reduce the level of replicating virus, but HIV-1 will rapidly re-emerge from the latent reservoir after stopping therapy. Antisense inhibitors of the latency-inducing miRNAs could counter latency and induce active virus production. Activation of the latent reservoir may eventually allow for purging of the viral reservoir, such that one can eradicate the virus in an infected individual and cure the patient.

Another example of a promising new target in antiviral drug design is miR-122. This miRNA is required for HCV translation and replication (41, 42). Inhibition of miR-122 by antisense oligonucleotides indeed results in a strong decrease of HCV replication in vitro (42, 87–89). Interestingly, Lanford et al. recently demonstrated that treatment of chronically HCV-infected chimpanzees with a locked nucleic acid (LNA)-modified oligonucleotide that targets miR-122 leads to long-lasting suppression of HCV viremia (90).

Besides direct targeting of specific cellular miRNAs to block viruses, researchers have also recognized the use of miRNAs as a novel tool in the design of viral vectors (36). Insertion of miRNA targets in vectors can be used to alter the cellular tropism of the vector or to generate attenuated virus variants that may be used for vaccine development. In this strategy, the inserted targets are usually fully complementary to the cellular miRNAs, thus resulting in cleavage of the viral/vector RNA. For example, insertion of miR-142-3p targets in a lentiviral vector restricted transgene expression in the hematopoietic cell lineage where this miRNA is highly expressed, whereas optimal transgene expression was maintained in non-hematopoietic cells (91). Insertion of the liver-specific miR-122 target in adenovirus severely attenuated virus replication in hepatocytes, but allowed normal replication in other cells (92). Insertion of targets for the muscle specific miR-133 and miR-206 in the 3′UTR of an oncolytic picornavirus, Coxsackie virus A21, strongly attenuated viral pathogenicity (93). Another example is VSV, which has been proposed as possible recombinant vaccine platform, but which causes encephalomyelitis in rodents and primates. By insertion of targets for neuron-specific miRNAs, virus replication in neurons could be inhibited without compromising replication in other tissues (94). A similar approach was used to construct an attenuated influenza A virus vaccine. Perez et al. inserted miRNA response elements into the open reading frame of the viral nucleoprotein. The resulting virus displayed a 2 log reduction in mortality and elicited a diverse antibody response, illustrating the potency of miRNA-mediated control of live-attenuated virus vaccines (95).

Finally, in addition to inhibition, overexpression of specific cellular miRNAs may also be used in therapeutic antiviral approaches. This strategy could be employed when the expression of a certain miRNA is down-regulated upon virus infection in order to increase the expression of a required host co-factor. Overexpression of this miRNA should inhibit the accumulation of the host protein and thus inhibit virus replication. So far, this approach has not been realized, but miR-198 and miR-20 form attractive candidates for such an anti-HIV approach (37, 47). A potential danger of this strategy is that miRNA overexpression affects the expression of many cellular genes. In the optimal scenario, miRNA expression is restored to the level of the uninfected cell.

11. Conclusions

In recent years much has been learned about miRNA biogenesis and function. However, because relatively few miRNA targets have been identified thus far, little is known about miRNA regulatory networks. Even more so, the way in which viruses interact with these systems remains largely unknown. It has gradually become apparent that an interaction between viruses and the host miRNA pathway can take place at multiple levels. To complicate matters further, cellular miRNAs can stimulate or inhibit virus replication, and sometimes a complex dual effect is observed. These RNAi-virus interactions vary significantly depending on the particular virus strain and the host cell type used for the infection, thus causing much variation among experimental systems. Many mechanistic details on how these processes interact and what the consequences are for viral replication, the host cell and virus-induced pathogenesis remains to be determined. Novel technology for high-throughput analysis of small virus-derived RNAs will help to answer some of these questions. It is likely that an increased understanding of these processes will ultimately lead to the identification of new targets for the development of antiviral therapeutics.

Acknowledgements

RNAi research in the Berkhout lab is sponsored by ZonMw (Vici grant and Translational Gene Therapy program), NWO-CW (Top grant), the European Union (LSHP-CT-2006-037301) and the Technology Foundation STW (grant AGT.7708).

References

1. Waterhouse, P. M., Wang, M. B., and Lough, T. (2001) Gene silencing as an adaptive defence against viruses. *Nature* **411**, 834–42.
2. Kim, V. N., Han, J., and Siomi, M. C. (2009) Biogenesis of small RNAs in animals. *Nat Rev Mol Cell Biol* **10**, 126–39.
3. Rana, T. M. (2007) Illuminating the silence: understanding the structure and function of small RNAs. *Nat Rev Mol Cell Biol* **8**, 23–36.
4. Kim, V. N. (2005) MicroRNA biogenesis: coordinated cropping and dicing. *Nat Rev Mol Cell Biol* **6**, 376–85.
5. Krek, A., Grun, D., Poy, M. N., Wolf, R., Rosenberg, L., Epstein, E. J., MacMenamin, P., da Piedade, I., Gunsalus, K. C., Stoffel, M., and Rajewsky, N. (2005) Combinatorial microRNA target predictions. *Nat Genet* **37**, 495–500.
6. Bartel, D. P. (2009) MicroRNAs: target recognition and regulatory functions. *Cell* **136**, 215–33.
7. Segers, G. C., Zhang, X., Deng, F., Sun, Q., and Nuss, D. L. (2007) Evidence that RNA silencing functions as an antiviral defense mechanism in fungi. *Proc Natl Acad Sci U S A* **104**, 12902–6.
8. Voinnet, O. (2001) RNA silencing as a plant immune system against viruses. *Trends Genet* **17**, 449–59.
9. Wilkins, C., Dishongh, R., Moore, S. C., Whitt, M. A., Chow, M., and Machaca, K. (2005) RNA interference is an antiviral

defence mechanism in *Caenorhabditis elegans*. *Nature* **436**, 1044–7.

10. Pfeffer, S., Sewer, A., Lagos-Quintana, M., Sheridan, R., Sander, C., Grasser, F. A., van Dyk, L. F., Ho, C. K., Shuman, S., Chien, M., Russo, J. J., Ju, J., Randall, G., Lindenbach, B. D., Rice, C. M., Simon, V., Ho, D. D., Zavolan, M., and Tuschl, T. (2005) Identification of microRNAs of the herpesvirus family. *Nat Methods* **2**, 269–76.
11. Cullen, B. R. (2006) Is RNA interference involved in intrinsic antiviral immunity in mammals? *Nat Immunol* **7**, 563–7.
12. Parameswaran, P., Sklan, E., Wilkins, C., Burgon, T., Samuel, M. A., Lu, R., Ansel, K. M., Heissmeyer, V., Einav, S., Jackson, W., Doukas, T., Paranjape, S., Polacek, C., dos Santos, F. B., Jalili, R., Babrzadeh, F., Gharizadeh, B., Grimm, D., Kay, M., Koike, S., Sarnow, P., Ronaghi, M., Ding, S. W., Harris, E., Chow, M., Diamond, M. S., Kirkegaard, K., Glenn, J. S., and Fire, A. Z. Six RNA viruses and forty-one hosts: viral small RNAs and modulation of small RNA repertoires in vertebrate and invertebrate systems. *PLoS Pathog* **6**, e1000764.
13. Yeung, M. L., Bennasser, Y., Watashi, K., Le, S. Y., Houzet, L., and Jeang, K. T. (2009) Pyrosequencing of small non-coding RNAs in HIV-1 infected cells: evidence for the processing of a viral-cellular double-stranded RNA hybrid. *Nucleic Acids Res* **37**, 6575–86.
14. Cullen, B. R. (2009) Viral and cellular messenger RNA targets of viral microRNAs. *Nature* **457**, 421–5.
15. Lee, Y., Kim, M., Han, J., Yeom, K. H., Lee, S., Baek, S. H., and Kim, V. N. (2004) MicroRNA genes are transcribed by RNA polymerase II. *EMBO J* **23**, 4051–60.
16. Han, J., Lee, Y., Yeom, K. H., Kim, Y. K., Jin, H., and Kim, V. N. (2004) The Drosha-DGCR8 complex in primary microRNA processing. *Genes Dev* **18**, 3016–27.
17. Berezikov, E., Chung, W. J., Willis, J., Cuppen, E., and Lai, E. C. (2007) Mammalian mirtron genes. *Mol Cell* **28**, 328–36.
18. Grishok, A., Pasquinelli, A. E., Conte, D., Li, N., Parrish, S., Ha, I., Baillie, D. L., Fire, A., Ruvkun, G., and Mello, C. C. (2001) Genes and mechanisms related to RNA interference regulate expression of the small temporal RNAs that control *C. elegans* developmental timing. *Cell* **106**, 23–34.
19. Kok, K. H., Ng, M. H., Ching, Y. P., and Jin, D. Y. (2007) Human TRBP and PACT directly interact with each other and associate with dicer to facilitate the production of small interfering RNA. *J Biol Chem* **282**, 17649–57.
20. Su, H., Trombly, M. I., Chen, J., and Wang, X. (2009) Essential and overlapping functions for mammalian Argonautes in microRNA silencing. *Genes Dev* **23**, 304–17.
21. Grimson, A., Farh, K. K., Johnston, W. K., Garrett-Engele, P., Lim, L. P., and Bartel, D. P. (2007) MicroRNA targeting specificity in mammals: determinants beyond seed pairing. *Mol Cell* **27**, 91–105.
22. Nilsen, T. W. (2007) Mechanisms of microRNA-mediated gene regulation in animal cells. *Trends Genet* **23**, 243–9.
23. Diaz-Pendon, J. A., and Ding, S. W. (2008) Direct and indirect roles of viral suppressors of RNA silencing in pathogenesis. *Annu Rev Phytopathol* **46**, 303–26.
24. Ding, S. W., and Voinnet, O. (2007) Antiviral immunity directed by small RNAs. *Cell* **130**, 413–26.
25. Bowie, A. G., and Unterholzner, L. (2008) Viral evasion and subversion of pattern-recognition receptor signalling. *Nat Rev Immunol* **8**, 911–22.
26. Lin, J., and Cullen, B. R. (2007) Analysis of the interaction of primate retroviruses with the human RNA interference machinery. *J Virol* **81**, 12218–26.
27. Haasnoot, J., Westerhout, E. M., and Berkhout, B. (2007) RNA interference against viruses: strike and counterstrike. *Nat Biotechnol* **25**, 1435–43.
28. Watanabe, T., Totoki, Y., Toyoda, A., Kaneda, M., Kuramochi-Miyagawa, S., Obata, Y., Chiba, H., Kohara, Y., Kono, T., Nakano, T., Surani, M. A., Sakaki, Y., and Sasaki, H. (2008) Endogenous siRNAs from naturally formed dsRNAs regulate transcripts in mouse oocytes. *Nature* **453**, 539–43.
29. Tam, O. H., Aravin, A. A., Stein, P., Girard, A., Murchison, E. P., Cheloufi, S., Hodges, E., Anger, M., Sachidanandam, R., Schultz, R. M., and Hannon, G. J. (2008) Pseudogene-derived small interfering RNAs regulate gene expression in mouse oocytes. *Nature* **453**, 534–8.
30. Yang, N., and Kazazian, H. H., Jr. (2006) L1 retrotransposition is suppressed by endogenously encoded small interfering RNAs in human cultured cells. *Nat Struct Mol Biol* **13**, 763–71.
31. Haussecker, D., Huang, Y., Lau, A., Parameswaran, P., Fire, A. Z., and Kay, M. A. Human tRNA-derived small RNAs in the global regulation of RNA silencing. *RNA* **16**, 673–95.
32. Huang, J., Wang, F., Argyris, E., Chen, K., Liang, Z., Tian, H., Huang, W., Squires, K., Verlinghieri, G., and Zhang, H. (2007) Cellular microRNAs contribute to HIV-1

latency in resting primary CD4+ T lymphocytes. *Nat Med* **13**, 1241–7.

33. Lecellier, C. H., Dunoyer, P., Arar, K., Lehmann-Che, J., Eyquem, S., Himber, C., Saib, A., and Voinnet, O. (2005) A cellular microRNA mediates antiviral defense in human cells. *Science* **308**, 557–60.
34. Otsuka, M., Jing, Q., Georgel, P., New, L., Chen, J., Mols, J., Kang, Y. J., Jiang, Z., Du, X., Cook, R., Das, S. C., Pattnaik, A. K., Beutler, B., and Han, J. (2007) Hypersusceptibility to vesicular stomatitis virus infection in Dicer1-deficient mice is due to impaired miR24 and miR93 expression. *Immunity* **27**, 123–34.
35. Pedersen, I. M., Cheng, G., Wieland, S., Volinia, S., Croce, C. M., Chisari, F. V., and David, M. (2007) Interferon modulation of cellular microRNAs as an antiviral mechanism. *Nature* **449**, 919–22.
36. Kelly, E. J., and Russell, S. J. (2009) MicroRNAs and the regulation of vector tropism. *Mol Ther* **17**, 409–16.
37. Wang, X., Ye, L., Hou, W., Zhou, Y., Wang, Y. J., Metzger, D. S., and Ho, W. Z. (2009) Cellular microRNA expression correlates with susceptibility of monocytes/macrophages to HIV-1 infection. *Blood* **113**, 671–4.
38. Hariharan, M., Scaria, V., Pillai, B., and Brahmachari, S. K. (2005) Targets for human encoded microRNAs in HIV genes. *Biochem Biophys Res Commun* **337**, 1214–8.
39. Ahluwalia, J. K., Khan, S. Z., Soni, K., Rawat, P., Gupta, A., Hariharan, M., Scaria, V., Lalwani, M., Pillai, B., Mitra, D., and Brahmachari, S. K. (2008) Human cellular microRNA hsa-miR-29a interferes with viral nef protein expression and HIV-1 replication. *Retrovirology* **5**, 117.
40. Nathans, R., Chu, C. Y., Serquina, A. K., Lu, C. C., Cao, H., and Rana, T. M. (2009) Cellular microRNA and P bodies modulate host-HIV-1 interactions. *Mol Cell* **34**, 696–709.
41. Jopling, C. L., Schutz, S., and Sarnow, P. (2008) Position-dependent function for a tandem microRNA miR-122-binding site located in the hepatitis C virus RNA genome. *Cell Host Microbe* **4**, 77–85.
42. Jopling, C. L., Yi, M., Lancaster, A. M., Lemon, S. M., and Sarnow, P. (2005) Modulation of hepatitis C virus RNA abundance by a liver-specific MicroRNA. *Science* **309**, 1577–81.
43. Jangra, R. K., Yi, M., and Lemon, S. M. Regulation of hepatitis C virus translation and infectious virus production by the microRNA miR-122. *J Virol* **84**, 6615–25.
44. Sarasin-Filipowicz, M., Krol, J., Markiewicz, I., Heim, M. H., and Filipowicz, W. (2009) Decreased levels of microRNA miR-122 in individuals with hepatitis C responding poorly to interferon therapy. *Nat Med* **15**, 31–3.
45. Yeung, M. L., Bennasser, Y., Myers, T. G., Jiang, G., Benkirane, M., and Jeang, K. T. (2005) Changes in microRNA expression profiles in HIV-1-transfected human cells. *Retrovirology* **2**, 81.
46. Houzet, L., Yeung, M. L., de Lame, V., Desai, D., Smith, S. M., and Jeang, K. T. (2008) MicroRNA profile changes in human immunodeficiency virus type 1 (HIV-1) seropositive individuals. *Retrovirology* **5**, 118.
47. Triboulet, R., Mari, B., Lin, Y. L., Chable-Bessia, C., Bennasser, Y., Lebrigand, K., Cardinaud, B., Maurin, T., Barbry, P., Baillat, V., Reynes, J., Corbeau, P., Jeang, K. T., and Benkirane, M. (2007) Suppression of microRNA-silencing pathway by HIV-1 during virus replication. *Science* **315**, 1579–82.
48. van Haaften, G., and Agami, R. Tumorigenicity of the miR-17-92 cluster distilled. *Genes Dev* **24**, 1–4.
49. Mallick, B., Ghosh, Z., and Chakrabarti, J. (2009) MicroRNome analysis unravels the molecular basis of SARS infection in broncho-alveolar stem cells. *PLoS One* **4**, e7837.
50. Li, Y., Chan, E. Y., Li, J., Ni, C., Peng, X., Rosenzweig, E., Tumpey, T. M., and Katze, M. G. MicroRNA expression and virulence in pandemic influenza virus-infected mice. *J Virol* **84**, 3023–32.
51. Wang, Y., Brahmakshatriya, V., Zhu, H., Lupiani, B., Reddy, S. M., Yoon, B. J., Gunaratne, P. H., Kim, J. H., Chen, R., Wang, J., and Zhou, H. (2009) Identification of differentially expressed miRNAs in chicken lung and trachea with avian influenza virus infection by a deep sequencing approach. *BMC Genomics* **10**, 512.
52. Taganov, K. D., Boldin, M. P., Chang, K. J., and Baltimore, D. (2006) NF-kappaB-dependent induction of microRNA miR-146, an inhibitor targeted to signaling proteins of innate immune responses. *Proc Natl Acad Sci U S A* **103**, 12481–6.
53. Buck, A. H., Perot, J., Chisholm, M. A., Kumar, D. S., Tuddenham, L., Cognat, V., Marcinowski, L., Dolken, L., and Pfeffer, S. Post-transcriptional regulation of miR-27 in murine cytomegalovirus infection. *RNA* **16**, 307–15.
54. Croce, C. M. (2009) Causes and consequences of microRNA dysregulation in cancer. *Nat Rev Genet* **10**, 704–14.
55. Tomita, M., Tanaka, Y., and Mori, N. (2009) MicroRNA miR-146a is induced by HTLV-1 tax and increases the growth of HTLV-1-infected T-cells. *Int J Cancer*. Dec. 16.

56. Wang, X., Tang, S., Le, S. Y., Lu, R., Rader, J. S., Meyers, C., and Zheng, Z. M. (2008) Aberrant expression of oncogenic and tumor-suppressive microRNAs in cervical cancer is required for cancer cell growth. *PLoS One* **3**, e2557.
57. Bolisetty, M. T., Dy, G., Tam, W., and Beemon, K. L. (2009) Reticuloendotheliosis virus strain T induces miR-155, which targets JARID2 and promotes cell survival. *J Virol* **83**, 12009–17.
58. Lambeth, L. S., Yao, Y., Smith, L. P., Zhao, Y., and Nair, V. (2009) MicroRNAs 221 and 222 target p27Kip1 in Marek's disease virus-transformed tumour cell line MSB-1. *J Gen Virol* **90**, 1164–71.
59. Li, W. X., and Ding, S. W. (2001) Viral suppressors of RNA silencing. *Curr Opin Biotechnol* **12**, 150–4.
60. de Vries, W., and Berkhout, B. (2008) RNAi suppressors encoded by pathogenic human viruses. *Int J Biochem Cell Biol* **40**, 2007–12.
61. Bucher, E., Hemmes, H., de Haan, P., Goldbach, R., and Prins, M. (2004) The influenza A virus NS1 protein binds small interfering RNAs and suppresses RNA silencing in plants. *J Gen Virol* **85**, 983–91.
62. Li, H., Li, W. X., and Ding, S. W. (2002) Induction and suppression of RNA silencing by an animal virus. *Science* **296**, 1319–21.
63. Haasnoot, J., de Vries, W., Geutjes, E. J., Prins, M., de Haan, P., and Berkhout, B. (2007) The Ebola virus VP35 protein is a suppressor of RNA silencing. *PLoS Pathog* **3**, e86.
64. Bennasser, Y., Le, S. Y., Benkirane, M., and Jeang, K. T. (2005) Evidence that HIV-1 encodes an siRNA and a suppressor of RNA silencing. *Immunity* **22**, 607–19.
65. Qian, S., Zhong, X., Yu, L., Ding, B., de Haan, P., and Boris-Lawrie, K. (2009) HIV-1 Tat RNA silencing suppressor activity is conserved across kingdoms and counteracts translational repression of HIV-1. *Proc Natl Acad Sci U S A* **106**, 605–10.
66. Chen, W., Zhang, Z., Chen, J., Zhang, J., Zhang, J., Wu, Y., Huang, Y., Cai, X., and Huang, A. (2008) HCV core protein interacts with Dicer to antagonize RNA silencing. *Virus Res* **133**, 250–8.
67. Ji, J., Glaser, A., Wernli, M., Berke, J. M., Moradpour, D., and Erb, P. (2008) Suppression of short interfering RNA-mediated gene silencing by the structural proteins of hepatitis C virus. *J Gen Virol* **89**, 2761–6.
68. Wang, Y., Kato, N., Jazag, A., Dharel, N., Otsuka, M., Taniguchi, H., Kawabe, T., and Omata, M. (2006) Hepatitis C virus core protein is a potent inhibitor of RNA silencing-based antiviral response. *Gastroenterology* **130**, 883–92.
69. Berkhout, B., Silverman, R. H., and Jeang, K. T. (1989) Tat trans-activates the human immunodeficiency virus through a nascent RNA target. *Cell* **59**, 273–82.
70. Schnettler, E., de Vries, W., Hemmes, H., Haasnoot, J., Kormelink, R., Goldbach, R., and Berkhout, B. (2009) The NS3 protein of rice hoja blanca virus complements the RNAi suppressor function of HIV-1 Tat. *EMBO Rep* **10**, 258–63.
71. Liao, H. J., Kobayashi, R., and Mathews, M. B. (1998) Activities of adenovirus virus-associated RNAs: purification and characterization of RNA binding proteins. *Proc Natl Acad Sci U S A* **95**, 8514–9.
72. Andersson, M. G., Haasnoot, P. C., Xu, N., Berenjian, S., Berkhout, B., and Akusjarvi, G. (2005) Suppression of RNA interference by adenovirus virus-associated RNA. *J Virol* **79**, 9556–65.
73. Aparicio, O., Razquin, N., Zaratiegui, M., Narvaiza, I., and Fortes, P. (2006) Adenovirus virus-associated RNA is processed to functional interfering RNAs involved in virus production. *J Virol* **80**, 1376–84.
74. Lu, S., and Cullen, B. R. (2004) Adenovirus VA1 noncoding RNA can inhibit small interfering RNA and MicroRNA biogenesis. *J Virol* **78**, 12868–76.
75. Sano, M., Kato, Y., and Taira, K. (2006) Sequence-specific interference by small RNAs derived from adenovirus VAI RNA. *FEBS Lett* **580**, 1553–64.
76. Xu, N., Segerman, B., Zhou, X., and Akusjarvi, G. (2007) Adenovirus virus-associated RNAII-derived small RNAs are efficiently incorporated into the rna-induced silencing complex and associate with polyribosomes. *J Virol* **81**, 10540–9.
77. Cardenas, W. B., Loo, Y. M., Gale, M., Jr., Hartman, A. L., Kimberlin, C. R., Martinez-Sobrido, L., Saphire, E. O., and Basler, C. F. (2006) Ebola virus VP35 protein binds double-stranded RNA and inhibits alpha/beta interferon production induced by RIG-I signaling. *J Virol* **80**, 5168–78.
78. Chang, H. W., Watson, J. C., and Jacobs, B. L. (1992) The E3L gene of vaccinia virus encodes an inhibitor of the interferon-induced, double-stranded RNA-dependent protein kinase. *Proc Natl Acad Sci U S A* **89**, 4825–9.
79. Hartman, A. L., Bird, B. H., Towner, J. S., Antoniadou, Z. A., Zaki, S. R., and Nichol, S. T. (2008) Inhibition of IRF-3 activation by VP35 is critical for the high level of virulence of ebola virus. *J Virol* **82**, 2699–704.

80. Katze, M. G., He, Y., and Gale, M., Jr. (2002) Viruses and interferon: a fight for supremacy. *Nat Rev Immunol* **2**, 675–87.
81. Lu, Y., Wambach, M., Katze, M. G., and Krug, R. M. (1995) Binding of the influenza virus NS1 protein to double-stranded RNA inhibits the activation of the protein kinase that phosphorylates the elF-2 translation initiation factor. *Virology* **214**, 222–8.
82. Mibayashi, M., Martinez-Sobrido, L., Loo, Y. M., Cardenas, W. B., Gale, M., Jr., and Garcia-Sastre, A. (2007) Inhibition of retinoic acid-inducible gene I-mediated induction of beta interferon by the NS1 protein of influenza A virus. *J Virol* **81**, 514–24.
83. Min, J. Y., and Krug, R. M. (2006) The primary function of RNA binding by the influenza A virus NS1 protein in infected cells: Inhibiting the 2′-5′ oligo (A) synthetase/RNase L pathway. *Proc Natl Acad Sci U S A* **103**, 7100–5.
84. Rivas, C., Gil, J., Melkova, Z., Esteban, M., and Diaz-Guerra, M. (1998) Vaccinia virus E3L protein is an inhibitor of the interferon (i.f.n.)-induced 2-5A synthetase enzyme. *Virology* **243**, 406–14.
85. Xiang, Y., Condit, R. C., Vijaysri, S., Jacobs, B., Williams, B. R., and Silverman, R. H. (2002) Blockade of interferon induction and action by the E3L double-stranded RNA binding proteins of vaccinia virus. *J Virol* **76**, 5251–9.
86. Zhang, H. (2009) Reversal of HIV-1 latency with anti-microRNA inhibitors. *Int J Biochem Cell Biol* **41**, 451–4.
87. Chang, J., Guo, J. T., Jiang, D., Guo, H., Taylor, J. M., and Block, T. M. (2008) Liver-specific microRNA miR-122 enhances the replication of hepatitis C virus in nonhepatic cells. *J Virol* **82**, 8215–23.
88. Jopling, C. L. (2008) Regulation of hepatitis C virus by microRNA-122. *Biochem Soc Trans* **36**, 1220–3.
89. Randall, G., Panis, M., Cooper, J. D., Tellinghuisen, T. L., Sukhodolets, K. E., Pfeffer, S., Landthaler, M., Landgraf, P., Kan, S., Lindenbach, B. D., Chien, M., Weir, D. B., Russo, J. J., Ju, J., Brownstein, M. J., Sheridan, R., Sander, C., Zavolan, M., Tuschl, T., and Rice, C. M. (2007) Cellular cofactors affecting hepatitis C virus infection and replication. *Proc Natl Acad Sci U S A* **104**, 12884–9.
90. Lanford, R. E., Hildebrandt-Eriksen, E. S., Petri, A., Persson, R., Lindow, M., Munk, M. E., Kauppinen, S., and Orum, H. Therapeutic silencing of microRNA-122 in primates with chronic hepatitis C virus infection. *Science* **327**, 198–201.
91. Brown, B. D., Venneri, M. A., Zingale, A., Sergi Sergi, L., and Naldini, L. (2006) Endogenous microRNA regulation suppresses transgene expression in hematopoietic lineages and enables stable gene transfer. *Nat Med* **12**, 585–91.
92. Cawood, R., Chen, H. H., Carroll, F., Bazan-Peregrino, M., van Rooijen, N., and Seymour, L. W. (2009) Use of tissue-specific microRNA to control pathology of wild-type adenovirus without attenuation of its ability to kill cancer cells. *PLoS Pathog* **5**, e1000440.
93. Kelly, E. J., Hadac, E. M., Greiner, S., and Russell, S. J. (2008) Engineering microRNA responsiveness to decrease virus pathogenicity. *Nat Med* **14**, 1278–83.
94. Kelly, E. J., Nace, R., Barber, G. N., and Russell, S. J. (2010) Attenuation of vesicular stomatitis virus encephalitis through microRNA targeting. *J Virol* **84**, 1550–62.
95. Perez, J. T., Pham, A. M., Lorini, M. H., Chua, M. A., Steel, J., and tenOever, B. R. (2009) MicroRNA-mediated species-specific attenuation of influenza A virus. *Nat Biotechnol* **27**, 572–6.
96. Murakami Y., Aly H.H., Tajima A., Inoue I., Shimotohno K. (2009) Regulation of the hepatitis C virus genome replication by miR-199a. *J Hepatol* **50**, 453–60.

Chapter 3

Viral miRNAs

Karlie Plaisance-Bonstaff and Rolf Renne

Abstract

Since 2004, more than 200 microRNAs (miRNAs) have been discovered in double-stranded DNA viruses, mainly herpesviruses and polyomaviruses (Nucleic Acids Res 32:D109–D111, 2004). miRNAs are short 22 ± 3 nt RNA molecules that posttranscriptionally regulate gene expression by binding to 3′-untranslated regions (3′UTR) of target mRNAs, thereby inducing translational silencing and/or transcript degradation (Nature 431:350–355, 2004; Cell 116:281–297, 2004). Since miRNAs require only limited complementarity for binding, miRNA targets are difficult to determine (Mol Cell 27:91–105, 2007). To date, targets have only been experimentally verified for relatively few viral miRNAs, which either target viral or host cellular gene expression: For example, SV40 and related polyomaviruses encode miRNAs which target viral large T antigen expression (Nature 435:682–686, 2005; J Virol 79:13094–13104, 2005; Virology 383:183–187, 2009; J Virol 82:9823–9828, 2008) and miRNAs of α-, β-, and γ-herpesviruses have been implicated in regulating the transition from latent to lytic gene expression, a key step in the herpesvirus life cycle. Viral miRNAs have also been shown to target various host cellular genes. Although this field is just beginning to unravel the multiple roles of viral miRNA in biology and pathogenesis, the current data strongly suggest that virally encoded miRNAs are able to regulate fundamental biological processes such as immune recognition, promotion of cell survival, angiogenesis, proliferation, and cell differentiation. This chapter aims to summarize our current knowledge of viral miRNAs, their targets and function, and the challenges lying ahead to decipher their role in viral biology, pathogenesis, and for γ-herepsvirus-encoded miRNAs, potentially tumorigenesis.

Key words: Viral microRNAs, Herpesvirus-encoded microRNAs, microRNA maturation, microRNA targets, microRNA regulation of latency, microRNAs and immune evasion, microRNAs and apoptosis

1. Introduction

1.1. MicroRNAs Regulate Fundamental Cell Processes in All Metazoans

The first miRNA, lin-4 of *C. elegans,* was found through analysis of a strong developmental timing defect. The responsible gene, lin-4, did not contain an ORF, and instead only expressed two short transcripts of 60 and 24 nucleotides in length. It was subsequently shown that the lin-4 RNA was involved in translationally

Ronald P. van Rij (ed.), *Antiviral RNAi: Concepts, Methods, and Applications*, Methods in Molecular Biology, vol. 721, DOI 10.1007/978-1-61779-037-9_3, © Springer Science+Business Media, LLC 2011

silencing the lin-14 transcript by binding to complementary sequences within the lin-14 3′UTR (1–4). This novel RNA-based inhibition was specific to *C. elegans* until the discovery of the let-7 miRNA, which was found to be conserved in metazoans, including humans and flies (5–7). miRNAs have been discovered in every metazoan and plant species tested thus far, and according to estimates, around 30% of all metazoan miRNAs are conserved in all species (8). There are currently 1,048 human miRNAs known and this number is predicted to expand as more sensitive techniques for discovering miRNAs are developed (http://microrna.org) (9). The major function of miRNAs appears to be regulation of cellular gene expression through translational inhibition and mRNA degradation; however, there are classes of miRNAs discovered which have different mechanisms of action. miRNAs are an integral part of innate immunity against viruses in plants, and in many organisms, miRNAs are also involved in chromatin silencing (for review see refs. (10–12)).

miRNAs participate in the regulation of apoptosis, cell fate decisions, cell differentiation, stress response, and development (8, 13). However, those miRNAs where targets are known and the rapid identification of new targets showcase that miRNAs regulate fundamental processes during development and differentiation and that miRNA expression is tightly regulated in both spatial and temporal manners (8, 13). This is further supported by the finding that aberrant miRNA expression is associated with pathogenesis including many human malignancies (for review see ref. (14)).

1.2. MicroRNA Biogenesis and Function

Viral miRNA genes are expressed from pol II transcripts or in the case of murine γ-herpesvirus type 68 (MHV-68), in which miRNA genes are embedded in tRNA-like genes by pol III (8, 13, 15, 16). miRNAs can occur individually or be organized into clusters and they can exist as stand-alone genes or located within the introns and exons of protein-coding genes (8, 13). Like metazoan miRNAs, viral miRNA processing begins with the formation of an imperfect stem-loop with a hairpin bulge that forms in a RNA transcript termed the pri-miRNA (Fig. 1). The dsRNA region of the pri-miRNA is recognized by DGCR8 (Pasha in flies), which recruits the endonuclease Drosha to cleave and release a 60–80-nt long hairpin. This pre-miRNA is then exported to the cytoplasm by the Exportin 5/RAN-GTPase pathway, where it is recognized by Dicer and cleaved to leave a short dsRNA molecule. One strand of this dsRNA product is loaded by Dicer and the dsRNA-binding protein TRBP (also known as R2D2 in flies and RDE-4 in nematodes) into the RNA-induced silencing complex (RISC). The other strand, known as the star (*) strand, is often degraded, but in many cases can also be loaded into RISC with variable efficiency (for review see refs. (8, 13)).

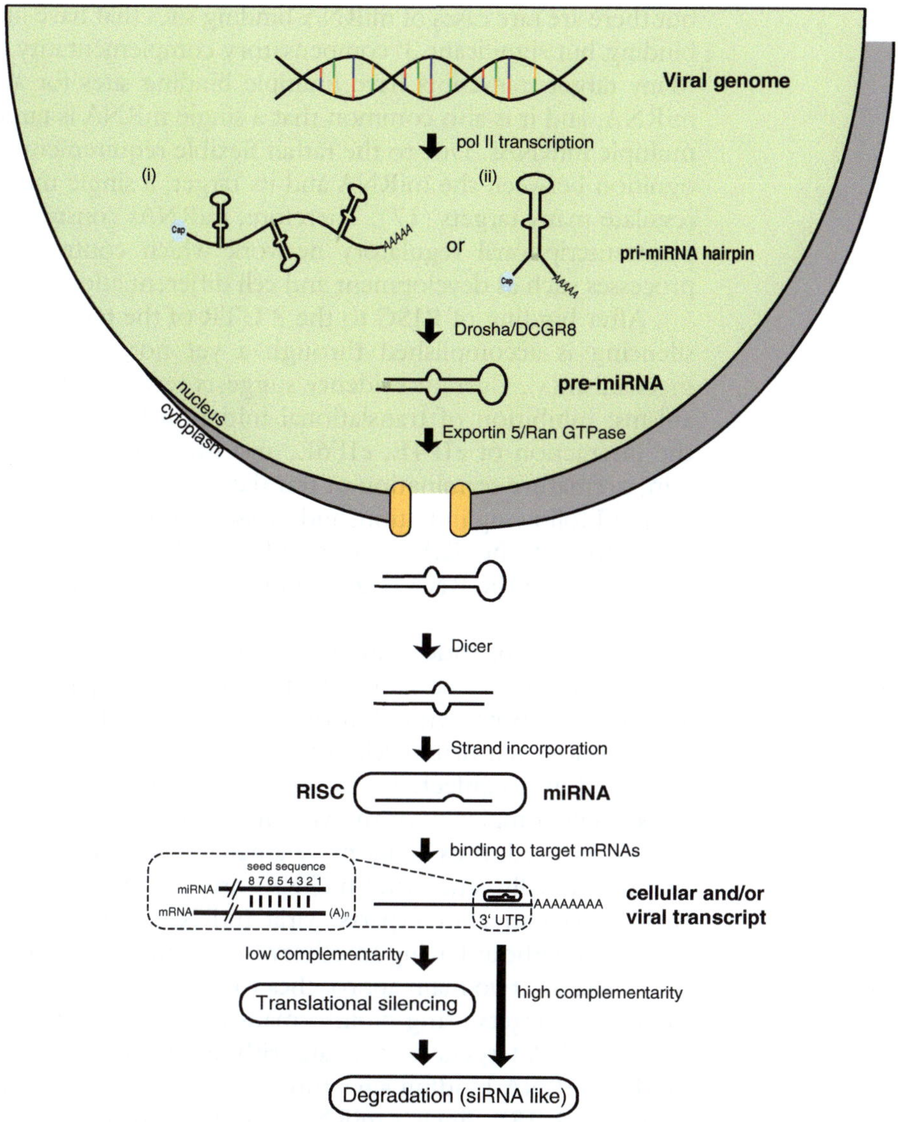

Fig. 1. Biogenesis pathway for metazoan miRNAs. miRNA precursors begin as hairpin loops in pol II or pol III transcripts in introns or exons. Drosha cleaves the pri-miRNA transcript leaving a ~80 bp stem-loop which is exported into the cytoplasm. Dicer cleaves off the loop structure leaving a 21–24 nt dsRNA molecule. The miRNA is incorporated into the RISC where it binds to the 3′UTR of target transcripts and induces either translational silencing or transcriptional degradation depending on the level of complementarity. The seed sequence of the miRNA, nts 2 through 8, is known to be a critical component of target recognition and binding (8, 13).

RISC functions by guiding miRNAs to semicomplementary sites within the 3′UTRs of target transcripts and induces translational silencing. For targets that are completely complementary, siRNA-like degradation occurs rather than translational silencing. The 5′ end of the miRNA, specifically nts 2 through 8 termed the seed sequence, is critical for determining mRNA target binding, but there are rare cases of miRNA-binding sites that have little seed binding but significant 3′ compensatory complementarity instead. Many target transcripts have multiple binding sites for a specific miRNA, and it is also common that a single mRNA is targeted by multiple miRNAs. Due to the rather flexible requirements for recognition between the miRNA and its target, a single miRNA can regulate many targets (17). Therefore, miRNAs constitute a large posttranscriptional regulatory network which controls complex processes such as development and cell differentiation.

After binding of RISC to the 3′UTR of the target transcript, silencing is accomplished through a yet not fully deciphered mechanism(s). Current evidence suggests several different mechanisms: inhibition of translational initiation by interfering with the interaction of eIF4E, eIF6E, and the poly(A)-binding protein, premature termination of translation by inducing ribosomal drop off following initiation, and messenger RNA degradation by relocation of the RISC to cytoplasmic Processing (P)-bodies which contain the RNA degradation machinery (18).

1.3. Techniques to Determine miRNA Targets

Viral miRNAs provide a unique opportunity to examine miRNA targets and functions in mammalian cells. Compared with the 1,048 miRNAs known in humans, the number of virally encoded miRNAs for each virus is relatively small. The fact that they are not expressed in noninfected cells provides a model system in which the overall complexity of the viral miRNA regulatory network is greatly reduced, thus being more amenable to biochemical analysis. Additionally, since viral miRNAs are not sequence-related to their metazoan counterparts, viral miRNAs can be ectopically expressed without having to otherwise genetically modify cells.

There are two main approaches to determine miRNA targets: predicting targets using bioinformatics, or experimentally identifying miRNA targets. Several algorithms have been developed to predict miRNA-binding sites through scanning of 3′UTR libraries (17, 19–23). Since a miRNA is not completely complementary to its target, the strength of a potential target site has to be calculated by weighing different factors such as seed binding, overall thermodynamic stability of the miRNA/mRNA duplex, and for some algorithms conservation of target sites between species (17). John et al. used miRanda with 218 human miRNAs and predicted over 2,000 potential targets (24). Thus, computer-based target predictions provide a useful start, but clearly need to be experimentally validated by directly demonstrating either mRNA degradation or translational silencing.

Using microarrays for gene expression profiling provides an efficient method to determine which cellular transcripts are up- or downregulated in response to miRNA expression. In addition, either microarrays containing viral probe sets or genome wide qRT-PCR assays, which have been developed for all human herpesviruses (25, 26), are available to investigate whether viral miRNAs target viral gene expression. Observed transcriptional repression or decreased protein levels may be due to secondary effects. Therefore, for target verification, miRNA binding should be functionally confirmed within a 3′UTR of each candidate target gene. The most commonly used method is to introduce 3′UTRs downstream of a luciferase reporter and demonstrate miRNA-dependent repression. This can be confirmed through mutational analysis of identified binding sites. More recently, both proteomics and genomics approaches in combination with a high-throughput sequencing (HTS) assays have been utilized to determine global impact of miRNA-induced expression changes. Two groups successfully analyzed total proteomes in cells engineered to ectopically express specific miRNAs by HTS mass spectrometry and found that while many proteins were changed, most changes were moderate (27, 28). In addition, UV cross-linking immune precipitation (CLIP) assays for RISC resident proteins such as Ago2 have been combined with HTS sequencing (CLIP-Seq) to analyze the total repertoire of miRNA-bound mRNAs within cells (29, 30). These novel methods will greatly increase our understanding of how different miRNA repertoires and their alteration in response to viral infection will influence host cellular transcriptomes and proteomes. The first CLIP-Seq data set has recently been reported for EBV- and KSHV-infected lymphoma cells and will be discussed in Subheading 3 (31).

It is important to note that determining specific target genes alone does not necessarily define biological function and vice versa. Functions of metazoan miRNAs have mostly been defined by genetic approaches (32–34). The ability of cellular as well as viral miRNAs to target specific genes depends on the complexity of the cellular transcriptome and the miRNA repertoire within a specific cell type, which may not be accurately represented by current miRNA target determination methods. For example, most herpesviruses infect several different cell lineages, and both miRNA expression and the presence of specific target mRNAs are likely different in each infected cell type.

2. Herpesviruses and Polyomaviruses Encode miRNAs

In 2004, Pfeffer et al. reported the cloning and identification of five miRNAs from Epstein Barr virus (EBV). Initially, three miRNAs were found to be located in the BHRF region and two miRNAs in the BART region of EBV (35). This first report of virally encoded miRNAs opened a new field of virology. The possibility that viral miRNAs can regulate hundreds of target genes suggests a novel and extremely complex level of host/virus interaction. A number of herpesvirus proteins, often pirated from host genomes, target specific cellular processes such as immune surveillance, apoptosis, and proliferation, and in retrospect, it seems obvious that viruses would also utilize miRNAs to regulate these pathways. To date, more than 200 miRNAs have been identified in 20 different DNA viruses (Table 1, Fig. 2).

2.1. γ-Herpesvirus-Encoded miRNAs

After EBV was shown to encode miRNAs, four independent groups cloned miRNAs from Kaposi's sarcoma-associated herpesvirus (KSHV)-infected primary effusion lymphoma cells (PEL) and identified a total of 12 miRNA genes giving rise to 18 mature miRNAs (36–40). Surprisingly, all KSHV miRNAs are located

Table 1
Verified known viral miRNAs

Virus	Number of miRNA genes	References
HSV-1	16	(49, 51, 53)
HSV-2	17	(50, 51, 54, 55)
MDV-1	13	(56, 59)
MDV-2	17	(52, 57, 58)
HCMV	11	(46, 47)
MCMV	18	(35, 102)
EBV	25	(35, 37, 41, 42)
rLCV	36	(41, 43, 44)
RRV	7	(45)
KSHV	12	(36–40)
MHV-68	9	(40)
SV40	1	(60)
SA12	1	(61)
BKV	1	(62)
JCV	1	(62)
MCV	1	(63)
PyV	1	(64)

HSV-1 and *-2* herpes simplex virus; *MDV*-1 and -2 Marek's disease virus; *HCMV* and *MCMV* human and mouse cytomegalovirus; *EBV* Epstein Barr virus, *rLCV* Rhesus lymphocryptovirus; *RRV* Rhesus monkey rhadinovirus; *KSHV* Kaposi's sarcoma-associated herpesvirus, *MHV*-68 Murine gammaherpesvirus-68; *SV40* Simian virus 40; *SA12* Simian agent 12, *BKV* Polyomavirus BK, *JCV* polyomavirus JC, *MCV* Merkel cell polyomavirus, *PyV* murine polyomavirus

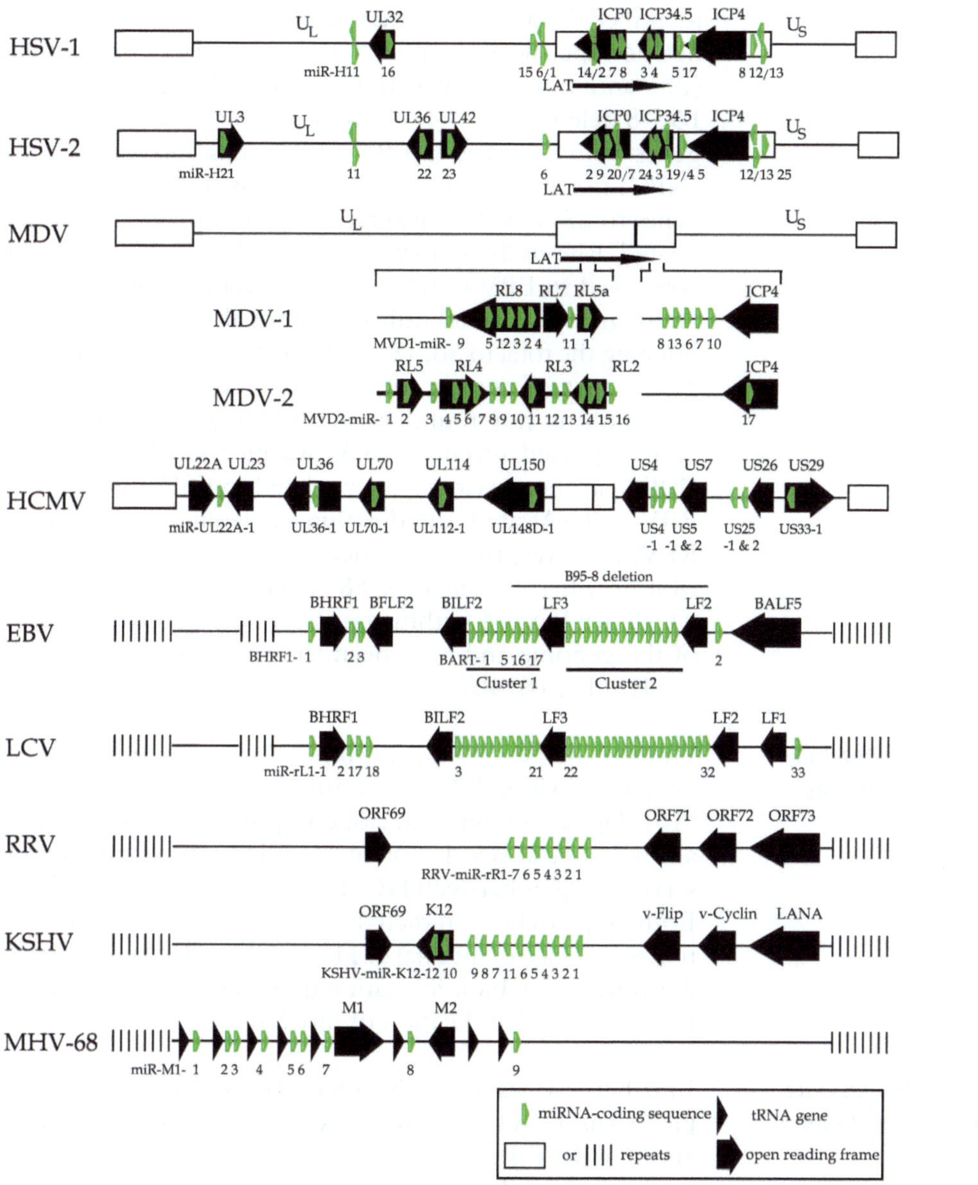

Fig. 2. Schematic representation of miRNAs found for several herpesviruses. Genomes are represented for HSV-1, -2, MDV-1, -2, HCMV, EBV, LCV, RRV, KSHV, and MHV-68 with *black arrows* for ORFs, *black triangles* for tRNA genes, and *black bars* or rectangles for repeat sequences. miRNA locations are indicated with *green arrows*. Genomes are not drawn to scale. MDV-1 and -2 are drawn as one complete genome with the respective miRNA-coding regions shown in more detail. Figure was compiled from (35–37, 39–41, 46–48, 57, 59). U_S unique short; U_L unique long; *LAT* latency-associated transcript.

within the major latency-associated region of the genome with 10 of the 12 miRNAs organized in a cluster in the intragenic region between v-Flip and the K12/Kaposin gene. Two additional miRNAs were found to be located within K12 open reading frame. A combination of tiled arrays, cloning, and bioinformatic approaches identified 18 additional EBV miRNAs, located within the 12 kb deletion specific to the B95-8 strain analyzed in the original report (35) and three more within the BART region outside of the B95-8 deletion (37, 41). Recently, two additional BART miRNA genes were identified in EBV-positive nasopharyngeal carcinomas (NPC) tissue samples (42). This brings the total of EBV miRNA genes to 25. Cai et al. also reported 16 miRNAs within the EBV-related Rhesus lymphocryptovirus (rLCV), eight of which are conserved to EBV miRNAs (41). Recently, both Grundhoff and Steitz groups have identified additional rLCV-encoded miRNAs bringing the total to 36, 18 of which are conserved to EBV (43, 44). The extent of conservation between EBV and rLCV miRNAs sequences is not found elsewhere in γ-herpesviruses. Schaefer et al. reported seven miRNAs within the Rhesus Rhadinovirus (RRV), a γ-herpesvirus closely related to KSHV (45). Like KSHV, RRV miRNAs are located within the latency-associated region of RRV; however, their sequences are not evolutionary conserved. Murine γ-herpesvirus type 68 (MHV68) encodes nine miRNAs which are located within transfer RNA-like genes at the 5′ end of the genome and have been shown to be transcribed by RNA pol III (15, 16, 40).

2.2. β-Herpesvirus-Encoded miRNAs

Within β-herpesviruses, nine miRNAs were identified from human cytomegalovirus (HCMV), scattered throughout the viral genome (40) (Fig. 2). Dunn et al. cloned a previously unreported miRNA and Grey et al. used a bioinformatics approach to predict conserved hairpins between HCMV and chimpanzee CCMV (46, 47). Both approaches predicted and confirmed two new HCMV miRNAs for a current total of 11 HCMV miRNAs. These studies also illustrated that bioinformatics approaches alone are not reliable tools for the identification of miRNAs.

2.3. α-Herpesvirus-Encoded miRNAs

In α-herpesviruses, miRNAs have been identified in Herpes simplex virus 1 and 2 (HSV-1 and -2) and Marek's disease virus 1 and 2 (MDV-1 and -2) (48–59). Latency of HSV is primarily established in the sensory neurons of the trigeminal and sacral ganglia. As seen in KSHV, α-herpesviruses also encode miRNAs during latency. In 2006, Cui et al. predicted several miRNAs within and upstream of the HSV-1 and HSV-2 latency-associated transcript (LAT) (48). LAT is a noncoding mRNA and is believed to be the only transcript expressed during latency. Subsequently, Umbach et al. showed that the HSV-1 LAT functions as a pri-miRNA giving rise to four miRNAs (53). Additionally, one miRNA is located directly upstream of LAT and was found to be expressed in latently infected mouse trigeminal ganglia (53). Tang et al. first reported LAT-encoded miRNAs of HSV-2 (54, 55). Umbach and colleagues validated the expression of the HSV-1 and HSV-2 miRNAs in human trigeminal ganglia and sacral ganglia by using sensitive deep sequencing methods. Additionally, two more miRNAs located within LAT of HSV-1 and three novel HSV-2 miRNAs were identified (49, 50). Recently, Jurak et al. utilized deep sequencing to confirm previously identified and predicted miRNAs by Cui, Umbach, and Tang, along with identifying 19 new HSV-1 and HSV-2-encoded miRNAs, bringing the total number of HSV-1 miRNAs and HSV-2 miRNAs to 16 and 17, respectively (Table 1, Fig. 2). Interestingly, HSV-1 and HSV-2 share nine miRNAs that are either positional conserved and/or show limited sequence homology (51).

Burnside et al. used 454 deep sequencing to identify 13 miRNAs from Marek's disease virus type 1 (MDV-1), which map to the inverted repeat short and long regions (IR_s and IR_L) (Fig. 2). Eight of these miRNAs are located within the meq oncogene and the remaining map to the LAT region of MDV-1 (56, 59). Conventional miRNA cloning, and recently deep sequencing, revealed 18 miRNAs within the closely related MDV-2 virus, 17 of which were clustered within IR_L with one additional located in IR_S (52, 57, 58).

2.4. miRNAs Encoded by Other Viruses

Outside the herpesvirus family, two miRNAs resulting from a single hairpin were identified within the 3′UTR of the SV40 late transcript (60). Positional homologs of these SV40 miRNAs were also isolated from SA12-infected cells (61). Human polyomaviruses BKV and JCV along with Merkel cell polyomavirus were also shown to encode one miRNA each (62, 63). Recently, Sullivan et al. also showed that murine polyomavirus (PyV) encodes one miRNA gene (64). All polyomavirus miRNAs are located antisense to the viral mRNAs encoding large T antigen. Hence, while common in herpesviruses and polyomaviruses, miRNA-encoding genes appear to be rare in other virus families.

It appears that no RNA virus investigated to date encodes miRNAs. Pfeffer et al. were unable to clone miRNAs from cells infected with either hepatitis C virus, yellow fever virus, or HIV (40), and there have been no other reports of miRNAs being predicted and verified in other RNA viruses. Thus, it appears as if miRNAs are a DNA virus-specific phenomenon. Since all RNA viruses (except for retroviruses) progress through a dsRNA intermediate, miRNAs would ultimately act as siRNA and inhibit viral replication. Additionally, transcripts from RNA viruses that often replicate exclusively in the cytoplasm would have to be shuttled into the nucleus to be processed by Drosha. This same issue arises with poxviruses replicating solely in the cytoplasm, which may explain why only very few miRNAs have been predicted within the more than 300 kbp large double-stranded DNA genomes of poxviruses and none have been experimentally confirmed (40). At this admittedly early stage in the field, it appears that predominantly herpesviruses encode miRNAs, which may be just another example for host cellular genes that have been successfully captured and subsequently coevolved into herpesvirus genomes.

2.5. Viral miRNAs Show Only Limited Sequence Conservation

With the discovery of the let-7 miRNA, it became clear that miRNAs are highly conserved between metazoan organisms. For human miRNAs, it is estimated that around 30% are conserved (8). The degree of miRNA conservation is astonishing. For example, let-7 is 100% identical between humans, flies, and worms (5–7). In contrast, viral miRNAs appear to be poorly conserved, as they show no sequence homology to metazoan miRNAs and little conservation between viruses. Only highly related viruses appear to have conserved miRNAs, and thus far, there are only three examples of miRNA homology between viruses. First, there are miRNAs that are hypothesized to be conserved between HCMV and CCMV (chimpanzee CMV) (47). In an effort to identify candidate CMV miRNAs, Grey et al. scanned the closely related HCMV and CCMV genomes for conserved hairpins. Several hits were previously known to encode HCMV miRNAs, strongly suggesting that miRNAs in CCMV and HCMV are conserved (47). Second, seven miRNA stem-loops were found to be conserved between EBV and rLCV (41). Whether these miRNAs have conserved functions remains to be determined. Interestingly, there is no miRNA conservation between KSHV and RRV, two closely related rhadinoviruses; yet miRNAs for both viruses are located within their respective latency-associated regions (45). Similarly, MDV-1 and MDV-2 miRNAs, which are also clustered, are positionally conserved, but lack significant sequence conservation (57, 58, 65). HSV-1 and -2 LAT miRNAs that are antisense to ORFs are conserved in location, but show only limited sequence homology, while miRNAs that presumably target through seed sequences show limited sequence homology (66).

These observations suggest that miRNAs have existed in viruses for millions of years and have slowly evolved and adapted within their specific hosts. Therefore, we only observe miRNA conservation in viruses that have the least amount of evolutionary divergence and the greatest similarity in viral biology. Accordingly, polyomavirus miRNA genes that target large T antigen are conserved among all SV40-related polyomaviridae (BKV, JCV and SA12), suggesting a common siRNA-based immune evasion strategy as discussed below (60, 61, 63).

Are viral miRNAs conserved within different viral strains or clinical isolates? To address this question, Marshall et al. analyzed clinical samples from AIDS-KS and KS patients of different geographical origin and found that most miRNAs were highly conserved, suggesting in vivo selection for functional miRNA genes (67). However, this study also identified a number of miRNA polymorphisms and a recent study suggests a linkage of such polymorphisms with pathogenesis (Whitby in press, Journal of Infectious Disease). Molecular studies reveal that miRNA polymorphisms can affect both miRNA maturation and targeting in infected cells (68) (Whitby and Renne, unpublished results).

3. Identifying Targets and Functions of Viral miRNAs

Understanding viral miRNA function ultimately will require comprehensive target gene identification. Viral miRNAs are unique in that there are two potential groups of targets. Viral miRNAs can target cellular transcripts to modulate the host environment and/or target viral transcripts to regulate viral gene expression. Indeed, cellular and/or viral mRNA targets have been found for at least one viral miRNA from all of the human viruses known to encode miRNAs. Although this field is relatively new, studies over the past 6 years strongly suggest that these novel viral posttranscriptional regulators play a role in many processes including immune evasion, cell survival, cell proliferation, and maintenance of latency.

3.1. Viral miRNAs Targeting Viral Genes

The SV40 miRNAs, located within the 3′ UTR of the late transcript, target and efficiently degrade early transcripts encoding the large T antigen (60). As a result, large T antigen expression is downregulated after DNA replication has been completed. Using in vitro studies, Sullivan et al. showed that SV40-infected cells were more efficiently lysed in cytotoxic T-cell assays when infected with a genetically engineered miRNA knock-out virus, strongly suggesting that these miRNAs function to reduce immune recognition. However, using the murine PyV to perform similar knock-out studies but analyze the miRNA minus virus in vivo did not show any replication differences in infected immunocompetent mice. Therefore, the biological significance of miRNA-dependent downregulation of SV40 large T, although conserved for all polyomaviruses, is not fully understood (60, 64).

HSV-1 and HSV-2 miRNAs are located antisense to viral genes due to perfect sequence complimentarity between miRNA and mRNA target function siRNA-like. The LAT of HSV-1 and HSV-2 is located antisense of ICP0 and ICP34.5, which are both important regulators of lytic reactivation. MiR-H2 of both HSV-1 and HSV-2 are antisense to ICP0, resulting in downregulation of ICP0 protein expression (53–55). Similarly, ICP34.5, a neurovirulence factor, is transcribed antisense to miR-H3 and miR-H4. Targeting of ICP34.5 has been confirmed for HSV-2 miRNAs (53–55). In contrast, HSV-1 miR-H6 is located upstream of LAT and bioinformatic analysis revealed a punitive binding site in the 3′UTR of ICP4, another immediate-early viral transactivator. Umbach et al. demonstrated miR-H6 targeting of ICP4 which significantly reduced protein levels (53). Targeting of two viral transactivators, both of which are crucial for reactivation, by viral miRNAs suggests that the primary function of HSV LAT is to express miRNAs, which contribute to the establishment and maintenance of latency.

HCMV miRNAs are not located in clusters, but are dispersed across the entire HCMV genome (Fig. 2). However, common to HSV-1, HCMV miRNAs also target immediate-early viral transactivators. Bioinformatics analysis suggested miR-UL112-1 targets sites within the 3′UTRs of UL112/113 and IE72, viral proteins that promote viral DNA replication, and UL120/121 whose function is unknown. Luciferase assays and transfection of miR-UL112-1 mimics into HCMV-infected cells experimentally validated miR-UL112-1 targeting of IE72 (69, 70). Murphy et al. developed an algorithm to predict miRNA-binding sites within their respective immediate-early transactivators for all herpesviruses and also predicted that miR-UL112-1 targets IE72. Using recombinant viruses with either the miR-UL112-1 gene removed or encoding an IE72 mRNA lacking the miR-UL112-1 target site showed specific IE72 protein regulation and moderate effects on viral replication, thereby confirming IE72 as a target (70). As seen in HSV-1 and HSV-2, miR-UL112-1 is also located antisense to the lytic viral gene UL114, a viral uracil DNA glycosylase, and was recently shown to downregulate UL114 expression (71).

EBV-encoded miR-BART2 is located antisense to BALF5, a transcript encoding the viral DNA polymerase, and was initially hypothesized to degrade its mRNA (40). Barth et al. later confirmed that miR-BART2 does cleave BALF5 in a siRNA-like manner in EBV-infected cells (72). The first example of a viral gene targeted by viral miRNAs through seed sequence binding was LMP1 (73). The 3′UTR of latent membrane protein 1 (LMP1) contains potential binding sites for several BART miRNAs. LMP1 is a viral oncoprotein required for immortalization of human B cells. However, overexpression of LMP1 leads to induction of apoptosis (74). BART cluster 1 miRNAs, miR-BART16, miR-BART17-5p, and miR-BART1-5p, were shown to decrease LMP1 protein expression. Therefore, these miRNAs provide a mechanism of fine-tuning LMP1 levels in latently infected cells, thereby promoting proliferation and inhibiting apoptosis (73). Recently, novel miRNAs including miR-BART22 were identified in EBV-associated NPCs. MiR-BART22, which is expressed at high copy numbers in NPCs, was shown to target the latent membrane protein 2A (LMP2A) and reduce its expression in NPC-derived cell lines. The regulation of LMP2a by miR-BART22 may facilitate NPC carcinogenesis by evading host immune responses (75).

To investigate whether KSHV miRNAs target KSHV immediate-early transactivators, Bellare and colleagues utilized luciferase reporter assays in which the 3′UTR of the KSHV reactivation and transcriptional activator gene (RTA) was cotransfected with individual KSHV miRNA mimics. Further analysis of miRNA knockdown using antagomirs (sequence-specific miRNA inhibitors) in latently infected PEL cells showed that miR-K12-9* modulates RTA expression at the protein level (76). Lu et al. also found that a KSHV miR-K12-5 can inhibit RTA expression. However, this may reflect an indirect effect rather than direct targeting, since the 3′UTR of RTA does not contain a favorable miR-K12-5 seed sequence (77).

In summary, with the mechanisms by which herpesvirus-encoded miRNA function varies (miRNA-like vs. siRNA-like for antisense control), it becomes clear that all herpesviruses have evolved to utilize miRNAs as means to regulate immediate-early transcriptional regulators. Future work using appropriate in vivo models will tell how important these regulatory pathways are in the context of viral latency and persistence.

3.2. Viral miRNAs Targeting Host Cellular Genes

Up to 25% of all herpesvirus genes modulate host cellular functions during both latent and lytic infection and it was hypothesized early on that viral miRNAs will greatly increase the complexity of virus/host interaction (78). To date, most efforts on identifying cellular genes targeted by viral miRNAs have been focused on KSHV and EBV. In summary, these studies reveal that viral miRNAs target key cellular pathways, including immunity, proliferation, angiogenesis, and apoptosis (Table 2, Fig. 3).

The first cellular target genes for viral miRNAs were identified by gene expression profiling of HEK 293 cells stably expressing KSHV miRNA cluster containing 10 miRNAs (79). A total of 65 genes were downregulated in the presence of the miRNAs. SPP1, PRG1, and THBS1 were verified as miRNA targets using luciferase reporter constructs containing the 3′UTRs. Additionally, protein levels of THBS1 were decreased >10-fold in KSHV miRNA expressing cells. This was significant since THBS, a strong tumor suppressor and antiangiogenic factor, had previously been reported to be downregulated in KS lesions (80). Interestingly, the 3′UTR of THBS contained seed sequence binding sites for multiple KSHV miRNAs suggesting that viral miRNAs in clusters coordinately regulate host cellular target genes. SPP1 and PRG1 are involved in cell-mediated immunity and apoptosis, respectively. These initial findings, albeit obtained in 293 cells, suggest that KSHV-encoded miRNAs contribute to viral pathogenesis by promoting angiogenesis (a hallmark of KS tumors) and by inhibiting cellular immunity and apoptosis (79).

The Ganem group devised an elegant tandem-array approach to identify KSHV miRNA targets that were either induced by miRNA knockdown in latently infected PEL cells or inhibited in uninfected B cells ectopically expressing the corresponding KSHV miRNA (81). This analysis revealed that miR-K12-5, along with

Table 2
Experimentally verified viral miRNA targets

Viral targets

				References
HSV-1	miR-H2-3p	ICP0	Immediate-early transactivator	(53)
	miR-H3	ICP34.5	Neuro-virulence factor	(53)
	miR-H4			(53)
	miR-H6	ICP4	Immediate-early transactivator	(53)
HSV-2	miR-H2	ICP0	Immediate-early transactivator	(55)
	miR-H3	ICP34.5	Immediate-early transactivator	(54)
	miR-H4			(55)
HCMV	miR-UL112-1	IE72	Immediate-early transactivator	(69, 70)
		UL114	Viral uracil DNA glycosylase	(71)
KSHV	miR-K12-9*	RTA	Replication and transcriptional activator	(76)
EBV	miR-BART2	BALF5	DNA polymerase	(72)
	miR-BART22	LMP2A	Viral oncogene in NPC	(75)
	miR-Bart 1-5p	LMP1	Viral oncogene	(73)
	miR-BART16			(73)
	miR-BART17-5p			(73)
Polyomaviruses				
SV40	miR-M1	T Antigens	Early genes	(60)
SA12	miR-S1			(61)
BKV	miR-B1			(63)
JCV	miR-J1			(63)
MCV	miR-S1			(62)

Cellular targets

HCMV	miR-UL112-1	MICB	NK cell ligand	(95)
KSHV	miR-Cluster	THBS1	Angiogenesis inhibitor	(79)
		EXOC6	SEC15 gene family	(31)
		ZNF684	Zinc finger protein	(31)
		CDK5RAP1	Regulation of neuronal differentiation	(31)
	miR-K12-1	IκBα	NF-κB inhibitor	(90)
		p21	Inducer of cell cycle arrest	(91)
	miR-K12-3	LRRC8D	Immune cell activator	(31)
		NHP2L1	U4 snRNA nuclear binding protein	(31)
	miR-K12-3	C/EBPβ (LIP)	Transcriptional activator	(86)
	miR-K12-7			
	miR-K12-4-3p	GEMIN8	Required for splicing	(31)
	miR-K12-5	BCLAF1	Proapoptotic factor	(81)
		Rbl-2	Rb-like protein	(77)

(continued)

Table 2 (continued)

Cellular targets

	miR-K12-6 miR-K12-11	MAF	Transcription factor	(89)
	miR-K12-7	MICB	NK cell ligand	(94)
	miR-K12-11[a]	BACH1	Transcriptional suppressor	(82, 83)
EBV	miR-BHRF1-3	CXCL11	Chemokine, T-cell attractant	(93)
	miR-BART2	MICB	NK cell ligand	(94)
	miR-BART3	IPO7	Nuclear import protein	(31)
	miR-BART5	PUMA	Proapoptotic factor	(92)
	miR-BART16	TOMM22	Mitochondrial membrane protein	(31)

[a]Shown to have seed sequence homology with human miR-155

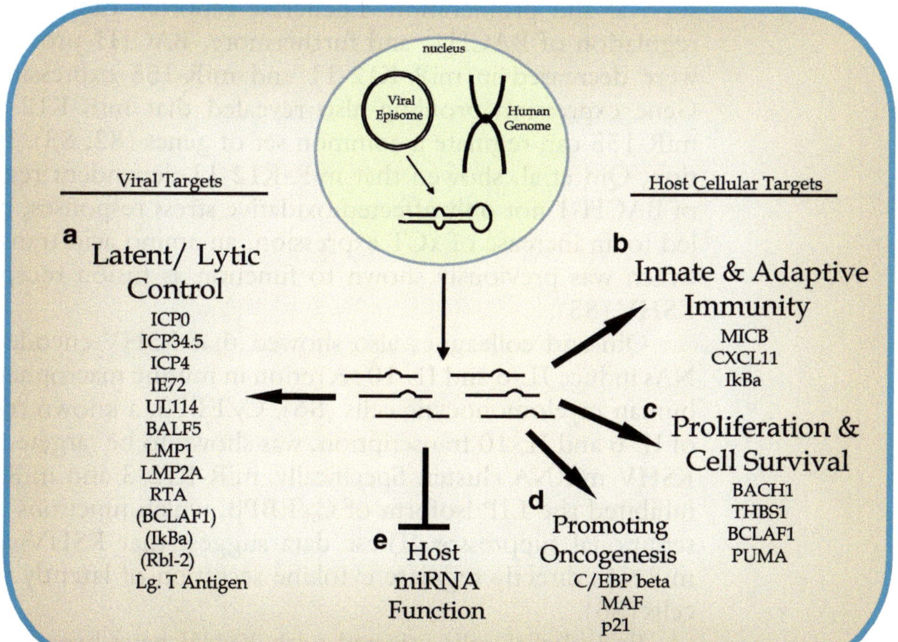

Fig. 3. Themes of viral miRNA gene regulation. Virus-infected cells can produce miRNAs which target both viral and cellular genes. (**a**) The common function of viral mRNA targets is the control of latency by targeting immediate-early or early genes. BCLAF1, IκBα, and Rbl-2 are host genes that contribute to the latent/lytic switch by either sensitizing cells to reactivate or functioning to maintain latency. (**b**) Targeting of antiviral responses to allow infected cell to evade immune response. (**c**) Targets involved in promoting proliferation and cell survival. (**d**) Inhibition of genes that may promote viral oncogenesis. (**e**) Modulating host cell miRNA expression and function. Experiments in MCMV show that viral miRNA synthesis completely overtakes host cell miRNA production early after infection (101). Viral miRNAs may hijack Drosha processing or RISC loading leading to impaired function of host miRNA, leading to global derepression of cellular miRNA targets.

K12-9 and miR-K12-10b, targets Bcl-2-associated factor (BCLAF1). BCLAF1 is a transcriptional repressor and overexpression can promote apoptosis. However, BCLAF1 expression in latently infected PEL cells can inhibit viral replication. Antagomir-based inhibition of KSHV miRNAs targeting BCLAF1 resulted in sensitizing latently infected endothelial cells for lytic reactivation. This data suggest that KSHV miRNAs can contribute to latency control by both targeting the viral RTA gene as discussed above and cellular genes like BCLAF1 (81).

Viral miRNAs can mimic cellular miRNA function. We and the Cullen group showed that miR-K12-11 and human miR-155 shared complete seed sequence identity (82, 83). Mir-155 is aberrantly expressed in many human malignancies, and when overexpressed in mice, causes lymphoproliferative disease (84). This led to question if miR-K12-11 and miR-155 can target a common set of genes. Bioinformatics identified the BACH1 gene, which has four binding sites for both miR-K12-11 and miR-155 within its 3′UTR. BACH1 is a transcriptional repressor affecting expression of heme-oxygenase 1 (HMOX1), a protein that promotes cell survival and proliferation. Luciferase reporter assays confirmed regulation of BACH1, and furthermore, BACH1 protein levels were decreased in miR-K12-11 and miR-155 expressing cells. Gene expression profiling also revealed that miR-K12-11 and miR-155 can regulate a common set of genes (82, 83). In addition, Qin et al. showed that miR-K12-11-dependent regulation of BACH-1 not only affected oxidative stress responses, but also led to an increase of xCT expression, an amino acid transporter, which was previously shown to function as fusion receptor for KSHV (85).

Qin and colleagues also showed that KSHV-encoded miRNAs induce IL-6 and IL-10 secretion in murine macrophages and human myelomonocytic cells (86). C/EBPβ, a known regulator of IL-6 and IL-10 transcription, was shown to be targeted by the KSHV miRNA cluster. Specifically, miR-K12-3 and miR-K12-7 inhibited the LIP isoform of C/EBPβ, which functions as transcriptional suppressor. These data suggest that KSHV-encoded miRNAs directly regulate cytokine secretion of latently infected cells (86).

Endothelial cells infected with KSHV have been shown to undergo transcriptional reprogramming, expressing markers for both lymphatic (LECs) and blood endothelial cells (BECs) (87, 88). Hansen et al. recently showed that KSHV miRNAs directly contribute to this reprogramming by targeting the cellular transcription factor musculoaponeurotic fibrosarcoma oncogene homolog (MAF). MiR-K12-6 and miR-K12-11 together target the 3′UTR of MAF, thereby inducing endothelial cell differentiation and possibly contribute to KSHV oncogenesis (89). Interestingly, miR-K12-11, the ortholog of miR-155, is also involved in B cell differentiation and proliferation in vivo (Boss and Renne, unpublished).

The above studies on KSHV were based on either latently infected cell lines or cell lines engineered to express viral miRNAs. Two recent studies addressed the role of KSHV miRNA within the context of the viral genome (77, 90). Both developed a recombinant KSHVΔmiRNA virus to study viral replication. Lei et al. observed a marked inhibition of NF-κB in 293 cells infected with the virus mutant, which was accompanied by a moderate increase in lytic replication. IκBα, the NF-κB repressor, was subsequently shown to be targeted by miR-K12-1 (90). This is the second example of a KSHV miRNA that contributes to latency by targeting cellular genes. Gottwein and colleagues reported another target for miR-K12-1 which like NF-κB is crucial for cell survival and proliferation. Using bioinformatic tools and luciferase assays and mutagenesis, it was shown that miR-K12-1 directly targets p21, a key inducer of cell cycle arrest and tumor suppressor (91).

Using a very similar miRNA knock-out virus, Lu et al. also observed a moderate increase in lytic replication, but identified entirely different mechanisms. In addition to inhibiting RTA through miR-K12-5 (see Subheading 3.1), Lu et al. observed a drastic inhibition of DNA methylation throughout the KSHV genome after deleting the miRNA cluster. Subsequently, it was shown that Retinoblastoma (Rb)-like protein 2 (Rbl-2), a potent inhibitor of DNA (cytosine-5-)-methyltransferases (DnmT1, 3a, and 3b), was targeted by several KSHV miRNAs. These data showed for the first time a role of viral miRNAs in epigenetic regulation of latency (77).

The latest technique to identify miRNA targets utilizes immunoprecipitation of RISCs followed by microarray or HTS sequencing analysis of the RISC-bound miRNA targets (29, 30). Using this technique, Dolken et al. was able to confirm a significant number of the above-discussed targets, and in addition, determined six novel targets of KSHV miRNAs and two targets of EBV miRNAs (31). KSHV miR-K12-3 was shown to target LRRC8D, thought to be involved in proliferation and activation of lymphocytes and macrophages, and NHP2L1, a nuclear protein that binds to U4 snRNA. MiR-K12-4-3p targets GEMIN8, which is required for spliceosomal snRNP assembly in the cytoplasm and pre-mRNA splicing in the nucleus. Also, the KSHV miR-cluster was found to target EXOC6, ZNF684, and CDK5RAP1; however, no functional studies have been presented on these novel target genes (31).

For the more than 25 EBV-encoded miRNAs, only a handful of targets have been described to date. Using the RISC-Chip, Dolken et al. showed that EBV miR-BART16 targets TOMM22 and miR-BART3 targets IP07, both involved in cellular transport processes. However, their role in EBV biology has not been determined (31). Previously, EBV miRNAs were shown to regulate PUMA and CXCL11. PUMA modulates apoptosis through p53 upregulation and is targeted by miR-BART5 (92). CXCL11 is an IFN-inducible T-cell chemoattractant and is targeted by miR-BHRF1-3, thereby inhibiting T-cell recognition (93). In this context, EBV miR-BART2 also suppresses the major histocompatibility complex 1-related chain B (MICB) as discussed below (94).

The first viral miRNA targeting a host gene involved in immune response was reported by Stern-Ginossar and colleagues who used a novel algorithm to identify miRNA targets for HCMV. MICB was predicted to be targeted by miR-UL112-1; interestingly, this MICB does not contain a conventional seed sequence site (95). MICB is a stress-induced ligand that is essential for natural killer (NK) cell recognition of virus-infected cells. Using elegant genetics approaches demonstrated that targeting MICB significantly reduced NK cell killing of HCMV-infected cells (95). Two years after this discovery, two more herpesvirus-encoded miRNAs were found to target MICB. Nachmani et al. showed that KSHV miR-K12-7 and EBV miR-BART2 both directly target MICB mRNA and reduce its expression by utilizing three different sites within the MICB 3′UTR (94). Thus, this coevolution suggests that targeting MICB to prevent virus-infected cells from being recognized by NK cells is a critical step for viral persistence in vivo.

4. Summary, Open Questions, and Future Perspectives

Since miRNAs are present in all herpesvirus genomes investigated thus far, it is likely that they play important roles in virus biology and possibly pathogenesis. Several common themes have emerged already from the limited number of validated viral and cellular miRNA targets (Fig. 3). Latent/lytic control either by regulating immediate-early transactivators or through modulation of cellular targets has been shown for all herpesvirus families. The striking observation that three different viral miRNAs from HCMV (95), EBV, and KSHV (94) that are unrelated in sequence have evolved to target MICB to evade NK immune surveillance further supports the idea that herpesvirus genomes have acquired miRNA genes a long time ago. The importance of miRNAs targeting antiviral responses is further supported by the observation that MICB is also targeted by the HCMV-encoded UL16 protein and the KSHV MIR3 and MIR5 proteins (78).

An important question is whether KSHV and EBV miRNAs directly contribute to viral tumorigenesis. A strong candidate for such role in KSHV is the discovery of miR-K12-11, mimicking human miR-155, which was shown to have oncogene activity

(84). This mechanism is also conserved since MDV-1, an avian tumorigenic virus, also encodes a miR-155 mimic (96, 97) and EBV strongly upregulates miR-155 (98). In addition, targeting pathways involved in apoptosis, proliferation, angiogenesis, and differentiation have the potential to contribute to immortalization and tumorgenesis. However, answering these important questions will depend on the availability of robust genetic systems and appropriate in vitro and in vivo models. One additional layer of complexity stems from recent discoveries in the miRNA field, suggesting that miRNAs can be secreted by exosomes; hence, the regulatory effects of viral-encoded miRNAs may not be limited to herpesvirus-infected cells as was recently reported for EBV (99, 100). Finally, while the discovery of these novel regulatory genes and their large number clearly increases the complexity of host/virus interactions, they may also point to attractive novel therapeutic approaches in the future.

References

1. Lee, R. C., Feinbaum, R. L., and Ambros, V. (1993) The *C. elegans* heterochronic gene lin-4 encodes small RNAs with antisense complementarity to lin-14, *Cell* **75**, 843–854.
2. Ruvkun, G., Wightman, B., and Ha, I. (2004) The 20 years it took to recognize the importance of tiny RNAs, *Cell* **116**, S93–S96, 92 p following S96.
3. Wightman, B., Burglin, T. R., Gatto, J., Arasu, P., and Ruvkun, G. (1991) Negative regulatory sequences in the lin-14 3 -untranslated region are necessary to generate a temporal switch during *Caenorhabditis elegans* development, *Genes Dev* **5**, 1813–1824.
4. Wightman, B., Ha, I., and Ruvkun, G. (1993) Posttranscriptional regulation of the heterochronic gene lin-14 by lin-4 mediates temporal pattern formation in *C. elegans*, *Cell* **75**, 855–862.
5. Pasquinelli, A. E., Reinhart, B. J., Slack, F., Martindale, M. Q., Kuroda, M. I., Maller, B., Hayward, D. C., Ball, E. E., Degnan, B., Muller, P., Spring, J., Srinivasan, A., Fishman, M., Finnerty, J., Corbo, J., Levine, M., Leahy, P., Davidson, E., and Ruvkun, G. (2000) Conservation of the sequence and temporal expression of let-7 heterochronic regulatory RNA, *Nature* **408**, 86–89.
6. Reinhart, B. J., Slack, F. J., Basson, M., Pasquinelli, A. E., Bettinger, J. C., Rougvie, A. E., Horvitz, H. R., and Ruvkun, G. (2000) The 21-nucleotide let-7 RNA regulates developmental timing in *Caenorhabditis elegans*, *Nature* **403**, 901–906.
7. Slack, F. J., Basson, M., Liu, Z., Ambros, V., Horvitz, H. R., and Ruvkun, G. (2000) The lin-41 RBCC gene acts in the *C. elegans* heterochronic pathway between the let-7 regulatory RNA and the LIN-29 transcription factor, *Mol Cell* **5**, 659–669.
8. Ambros, V. (2004) The functions of animal microRNAs, *Nature* **431**, 350–355.
9. Griffiths-Jones, S. (2004) The microRNA Registry, *Nucleic Acids Res* **32**, D109–D111.
10. Li, H., Li, W. X., and Ding, S. W. (2002) Induction and suppression of RNA silencing by an animal virus, *Science* **296**, 1319–1321.
11. Lippman, Z., and Martienssen, R. (2004) The role of RNA interference in heterochromatic silencing, *Nature* **431**, 364–370.
12. Iorio, M. V., Piovan, C., and Croce, C. M. (2010) Interplay between microRNAs and the epigenetic machinery: An intricate network, *Biochimica et biophysica acta* **1799**, 694–701.
13. Bartel, D. P. (2004) MicroRNAs: genomics, biogenesis, mechanism, and function, *Cell* **116**, 281–297.
14. Calin, G. A., and Croce, C. M. (2006) MicroRNA signatures in human cancers, *Nat Rev Cancer* **6**, 857–866.
15. Bogerd, H. P., Karnowski, H. W., Cai, X., Shin, J., Pohlers, M., and Cullen, B. R. (2010) A mammalian herpesvirus uses noncanonical expression and processing mechanisms to generate viral MicroRNAs, *Mol Cell* **37**, 135–142.

16. Diebel, K. W., Smith, A. L., and van Dyk, L. F. (2010) Mature and functional viral miRNAs transcribed from novel RNA polymerase III promoters, *RNA* **16**, 170–185.
17. Grimson, A., Farh, K. K., Johnston, W. K., Garrett-Engele, P., Lim, L. P., and Bartel, D. P. (2007) MicroRNA targeting specificity in mammals: determinants beyond seed pairing, *Mol Cell* **27**, 91–105.
18. Filipowicz, W., Bhattacharyya, S. N., and Sonenberg, N. (2008) Mechanisms of post-transcriptional regulation by microRNAs: are the answers in sight? *Nat Rev* **9**, 102–114.
19. Yue, D., Liu, H., and Huang, Y. (2009) Survey of computational algorithms for MicroRNA target prediction, *Curr Genomics* **10**, 478–492.
20. Lewis, B. P., Burge, C. B., and Bartel, D. P. (2005) Conserved seed pairing, often flanked by adenosines, indicates that thousands of human genes are microRNA targets, *Cell* **120**, 15–20.
21. Lewis, B. P., Shih, I. H., Jones-Rhoades, M. W., Bartel, D. P., and Burge, C. B. (2003) Prediction of mammalian microRNA targets, *Cell* **115**, 787–798.
22. Burgler, C., and Macdonald, P. M. (2005) Prediction and verification of microRNA targets by MovingTargets, a highly adaptable prediction method, *BMC Genomics* **6**, 88.
23. Saetrom, O., Snove, O., Jr., and Saetrom, P. (2005) Weighted sequence motifs as an improved seeding step in microRNA target prediction algorithms, *RNA* **11**, 995–1003.
24. John, B., Enright, A. J., Aravin, A., Tuschl, T., Sander, C., and Marks, D. S. (2004) Human MicroRNA targets, *PLoS Biol* **2**, e363.
25. Dittmer, D. P. (2003) Transcription profile of Kaposi's sarcoma-associated herpesvirus in primary Kaposi's sarcoma lesions as determined by real-time PCR arrays, *Cancer Res* **63**, 2010–2015.
26. Dittmer, D. P., Gonzalez, C. M., Vahrson, W., DeWire, S. M., Hines-Boykin, R., and Damania, B. (2005) Whole-genome transcription profiling of rhesus monkey rhadinovirus, *J Virol* **79**, 8637–8650.
27. Baek, D., Villen, J., Shin, C., Camargo, F. D., Gygi, S. P., and Bartel, D. P. (2008) The impact of microRNAs on protein output, *Nature* **455**, 64–71.
28. Selbach, M., Schwanhausser, B., Thierfelder, N., Fang, Z., Khanin, R., and Rajewsky, N. (2008) Widespread changes in protein synthesis induced by microRNAs, *Nature* **455**, 58–63.
29. Chi, S. W., Zang, J. B., Mele, A., and Darnell, R. B. (2009) Argonaute HITS-CLIP decodes microRNA-mRNA interaction maps, *Nature* **460**, 479–486.
30. Hafner, M., Landthaler, M., Burger, L., Khorshid, M., Hausser, J., Berninger, P., Rothballer, A., Ascano, M., Jr., Jungkamp, A. C., Munschauer, M., Ulrich, A., Wardle, G. S., Dewell, S., Zavolan, M., and Tuschl, T. (2010) Transcriptome-wide identification of RNA-binding protein and microRNA target sites by PAR-CLIP, *Cell* **141**, 129–141.
31. Dolken, L., Malterer, G., Erhard, F., Kothe, S., Friedel, C. C., Suffert, G., Marcinowski, L., Motsch, N., Barth, S., Beitzinger, M., Lieber, D., Bailer, S. M., Hoffmann, R., Ruzsics, Z., Kremmer, E., Pfeffer, S., Zimmer, R., Koszinowski, U. H., Grasser, F., Meister, G., and Haas, J. (2010) Systematic analysis of viral and cellular microRNA targets in cells latently infected with human gamma-herpesviruses by RISC immunoprecipitation assay, *Cell Host Microbe* **7**, 324–334.
32. Monticelli, S., Ansel, K. M., Xiao, C., Socci, N. D., Krichevsky, A. M., Thai, T. H., Rajewsky, N., Marks, D. S., Sander, C., Rajewsky, K., Rao, A., and Kosik, K. S. (2005) MicroRNA profiling of the murine hematopoietic system, *Genome Biol* **6**, R71.
33. Rodriguez, A., Vigorito, E., Clare, S., Warren, M. V., Couttet, P., Soond, D. R., van Dongen, S., Grocock, R. J., Das, P. P., Miska, E. A., Vetrie, D., Okkenhaug, K., Enright, A. J., Dougan, G., Turner, M., and Bradley, A. (2007) Requirement of bic/microRNA-155 for normal immune function, *Science* **316**, 608–611.
34. Thai, T. H., Calado, D. P., Casola, S., Ansel, K. M., Xiao, C., Xue, Y., Murphy, A., Frendewey, D., Valenzuela, D., Kutok, J. L., Schmidt-Supprian, M., Rajewsky, N., Yancopoulos, G., Rao, A., and Rajewsky, K. (2007) Regulation of the germinal center response by microRNA-155, *Science* **316**, 604–608.
35. Pfeffer, S., Zavolan, M., Grasser, F. A., Chien, M., Russo, J. J., Ju, J., John, B., Enright, A. J., Marks, D., Sander, C., and Tuschl, T. (2004) Identification of virus-encoded microRNAs, *Science* **304**, 734–736.
36. Samols, M. A., Hu, J., Skalsky, R. L., and Renne, R. (2005) Cloning and identification of a microRNA cluster within the latency-associated region of Kaposi's sarcoma-associated herpesvirus, *J Virol* **79**, 9301–9305.
37. Grundhoff, A., Sullivan, C. S., and Ganem, D. (2006) A combined computational and

microarray-based approach identifies novel microRNAs encoded by human gammaherpesviruses, *RNA* **12**, 733–750.

38. Umbach, J. L., and Cullen, B. R. (2010) In-depth analysis of Kaposi's sarcoma-associated herpesvirus microRNA expression provides insights into the mammalian microRNA-processing machinery, *J Virol* **84**, 695–703.
39. Cai, X., Lu, S., Zhang, Z., Gonzalez, C. M., Damania, B., and Cullen, B. R. (2005) Kaposi's sarcoma-associated herpesvirus expresses an array of viral microRNAs in latently infected cells, *Proc Natl Acad Sci U S A* **102**, 5570–5575.
40. Pfeffer, S., Sewer, A., Lagos-Quintana, M., Sheridan, R., Sander, C., Grasser, F. A., van Dyk, L. F., Ho, C. K., Shuman, S., Chien, M., Russo, J. J., Ju, J., Randall, G., Lindenbach, B. D., Rice, C. M., Simon, V., Ho, D. D., Zavolan, M., and Tuschl, T. (2005) Identification of microRNAs of the herpesvirus family, *Nat Methods* **2**, 269–276.
41. Cai, X., Schafer, A., Lu, S., Bilello, J. P., Desrosiers, R. C., Edwards, R., Raab-Traub, N., and Cullen, B. R. (2006) Epstein-Barr virus microRNAs are evolutionarily conserved and differentially expressed, *PLoS Pathog* **2**, e23.
42. Zhu, J. Y., Pfuhl, T., Motsch, N., Barth, S., Nicholls, J., Grasser, F., and Meister, G. (2009) Identification of novel Epstein-Barr virus microRNA genes from nasopharyngeal carcinomas, *J Virol* **83**, 3333–3341.
43. Riley, K. J., Rabinowitz, G. S., and Steitz, J. A. (2010) Comprehensive analysis of Rhesus lymphocryptovirus microRNA expression, *J Virol* **84**, 5148–5157.
44. Walz, N., Christalla, T., Tessmer, U., and Grundhoff, A. (2010) A global analysis of evolutionary conservation among known and predicted gammaherpesvirus microRNAs, *J Virol* **84**, 716–728.
45. Schafer, A., Cai, X., Bilello, J. P., Desrosiers, R. C., and Cullen, B. R. (2007) Cloning and analysis of microRNAs encoded by the primate gamma-herpesvirus rhesus monkey rhadinovirus, *Virology* **364**, 21–27.
46. Dunn, W., Trang, P., Zhong, Q., Yang, E., van Belle, C., and Liu, F. (2005) Human cytomegalovirus expresses novel microRNAs during productive viral infection, *Cell Microbiol* **7**, 1684–1695.
47. Grey, F., Antoniewicz, A., Allen, E., Saugstad, J., McShea, A., Carrington, J. C., and Nelson, J. (2005) Identification and characterization of human cytomegalovirus-encoded microRNAs, *J Virol* **79**, 12095–12099.
48. Cui, C., Griffiths, A., Li, G., Silva, L. M., Kramer, M. F., Gaasterland, T., Wang, X. J., and Coen, D. M. (2006) Prediction and identification of herpes simplex virus 1-encoded microRNAs, *J Virol* **80**, 5499–5508.
49. Umbach, J. L., Nagel, M. A., Cohrs, R. J., Gilden, D. H., and Cullen, B. R. (2009) Analysis of human {alpha}-herpesvirus microRNA expression in latently infected human trigeminal ganglia, *J Virol* **79**, 6162–6171.
50. Umbach, J. L., Wang, K., Tang, S., Krause, P. R., Mont, E. K., Cohen, J. I., and Cullen, B. R. (2010) Identification of viral microRNAs expressed in human sacral ganglia latently infected with herpes simplex virus 2, *J Virol* **84**, 1189–1192.
51. Jurak, I., Kramer, M. F., Mellor, J. C., van Lint, A. L., Roth, F. P., Knipe, D. M., and Coen, D. M. (2010) Numerous conserved and divergent microRNAs expressed by herpes simplex viruses 1 and 2, *J Virol* **84**, 4659–4672.
52. Waidner, L. A., Morgan, R. W., Anderson, A. S., Bernberg, E. L., Kamboj, S., Garcia, M., Riblet, S. M., Ouyang, M., Isaacs, G. K., Markis, M., Meyers, B. C., Green, P. J., and Burnside, J. (2009) MicroRNAs of Gallid and Meleagrid herpesviruses show generally conserved genomic locations and are virus-specific, *Virology* **388**, 128–136.
53. Umbach, J. L., Kramer, M. F., Jurak, I., Karnowski, H. W., Coen, D. M., and Cullen, B. R. (2008) MicroRNAs expressed by herpes simplex virus 1 during latent infection regulate viral mRNAs, *Nature* **454**, 780–783.
54. Tang, S., Bertke, A. S., Patel, A., Wang, K., Cohen, J. I., and Krause, P. R. (2008) An acutely and latently expressed herpes simplex virus 2 viral microRNA inhibits expression of ICP34.5, a viral neurovirulence factor, Proc Natl Acad Sci U S A **105**, 10931–10936.
55. Tang, S., Patel, A., and Krause, P. R. (2009) Novel less-abundant viral microRNAs encoded by herpes simplex virus 2 latency-associated transcript and their roles in regulating ICP34.5 and ICP0 mRNAs, *J Virol* **83**, 1433–1442.
56. Burnside, J., Ouyang, M., Anderson, A., Bernberg, E., Lu, C., Meyers, B. C., Green, P. J., Markis, M., Isaacs, G., Huang, E., and Morgan, R. W. (2008) Deep sequencing

of chicken microRNAs, *BMC Genomics* **9**, 185.

57. Yao, Y., Zhao, Y., Xu, H., Smith, L. P., Lawrie, C. H., Sewer, A., Zavolan, M., and Nair, V. (2007) Marek's disease virus type 2 (MDV-2)-encoded microRNAs show no sequence conservation with those encoded by MDV-1, *J Virol* **81**, 7164–7170.
58. Yao, Y., Zhao, Y., Xu, H., Smith, L. P., Lawrie, C. H., Watson, M., and Nair, V. (2008) MicroRNA profile of Marek's disease virus-transformed T-cell line MSB-1: predominance of virus-encoded microRNAs, *J Virol* **82**, 4007–4015.
59. Burnside, J., Bernberg, E., Anderson, A., Lu, C., Meyers, B. C., Green, P. J., Jain, N., Isaacs, G., and Morgan, R. W. (2006) Marek's disease virus encodes MicroRNAs that map to meq and the latency-associated transcript, *J Virol* **80**, 8778–8786.
60. Sullivan, C. S., Grundhoff, A. T., Tevethia, S., Pipas, J. M., and Ganem, D. (2005) SV40-encoded microRNAs regulate viral gene expression and reduce susceptibility to cytotoxic T cells, *Nature* **435**, 682–686.
61. Cantalupo, P., Doering, A., Sullivan, C. S., Pal, A., Peden, K. W., Lewis, A. M., and Pipas, J. M. (2005) Complete nucleotide sequence of polyomavirus SA12, *J Virol* **79**, 13094–13104.
62. Seo, G. J., Chen, C. J., and Sullivan, C. S. (2009) Merkel cell polyomavirus encodes a microRNA with the ability to autoregulate viral gene expression, *Virology* **383**, 183–187.
63. Seo, G. J., Fink, L. H., O'Hara, B., Atwood, W. J., and Sullivan, C. S. (2008) Evolutionarily conserved function of a viral microRNA, *J Virol* **82**, 9823–9828.
64. Sullivan, C. S., Sung, C. K., Pack, C. D., Grundhoff, A., Lukacher, A. E., Benjamin, T. L., and Ganem, D. (2009) Murine polyomavirus encodes a microRNA that cleaves early RNA transcripts but is not essential for experimental infection, *Virology* **387**, 157–167.
65. Yao, Y., Zhao, Y., Smith, L. P., Lawrie, C. H., Saunders, N. J., Watson, M., and Nair, V. K. (2009) Differential expression of miRNAs in Marek's disease virus-transformed T-lymphoma cell lines, *J Gen Virol* **90**, 1551–1559.
66. Umbach, J. L., and Cullen, B. R. (2009) The role of RNAi and microRNAs in animal virus replication and antiviral immunity, *Genes Dev* **23**, 1151–1164.
67. Marshall, V., Parks, T., Bagni, R., Wang, C. D., Samols, M. A., Hu, J., Wyvil, K. M., Aleman, K., Little, R. F., Yarchoan, R., Renne, R., and Whitby, D. (2007) Conservation of virally encoded micrornas in Kaposi sarcoma--associated herpesvirus in primary effusion lymphoma cell lines and in patients with Kaposi sarcoma or multicentric castleman disease, *J Infect Dis* **195**, 645–659.
68. Gottwein, E., Cai, X., and Cullen, B. R. (2006) A novel assay for viral microRNA function identifies a single nucleotide polymorphism that affects Drosha processing, *J Virol* **80**, 5321–5326.
69. Grey, F., Meyers, H., White, E. A., Spector, D. H., and Nelson, J. (2007) A human cytomegalovirus-encoded microRNA regulates expression of multiple viral genes involved in replication, *PLoS Pathog* **3**, e163.
70. Murphy, E., Vanicek, J., Robins, H., Shenk, T., and Levine, A. J. (2008) Suppression of immediate-early viral gene expression by herpesvirus-coded microRNAs: implications for latency, *Proc Natl Acad Sci U S A* **105**, 5453–5458.
71. Stern-Ginossar, N., Saleh, N., Goldberg, M. D., Prichard, M., Wolf, D. G., and Mandelboim, O. (2009) Analysis of human cytomegalovirus-encoded microRNA activity during infection, *J Virol* **83**, 10684–10693.
72. Barth, S., Pfuhl, T., Mamiani, A., Ehses, C., Roemer, K., Kremmer, E., Jaker, C., Hock, J., Meister, G., and Grasser, F. A. (2008) Epstein-Barr virus-encoded microRNA miR-BART2 down-regulates the viral DNA polymerase BALF5, *Nucleic Acids Res* **36**, 666–675.
73. Lo, A. K., To, K. F., Lo, K. W., Lung, R. W., Hui, J. W., Liao, G., and Hayward, S. D. (2007) Modulation of LMP1 protein expression by EBV-encoded microRNAs, *Proc Natl Acad Sci U S A* **104**, 16164–16169.
74. Izumi, K. M., and Kieff, E. D. (1997) The Epstein-Barr virus oncogene product latent membrane protein 1 engages the tumor necrosis factor receptor-associated death domain protein to mediate B lymphocyte growth transformation and activate NF-kappaB, *Proc Natl Acad Sci U S A* **94**, 12592–12597.
75. Lung, R. W., Tong, J. H., Sung, Y. M., Leung, P. S., Ng, D. C., Chau, S. L., Chan, A. W., Ng, E. K., Lo, K. W., and To, K. F. (2009) Modulation of LMP2A expression by a newly identified Epstein-Barr virus-encoded microRNA miR-BART22, *Neoplasia* **11**, 1174–1184.

76. Bellare, P., and Ganem, D. (2009) Regulation of KSHV lytic switch protein expression by a virus-encoded microRNA: an evolutionary adaptation that fine-tunes lytic reactivation, *Cell Host Microbe* **6**, 570–575.
77. Lu, F., Stedman, W., Yousef, M., Renne, R., and Lieberman, P. M. (2010) Epigenetic regulation of Kaposi's sarcoma-associated herpesvirus latency by virus-encoded microRNAs that target Rta and the cellular Rbl2-DNMT pathway, *J Virol* **84**, 2697–2706.
78. Areste, C., and Blackbourn, D. J. (2009) Modulation of the immune system by Kaposi's sarcoma-associated herpesvirus, *Trends Microbiol* **17**, 119–129.
79. Samols, M. A., Hu, J., Skalsky, R.L., Maldonado, A.M., Riva, A., Lopez, M.C., Baker, H.V., and R. Renne. (2007) Identification of cellular genes targeted by KSHV-encoded microRNAs, *PLoS Pathog* **3**, e65.
80. Taraboletti, G., Benelli, R., Borsotti, P., Rusnati, M., Presta, M., Giavazzi, R., Ruco, L., and Albini, A. (1999) Thrombospondin-1 inhibits Kaposi's sarcoma (KS) cell and HIV-1 Tat-induced angiogenesis and is poorly expressed in KS lesions, *J Pathol* **188**, 76–81.
81. Ziegelbauer, J. M., Sullivan, C. S., and Ganem, D. (2009) Tandem array-based expression screens identify host mRNA targets of virus-encoded microRNAs, *Nat Genet* **41**, 130–134.
82. Gottwein, E., Mukherjee, N., Sachse, C., Frenzel, C., Majoros, W. H., Chi, J. T., Braich, R., Manoharan, M., Soutschek, J., Ohler, U., and Cullen, B. R. (2007) A viral microRNA functions as an orthologue of cellular miR-155, *Nature* **450**, 1096–1099.
83. Skalsky, R. L., Samols, M. A., Plaisance, K. B., Boss, I. W., Riva, A., Lopez, M. C., Baker, H. V., and Renne, R. (2007) Kaposi's sarcoma-associated herpesvirus encodes an ortholog of miR-155, *J Virol* **81**, 12836–12845.
84. Garzon, R., and Croce, C. M. (2008) MicroRNAs in normal and malignant hematopoiesis, *Curr Opin Hematol* **15**, 352–358.
85. Qin, Z., Freitas, E., Sullivan, R., Mohan, S., Bacelieri, R., Branch, D., Romano, M., Kearney, P., Oates, J., Plaisance, K., Renne, R., Kaleeba, J., and Parsons, C. (2010) Upregulation of xCT by KSHV-encoded microRNAs facilitates KSHV dissemination and persistence in an environment of oxidative stress, *PLoS Pathog* **6**, e1000742.
86. Qin, Z., Kearney, P., Plaisance, K., and Parsons, C. H. (2009) Pivotal Advance: Kaposi's sarcoma-associated herpesvirus (KSHV)-encoded microRNA specifically induce IL-6 and IL-10 secretion by macrophages and monocytes, *J Leukoc Biol* **87**, 25–34.
87. Wang, H. W., Trotter, M. W., Lagos, D., Bourboulia, D., Henderson, S., Makinen, T., Elliman, S., Flanagan, A. M., Alitalo, K., and Boshoff, C. (2004) Kaposi sarcoma herpesvirus-induced cellular reprogramming contributes to the lymphatic endothelial gene expression in Kaposi sarcoma, *Nat Genet* **36**, 687–693.
88. Carroll, P. A., Brazeau, E., and Lagunoff, M. (2004) Kaposi's sarcoma-associated herpesvirus infection of blood endothelial cells induces lymphatic differentiation, *Virology* **328**, 7–18.
89. Hansen, A., Henderson, S., Lagos, D., Nikitenko, L., Coulter, E., Roberts, S., Gratrix, F., Plaisance, K., Renne, R., Bower, M., Kellam, P., and Boshoff, C. (2010) KSHV-encoded miRNAs target MAF to induce endothelial cell reprogramming, *Genes Dev* **24**, 195–205.
90. Lei, X., Bai, Z., Ye, F., Xie, J., Kim, C. G., Huang, Y., and Gao, S. J. (2010) Regulation of NF-kappaB inhibitor IkappaBalpha and viral replication by a KSHV microRNA, *Nat Cell Biol* **12**, 625.
91. Gottwein, E., and Cullen, B. R. (2010) A human herpesvirus microRNA inhibits p21 expression and attenuates p21-mediated cell cycle arrest, *J Virol* **84**, 5229–5237.
92. Choy, E. Y., Siu, K. L., Kok, K. H., Lung, R. W., Tsang, C. M., To, K. F., Kwong, D. L., Tsao, S. W., and Jin, D. Y. (2008) An Epstein-Barr virus-encoded microRNA targets PUMA to promote host cell survival, *J Exp Med* **205**, 2551–2560.
93. Xia, T., O'Hara, A., Araujo, I., Barreto, J., Carvalho, E., Sapucaia, J. B., Ramos, J. C., Luz, E., Pedroso, C., Manrique, M., Toomey, N. L., Brites, C., Dittmer, D. P., and Harrington, W. J., Jr. (2008) EBV micro RNAs in primary lymphomas and targeting of CXCL-11 by ebv-mir-BHRF1-3, *Cancer Res* **68**, 1436–1442.
94. Nachmani, D., Stern-Ginossar, N., Sarid, R., and Mandelboim, O. (2009) Diverse herpesvirus microRNAs target the stress-induced immune ligand MICB to escape recognition by natural killer cells, *Cell Host Microbe* **5**, 376–385.
95. Stern-Ginossar, N., Elefant, N., Zimmermann, A., Wolf, D. G., Saleh, N.,

Biton, M., Horwitz, E., Prokocimer, Z., Prichard, M., Hahn, G., Goldman-Wohl, D., Greenfield, C., Yagel, S., Hengel, H., Altuvia, Y., Margalit, H., and Mandelboim, O. (2007) Host immune system gene targeting by a viral miRNA, *Science* **317**, 376–381.

96. Morgan, R., Anderson, A., Bernberg, E., Kamboj, S., Huang, E., Lagasse, G., Isaacs, G., Parcells, M., Meyers, B. C., Green, P. J., and Burnside, J. (2008) Sequence conservation and differential expression of Marek's disease virus microRNAs, *J Virol* **82**, 12213–12220.
97. Zhao, Y., Yao, Y., Xu, H., Lambeth, L., Smith, L. P., Kgosana, L., Wang, X., and Nair, V. (2009) A functional MicroRNA-155 ortholog encoded by the oncogenic Marek's disease virus, *J Virol* **83**, 489–492.
98. Cameron, J. E., Fewell, C., Yin, Q., McBride, J., Wang, X., Lin, Z., and Flemington, E. K. (2008) Epstein-Barr virus growth/latency III program alters cellular microRNA expression, *Virology* **382**, 257–266.
99. Pegtel, D. M., Cosmopoulos, K., Thorley-Lawson, D. A., van Eijndhoven, M. A., Hopmans, E. S., Lindenberg, J. L., de Gruijl, T. D., Wurdinger, T., and Middeldorp, J. M. (2010) Functional delivery of viral miRNAs via exosomes, *Proc Natl Acad Sci U S A* **107**, 6328–6333.
100. Valadi, H., Ekstrom, K., Bossios, A., Sjostrand, M., Lee, J. J., and Lotvall, J. O. (2007) Exosome-mediated transfer of mRNAs and microRNAs is a novel mechanism of genetic exchange between cells, *Nat Cell Biol* **9**, 654–659.
101. Dolken, L., Perot, J., Cognat, V., Alioua, A., John, M., Soutschek, J., Ruzsics, Z., Koszinowski, U., Voinnet, O., and Pfeffer, S. (2007) Mouse cytomegalovirus microRNAs dominate the cellular small RNA profile during lytic infection and show features of posttranscriptional regulation, *J Virol* **81**, 13771–13782.
102. Pfeffer, S. (2007) Identification of virally encoded microRNAs, *Methods Enzymol* **427**, 51–63.

Chapter 4

Progress in RNAi-Based Antiviral Therapeutics

Jiehua Zhou and John J. Rossi

Abstract

RNA interference (RNAi) refers to the conserved sequence-specific degradation of message RNA mediated by small interfering (si)RNA duplexes 21–25 nucleotides in length. Given the ability to specifically silence any gene of interest, siRNAs offers several advantages over conventional drugs as potential therapeutic agents for the treatment of human maladies including cancers, genetic disorders, and infectious diseases. Antiviral RNAi strategies have received much attention and several compounds are currently being tested in clinical trials. In particular, the development of siRNA-based HIV (human immunodeficiency virus) therapeutics has progressed rapidly and many recent studies have shown that the use of RNAi could inhibit HIV-1 replication by targeting a number of viral or cellular genes. Therefore, the present chapter mainly focuses on the recent progress of RNAi-based anti-HIV gene therapeutics, with particular attention to molecular targets and delivery strategies of the siRNAs.

Key words: RNA interference (RNAi), Small interfering RNA (siRNA), Antiviral RNAi, Anti-HIV gene therapeutics, siRNA delivery

1. Introduction

Since the first description of RNA interference (RNAi) in 1998 (1), RNAi has rapidly become one of the methods of choice for gene function analyses (2), and has also rapidly been exploited for therapeutic applications (3–6). Synthetic small interfering RNA (siRNA) was first demonstrated to achieve sequence-specific gene knockdown in a mammalian cell line by Tuschl and coworkers (7). Soon thereafter, Song et al. successfully showed the first *in vivo* evidence of siRNA-mediated gene silencing in an animal by demonstrating inhibition of a cell death receptor to recover liver function in a mouse model of hepatitis (8). To date, several RNAi-based drugs against a variety of targets in humans have been developed or are in the process of development. These include age-related

Ronald P. van Rij (ed.), *Antiviral RNAi: Concepts, Methods, and Applications*, Methods in Molecular Biology, vol. 721,
DOI 10.1007/978-1-61779-037-9_4, © Springer Science+Business Media, LLC 2011

macular degeneration (AMD), respiratory syncytial virus (RSV) infection, Hepatitis B (HBV), Hepatitis C (HCV), HIV/AIDS, Pachyonychia congenita, solid tumors, and acute renal failure (4, 9). Several of these siRNAs have already been evaluated in early stage clinical trials. Acuity Pharmaceuticals in 2004 conducted the first clinical trials involving siRNA against vascular endothelial growth factor (VEGF) for the treatment of AMD, a leading cause of blindness. Currently, phase III trials are being performed by Opko Health Inc (9). Moreover, RNAi has been demonstrated to inhibit replication of a number of viral infectious agents, including HIV-1 (10, 11).

Although current highly active antiretroviral therapy (HAART) treatment for HIV has been therapeutically effective in the majority of patients, drug resistance and toxicity issues still remain a concern for some individuals (12). Because the essential feature of the RNAi mechanism is the sequence-specificity, deriving from complementary Watson-Crick base pairing of a target messenger RNA (mRNA) and the guide strand of the siRNA, RNAi is considered to have some unique therapeutic attributes for the treatment of HIV-1 infection (13, 14). RNAi-based therapeutics can be multiplexed by combining different siRNAs for different targets or by combining siRNAs with other small RNAs or antiviral proteins. HIV dependency factors (HDFs) of host cells, such as the chemokine receptor CCR5 can be targeted alone or in combination with viral sequences in a tailored fashion to reduce the possibility of viral escape mutants and to avoid toxicity risks associated with long-term HAART treatment (11, 14). For example (15), a triple combination lentiviral construct comprised of an anti-*tat/rev* shRNA, a U6-driven nucleolar localizing TAR RNA decoy, and an anti-CCR5 ribozyme has been approved by the US Food and Drug Administration (FDA) and is currently being tested in clinical trials for AIDS/lymphoma patients at the City of Hope Medical Center (16).

Although siRNAs hold great therapeutic promise for HIV/AIDS, getting the technique to work in the clinic has been limited by difficulties of delivering siRNAs in the bloodstream and to the right target cells (9, 17). From a practical perspective, all HIV-susceptible cells (such as, human T cells and primary peripheral blood mononuclear cells) are difficult to transfect by nonviral agents, such as liposomes. Therefore, efficient, systemic delivery of siRNAs *in vivo*, especially targeted siRNA delivery to HIV-1 susceptible or infected cells, remains a principal challenge for successful anti-HIV therapeutic application and future clinical translation. Several recent studies have described methods or agents for siRNA delivery *in vivo* using different nonviral delivery approaches. For example, single chain antibody-siRNA chimeras targeting either gp120 expressing cells or the T-cell CD7 receptor have been shown to functionally deliver anti-HIV siRNAs *in vivo* (see Chapter 21 in this volume, and ref. (18). This chapter primarily discusses the potential of RNAi-based anti-HIV therapeutics. We also place particular emphasis on discussing development of various molecular targets and delivery strategies for siRNAs.

2. Potential Molecular Targets for Anti-HIV RNAi Therapeutics

To date, numerous siRNAs targeted to a number of HIV-1 or cellular HDF transcripts have been demonstrated to achieve viral inhibition both *in vitro* and *in vivo*. Indeed, all the HIV-1 encoded genes (*tat*, *rev*, *gag*, *pol*, *nef*, *vif*, *env*, *vpr*, and the long terminal repeat (LTR)) are susceptible to RNAi-induced gene silencing in cell lines (19). For example, the Tat and Rev proteins are essential for subsequent expression of HIV-1 structural genes (*gag*, *pol*, and *env*) and for the synthesis of full length viral genomic RNA (20). SiRNAs designed to destroy the *tat/rev* transcripts were found to be highly effective in viral suppression (15).

One of crucial concerns in achieving the desired RNAi efficacy is mitigating viral escape from RNAi (11). In addition, as opposed to acute virus infections, HIV-1 chronically infects individuals, thus requiring long-term RNAi treatment. However, development of viral resistance is a common setback with HIV therapies due to the generation of viral escape mutants. It has been demonstrated that prolonged culturing of cells with gene-based expression of anti-HIV siRNAs can result in the evolution of escape variants that are resistant to the expressed siRNA (21, 22). A single nucleotide mutation in a critical position within the target sequence relative to the siRNA site of interaction can diminish and even eliminate RNAi inhibition. Therefore, it is important to target sequences that are conserved among different virus strains to reduce the chance of mutant escape. Similar to the conventional HAART strategy, using multiple siRNAs targeting separate conserved sites in HIV or, alternatively, targeting HIV host dependency factors (HDFs) has been shown to minimize and even prevent escape while achieving prolonged inhibition (23). In this regard, host factors that are essential for viral entry and replication represent attractive molecular targets.

Several well-known host factors, including NF kappa Beta (NF-κB), the receptor CD4 and the coreceptors CCR5, and CXCR4 (C-X-C motif receptor 4) have been successfully targeted by siRNAs, thereby suppressing viral replication or entry (14). Recently, three siRNA-screening studies have been conducted to identify hundreds of host factors that are critical for HIV replication (24–26), which were not previously known to be commandeered by the virus during infection. For example, in the first screening (24), Brass et al., using a large scale siRNA screen identified 273 genes whose depletion inhibited either HIV p24 production or viral gene activities. There were some surprises among these targets, such as Rab6 (a regulator of retrograde protein transport to the Golgi), Transportin 3-SR2 (TNPO3, a nuclear import factor for serine/arginine-rich (SR) substrates), and Med28 (the Mediator transcription activation complex component). Specifically, depletion of Rab6 or TNPO3 potently blocked early stages of virus infection. The hits in the three screens differed due to the conditions and readouts of the screens. Nevertheless, these screening approaches have opened up a new landscape of viral–host interactions that are potential targets for drug development, including RNAi (27).

Despite the emergence of these new cellular genes as molecular targets for RNAi-based antiviral drugs, it should be considered that these targets may represent a double edge sword with respect to efficacy vs. the safety and potential toxicity associated with downregulating cellular targets in uninfected cells.

3. Delivery Strategies for Anti-HIV RNAi Therapeutic

In addition to identifying new anti-HIV targets, there are ongoing efforts to develop new innovative approaches for siRNA delivery. However, safe and effective siRNA drug delivery remains a principal challenge to achieve the desired RNAi potency for successful disease prevention and treatment. The feasibility of siRNA delivery is largely dependent on the accessibility of the target organ or tissue within the body. Local siRNA delivery is particularly well-suited for the treatment of lung diseases and local infections (9). For example, several tissues, including the eye, skin, mucus membranes, and some tumors, are amenable to topical or localized therapy. However, the HIV-1 infected cells in HIV/AIDS patients can only be reached through systemic administration of delivery vehicles in the bloodstream. Various nonviral or viral delivery systems have been developed to facilitate siRNA cellular uptake and promote anti-HIV efficacy, some of which will be highlighted in the following sections.

3.1. Nonviral siRNA Delivery

Nonviral siRNA delivery for HIV-1 treatment is an attractive alternative for patients that have run out of treatment options. Several recent studies have described approaches for achieving siRNA delivery in cell culture and *in vivo* through different nonviral delivery approaches, thereby successfully inhibiting HIV-1 replication.

Chemical synthetic materials or polymers including carbon nanotubes and carbosilane dendrimers have been explored for delivering siRNAs in cell culture. In 2007, Liu et al. showed that single walled nanotubes (SWNTs) were capable of siRNA delivery resulting in efficient RNAi-mediated inhibition of the CXCR4 and CD4 receptors on human T cells and peripheral blood mononuclear cells (PBMCs) (28). They observed up to 90% knockdown of the CXCR4 receptor and up to 60% knockdown of CD4 expression in cultured T cells, and up to 60% downregulation of CXCR4 in primary PBMCs. Additionally, amino terminated carbosilane dendrimers (with interior carbon-silicon bonds) were also used for delivery of siRNAs to HIV-infected lymphocytes (29). The dendrimer/siRNA complex silenced GAPDH expression and reduced HIV replication. However, the complicated formulation (nanotube functionalization and nanotube-siRNA conjugation) and suboptimal RNAi activity makes this system less attractive for clinical applications in the treatment of HIV infection.

Fusion proteins, such as: cell-penetrating peptide-dsRNA-binding domain fusion proteins and an antibody Fab fragment-protamine fusion protein, and a CD7-antibody-polyarginine conjugate have been used to functionally deliver siRNAs and induce RNAi responses *in vivo* in the absence of cytotoxicity. Eguchi et al. reported an efficient siRNA delivery approach that uses a peptide transduction domain-double stranded RNA-binding domain (PTD-DRBD) fusion protein (30). PTD-DRBD-mediated CD4 and CD8 specific siRNA delivery induced a rapid RNAi response in difficult-to-transfect primary cell types (e.g., various primary and transformed cells and human embryonic stem cells) without cytotoxicity and no innate immune responses. A targeted intracellular delivery approach for siRNAs to specific cell populations or tissues is highly desirable for the safety and efficacy of RNAi-based therapeutics. In 2005, Song et al. showed an antibody against HIV-1 envelope-mediated *in vivo* delivery of siRNA *via* a cell-surface receptor, in which a protamine-antibody F105 fusion protein was able to specifically deliver siRNAs to HIV-infected primary T cells or HIV envelope-expressing cells (31). Their results demonstrated a siRNA targeted against the HIV-1 *gag* capsid gene delivered by such a fusion protein could inhibit HIV-1 replication only in cells expressing the HIV-1 envelope. In a similar type of study, Kumar et al. conjugated a CD7 specific antibody with an oligo-9-arginine peptide to obtain a peptide-antibody based cell-specific siRNA delivery system (scFvCD7-9R) (18). A combination of siRNAs targeting the cellular CCR5 coreceptor and two conserved HIV-1 genes (*vif* and *tat*) were complexed to scFvCD7-9R prior to weekly systemic injection and resulted in suppression of HIV-1 infection and protection of CD4+T cell depletion in humanized mice. Furthermore, a comparison of a single anti-CCR5 siRNA and a triple siRNA combination revealed that more robust suppression of viral infection required virus-specific siRNAs along with the anti-CCR5 siRNA, although anti-CCR5 siRNA definitely contributed to protection. These results demonstrate that targeting combinations of cellular and viral genes may be the most effective strategy for RNAi-based treatment of HIV-1 infection.

Despite these advancements, the time-consuming protein fusion and complicated formulations limit their practical feasibility for clinical translation. Expensive production processes with batch-to-batch variability and potential immunogenicity may limit the utility of these antibody-based delivery approaches. In this regard, nucleic acid-based aptamers may offer advantages over protein vectors, such as the potential for chemical modification, enhanced stability, and no or minimal immunogenicity (32). HIV-1 gp120 expressed on the surface of HIV-1 infected cells represents a unique target for aptamer-mediated siRNA delivery. It has been demonstrated that HIV-1 gp120 RNA specific aptamers can specifically bind to and be internalized into gp120 expressing cells *via* receptor-mediated endocytosis (see Chapter 22 and refs. (33, 34)). These aptamer-siRNA conjugates provide a dual inhibitory approach for cell type-specific delivery of the siRNAs as well as aptamer neutralizing activity of HIV-1. Although the aptamer alone provides HIV inhibitory function, the aptamer-siRNA chimeras provide a prolonged inhibition of HIV, suggesting cooperativity between the siRNA and aptamer portions in inhibiting HIV replication and spread. In particular, these dual action constructs might be useful for treatment of patients who do not respond to HAART. In the future, continued efforts to improve the therapeutic efficacy by increasing circulation time and enhancing endosomal escape are challenges for making these approaches better therapeutics.

3.2. Lentiviral Vectors-Mediated shRNA Delivery

Different viral vectors have been used to stably transduce cells with shRNA expression constructs. Currently, lentiviral vectors represent one of the most popular choices due to their capability of transducing nondividing hematopoietic stem cells. For example, in a proof-of-concept study, viral infection and replication was inhibited by using lentiviral vector-mediated delivery of truncated CD4 molecules to CD4+ cell lines and primary lymphocytes (35). Similarly, Berkhout and coworkers have shown that HIV-1 replication can be strongly suppressed in cells transduced with a lentiviral shRNA vector (23).

As mentioned above, the capacity of HIV to mutate at high frequency may enable the virus to rapidly escape the selective pressure of RNAi *via* simple mutations of the targeted mRNA sequences. Therefore, lentiviral vectors with multiple payloads that confer antiviral effects in T cells have been developed to overcome this problem. We previously described that coexpression of a U6 promoter transcribed anti-tat/rev shRNA, a U6 promoter driven nucleolar localizing TAR RNA decoy and VA1 promoted anti-CCR5 ribozyme in a single vector efficiently inhibited HIV-1 over 42 days and was more effective than a single anti-*tat/rev* shRNA or double combinations of the shRNA/ribozyme or decoy (15). The transduced cells are created *ex vivo* by mobilizing and removing hematopoietic progenitor cells from HIV infected patients, transducing the cells *ex vivo*, and reinfusing the genetically modified cells in patients who have also undergone full marrow preparation for stem cell transplantation. The triple combination lentiviral construct has been approved by the US FDA and has presently entered in a human clinical trial for AIDS/lymphoma patients at the City of Hope Medical Center. This approach may provide the impetus for future trials with combinations of shRNAs in stem cell-based gene therapies.

4. Conclusions

Within the past decade, RNAi has show great potential as a therapeutic modality for various diseases. HIV-1 became one of the first infectious agents targeted by RNAi due to its well-understood life cycle and pattern of gene expression. However, two major obstacles for the long-term use of RNAi against chronic HIV-1 infection are viral escape and poor cellular uptake/stability of siRNA.

Currently, significant progress has been made to identify therapeutic targets and in the development of efficient and safe siRNA delivery approaches. Efforts have been undertaken to identify hundreds of cellular targets and to develop combination therapies. Moreover, several examples discussed in this chapter demonstrate cell type-specific, nonviral vector-mediated siRNA delivery into HIV infected and target cells. Moreover viral vector-mediated multi-RNAi therapeutic approaches may provide a complementary mechanism for combining the power of RNAi with other nucleic acid therapeutics, thereby providing a versatile technology platform for the treatment of various diseases.

Despite of these advances, RNAi technology still requires refinement before its full potential can be utilized for the routine clinical treatment of HIV. For example when targeting HDFs the safety and toxicity profiles of downregulating these host targets must be carefully investigated in long-term knockdown studies. In addition, further efforts should be conducted to refine siRNA delivery schemes and to reduce or avoid unwanted immunogenicity and unwanted side-effects.

References

1. Fire, A., Xu, S., Montgomery, M. K., Kostas, S. A., Driver, S. E., and Mello, C. C. (1998) Potent and specific genetic interference by double-stranded RNA in Caenorhabditis elegans. *Nature 391*, 806–11.
2. Zamore, P. D. (2006) RNA interference: big applause for silencing in Stockholm. *Cell 127*, 1083–6.
3. Castanotto, D., and Rossi, J. J. (2009) The promises and pitfalls of RNA-interference-based therapeutics. *Nature 457*, 426–33.
4. Kim, D. H., and Rossi, J. J. (2007) Strategies for silencing human disease using RNA interference. *Nat Rev Genet 8*, 173–84.
5. de Fougerolles, A., Vornlocher, H. P., Maraganore, J., and Lieberman, J. (2007) Interfering with disease: a progress report on siRNA-based therapeutics. *Nat Rev Drug Discov 6*, 443–53.
6. Kurreck, J. (2009) RNA interference: from basic research to therapeutic applications. *Angew Chem Int Ed Engl 48*, 1378–98.
7. Zamore, P. D., Tuschl, T., Sharp, P. A., and Bartel, D. P. (2000) RNAi: double-stranded RNA directs the ATP-dependent cleavage of mRNA at 21 to 23 nucleotide intervals. *Cell 101*, 25–33.
8. Song, E., Lee, S. K., Wang, J., Ince, N., Ouyang, N., Min, J., Chen, J., Shankar, P., and Lieberman, J. (2003) RNA interference targeting Fas protects mice from fulminant hepatitis. *Nat Med 9*, 347–51.
9. Whitehead, K. A., Langer, R., and Anderson, D. G. (2009) Knocking down barriers: advances in siRNA delivery. *Nat Rev Drug Discov 8*, 129–38.
10. Haasnoot, J., Westerhout, E. M., and Berkhout, B. (2007) RNA interference against viruses: strike and counterstrike. *Nat Biotechnol 25*, 1435–43.
11. Rossi, J. J., June, C. H., and Kohn, D. B. (2007) Genetic therapies against HIV. *Nat Biotechnol 25*, 1444–54.
12. Richman, D. D., Margolis, D. M., Delaney, M., Greene, W. C., Hazuda, D., and Pomerantz, R. J. (2009) The challenge of finding a cure for HIV infection. *Science 323*, 1304–7.
13. Scherer, L., Rossi, J. J., and Weinberg, M. S. (2007) Progress and prospects: RNA-based therapies for treatment of HIV infection. *Gene Ther 14*, 1057–64.
14. Singh, S. K., and Gaur, R. K. (2009) Progress towards therapeutic application of RNA interference for HIV infection. *BioDrugs 23*, 269–76.
15. Li, M. J., Kim, J., Li, S., Zaia, J., Yee, J. K., Anderson, J., Akkina, R., and Rossi, J. J. (2005) Long-term inhibition of HIV-1 infection in primary hematopoietic cells by lentiviral vector delivery of a triple combination of anti-HIV shRNA, anti-CCR5 ribozyme, and a nucleolar-localizing TAR decoy. *Mol Ther 12*, 900–9.
16. DiGiusto, D. L., Krishnan, A., Li, L., Li, H., Li, S., Rao, A., Mi, S., Yam, P., Stinson, S., Kalos, M., Alvarnas, J., Lacey, S. F., Yee, J. K., Li, M., Couture, L., Hsu, D., Forman, S. J., Rossi, J. J., and Zaia, J. A. (2010) RNA-based gene therapy for HIV with lentiviral vector-modified CD34(+) cells in patients undergoing transplantation for AIDS-related lymphoma. *Sci Transl Med 2*, 36ra43.
17. Berkhout, B., and ter Brake, O. (2009) Towards a durable RNAi gene therapy for HIV-AIDS. *Expert Opin Biol Ther 9*, 161–70.
18. Kumar, P., Ban, H. S., Kim, S. S., Wu, H., Pearson, T., Greiner, D. L., Laouar, A., Yao, J., Haridas, V., Habiro, K., Yang, Y. G., Jeong, J. H., Lee, K. Y., Kim, Y. H., Kim, S. W., Peipp, M., Fey, G. H., Manjunath, N., Shultz, L. D., Lee, S. K., and Shankar, P. (2008) T cell-specific siRNA delivery suppresses HIV-1 infection in humanized mice. *Cell 134*, 577–86.
19. Tsygankov, A. Y. (2009) Current developments in anti-HIV/AIDS gene therapy. *Curr Opin Investig Drugs 10*, 137–49.
20. Podlekareva, D., Mocroft, A., Dragsted, U. B., Ledergerber, B., Beniowski, M., Lazzarin, A., Weber, J., Clumeck, N., Vetter, N., Phillips, A., and Lundgren, J. D. (2006) Factors associated with the development of opportunistic infections in HIV-1-infected adults with high CD4+ cell counts: a EuroSIDA study. *J Infect Dis 194*, 633–41.
21. Boden, D., Pusch, O., Lee, F., Tucker, L., and Ramratnam, B. (2003) Human immunodeficiency virus type 1 escape from RNA interference. *J Virol 77*, 11531–5.
22. Das, A. T., Brummelkamp, T. R., Westerhout, E. M., Vink, M., Madiredjo, M., Bernards, R., and Berkhout, B. (2004) Human immunodeficiency virus type 1 escapes from RNA interference-mediated inhibition. *J Virol 78*, 2601–5.
23. ter Brake, O., t Hooft, K., Liu, Y. P., Centlivre, M., von Eije, K. J., and Berkhout, B. (2008) Lentiviral vector design for multiple shRNA expression and durable HIV-1 inhibition. *Mol Ther 16*, 557–64.

24. Brass, A. L., Dykxhoorn, D. M., Benita, Y., Yan, N., Engelman, A., Xavier, R. J., Lieberman, J., and Elledge, S. J. (2008) Identification of host proteins required for HIV infection through a functional genomic screen. *Science* ***319***, 921–6.

25. Konig, R., Zhou, Y., Elleder, D., Diamond, T. L., Bonamy, G. M., Irelan, J. T., Chiang, C. Y., Tu, B. P., De Jesus, P. D., Lilley, C. E., Seidel, S., Opaluch, A. M., Caldwell, J. S., Weitzman, M. D., Kuhen, K. L., Bandyopadhyay, S., Ideker, T., Orth, A. P., Miraglia, L. J., Bushman, F. D., Young, J. A., and Chanda, S. K. (2008) Global analysis of host-pathogen interactions that regulate early-stage HIV-1 replication. *Cell* ***135***, 49–60.

26. Zhou, H., Xu, M., Huang, Q., Gates, A. T., Zhang, X. D., Castle, J. C., Stec, E., Ferrer, M., Strulovici, B., Hazuda, D. J., and Espeseth, A. S. (2008) Genome-scale RNAi screen for host factors required for HIV replication. *Cell Host Microbe* ***4***, 495–504.

27. Goff, S. P. (2008) Knockdown screens to knockout HIV-1. *Cell* ***135***, 417–20.

28. Liu, Z., Winters, M., Holodniy, M., and Dai, H. (2007) siRNA delivery into human T cells and primary cells with carbon-nanotube transporters. *Angew Chem Int Ed Engl* ***46***, 2023–7.

29. Weber, N., Ortega, P., Clemente, M. I., Shcharbin, D., Bryszewska, M., de la Mata, F. J., Gomez, R., and Munoz-Fernandez, M. A. (2008) Characterization of carbosilane dendrimers as effective carriers of siRNA to HIV-infected lymphocytes. *J Control Release* ***132***, 55–64.

30. Eguchi, A., Meade, B. R., Chang, Y. C., Fredrickson, C. T., Willert, K., Puri, N., and Dowdy, S. F. (2009) Efficient siRNA delivery into primary cells by a peptide transduction domain-dsRNA binding domain fusion protein. *Nat Biotechnol* ***27***, 567–71.

31. Song, E., Zhu, P., Lee, S. K., Chowdhury, D., Kussman, S., Dykxhoorn, D. M., Feng, Y., Palliser, D., Weiner, D. B., Shankar, P., Marasco, W. A., and Lieberman, J. (2005) Antibody mediated in vivo delivery of small interfering RNAs via cell-surface receptors. *Nat Biotechnol* ***23***, 709–17.

32. Zhou, J., and Rossi, J. J. (2009) The therapeutic potential of cell-internalizing aptamers. *Curr Top Med Chem* ***9***, 1144–57.

33. Zhou, J., Li, H., Li, S., Zaia, J., and Rossi, J. J. (2008) Novel dual inhibitory function aptamer-siRNA delivery system for HIV-1 therapy. *Mol Ther* ***16***, 1481–9.

34. Zhou, J., Swiderski, P., Li, H., Zhang, J., Neff, C. P., Akkina, R., and Rossi, J. J. (2009) Selection, characterization and application of new RNA HIV gp 120 aptamers for facile delivery of Dicer substrate siRNAs into HIV infected cells. *Nucleic Acids Res.* 37:3094–109

35. Pham, H. M., Arganaraz, E. R., Groschel, B., Trono, D., and Lama, J. (2004) Lentiviral vectors interfering with virus-induced CD4 down-modulation potently block human immunodeficiency virus type 1 replication in primary lymphocytes. *J Virol* ***78***, 13072–81.

Chapter 5

Chemical Modification of Small Interfering RNA

Jesper B. Bramsen and Jørgen Kjems

Abstract

Chemically synthesized siRNAs are widely used for gene silencing. For *in vitro* applications, stability, delivery, and immunological issues are rarely problematic, but for *in vivo* applications the situation is different. Limited stability, undesirable pharmacokinetic behaviour, and unanticipated side effects from the immune system call for more careful structural siRNA design and inclusion of chemical modifications at selected positions. Also the notion that siRNA induces significant off-target silencing of many non-related genes has promted new effective measures to enhance specificity. The scope of this review is to provide a simple guide to successful chemical and structural modification of siRNAs with improved activity, stability, specificity, and low toxicity.

Key words: RNAi, Small interfering RNA, Off-target effect, Gene silencing, Chemical modification

1. Introduction

The discovery of RNA interference (RNAi) by Fire et al. (1) not only triggered a paradigm change in our perception of RNA as a molecule taking active part in gene regulation, but also provided new RNA tools to experimentally reduce expression of specific genes in both basic research and therapeutics. Particularly, the demonstration by Tuschl et al. in 2001 that synthetic 21-mer double-stranded (ds) small interfering RNAs (siRNAs) sequence-specifically and efficiently silenced gene expression by RNAi in mammalian cells (2) sparked further optimization of siRNA designs and established siRNA as the preferred tool to silence gene expression today (3).

Synthetic siRNAs are typically designed to mimic the structure of RNA intermediates in the cellular RNAi pathway which process both endogenous microRNAs (miRNAs) from longer

Ronald P. van Rij (ed.), *Antiviral RNAi: Concepts, Methods, and Applications*, Methods in Molecular Biology, vol. 721,
DOI 10.1007/978-1-61779-037-9_5, © Springer Science+Business Media, LLC 2011

hairpin-folded RNA primary transcripts and siRNAs typically derived from longer exogenous dsRNAs (reviewed in refs. (4, 5)). In the cell cytoplasm, the siRNA or miRNA precursors, dsRNA, and precursor miRNAs (pre-miRNAs), respectively, are cleaved by the riboendonuclease Dicer and loaded into the RNA-induced silencing complex (RISC) containing a core of one of four Argonaute 1-4 (Ago1-4) proteins. In Ago2-directed RISC, one siRNA strand is cleaved and released (denoted passenger strand or sense strand (SS)), enabling RISC to subsequently bind and cleave any cellular RNA having (near-)perfect sequence complementary to the incorporated siRNA strand (therefore denoted guide strand or antisense strand (AS)), thereby resulting in sequence-specific gene knockdown (KD).

The current success and future promises of siRNAs lie in their ease of use, broad applicability, and strong reproducible potency as compared to other silencing nucleic acids such as antisense oligonucleotides (AONs) or ribozymes (6); the exploitation of endogenous RNAi pathways by introducing artificial RNAi substrates will not only direct multiple rounds of RNA cleavage by each RISC (7), but may also allow the rapid incorporation of silencing RNAs into RNAi protein complexes to protect them from nuclease degradation (8, 9). Yet, the experimental harnessing of endogenous RNAi pathways poses new challenges and requires careful design, e.g. in order to not disturb gene regulation by miRNAs as reported in mice (10).

2. Design of Unmodified siRNA

2.1. A Variety of siRNA Designs

The most widely used siRNA design today mimics natural Dicer cleavage products and comprises a 21 nucleotide (nt) guiding strand antisense to a given RNA target and a complementary passenger SS annealed to form a siRNA duplex with a 19 base-pair (bp) dsRNA stem and 2-nt 3′ overhangs at both ends (2, 11). Other siRNA designs mimic Dicer substrates to enhance incorporation into RNAi pathways and thereby siRNA potency; both synthetic 25–27mer siRNAs (12, 13) and short hairpin RNAs (shRNAs) (14) are Dicer-substrate siRNAs (DsiRNAs) that are reported to be more potent than corresponding conventional 21-mer siRNAs presumably due to enhanced uptake by the RNAi machinery (15). Yet, 27mer siRNA has a higher cost of synthesis and reports of cellular interferon (IFN) induction by blunt 27mers have posed concerns (16). Other efficient, yet less widely used, siRNA designs are the Dicer-independent short shRNAs (sshRNAs) (17–19), blunt 19mers (20, 21), blunt fork-siRNA (22, 23), and siRNAs using segmented passenger strands (sisiRNAs) (24).

2.2. siRNA Target Sequence Considerations

A number of siRNA design rules have so far been deduced from the experimental testing of large siRNA sets (25–31). One line of rules reflects the requirement for high target site accessibility; RNA target sites should not overlap with protein-occupied region (32) and should preferentially be found in AU-rich region (33) to avoid secondary structures in the target RNA from blocking RISC-binding (34–37). In effect, effective siRNAs have relatively low GC-content (30–50%) (27, 38), especially in the 2–8 nucleotide region of the AS guide strand denoted the "seed" region (27, 30) that serves as nucleation point for RISC-target interactions (39). As siRNAs can trigger both stimulation of innate immune responses and miRNA-like off-target effects (discussed in Subheadings 6 and 7), siRNAs should be designed to avoid known immunostimulatory sequence motifs and target sequences should be unique in the transcriptome.

2.3. siRNA Sequence Considerations

Another important criterion is the efficiency by which the siRNA is recognized and adopted by the RNAi machinery, particularly during RISC-loading and subsequent target cleavage. A major determinant of siRNA potency is strand selection from the symmetrical siRNA during RISC loading. Here the relative difference in thermodynamic stability of the siRNA duplex termini affects the outcome (27, 40); The siRNA strand having the 5′ end engaged in the thermodynamically least stable part of the duplex will preferentially be utilized as guiding strand in RISC (41, 42). In effect, potent guide strands should be thermodynamically asymmetric and relative AU-rich in the 5′ end and/or more GC-rich in the 3′ end (27, 28, 30). In fact, an inefficient siRNA can be converted into a potent silencer by altering the thermodynamic properties of the 5′ ends of either siRNA strands (43). Similarly, the siRNA overhangs can be designed to favour loading of the intended guide strand into RISC; albeit blunt-ended siRNAs do support high levels of RNAi (21), they are slightly less efficiently processed by the RNAi machinery (44). Therefore, the use of asymmetric siRNA with a blunt end at the passenger strand 3′ end and a normal guide strand 2 nt overhang will favour guide strand loading into RISC irrespectively of the thermodynamic asymmetry of siRNA duplex ends (45–47). Some studies reported that low internal stability is found in potent siRNAs and that the tenth base of the guide strand should be an A or U (27, 29, 30), which likely ease either sense or the target strand cleavage by RISC or subsequent strand release. Other nucleotide preferences at specific positions within the siRNA duplex have been reported (25–30), but seem not to be widely used in siRNA design.

3. Strategies for Chemical Modification of siRNAs

3.1. Motivations for Chemical Modification of siRNA

The success of siRNAs in the laboratory has inspired researches to bring siRNA-mediated gene silencing into animal studies with the scope of developing a new class of therapeutics for human diseases. The performance of well-designed, unmodified siRNAs are in most cases adequate for short-term KD experiments in mammalian cell cultures, yet siRNA applications *in vivo* require even higher standards for siRNA potency, specificity, and safety. Particularly, the high susceptibility of RNA to ribonuclease degradation in biological fluids (48), the potential immunogenic properties of (some) siRNAs (49), the non-intended off-target regulation of genes sharing only partial sequence complementarity to either siRNA strands (50), and poor pharmacokinetic properties and biodistribution of siRNAs (51, 52) are key concerns that are currently being addressed. The chemical synthesis of siRNAs allows the incorporation of chemically modified nucleoside phosphoramidites into discrete positions in the siRNA to modulate its biochemical properties. Great efforts have been undertaken to improve siRNA performance by chemical modification benefitting from an arsenal of well-described nucleotide modifications that has been employed in AON designs for many years (53, 54). As described below, a number of different types of modifications have been tested in siRNA designs; modification of the phosphodiester backbone has been utilized primarily to enhance siRNA stability, yet may also affect RNA biodistribution and cellular uptake; modifications of the ribose 2′-OH were originally used to enhance siRNA stability, yet are now widely used to modulate siRNA potency, specificity of silencing, and to reduce siRNA immunogenicity; conjugation of lipophilic and cationic molecules to the termini of siRNA has been utilized primarily to facilitate cellular uptake and alter the biodistribution of siRNAs *in vivo*.

3.2. Phosphodiester Backbone Modifications

Phosphothioate (PS) modifications have been used extensively to increase the stability of AONs (55) and were therefore immediately tested in siRNA designs. Although moderately PS-modified siRNAs support efficient RNAi (48, 56–60), extensive PS modification reduces silencing (48, 61) and has toxic side-effects (56, 58). Furthermore, PS-substitution does not dramatically improve serum stability (48, 59) and does not alter the biodistribution *in vivo* (62), although PS-modified oligonucleotides do nonspecifically bind to cellular proteins (63). Notably, PS-modified AONs have been shown to dramatically enhance cellular uptake of naked siRNAs in trans via caveolin-mediated endocytosis in mammalian cell culture (64), however, gene silencing is very limited, likely due to siRNA entrapment in vesicles in the vicinity of the cell nucleus (64, 65). Also substitutions of the native phosphodiester linkage with either a boranophosphate linkage (61), amide linkage (66), or 2′,5′-linkage (67) has been found to enhance nuclease resistance of siRNA, yet are not widely used.

3.3. Ribose 2′-OH Modifications

Since the ribose 2′-OH group is not required for siRNA function (48), this position has been extensively modified in siRNA design both by the substitution of the 2′OH group with, for example, 2′*O*-Methyl (2′-OMe), 2′-Fluoro (2′F), and 2′-methoxyethyl (2′-O-MOE) or by locking the 2′OH via intermolecular linkages, e.g. locked nucleic acid (LNA) and ethylene-bridged nucleic acid (ENA).

The naturally occurring 2′-OMe is among the most extensively tested 2′ substitutions (20, 21, 48, 56, 57, 60, 68) and a 2′OMe/PS-modified siRNA was the first to successfully silence an endogenous gene *in vivo* (52). 2′-OMe has a *C3′*-endo sugar pucker, slightly enhances thermostability of the siRNA duplex and is well-tolerated at most duplex positions. Yet, extensive or full modification, particularly of the AS, can reduce siRNA potency (11, 21, 48, 59), albeit conflicting results are reported (57, 69). 2′OMe modifications have been successfully combined with other 2′-modifications, e.g. 2′Flouro (see Subheading 4), to generate fully substituted, nuclease resistant, yet potent siRNAs (70). Interestingly, 2′OMe modification may also reduce the immunogenic potential of siRNAs (described in Subheading 6.2). Fluorine substitution (2′-F) of the 2′-OH is another well-characterized modification tolerated in both siRNA strands except for very extensively modified duplexes (20, 48, 57–59) and certain siRNA sequences (71); 2′-F modifications on all siRNA pyrimidines were reported to preserve siRNA potency while greatly enhancing stability and supported effective silencing *in vitro* and *in vivo* (72–74). Alternating modifications using 2′F and DNA (75) or 2′OMe (70) also produce potent siRNAs with increased nuclease resistance. DNA has been utilized in siRNA designs from the very beginning where siRNAs have been synthesized with DNA overhangs, typically dTdT, to reduce cost and infer nuclease resistance (2), albeit this may slightly reduce silencing efficiency (48, 59). siRNAs are relatively tolerant to 2′-deoxy modification and the passenger strand has been fully modified with little loss of siRNA function (76, 77). In contrast, only partial substitution of the guide strand is tolerated (48, 71); however, alternating modification with 2′F has created fully substituted, active guide strands (48). Notably, DNA substitution of the guide strand seed region may even reduce off-target effects by some siRNA sequences (discussed in Subheading 7.2) (78).

More bulky 2′-modifications such as 2′-O-MOE and 2′-O-allyl modifications have been tested in siRNA design. They are only very position-specifically tolerated within the base-pairing part of the siRNA (20, 56, 79, 80) and do not immediately confer unique beneficial properties to siRNA designs (in addition to the enhanced nuclease resistance provided by most investigated chemistries). Yet, 2′-aminoethyl modification has been reported to enhance siRNA function when inserted into the passenger strand 3′ end, likely by altering the thermodynamic asymmetry of the siRNA duplex (80). Most bulky 2′ modifications, however, are not widely used and mainly restricted to siRNA overhangs to enhance exonuclease resistance (20, 56).

A more radical approach to modifying 2′OH groups is by using conformationally locked nucleic acids where the 2′-oxygen is connected to, e.g. the 4′ carbon via a methylene bridge as in LNA (81) and carbocyclic-LNA (80, 82) or ethylene bridge as in ENA (83), and carbocyclic-ENA (80, 82) or to the 1′ carbon as in oxetane (OXE) (80, 84). Among these, LNA has been most extensively used to enhance the performance of both AON (85, 86) and siRNAs (24, 59, 60, 80, 87). The intermolecular methylene bridge locks the furanose ring of LNA in a 3′-endo RNA-like conformation, which will lead to an additive increase in thermal stability by 2–10°C per LNA monomer upon incorporation into RNA duplexes (88). This dramatic increase in thermostability severely limits the number of LNAs that are tolerated in siRNA design (59, 60, 87); however, limited LNA modification allows the modulating of the local thermodynamic profile within the siRNA to increase nuclease resistance *in vitro* (59) and *in vivo* (89, 90), it can reduce the immunogenic properties of certain siRNA sequences (91), enhances siRNA potency and specificity by altering strand selection during RISC loading (24, 80, 87), and has even allowed the development of novel siRNA designs (24).

3.4. Ribose Modifications

Substituting the ribose sugar moiety of nucleotides has also been employed in siRNA design such as in ANA, HNA, and FANA nucleotides which are based on anitrol, hexitol, and arabinose, respectively (80, 92–94). These nucleotides are generally well-tolerated in the passenger strand and in the 3′ end of the guide strand and have been reported to enhance siRNA stability, potency and KD duration in some studies (92, 93). A more radical modification of the nucleotide ribose is found in unlocked nucleic acid (UNA) monomers which are acyclic derivatives of RNA lacking the C2′–C3′-bond of the RNA ribose ring, yet still structurally mimic unmodified RNA upon incorporation into RNA duplexes. Incorporation of UNA monomers induces additive destabilization by 5–8°C per UNA monomer, thereby allowing local destabilization of the siRNA duplex (95). Extensive UNA modification will not allow annealing of siRNA strands (96) and UNA is not well-tolerated at the 5′-most positions of the guide stand (97). However, limited UNA modification can be strategically used in both siRNA strands to alter siRNA strand selection and improve the potency of extensively modified siRNAs, e.g. by LNAs that are otherwise too stable or rigid to support RNAi (98). Notably, a siRNA modified by UNA only in its overhangs has prolonged biostability *in vivo*, even compared to extensively LNA-modified siRNA, and produces efficient gene KD in contrast to unmodified siRNAs (96). Furthermore, the inclusion of single UNA modification at position 7 in the guide strand can efficiently reduce siRNA off-targeting by weakening interactions between guide strands and potential off-target mRNAs (99). Finally, 4′ thio-modified nucleotides contain a sulphur atom instead of the 4′ carbon of the ribose ring, which has been shown to enhance nuclease resistance and potency of siRNAs upon insertion into the siRNA duplex; however, extensive modification of the guide strand is not well-tolerated (100, 101) and sequence-specificity of 4′ thio modification effects have been reported (102).

3.5. Base Modifications

A number of modified nucleotide bases, such as 5-bromo-, 5-iodo-, 2-thio-, 4-thio, dihydro, and pseudo-uracil, have been tested in siRNA design, yet are not widely utilized. Modified bases are generally utilized to enhance RNA-binding affinity to stabilize base-pairing potential; whereas 5-bromo- and 5-iodo uracil slightly reduce siRNA potency slightly (48), 2-thio- and pseudouracil have been reported to enhance siRNA potency (103) and reduce cellular immune responses (104).

4. Improving siRNA Potency by Chemical Modification

Maximizing siRNA potency by chemical modification is desirable to minimize the dose of delivered siRNA required for efficient RNAi or when target sequences are suboptimal and cannot be freely chosen, e.g. when targeting fusion oncogenes, highly repetitive sequences, or particular splice variants. Although most types of chemical modifications have been shown to negatively impact siRNA potency, a few studies have found an increase in siRNA potency upon chemical modification (70, 80, 87). Notably, Allerson et al. reported an up to 500-fold increase in siRNA potency using fully modified siRNA with alternating 2′OMe/2′F-modifications (70), albeit this effect seems sequence-specific (105). The reason for this dramatic increase in potency is not fully understood, but 2′OMe/2F′-modified siRNAs are preferentially taken up by RISC as compared to unmodified siRNA (105). Other studies have observed enhanced siRNA potency by chemical modification that favours guide strand selection during RISC loading. One strategy is to alter the thermodynamic asymmetry of base-pairing region, e.g. by incorporating stabilizing LNA in the passenger strand 5′ end (87) or 2-thiouracil in the guide stand 3′ end (103) or inserting destabilizing modifications such as OXE, ethylamino, UNA, or dihydrouracil in the passenger strand 3′ end (80, 87). Another strategy involves modification of the siRNA 3′ overhangs to modulate strand selection by RISC; both chemically modified overhangs that are favoured and disfavoured during strand selection by RISC have been identified and can be incorporated into the guide and passenger strands of the siRNA, respectively (80).

5. Improving siRNA Nuclease Resistance by Chemical Modification

5.1. siRNA Degradation in Biological Fluids

RNA is highly labile in extracellular compartments due to degradation by ribonucleases, e.g. >99% of exogenous ssRNA is degraded in human blood within seconds of incubation (106). Double-stranded RNAs, such as siRNAs, are more resistant than their single-stranded counterparts, yet are still degraded within minutes in mammalian serum (48, 52, 59, 80, 107). A number of RNases have been ascribed as key mediators of siRNA degradation; levels of the 3′ exonuclease ERI-1 have been shown to negatively correlate with duration of siRNA silencing (108, 109) and also RNase A-like, yet unidentified, endonuclease activities have been described in human serum (110, 111). The RNase composition, especially the relative activities of exonuclease and pyrimidine-specific endonuclease, differs between biological fluids and species (112–116) and siRNA modification should therefore address the particular siRNA application, e.g. species, delivery vehicle, and entry route. Once inside cells, however, siRNAs seem relatively stable and enhancing siRNA stability does not immediately influence KD potency or persistence (73) and silencing can last for several weeks in terminally differentiated non-dividing cells, such as macrophages (117).

5.2. Strategies for Enhancing siRNA Nuclease Resistance

The degradation of RNA by ribonucleases involves the nucleophilic attack and hydrolysis of the interphosphate linkage via a 2′,3′-cyclic phosphate intermediate (118). a popular strategy for increasing siRNA nuclease resistance has been chemical modification of internucleotide phosphate linkages or ribose 2′ OH groups (or more simply by the physical separation of siRNA and nucleases by utilizing shielding delivery agent which is not discussed further here). Replacement of an inter-ribonucleotide non-bridging oxygen by sulphur to create a PS linkage has been used extensively to increase the stability of AONs (55) and was therefore immediately tested in siRNA designs; albeit moderately PS-modified siRNAs support efficient RNAi (48, 56–60), extensive PS modification does not dramatically improve serum stability (48, 59) and was reported to reduce silencing and to have toxic side-effects (56, 58). However, combining moderate PS modification with various 2′ substitutions and end-conjugations has been very successful in creating highly stable and potent siRNA for applications *in vivo* (52, 119, 120). Also, 4′ thioribose has been reported to increase siRNA stability by more than 600 times (100, 101).

A great number of ribose 2′ modifications have been used to increase nuclease resistance by either full, partial, or 3′ overhang modification of the siRNA duplex. In most cases, full modification will dramatically reduce siRNA function, yet some fully modified siRNAs, especially using DNA, 2′OMe, and 2′F substitutions, are reported to be both highly stable and potent (70, 72, 75).

5.3. Selective siRNA Modifications Enhance Endonuclease Resistance

Other strategies modify only vulnerable nucleotide positions of the siRNAs to greatly improve siRNA stability; as most dsRNA-specific endoribonucleases are pyrimidine-specific (and preferentially recognize UpA, UpG, and CpA dinucleotide motifs (114, 121–123)), these can be specifically modified with minor impact on activity (57, 122, 123). Vulnerable base positions may alternatively be protected by lowering their accessibility to nucleases; several studies suggest that the thermodynamic stability of dsRNA greatly influences resistance toward endonuclease attack as even double-strand-specific riboendonuclease cleaves dsRNA by the preferential binding to short single-stranded regions which are transiently exposed by spontaneous thermal fluctuations (124). Thus, enhancing siRNA thermostability slightly, e.g. by the introduction of LNA (59, 60, 80, 87, 90) or 4′ thioribose (100) at selected positions within the duplex can enhance siRNA stability while generally preserving siRNA potency. Too extensive thermodynamic stabilization of the siRNA duplex by, e.g. LNA dramatically reduces silencing potency, yet can be compensated for by the simultaneous incorporation of destabilizing modifications such as UNA (96) or using the sisiRNA design containing a segmented passenger strand (24). Thus, ensuring a thermodynamic profile tolerated by the RNAi machinery, yet detrimental to endonuclease attack, will generate highly potent, stable siRNAs.

5.4. Overhang Modification Enhances Exonuclease Resistance

The siRNA 3′ overhangs are subject to degradation by 3′ riboexonucleases, and in effect, blunt-ended siRNAs are reported to have slightly higher stability in foetal calf serum (FCS) (21). Yet, the 3′ overhangs of siRNAs are very tolerant to chemical modification and have been very extensively modified to prevent degradation by exonucleases (48, 56–60, 80), which has been reported to be the main nuclease activity in serum (116, 125). In most cases, stability is only moderately enhanced; however, this may prove sufficient for some siRNA applications, even *in vivo*; UNA modification of siRNA overhangs dramatically increases siRNA stability and function in mice as compared to unmodified siRNA (yet stability was not dramatically increased in bovine serum (96)). Interestingly, specific chemical modification of 3′ overhangs may be simultaneously used to increase resistance to exonuclease attack, to enhance siRNA potency by maximizing incorporation of the guiding strand during RISC loading (80), or for conjugation to cell-targeting or penetrating molecules to enhance cellular delivery (described in Subheading 8.1).

6. Using Chemical Modification to Reduce siRNA Immunogenicity

Long dsRNA has been long known to trigger innate immune responses in mammalian cells resulting in release of proinflammatory cytokines and IFNs and trigger shutdown of transcriptional and translational activity (126). Therefore, the demonstration that shorter exogenous siRNAs, mimicking endogenous RNA species, are apparently non-immunogenic was very much appreciated (2). Upon closer inspection, however, numerous studies have established that siRNAs can indeed trigger innate immune responses *in vivo* and in peripheral blood mononuclear cells (PBMC) *in vitro* and induce high levels of inflammatory cytokines such as tumour necrosis factor alpha (TNFα), interleukin-6 (IL-6), and interferon-alpha (IFN-α) (16, 49, 91, 104, 127–129). This siRNA immunogenicity depends on siRNA sequence and structure, delivery route/vehicle, and cell type, thereby underscoring that the evaluation of the immunogenic properties of siRNAs must be evaluated in a relevant experimental setup *in vivo* or using primary immune cells *in vitro*.

6.1. Cellular Sensors of siRNA

The innate immune response is a non-adaptive response of immune cells which relies on a range of specialized patterns-recognition receptors (PRRs) recognizing so-called pathogen-associated molecular patterns (PAMPs) such as foreign polysaccharides, peptides, DNA, and (viral) RNA (130). Different PRRs exit in different cell types and cellular compartments; in the cytoplasm of most cells, foreign dsRNA is sequence-independently recognized by dsRNA responsive kinase R(PKR) (131), the helicases retinoic acid inducible geneI (RIG-I) (132, 133), and melanoma differentiation-associated gene 5 (Mda5) (134). PKR was originally described to respond only to dsRNA longer than 30 bp (135), yet recent studies found low levels of PKR activation by 21mer siRNAs (133, 136). However, the helicase RIG-I is now considered a main PPR of cytoplasmic dsRNA where it senses the nature of dsRNA ends; the standard 21mer siRNA design having two 2 nt 3′ overhangs is tolerated, whereas longer blunt-ended dsRNA, including blunt 27mer dsiRNAs, and 5′ end triphosphates can trigger RIG-I activation (104, 133, 137).

Toll-like receptors are a family of membrane-bound PRRs, which are found primarily in immune cells and are responsible for detecting foreign RNA primarily in endosomes or on the cell surface. The TLR family comprises 13 members among which TLR3 (129), TLR7 (91, 138, 139), and TLR8 (138) are key mediators of siRNA detection. Human TLR7 and TLR8 sequence-dependently recognize U-rich or GU-rich ssRNA and are primarily expressed in plasmacytoid dendritic cells (pDC) and B cells or myeloid DC (mDC), monocytes, and macrophages (140). The sequence-specificity of TLR7/8 remains to be fully established, yet certain nucleotide motifs have been reported as strong inducers of TLR7/8 activation (91, 127, 128, 141, 142) and even any U-rich RNA sequence may be subject to TLR7 recognition (141). Human TLR3 is primarily expressed on the surface of mature mDCs (143, 144) where it recognizes foreign dsRNAs (145, 146) and some exogenous siRNAs in a sequence-unspecific manner, leading to the production of, e.g. IFN-γ and Interleukin 12 (144).

Due to the specificity and localization of siRNA-responsive PPRs, the host immunological responses towards siRNA, and thereby the siRNA modifications needed to abrogate these, are very much dependent on the entry route and delivery method; siRNA delivery via endosomes using common delivery vehicles such as cationic lipids, polymers, and nanoparticles will expose siRNAs to potential TLR7/8 stimulation. In contrast, delivery of naked siRNA will likely not lead to TLR7/8 activation (91, 127, 147), yet can instead trigger immune stimulation via surface-bound TLR3 (144).

6.2. Reducing siRNA Immunogenicity by Chemical Modification

Originally, chemical modification of siRNAs has primarily aimed at abrogating their immunogenic potential by making immunostimulatory sequence motifs unrecognizable to the endosomal TLR7, the primary TLR responsible for siRNA immune stimulation (130). Perhaps not unexpectedly, the immunogenicity of siRNAs can be reduced by chemically modifying the immunostimulatory motif using, e.g. fully 2′OMe/2′F-modified siRNA (148) or selected LNA modification (91). Furthermore, a number of base modifications, 5-methylcytidine (m5C), 5-methyluracil (m5U), N6-methyladenosine (m6A), 2-thiouridine (s2U), or pseudouridine, have been shown to reduce TLR7/8 immunostimulation (149). Notably, recent studies suggest that even very limited but selective 2′OMe modifications can abrogate siRNA immunogenicity; replacement of only uridine residues with either 2′-F or 2′-OMe abrogated all immunogenicity upon delivery to PBMCs using a cationic lipid (150). Also, replacement of selected uridine or guanosine residues, even as few as two residues in the SS, abrogated siRNA immunogenicity upon liposomal delivery in human PBMCs and mice (151), and very recently, Hamm et al. proposed that alternating 2′OMe modification of the SS can be an universal approach to avoid TLR7 activation by siRNAs (152). It is unclear how the introduction of 2′OMe nucleotides prevents recognition of the duplex RNA by the immune system. Interestingly, naturally occurring 2′OMe-modified RNA has been reported as a potent antagonist of immunostimulatory RNA by preventing TLR7 activation (153) and it was suggested that 2′OMe modification of endogenous RNAs may allow immune receptors to distinguish pathogen-derived (unmodified) RNA from host cell RNA (149).

Only few studies have aimed at abrogating siRNA-mediated activation of RIG-I by chemical modification as the standard 21mer siRNA design having canonical 2nt 3′ overhangs generally does not trigger RIG-I activation (133). However, blunt-ended siRNAs, such as dsiRNAs, may be modified by 2′OMe (154, 155) or DNA (133) to reduce RIG-I activation.

7. Reducing siRNA Off-Target Effects by Chemical Modification

7.1. siRNA Off-Target Effects Are an Inherent Feature of RNAi

The cleavage of intended target RNAs by Ago2-RISC is highly sequence-specific and only few mismatches between the guide stand and the target are tolerated (156, 157). Yet, siRNAs still trigger unintended silencing of hundreds of endogenous genes upon introduction into cells (50, 158, 159), which can result in toxic phenotypes (160) and compromise the interpretation and outcome of the particular siRNA application. These sequence-specific off-target effects are triggered by the inherent miRNA-like behaviour of all investigated siRNA (50) and reflect the shared handling of miRNAs and siRNAs by the cellular RNAi pathway (161). Like miRNA-target regulation, siRNA off-targeting is primarily mediated by the interaction between the seed region of the RISC-associated guide strand (nucleotide position 2–8 counting from the 5′ end) and complementary sites in the 3′ UTR of the target mRNA (158, 162). Upon target binding, partially complementary (off-)targets are silenced through several mechanisms such as translational inhibition and mRNA destabilization (163–165). Off-target effects can be dramatic; during a large-scale siRNA library screen, Lin et al. found the highly efficient siRNAs to primarily function through such unintended off-targeting rather than the intended siRNA target cleavage (162).

7.2. Chemical Modification of the Guide Strand Seed Region Reduce Off-Targeting

As both siRNA and miRNA effects are concentration-dependent (50), success in reducing siRNA off-targeting has been achieved by utilizing siRNA pools to minimize the contribution of the individual siRNAs to off-targeting while preserving on-target activity. Yet chemical modification of siRNA to reduce off-targeting of individual siRNA species is still considered a paramount prerequisite for future applications of siRNA in therapeutics. The challenge here is identifying modification types that reduce off-target potentials by exerting a higher negative impact on target interactions relying on partial sequence complementarity (i.e. off-targeting) relative to fully complementary targets (i.e. siRNA effect). It is well-described that initial interactions between the siRNA guide strand and target RNA are mediated by the seed region exposed by RISC (39) and thermodynamic stability of this interaction correlate positively with off-targeting (78, 158). In agreement, a number of studies have aimed at reducing siRNA off-targeting by chemically modifying the seed region of the guide strand; Jackson et al. found 2′OMe modification of position 2 of the guide strand to site-specifically reduce off-targeting and proposed that the strict size constraints of Ago2 to accommodate a 2′-OMe group at this position would affect conformational adjustments in RISC and reduce off-targeting (68). Other studies have instead sought to destabilize seed-target interaction by substituting position 1–8 with DNA (78) or incorporating the strongly destabilizing UNA modification at position 7 with only limited negative impact of siRNA efficiency (99). Particularly, UNA modification allows position-specific destabilization of seed–target interactions to adjust the thermodynamic stability to a level that will require additional base pairing, ideally perfect complementarity, to support efficient RNAi. It remains to be fully established, however, how generally applicable these modifications are in siRNA designs and how they respond to differences in seed sequence and thermodynamic properties.

7.3. Chemical Modification Abrogates Passenger Strand Off-Targeting

Similarly the passenger strand can significantly contribute to off-targeting, e.g. Clark et al. found that the passenger strand of an ICAM-1-directed siRNA potently targeted the tumour necrosis factor receptor 1 (TNFR1) mRNA, thereby abrogating an expected TNF response (166). A number of studies have aimed at reducing the contribution of the passenger strand to off-targeting by abrogating its function or incorporation into RISC; the sisiRNA design utilizes two shorter sense strands incapable of RNAi function (24), LNA modification of the passenger strand 5′ duplex end disfavours passenger strand incorporation into RISC (87) and chemical blockage of the passenger strand 5′ phosphate (168) by, e.g. 5′-*O*-methyl modification (167) abrogates its silencing potential.

8. Improving siRNA Pharmacokinetics by Chemical Modification

siRNAs are structurally and biochemically very dissimilar to most drugs approved today; they are relatively big with a molecular weight of roughly 14 kDa, highly labile in biological fluids, and highly charged due to their phosphate backbone, which prevent siRNA from penetrating cellular membranes by diffusion. Furthermore, mammalian cells, even phagocytic macrophages and dendritic cells, do not immediately internalize siRNA (117, 169), and in effect, effective delivery into the target cell cytoplasm still poses the major obstacle of siRNA applications *in vivo* and therapeutics (although inefficient cellular uptake of oligonucleotides and siRNAs have been described in cell culture using high nucleic acids concentrations (170–173)).

8.1. Enhancing Cellular Delivery by siRNA Conjugation

A great number of delivery vehicles have been developed to facilitate siRNA delivery across the plasma membrane into the cytoplasma such as cationic lipids (such as RNAifect, oligoefectamine, lipofectamine, DOTAP, and TransIT TKO), cationic polymer (such as polyethylenimine (PEI), Chitosan, and cyclodextrin), and dendrimers. All these compounds electrostatically adsorb the anionic siRNA onto their surface and subsequently dock on the anionic cell membranes and allow cellular uptake via adsorptive endocytosis. These non-covalently siRNA-binding vehicles and their applications are described in detail elsewhere (171). Chemical modification of the siRNA itself has also been employed to enhance cellular uptake. As the integrity of the guide strand 5′ end is required for siRNA function (48, 168), the 3′ end of the guide strand and both ends of the passenger strand have been conjugated to various cationic cell-penetrating molecules or liposomes typically via acid-labile or reducible linkages, often thio-linkages. In particular, cell-penetrating peptides (CPP) such as penetratin (174–176), transportan (175), oligoarginine (177), and TAT (176, 178) have been conjugated to siRNA ends. Although not fully understood, these short cationic, hydrophobic, and/or amphipathic CPPs interact electrostatically with proteoglycans on the cell surface and are internalized by endocytosis, thereby bringing conjugated siRNAs to endosomes from which they are subsequently released to the cytoplasm (179). siRNA–TAT conjugates exhibit a dramatic increase in cellular uptake *in vitro* comparable to commercial transfection reagents and support-efficient RNAi (178), yet no improvement in efficiency of siRNA–TAT conjugates was seen *in vivo* upon intranasal delivery in mice and TAT alone seemed to induce unspecific side-effects (176). siRNA conjugated to penetratin (derived from the antennapedia protein) and transportin (a fusion peptide between the neuropeptide galanin and mastoparan, a peptide toxin from wasp venom) has been reported to be similarly efficient as cationic liposomes in a variety of cell lines (175), yet penetratin has been shown to trigger innate immune responses in mice upon intratracheal delivery (176). Chemically stabilized siRNA has been modified by cholesterol conjugation to the 3′ end of the passenger stand via a pyrrolidine linker to ensure efficient uptake into liver cells upon intravenous injection in mice, thereby resulting in 60% silencing of target apoB mRNA (52). Similarly, another lipophilic conjugate, alfa-tocopherol, was conjugated to the 5′ end of the guide strand in a dsiRNA design to successfully reduce apoB protein levels in mouse livers upon intravenous injection (120). Finally, siRNA delivery using commercial transfection reagent can be significantly improved, at least *in vitro*, by siRNA concatamerization using short complementary "sticky" overhangs (180) or end-conjugation of siRNAs via reducible disulphide-bridges (181).

8.2. Targeted Delivery by siRNA Conjugation

Targeted delivery of siRNAs to specific tissues or cell types is an attractive strategy, if not a prerequisite, to develop siRNAs into effective therapeutic drugs; it minimizes the amount of required siRNA and potential side-effects. Several studies have utilized cell-targeting ligands such as glycosylated molecules, peptides, antibodies, hormones, vitamines, and aptamers by conjugation to various carrier systems (reviewed in ref. (182)). Direct conjugations of siRNAs to cell-targeting ligands such as peptides, antibodies, aptamers, micelles (183), and nanoparticles (184, 185) have also been reported; this strategy not only confers specificity of targeting, but may also enhance cellular uptake through receptor-mediated endocytosis of the specific ligand. The conjugation of a peptide-mimicking insulin growth factor 1 (IGF1) to the 5′ end of the siRNA passenger strand resulted in a 60% KD of target gene expression in MCF7 cells that overexpresses the IGF1 receptor (186). Also, an antibody targeting the transferrin receptor expressed at the blood–brain barrier was conjugated to either passenger strand ends via biotin–streptavidin linkages leading to effective target gene silencing in a rat brain tumour model upon intravenous injection (187). To target siRNA delivery to hepatocytes, Oishi et al. constructed a lactose–PEG–siRNA conjugate with an acid-labile linker between siRNA and lactosylated PEG and delivered high amounts of siRNAs into hepatocytes in a receptor-mediated manner (188). An aptamer targeting prostate-specific membrane antigen (PMSA), a receptor expressed in prostate cancers, has been conjugated to the 5′ end of the siRNA passenger strand via a streptavidine linkage. Target gene KD was as efficient as when using a conventional lipid-based reagent in LNCaP cells, a prostate tumour cell lines expressing PMSA (189). A clever chimeric variant of the siRNA-PMSA aptamer has been generated by combining the PMSA aptamer and a dsiRNA design in a single T7 transcript that will be cleaved into effective siRNA by endogenous Dicer upon receptor-mediated endocytosis of the PMSA aptamer (190).

8.3. Altering Biodistribution by siRNA Conjugation

A major obstacle in systemic siRNA delivery *in vivo* is the rapid clearance of siRNAs typically observed upon delivery via passive or hydrodynamic intravenous injections (191). Naked siRNAs are very rapidly cleared from the bloodstream primarily due to their renal excretion (51, 52) and degradation by serum RNases (48, 52, 59, 80, 107). As naked siRNAs are smaller than the size threshold for glomerular filtration, incorporating siRNAs into several types of particles has allowed prolonged siRNA circulation and represents the most efficient strategy to avoid renal excretion (148, 183, 188, 192–197). siRNA bioavailability may be enhanced by using nuclease-resistant siRNAs, albeit this strategy will principally not prevent renal excretion and instead lead to excretion of intact siRNA. Several studies report on the improved efficiency of naked, chemically stabilized siRNAs upon introduction *in vivo* (198) using LNA (89), UNA (96, 198), PS/2′OMe (52), DNA/2′F/2′OMe/PS (72), inverted abasic moieties, and PS-modified siRNAs (199), whereas other studies did find enhancement of siRNA efficacy by 2′-F substitution upon hydrodynamic injection in mice (73).

Instead, chemical modification or conjugation of the siRNA itself has also been employed to enhance siRNA biodistribution; cholesterol conjugations have been reported to improve siRNA pharmacokinetics and exhibited a higher binding to blood serum albumin upon intravenous injection in mice resulting in siRNAs being detectable in liver, heart, lung, kidney, and fat tissue after 24 h in contrast to naked siRNAs (52). Also, conjugations to bile acids and various long chain fatty acids have been shown to influence siRNA tissue distribution upon intravenous injection as they bind to various lipoproteins, lipoprotein receptors, and transmembrane proteins in blood facilitating cellular uptake (200). PS modification of the siRNA backbone may be expected to alter siRNA biodistribution as PS oligonucleotide non-specifically binds to cellular proteins (63); however, no alteration in the biodistribution of modified siRNAs is seen *in vivo* (62).

Delivery strategies, and thereby the chemical modifications needed in siRNA design, may well focus on local rather than systemic delivery and indeed the first clinical trials have relied on intraocular injection of siRNA to treat macular degeneration and inhalation of siRNA to treat respiratory syncytial virus (RSV) infection (201).

9. Summary: The Creation of Superior siRNAs

Chemical modification studies are still in the stage of identifying modifications that are beneficial in siRNA design and no universally favorable siRNA design has, so far, been deduced (202). The types and extent of chemical modification employed in siRNA design will likely be tailored to fit the particular application to balance potency, stability, and pharmacokinetics depending on the

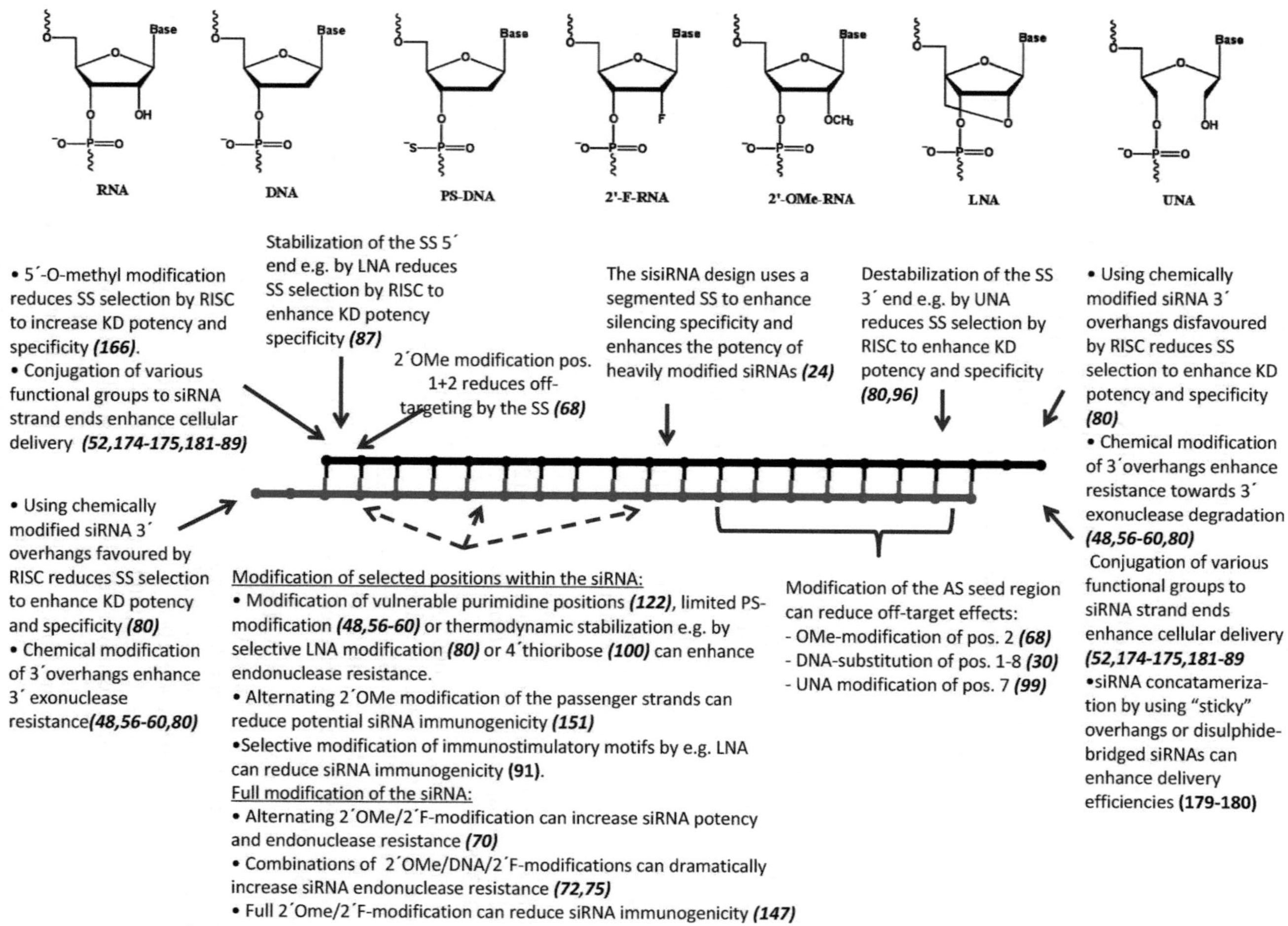

Fig. 1. Overview of chemical modification types and positions reported to enhance siRNA performance. *Upper panel*: examples of nucleoside analogues widely used in the design of chemically modified siRNAs. Abbreviations as found in the main text. *Lower panel*: Overview of chemical modifications used at various positions throughout the canonical siRNA to enhance siRNA potency, nuclease resistance, and silencing specificity. The passenger sense strand (SS) and guiding antisense strand (AS) are shown in *black* and *grey*, respectively.

delivery methods and target cell/tissue. Even so, important insight into optimizing the performance of siRNAs by chemical modification resulted in the identification of combinations of several types of chemical modifications that synergistically improve siRNA performance (see main text and Fig. 1 for an overview). In essence, modification of single positions can serve multiple purposes; siRNA 3′ overhang modification can enhance 3′ exonuclease resistance, boost siRNA efficiency, or enhance cellular delivery by various end-conjugations. Also, site-specific modifications of the base-pairing siRNA stem may simultaneously enhance endonuclease resistance and reduce siRNA immunogenicity and off-target effects. It is very likely that the combinations of such modifications will allow the generation of superior siRNAs that can form the basis for potent and safe siRNA therapeutics in the near future.

References

1. Fire, A., Xu, S., Montgomery, M. K., Kostas, S. A., Driver, S. E. and Mello, C. C. (1998) Potent and specific genetic interference by double-stranded RNA in *Caenorhabditis elegans*. *Nature*, **391**, 806–811.
2. Elbashir, S. M., Harborth, J., Lendeckel, W., Yalcin, A., Weber, K. and Tuschl, T. (2001) Duplexes of 21-nucleotide RNAs mediate RNA interference in cultured mammalian cells. *Nature*, **411**, 494–498.
3. Dorsett, Y. and Tuschl, T. (2004) siRNAs: applications in functional genomics and potential as therapeutics. *Nat Rev Drug Discov*, **3**, 318–329.
4. Jinek, M. and Doudna, J. A. (2009) A three-dimensional view of the molecular machinery of RNA interference. *Nature*, **457**, 405–412.
5. Siomi, H. and Siomi, M. C. (2009) On the road to reading the RNA-interference code. *Nature*, **457**, 396–404.
6. Bertrand, J. R., Pottier, M., Vekris, A., Opolon, P., Maksimenko, A. and Malvy, C. (2002) Comparison of antisense oligonucleotides and siRNAs in cell culture and *in vivo*. *Biochem Biophys Res Commun*, **296**, 1000–1004.
7. Hutvagner, G. and Zamore, P. D. (2002) A microRNA in a multiple-turnover RNAi enzyme complex. *Science*, **297**, 2056–2060.
8. Yi, R., Qin, Y., Macara, I. G. and Cullen, B. R. (2003) Exportin-5 mediates the nuclear export of pre-microRNAs and short hairpin RNAs. *Genes Dev*, **17**, 3011–3016.
9. Zeng, Y. and Cullen, B. R. (2003) Sequence requirements for micro RNA processing and function in human cells. *RNA*, **9**, 112–123.
10. Grimm, D., Streetz, K. L., Jopling, C. L., Storm, T. A., Pandey, K., Davis, C. R., Marion, P., Salazar, F. and Kay, M. A. (2006) Fatality in mice due to oversaturation of cellular microRNA/short hairpin RNA pathways. *Nature*, **441**, 537–541.
11. Elbashir, S. M., Martinez, J., Patkaniowska, A., Lendeckel, W. and Tuschl, T. (2001) Functional anatomy of siRNAs for mediating efficient RNAi in Drosophila melanogaster embryo lysate. *EMBO J*, **20**, 6877–6888.
12. Kim, D. H., Behlke, M. A., Rose, S. D., Chang, M. S., Choi, S. and Rossi, J. J. (2005) Synthetic dsRNA Dicer substrates enhance RNAi potency and efficacy. *Nat Biotechnol*, **23**, 222–226.
13. Amarzguioui, M., Lundberg, P., Cantin, E., Hagstrom, J., Behlke, M. A. and Rossi, J. J. (2006) Rational design and *in vitro* and *in vivo* delivery of Dicer substrate siRNA. *Nat Protoc*, **1**, 508–517.
14. Siolas, D., Lerner, C., Burchard, J., Ge, W., Linsley, P. S., Paddison, P. J., Hannon, G. J. and Cleary, M. A. (2005) Synthetic shRNAs as potent RNAi triggers. *Nat Biotechnol*, **23**, 227–231.
15. Tanudji, M., Machalek, D., Arndt, G. M. and Rivory, L. (2010) Competition between siRNA duplexes: impact of RISC loading efficiency and comparison between conventional-21 bp and dicer-substrate siRNAs. *Oligonucleotides* 20, 27–32.
16. Reynolds, A., Anderson, E. M., Vermeulen, A., Fedorov, Y., Robinson, K., Leake, D., Karpilow, J., Marshall, W. S. and Khvorova, A.

(2006) Induction of the interferon response by siRNA is cell type- and duplex length-dependent. *RNA*, **12**, 988–993.

17. Ge, Q., Dallas, A., Ilves, H., Shorenstein, J., Behlke, M. A. and Johnston, B. H. (2010) Effects of chemical modification on the potency, serum stability, and immunostimulatory properties of short shRNAs. *RNA*, **16**, 118–130.
18. Ge, Q., Ilves, H., Dallas, A., Kumar, P., Shorenstein, J., Kazakov, S. A. and Johnston, B. H. (2010) Minimal-length short hairpin RNAs: The relationship of structure and RNAi activity. *RNA*, **16**, 106–117.
19. Chu, C. Y. and Rana, T. M. (2008) Potent RNAi by short RNA triggers. *RNA*, **14**, 1714–1719.
20. Prakash, T. P., Allerson, C. R., Dande, P., Vickers, T. A., Sioufi, N., Jarres, R., Baker, B. F., Swayze, E. E., Griffey, R. H. and Bhat, B. (2005) Positional effect of chemical modifications on short interference RNA activity in mammalian cells. *J Med Chem*, **48**, 4247–4253.
21. Czauderna, F., Fechtner, M., Dames, S., Aygun, H., Klippel, A., Pronk, G. J., Giese, K. and Kaufmann, J. (2003) Structural variations and stabilising modifications of synthetic siRNAs in mammalian cells. *Nucleic Acids Res*, **31**, 2705–2716.
22. Hohjoh, H. (2002) RNA interference (RNA(i)) induction with various types of synthetic oligonucleotide duplexes in cultured human cells. *FEBS Lett*, **521**, 195–199.
23. Hohjoh, H. (2004) Enhancement of RNAi activity by improved siRNA duplexes. *FEBS Lett*, **557**, 193–198.
24. Bramsen, J. B., Laursen, M. B., Damgaard, C. K., Lena, S. W., Babu, B. R., Wengel, J. and Kjems, J. (2007) Improved silencing properties using small internally segmented interfering RNAs. *Nucleic Acids Res*, **35**, 5886–5897.
25. Ladunga, I. (2007) More complete gene silencing by fewer siRNAs: transparent optimized design and biophysical signature. *Nucleic Acids Res*, **35**, 433–440.
26. Huesken, D., Lange, J., Mickanin, C., Weiler, J., Asselbergs, F., Warner, J., Meloon, B., Engel, S., Rosenberg, A., Cohen, D. et al. (2005) Design of a genome-wide siRNA library using an artificial neural network. *Nat Biotechnol*, **23**, 995–1001.
27. Reynolds, A., Leake, D., Boese, Q., Scaringe, S., Marshall, W. S. and Khvorova, A. (2004) Rational siRNA design for RNA interference. *Nat Biotechnol*, **22**, 326–330.
28. Shabalina, S. A., Spiridonov, A. N. and Ogurtsov, A. Y. (2006) Computational models with thermodynamic and composition features improve siRNA design. *BMC Bioinformatics*, **7**, 65.
29. Jagla, B., Aulner, N., Kelly, P. D., Song, D., Volchuk, A., Zatorski, A., Shum, D., Mayer, T., De Angelis, D. A., Ouerfelli, O. et al. (2005) Sequence characteristics of functional siRNAs. *RNA*, **11**, 864–872.
30. Ui-Tei, K., Naito, Y., Takahashi, F., Haraguchi, T., Ohki-Hamazaki, H., Juni, A., Ueda, R. and Saigo, K. (2004) Guidelines for the selection of highly effective siRNA sequences for mammalian and chick RNA interference. *Nucleic Acids Res*, **32**, 936–948.
31. Li, W. and Cha, L. (2007) Predicting siRNA efficiency. *Cell Mol Life Sci*, **64**, 1785–1792.
32. Kretschmer-Kazemi Far, R. and Sczakiel, G. (2003) The activity of siRNA in mammalian cells is related to structural target accessibility: a comparison with antisense oligonucleotides. *Nucleic Acids Res*, **31**, 4417–4424.
33. Yuan, B., Latek, R., Hossbach, M., Tuschl, T. and Lewitter, F. (2004) siRNA Selection Server: an automated siRNA oligonucleotide prediction server. *Nucleic Acids Res*, **32**, W130–W134.
34. Brown, K. M., Chu, C. Y. and Rana, T. M. (2005) Target accessibility dictates the potency of human RISC. *Nat Struct Mol Biol*, **12**, 469–470.
35. Schubert, S., Grunweller, A., Erdmann, V. A. and Kurreck, J. (2005) Local RNA target structure influences siRNA efficacy: systematic analysis of intentionally designed binding regions. *J Mol Biol*, **348**, 883–893.
36. Overhoff, M., Alken, M., Far, R. K., Lemaitre, M., Lebleu, B., Sczakiel, G. and Robbins, I. (2005) Local RNA target structure influences siRNA efficacy: a systematic global analysis. *J Mol Biol*, **348**, 871–881.
37. Tafer, H., Ameres, S. L., Obernosterer, G., Gebeshuber, C. A., Schroeder, R., Martinez, J. and Hofacker, I. L. (2008) The impact of target site accessibility on the design of effective siRNAs. *Nat Biotechnol*, **26**, 578–583.
38. Holen, T. (2005) Mechanisms of RNAi: mRNA cleavage fragments may indicate stalled RISC. *J RNAi Gene Silencing*, **1**, 21–25.
39. Wang, Y., Sheng, G., Juranek, S., Tuschl, T. and Patel, D. J. (2008) Structure of the guide-strand-containing argonaute silencing complex. *Nature*, **456**, 209–213.
40. Aza-Blanc, P., Cooper, C. L., Wagner, K., Batalov, S., Deveraux, Q. L. and Cooke, M. P.

(2003) Identification of modulators of TRAIL-induced apoptosis via RNAi-based phenotypic screening. *Mol Cell*, **12**, 627–637.

41. Khvorova, A., Reynolds, A. and Jayasena, S. D. (2003) Functional siRNAs and miRNAs exhibit strand bias. *Cell*, **115**, 209–216.
42. Schwarz, D. S., Hutvagner, G., Du, T., Xu, Z., Aronin, N. and Zamore, P. D. (2003) Asymmetry in the assembly of the RNAi enzyme complex. *Cell*, **115**, 199–208.
43. Dykxhoorn, D. M., Schlehuber, L. D., London, I. M. and Lieberman, J. (2006) Determinants of specific RNA interference-mediated silencing of human beta-globin alleles differing by a single nucleotide polymorphism. *Proc Natl Acad Sci U S A*, **103**, 5953–5958.
44. Ghosh, P., Dullea, R., Fischer, J. E., Turi, T. G., Sarver, R. W., Zhang, C., Basu, K., Das, S. K. and Poland, B. W. (2009) Comparing 2-nt 3′ overhangs against blunt-ended siRNAs: a systems biology based study. *BMC Genomics*, **10 Suppl 1**, S17.
45. Sano, M., Sierant, M., Miyagishi, M., Nakanishi, M., Takagi, Y. and Sutou, S. (2008) Effect of asymmetric terminal structures of short RNA duplexes on the RNA interference activity and strand selection. *Nucleic Acids Res*, **36**, 5812–5821.
46. Vermeulen, A., Behlen, L., Reynolds, A., Wolfson, A., Marshall, W. S., Karpilow, J. and Khvorova, A. (2005) The contributions of dsRNA structure to Dicer specificity and efficiency. *RNA*, **11**, 674–682.
47. Rose, S. D., Kim, D. H., Amarzguioui, M., Heidel, J. D., Collingwood, M. A., Davis, M. E., Rossi, J. J. and Behlke, M. A. (2005) Functional polarity is introduced by Dicer processing of short substrate RNAs. *Nucleic Acids Res*, **33**, 4140–4156.
48. Chiu, Y. L. and Rana, T. M. (2003) siRNA function in RNAi: a chemical modification analysis. *RNA*, **9**, 1034–1048.
49. Sledz, C. A., Holko, M., de Veer, M. J., Silverman, R. H. and Williams, B. R. (2003) Activation of the interferon system by short-interfering RNAs. *Nat Cell Biol*, **5**, 834–839.
50. Jackson, A. L., Bartz, S. R., Schelter, J., Kobayashi, S. V., Burchard, J., Mao, M., Li, B., Cavet, G. and Linsley, P. S. (2003) Expression profiling reveals off-target gene regulation by RNAi. *Nat Biotechnol*, **21**, 635–637.
51. Gao, S., Dagnaes-Hansen, F., Nielsen, E. J., Wengel, J., Besenbacher, F., Howard, K. A. and Kjems, J. (2009) The effect of chemical modification and nanoparticle formulation on stability and biodistribution of siRNA in mice. *Mol Ther*, **17**, 1225–1233.
52. Soutschek, J., Akinc, A., Bramlage, B., Charisse, K., Constien, R., Donoghue, M., Elbashir, S., Geick, A., Hadwiger, P., Harborth, J. et al. (2004) Therapeutic silencing of an endogenous gene by systemic administration of modified siRNAs. *Nature*, **432**, 173–178.
53. Wilson, C. and Keefe, A. D. (2006) Building oligonucleotide therapeutics using non-natural chemistries. *Curr Opin Chem Biol*, **10**, 607–614.
54. Kurreck, J. (2003) Antisense technologies. Improvement through novel chemical modifications. *Eur J Biochem*, **270**, 1628–1644.
55. Aboul-Fadl, T. (2005) Antisense oligonucleotides: the state of the art. *Curr Med Chem*, **12**, 2193–2214.
56. Amarzguioui, M., Holen, T., Babaie, E. and Prydz, H. (2003) Tolerance for mutations and chemical modifications in a siRNA. *Nucleic Acids Res*, **31**, 589–595.
57. Choung, S., Kim, Y. J., Kim, S., Park, H. O. and Choi, Y. C. (2006) Chemical modification of siRNAs to improve serum stability without loss of efficacy. *Biochem Biophys Res Commun*, **342**, 919–927.
58. Harborth, J., Elbashir, S. M., Vandenburgh, K., Manninga, H., Scaringe, S. A., Weber, K. and Tuschl, T. (2003) Sequence, chemical, and structural variation of small interfering RNAs and short hairpin RNAs and the effect on mammalian gene silencing. *Antisense Nucleic Acid Drug Dev*, **13**, 83–105.
59. Braasch, D. A., Jensen, S., Liu, Y., Kaur, K., Arar, K., White, M. A. and Corey, D. R. (2003) RNA interference in mammalian cells by chemically-modified RNA. *Biochemistry*, **42**, 7967–7975.
60. Grunweller, A., Wyszko, E., Bieber, B., Jahnel, R., Erdmann, V. A. and Kurreck, J. (2003) Comparison of different antisense strategies in mammalian cells using locked nucleic acids, 2′-O-methyl RNA, phosphorothioates and small interfering RNA. *Nucleic Acids Res*, **31**, 3185–3193.
61. Hall, A. H., Wan, J., Shaughnessy, E. E., Ramsay Shaw, B. and Alexander, K. A. (2004) RNA interference using boranophosphate siRNAs: structure-activity relationships. *Nucleic Acids Res*, **32**, 5991–6000.
62. Braasch, D. A., Paroo, Z., Constantinescu, A., Ren, G., Oz, O. K., Mason, R. P. and Corey, D. R. (2004) Biodistribution of phosphodiester and phosphorothioate siRNA. *Bioorg Med Chem Lett*, **14**, 1139–1143.

63. Krieg, A. M. and Stein, C. A. (1995) Phosphorothioate oligodeoxynucleotides: antisense or anti-protein? *Antisense Res Dev*, **5**, 241.
64. Overhoff, M. and Sczakiel, G. (2005) Phosphorothioate-stimulated uptake of short interfering RNA by human cells. *EMBO Rep*, **6**, 1176–1181.
65. Detzer, A. and Sczakiel, G. (2009) Phosphorothioate-stimulated uptake of siRNA by mammalian cells: a novel route for delivery. *Curr Top Med Chem*, **9**, 1109–1116.
66. Iwase, R., Toyama, T. and Nishimori, K. (2007) Solid-phase synthesis of modified RNAs containing amide-linked oligoribonucleosides at their 3′-end and their application to siRNA. *Nucleosides Nucleotides Nucleic Acids*, **26**, 1451–1454.
67. Prakash, T. P., Kraynack, B., Baker, B. F., Swayze, E. E. and Bhat, B. (2006) RNA interference by 2′,5′-linked nucleic acid duplexes in mammalian cells. *Bioorg Med Chem Lett*, **16**, 3238–3240.
68. Jackson, A. L., Burchard, J., Leake, D., Reynolds, A., Schelter, J., Guo, J., Johnson, J. M., Lim, L., Karpilow, J., Nichols, K. et al. (2006) Position-specific chemical modification of siRNAs reduces "off-target" transcript silencing. *RNA*, **12**, 1197–1205.
69. Kraynack, B. A. and Baker, B. F. (2006) Small interfering RNAs containing full 2′-O-methylribonucleotide-modified sense strands display Argonaute2/eIF2C2-dependent activity. *RNA*, **12**, 163–176.
70. Allerson, C. R., Sioufi, N., Jarres, R., Prakash, T. P., Naik, N., Berdeja, A., Wanders, L., Griffey, R. H., Swayze, E. E. and Bhat, B. (2005) Fully 2′-modified oligonucleotide duplexes with improved *in vitro* potency and stability compared to unmodified small interfering RNA. *J Med Chem*, **48**, 901–904.
71. Parrish, S., Fleenor, J., Xu, S., Mello, C. and Fire, A. (2000) Functional anatomy of a dsRNA trigger: differential requirement for the two trigger strands in RNA interference. *Mol Cell*, **6**, 1077–1087.
72. Morrissey, D. V., Blanchard, K., Shaw, L., Jensen, K., Lockridge, J. A., Dickinson, B., McSwiggen, J. A., Vargeese, C., Bowman, K., Shaffer, C. S. et al. (2005) Activity of stabilized short interfering RNA in a mouse model of hepatitis B virus replication. *Hepatology*, **41**, 1349–1356.
73. Layzer, J. M., McCaffrey, A. P., Tanner, A. K., Huang, Z., Kay, M. A. and Sullenger, B. A. (2004) In vivo activity of nuclease-resistant siRNAs. *RNA*, **10**, 766–771.
74. Capodici, J., Kariko, K. and Weissman, D. (2002) Inhibition of HIV-1 infection by small interfering RNA-mediated RNA interference. *J Immunol*, **169**, 5196–5201.
75. Blidner, R. A., Hammer, R. P., Lopez, M. J., Robinson, S. O. and Monroe, W. T. (2007) Fully 2′-deoxy-2′-fluoro substituted nucleic acids induce RNA interference in mammalian cell culture. *Chem Biol Drug Des*, **70**, 113–122.
76. Pirollo, K. F., Rait, A., Zhou, Q., Hwang, S. H., Dagata, J. A., Zon, G., Hogrefe, R. I., Palchik, G. and Chang, E. H. (2007) Materializing the potential of small interfering RNA via a tumor-targeting nanodelivery system. *Cancer Res*, **67**, 2938–2943.
77. Hogrefe, R. I., Lebedev, A. V., Zon, G., Pirollo, K. F., Rait, A., Zhou, Q., Yu, W. and Chang, E. H. (2006) Chemically modified short interfering hybrids (siHYBRIDS): nanoimmunoliposome delivery *in vitro* and *in vivo* for RNAi of HER-2. *Nucleosides Nucleotides Nucleic Acids*, **25**, 889–907.
78. Ui-Tei, K., Naito, Y., Zenno, S., Nishi, K., Yamato, K., Takahashi, F., Juni, A. and Saigo, K. (2008) Functional dissection of siRNA sequence by systematic DNA substitution: modified siRNA with a DNA seed arm is a powerful tool for mammalian gene silencing with significantly reduced off-target effect. *Nucleic Acids Res*, **36**, 2136–2151.
79. Odadzic, D., Bramsen, J. B., Smicius, R., Bus, C., Kjems, J. and Engels, J. W. (2008) Synthesis of 2′-O-modified adenosine building blocks and application for RNA interference. *Bioorg Med Chem*, **16**, 518–529.
80. Bramsen, J. B., Laursen, M. B., Nielsen, A. F., Hansen, T. B., Bus, C., Langkjær, N., Babu, B. R., Højland, T., Abramov, M., Van Aerschot, A. et al. (2009) A large-scale chemical modification screen identifies design rules to generate siRNAs with high activity, high stability and low toxicity. *Nucleic Acids Res*, **37**, 2867–2881.
81. Wengel, J., Petersen, M., Nielsen, K. E., Jensen, G. A., Hakansson, A. E., Kumar, R., Sorensen, M. D., Rajwanshi, V. K., Bryld, T. and Jacobsen, J. P. (2001) LNA (locked nucleic acid) and the diastereoisomeric alpha-L-LNA: conformational tuning and high-affinity recognition of DNA/RNA targets. *Nucleosides Nucleotides Nucleic Acids*, **20**, 389–396.
82. Srivastava, P., Barman, J., Pathmasiri, W., Plashkevych, O., Wenska, M. and Chattopadhyaya, J. (2007) Five- and six-membered conformationally locked 2′,4′-carbocyclic ribo-thymidines: synthesis, structure, and biochemical studies. *J Am Chem Soc*, **129**, 8362–8379.

83. Hamada, M., Ohtsuka, T., Kawaida, R., Koizumi, M., Morita, K., Furukawa, H., Imanishi, T., Miyagishi, M. and Taira, K. (2002) Effects on RNA interference in gene expression (RNAi) in cultured mammalian cells of mismatches and the introduction of chemical modifications at the 3′-ends of siRNAs. *Antisense Nucleic Acid Drug Dev*, **12**, 301–309.

84. Pradeepkumar, P. I., Amirkhanov, N. V. and Chattopadhyaya, J. (2003) Antisense oligonucleotides with oxetane-constrained cytidine enhance heteroduplex stability, and elicit satisfactory RNase H response as well as showing improved resistance to both exo and endonucleases. *Org Biomol Chem*, **1**, 81–92.

85. Kurreck, J., Wyszko, E., Gillen, C. and Erdmann, V. A. (2002) Design of antisense oligonucleotides stabilized by locked nucleic acids. *Nucleic Acids Res*, **30**, 1911–1918.

86. Fluiter, K., ten Asbroek, A. L., de Wissel, M. B., Jakobs, M. E., Wissenbach, M., Olsson, H., Olsen, O., Oerum, H. and Baas, F. (2003) In vivo tumor growth inhibition and biodistribution studies of locked nucleic acid (LNA) antisense oligonucleotides. *Nucleic Acids Res*, **31**, 953–962.

87. Elmén, J., Thonberg, H., Ljungberg, K., Frieden, M., Westergaard, M., Xu, Y., Wahren, B., Liang, Z., Ørum, H., Koch, T. et al. (2005) Locked nucleic acid (LNA) mediated improvements in siRNA stability and functionality. *Nucleic Acids Res*, **33**, 439–447.

88. Petersen, M. and Wengel, J. (2003) LNA: a versatile tool for therapeutics and genomics. *Trends Biotechnol*, **21**, 74–81.

89. Glud, S. Z., Bramsen, J. B., Dagnaes-Hansen, F., Wengel, J., Howard, K. A., Nyengaard, J. R. and Kjems, J. (2009) Naked siLNA-mediated gene silencing of lung bronchoepithelium EGFP expression after intravenous administration. *Oligonucleotides*, **19**, 163–168.

90. Mook, O. R., Baas, F., de Wissel, M. B. and Fluiter, K. (2007) Evaluation of locked nucleic acid-modified small interfering RNA *in vitro* and *in vivo*. *Mol Cancer Ther*, **6**, 833–843.

91. Hornung, V., Guenthner-Biller, M., Bourquin, C., Ablasser, A., Schlee, M., Uematsu, S., Noronha, A., Manoharan, M., Akira, S., de Fougerolles, A. et al. (2005) Sequence-specific potent induction of IFN-alpha by short interfering RNA in plasmacytoid dendritic cells through TLR7. *Nat Med*, **11**, 263–270.

92. Dowler, T., Bergeron, D., Tedeschi, A. L., Paquet, L., Ferrari, N. and Damha, M. J. (2006) Improvements in siRNA properties mediated by 2′-deoxy-2′-fluoro-beta-D-arabinonucleic acid (FANA). *Nucleic Acids Res*, **34**, 1669–1675.

93. Fisher, M., Abramov, M., Van Aerschot, A., Rozenski, J., Dixit, V., Juliano, R. L. and Herdewijn, P. (2009) Biological effects of hexitol and altritol-modified siRNAs targeting B-Raf. *Eur J Pharmacol*, **606**, 38–44.

94. Watts, J. K., Choubdar, N., Sadalapure, K., Robert, F., Wahba, A. S., Pelletier, J., Pinto, B. M. and Damha, M. J. (2007) 2′-fluoro-4′-thioarabino-modified oligonucleotides: conformational switches linked to siRNA activity. *Nucleic Acids Res*, **35**, 1441–1451.

95. Langkjær, N., Pasternak, A. and Wengel, J. (2009) UNA (unlocked nucleic acid): a flexible RNA mimic that allows engineering of nucleic acid duplex stability. *Bioorg Med Chem*, **17**, 5420–5425.

96. Laursen, M. B., Pakula, M. M., Gao, S., Fluiter, K., Mook, O. R., Baas, F., Langkjaer, N., Wengel, S. L., Wengel, J., Kjems, J. et al. (2010) Utilization of unlocked nucleic acid (UNA) to enhance siRNA performance *in vitro* and *in vivo*. *Mol BioSyst*, **6**, 862–870.

97. Kenski, D. M., Cooper, A. J., Li, J. J., Willingham, A. T., Haringsma, H. J., Young, T. A., Kuklin, N. A., Jones, J. J., Cancilla, M. T., McMasters, D. R. et al. (2009) Analysis of acyclic nucleoside modifications in siRNAs finds sensitivity at position 1 that is restored by 5′-terminal phosphorylation both *in vitro* and *in vivo*. *Nucleic Acids Res*, **38**, 660–671.

98. Werk, D., Wengel, J., Lena, S. W., Grunert, H. P., Zeichhardt, H. and Kurreck, J. (2009) Application of small interfering RNAs modified by unlocked nucleic acid (UNA) to inhibit the heart-pathogenic coxsackievirus B3. *FEBS Lett*, **584**, 591–598.

99. Bramsen, J. B., Pakula, M. M., Hansen, T. B., Bus, C., Langkjaer, N., Odadzic, D., Smicius, R., Wengel, S. L., Chattopadhyaya, J., Engels, J. W. et al. (2010) A screen of chemical modifications identifies position-specific modification by UNA to most potently reduce siRNA off-target effects. *Nucleic Acids Res*, **38**, 5761–5773.

100. Dande, P., Prakash, T. P., Sioufi, N., Gaus, H., Jarres, R., Berdeja, A., Swayze, E. E., Griffey, R. H. and Bhat, B. (2006) Improving RNA interference in mammalian cells by 4′-thio-modified small interfering RNA (siRNA): effect on siRNA activity and nuclease stability when used in combination with 2′-O-alkyl modifications. *J Med Chem*, **49**, 1624–1634.

101. Hoshika, S., Minakawa, N., Kamiya, H., Harashima, H. and Matsuda, A. (2005) RNA interference induced by siRNAs modified with 4′-thioribonucleosides in cultured mammalian cells. *FEBS Lett*, **579**, 3115–3118.
102. Hoshika, S., Minakawa, N., Shionoya, A., Imada, K., Ogawa, N. and Matsuda, A. (2007) Study of modification pattern-RNAi activity relationships by using siRNAs modified with 4′-thioribonucleosides. *Chembiocem*, **8**, 2133–2138.
103. Sipa, K., Sochacka, E., Kazmierczak-Baranska, J., Maszewska, M., Janicka, M., Nowak, G. and Nawrot, B. (2007) Effect of base modifications on structure, thermodynamic stability, and gene silencing activity of short interfering RNA. *RNA*, **13**, 1301–1316.
104. Hornung, V., Ellegast, J., Kim, S., Brzozka, K., Jung, A., Kato, H., Poeck, H., Akira, S., Conzelmann, K. K., Schlee, M. et al. (2006) 5′-Triphosphate RNA is the ligand for RIG-I. *Science*, **314**, 994–997.
105. Koller, E., Propp, S., Murray, H., Lima, W., Bhat, B., Prakash, T. P., Allerson, C. R., Swayze, E. E., Marcusson, E. G. and Dean, N. M. (2006) Competition for RISC binding predicts *in vitro* potency of siRNA. *Nucleic Acids Res*, **34**, 4467–4476.
106. Tsui, N. B., Ng, E. K. and Lo, Y. M. (2002) Stability of endogenous and added RNA in blood specimens, serum, and plasma. *Clin Chem*, **48**, 1647–1653.
107. Chiu, Y. L. and Rana, T. M. (2002) RNAi in human cells: basic structural and functional features of small interfering RNA. *Mol Cell*, **10**, 549–561.
108. Kennedy, S., Wang, D. and Ruvkun, G. (2004) A conserved siRNA-degrading RNase negatively regulates RNA interference in *C. elegans*. *Nature*, **427**, 645–649.
109. Takabatake, Y., Isaka, Y., Mizui, M., Kawachi, H., Takahara, S. and Imai, E. (2007) Chemically modified siRNA prolonged RNA interference in renal disease. *Biochem Biophys Res Commun*, **363**, 432–437.
110. Haupenthal, J., Baehr, C., Zeuzem, S. and Piiper, A. (2007) RNAse A-like enzymes in serum inhibit the anti-neoplastic activity of siRNA targeting polo-like kinase 1. *Int J Cancer*, **121**, 206–210.
111. Haupenthal, J., Baehr, C., Kiermayer, S., Zeuzem, S. and Piiper, A. (2006) Inhibition of RNAse A family enzymes prevents degradation and loss of silencing activity of siRNAs in serum. *Biochem Pharmacol*, **71**, 702–710.
112. Hickerson, R. P., Vlassov, A. V., Wang, Q., Leake, D., Ilves, H., Gonzalez-Gonzalez, E., Contag, C. H., Johnston, B. H. and Kaspar, R. L. (2008) Stability study of unmodified siRNA and relevance to clinical use. *Oligonucleotides*, **18**, 345–354.
113. Tourriere, H., Chebli, K. and Tazi, J. (2002) mRNA degradation machines in eukaryotic cells. *Biochimie*, **84**, 821–837.
114. Sorrentino, S. (1998) Human extracellular ribonucleases: multiplicity, molecular diversity and catalytic properties of the major RNase types. *Cell Mol Life Sci*, **54**, 785–794.
115. Probst, J., Brechtel, S., Scheel, B., Hoerr, I., Jung, G., Rammensee, H. G. and Pascolo, S. (2006) Characterization of the ribonuclease activity on the skin surface. *Genet Vaccines Ther*, **4**, 4.
116. Zou, Y., Tiller, P., Chen, I. W., Beverly, M. and Hochman, J. (2008) Metabolite identification of small interfering RNA duplex by high-resolution accurate mass spectrometry. *Rapid Commun Mass Spectrom*, **22**, 1871–1881.
117. Song, E., Lee, S. K., Dykxhoorn, D. M., Novina, C., Zhang, D., Crawford, K., Cerny, J., Sharp, P. A., Lieberman, J., Manjunath, N. et al. (2003) Sustained small interfering RNA-mediated human immunodeficiency virus type 1 inhibition in primary macrophages. *J Virol*, 77, 7174–7181.
118. Usher, D. A. (1969) On the mechanism of ribonuclease action. *Proc Natl Acad Sci U S A*, **62**, 661–667.
119. Zimmermann, T. S., Lee, A. C., Akinc, A., Bramlage, B., Bumcrot, D., Fedoruk, M. N., Harborth, J., Heyes, J. A., Jeffs, L. B., John, M. et al. (2006) RNAi-mediated gene silencing in non-human primates. *Nature*, **441**, 111–114.
120. Nishina, K., Unno, T., Uno, Y., Kubodera, T., Kanouchi, T., Mizusawa, H. and Yokota, T. (2008) Efficient *in vivo* delivery of siRNA to the liver by conjugation of alpha-tocopherol. *Mol Ther*, **16**, 734–740.
121. Turner, J. J., Jones, S. W., Moschos, S. A., Lindsay, M. A. and Gait, M. J. (2007) MALDI-TOF mass spectral analysis of siRNA degradation in serum confirms an RNAse A-like activity. *Mol Biosyst*, **3**, 43–50.
122. Qiu, L., Moreira, A., Kaplan, G., Levitz, R., Wang, J. Y., Xu, C. and Drlica, K. (1998) Degradation of hammerhead ribozymes by human ribonucleases. *Mol Gen Genet*, **258**, 352–362.
123. Volkov, A. A., Kruglova, N. S., Meschaninova, M. I., Venyaminova, A. G., Zenkova, M. A., Vlassov, V. V. and Chernolovskaya, E. L. (2009) Selective protection of nuclease-sensitive sites in siRNA prolongs silencing effect. *Oligonucleotides*, **19**, 191–202.

124. Libonati, M. and Sorrentino, S. (1992) Revisiting the action of bovine ribonuclease A and pancreatic-type ribonucleases on double-stranded RNA. *Mol Cell Biochem*, **117**, 139–151.

125. Eder, P. S., DeVine, R. J., Dagle, J. M. and Walder, J. A. (1991) Substrate specificity and kinetics of degradation of antisense oligonucleotides by a 3′ exonuclease in plasma. *Antisense Res Dev*, **1**, 141–151.

126. Minks, M. A., West, D. K., Benvin, S. and Baglioni, C. (1979) Structural requirements of double-stranded RNA for the activation of 2′,5′-oligo(A) polymerase and protein kinase of interferon-treated HeLa cells. *J Biol Chem*, **254**, 10180–10183.

127. Judge, A. D., Sood, V., Shaw, J. R., Fang, D., McClintock, K. and MacLachlan, I. (2005) Sequence-dependent stimulation of the mammalian innate immune response by synthetic siRNA. *Nat Biotechnol*, **23**, 457–462.

128. Sioud, M. (2005) Induction of inflammatory cytokines and interferon responses by double-stranded and single-stranded siRNAs is sequence-dependent and requires endosomal localization. *J Mol Biol*, **348**, 1079–1090.

129. Kariko, K., Bhuyan, P., Capodici, J. and Weissman, D. (2004) Small interfering RNAs mediate sequence-independent gene suppression and induce immune activation by signaling through toll-like receptor 3. *J Immunol*, **172**, 6545–6549.

130. Sioud, M. (2009) Deciphering the code of innate immunity recognition of siRNAs. *Methods Mol Biol*, **487**, 41–59.

131. Meurs, E., Chong, K., Galabru, J., Thomas, N. S., Kerr, I. M., Williams, B. R. and Hovanessian, A. G. (1990) Molecular cloning and characterization of the human double-stranded RNA-activated protein kinase induced by interferon. *Cell*, **62**, 379–390.

132. Yoneyama, M., Kikuchi, M., Natsukawa, T., Shinobu, N., Imaizumi, T., Miyagishi, M., Taira, K., Akira, S. and Fujita, T. (2004) The RNA helicase RIG-I has an essential function in double-stranded RNA-induced innate antiviral responses. *Nat Immunol*, **5**, 730–737.

133. Marques, J. T., Devosse, T., Wang, D., Zamanian-Daryoush, M., Serbinowski, P., Hartmann, R., Fujita, T., Behlke, M. A. and Williams, B. R. (2006) A structural basis for discriminating between self and nonself double-stranded RNAs in mammalian cells. *Nat Biotechnol*, **24**, 559–565.

134. Kang, D. C., Gopalkrishnan, R. V., Wu, Q., Jankowsky, E., Pyle, A. M. and Fisher, P. B. (2002) mda-5: An interferon-inducible putative RNA helicase with double-stranded RNA-dependent ATPase activity and melanoma growth-suppressive properties. *Proc Natl Acad Sci U S A*, **99**, 637–642.

135. Manche, L., Green, S. R., Schmedt, C. and Mathews, M. B. (1992) Interactions between double-stranded RNA regulators and the protein kinase DAI. *Mol Cell Biol*, **12**, 5238–5248.

136. Puthenveetil, S., Whitby, L., Ren, J., Kelnar, K., Krebs, J. F. and Beal, P. A. (2006) Controlling activation of the RNA-dependent protein kinase by siRNAs using site-specific chemical modification. *Nucleic Acids Res*, **34**, 4900–5011.

137. Kim, D. H., Longo, M., Han, Y., Lundberg, P., Cantin, E. and Rossi, J. J. (2004) Interferon induction by siRNAs and ssRNAs synthesized by phage polymerase. *Nat Biotechnol*, **22**, 321–325.

138. Heil, F., Hemmi, H., Hochrein, H., Ampenberger, F., Kirschning, C., Akira, S., Lipford, G., Wagner, H. and Bauer, S. (2004) Species-specific recognition of single-stranded RNA via toll-like receptor 7 and 8. *Science*, **303**, 1526–1529.

139. Gantier, M. P., Tong, S., Behlke, M. A., Xu, D., Phipps, S., Foster, P. S. and Williams, B. R. (2008) TLR7 is involved in sequence-specific sensing of single-stranded RNAs in human macrophages. *J Immunol*, **180**, 2117–2124.

140. Zarember, K. A. and Godowski, P. J. (2002) Tissue expression of human Toll-like receptors and differential regulation of Toll-like receptor mRNAs in leukocytes in response to microbes, their products, and cytokines. *J Immunol*, **168**, 554–561.

141. Diebold, S. S., Massacrier, C., Akira, S., Paturel, C., Morel, Y. and Reis e Sousa, C. (2006) Nucleic acid agonists for Toll-like receptor 7 are defined by the presence of uridine ribonucleotides. *Eur J Immunol*, **36**, 3256–3267.

142. Diebold, S. S., Kaisho, T., Hemmi, H., Akira, S. and Reis e Sousa, C. (2004) Innate antiviral responses by means of TLR7-mediated recognition of single-stranded RNA. *Science*, **303**, 1529–1531.

143. Muzio, M., Bosisio, D., Polentarutti, N., D'amico, G., Stoppacciaro, A., Mancinelli, R., van't Veer, C., Penton-Rol, G., Ruco, L. P., Allavena, P. et al. (2000) Differential expression and regulation of toll-like receptors (TLR) in human leukocytes: Selective expression of TLR3 in dendritic cells. *Journal of Immunology*, **164**, 5998–6004.

144. Kleinman, M. E., Yamada, K., Takeda, A., Chandrasekaran, V., Nozaki, M., Baffi, J. Z.,

Albuquerque, R. J. C., Yamasaki, S., Itaya, M., Pan, Y. Z. et al. (2008) Sequence- and target-independent angiogenesis suppression by siRNA via TLR3. *Nature*, **452**, 591–597.

145. Kariko, K., Bhuyan, P., Capodici, J., Ni, H., Lubinski, J., Friedman, H. and Weissman, D. (2004) Exogenous siRNA mediates sequence-independent gene suppression by signaling through toll-like receptor 3. *Cells Tissues Organs*, **177**, 132–138.
146. Alexopoulou, L., Holt, A. C., Medzhitov, R. and Flavell, R. A. (2001) Recognition of double-stranded RNA and activation of NF-kappaB by Toll-like receptor 3. *Nature*, **413**, 732–738.
147. Heidel, J. D., Hu, S., Liu, X. F., Triche, T. J. and Davis, M. E. (2004) Lack of interferon response in animals to naked siRNAs. *Nat Biotechnol*, **22**, 1579–1582.
148. Morrissey, D. V., Lockridge, J. A., Shaw, L., Blanchard, K., Jensen, K., Breen, W., Hartsough, K., Machemer, L., Radka, S., Jadhav, V. et al. (2005) Potent and persistent *in vivo* anti-HBV activity of chemically modified siRNAs. *Nat Biotechnol*, **23**, 1002–1007.
149. Kariko, K., Buckstein, M., Ni, H. and Weissman, D. (2005) Suppression of RNA recognition by Toll-like receptors: the impact of nucleoside modification and the evolutionary origin of RNA. *Immunity*, **23**, 165–175.
150. Cekaite, L., Furset, G., Hovig, E. and Sioud, M. (2007) Gene expression analysis in blood cells in response to unmodified and 2′-modified siRNAs reveals TLR-dependent and independent effects. *Journal of Molecular Biology*, **365**, 90–108.
151. Judge, A. D., Bola, G., Lee, A. C. and MacLachlan, I. (2006) Design of noninflammatory synthetic siRNA mediating potent gene silencing *in vivo*. *Mol Ther*, **13**, 494–505.
152. Hamm, S., Latz, E., Hangel, D., Muller, T., Yu, P., Golenbock, D., Sparwasser, T., Wagner, H. and Bauer, S. (2010) Alternating 2′-O-ribose methylation is a universal approach for generating non-stimulatory siRNA by acting as TLR7 antagonist. *Immunobiology*, **215**, 559–569.
153. Robbins, M., Judge, A., Liang, L., McClintock, K., Yaworski, E. and MacLachlan, I. (2007) 2′-O-methyl-modified RNAs act as TLR7 antagonists. *Mol Ther*, **15**, 1663–1669.
154. Zamanian-Daryoush, M., Marques, J. T., Gantier, M. P., Behlke, M. A., John, M., Rayman, P., Finke, J. and Williams, B. R. (2008) Determinants of cytokine induction by small interfering RNA in human peripheral blood mononuclear cells. *J Interferon Cytokine Res*, **28**, 221–233.
155. Collingwood, M. A., Rose, S. D., Huang, L., Hillier, C., Amarzguioui, M., Wiiger, M. T., Soifer, H. S., Rossi, J. J. and Behlke, M. A. (2008) Chemical modification patterns compatible with high potency dicer-substrate small interfering RNAs. *Oligonucleotides*, **18**, 187–200.
156. Du, Q., Thonberg, H., Wang, J., Wahlestedt, C. and Liang, Z. (2005) A systematic analysis of the silencing effects of an active siRNA at all single-nucleotide mismatched target sites. *Nucleic Acids Res*, **33**, 1671–1677.
157. Dahlgren, C., Zhang, H. Y., Du, Q., Grahn, M., Norstedt, G., Wahlestedt, C. and Liang, Z. (2008) Analysis of siRNA specificity on targets with double-nucleotide mismatches. *Nucleic Acids Res*, **36**, e53.
158. Birmingham, A., Anderson, E. M., Reynolds, A., Ilsley-Tyree, D., Leake, D., Fedorov, Y., Baskerville, S., Maksimova, E., Robinson, K., Karpilow, J. et al. (2006) 3′ UTR seed matches, but not overall identity, are associated with RNAi off-targets. *Nat Methods*, **3**, 199–204.
159. Lim, L. P., Lau, N. C., Garrett-Engele, P., Grimson, A., Schelter, J. M., Castle, J., Bartel, D. P., Linsley, P. S. and Johnson, J. M. (2005) Microarray analysis shows that some microRNAs downregulate large numbers of target mRNAs. *Nature*, **433**, 769–773.
160. Fedorov, Y., Anderson, E. M., Birmingham, A., Reynolds, A., Karpilow, J., Robinson, K., Leake, D., Marshall, W. S. and Khvorova, A. (2006) Off-target effects by siRNA can induce toxic phenotype. *RNA*, **12**, 1188–1196.
161. Doench, J. G., Petersen, C. P. and Sharp, P. A. (2003) siRNAs can function as miRNAs. *Genes Dev*, **17**, 438–442.
162. Lin, X., Ruan, X., Anderson, M. G., McDowell, J. A., Kroeger, P. E., Fesik, S. W. and Shen, Y. (2005) siRNA-mediated off-target gene silencing triggered by a 7 nt complementation. *Nucleic Acids Res*, **33**, 4527–4535.
163. Doench, J. G. and Sharp, P. A. (2004) Specificity of microRNA target selection in translational repression. *Genes Dev*, **18**, 504–511.
164. Williams, A. E. (2008) Functional aspects of animal microRNAs. *Cell Mol Life Sci*, **65**, 545–562.

165. Wu, L., Fan, J. and Belasco, J. G. (2006) MicroRNAs direct rapid deadenylation of mRNA. *Proc Natl Acad Sci U S A*, **103**, 4034–4039.
166. Clark, P. R., Pober, J. S. and Kluger, M. S. (2008) Knockdown of TNFR1 by the sense strand of an ICAM-1 siRNA: dissection of an off-target effect. *Nucleic Acids Res*, **36**, 1081–1097.
167. Chen, P. Y., Weinmann, L., Gaidatzis, D., Pei, Y., Zavolan, M., Tuschl, T. and Meister, G. (2008) Strand-specific 5′-O-methylation of siRNA duplexes controls guide strand selection and targeting specificity. *RNA*, **14**, 263–274.
168. Nykanen, A., Haley, B. and Zamore, P. D. (2001) ATP requirements and small interfering RNA structure in the RNA interference pathway. *Cell*, **107**, 309–321.
169. Stewart, S. A., Dykxhoorn, D. M., Palliser, D., Mizuno, H., Yu, E. Y., An, D. S., Sabatini, D. M., Chen, I. S., Hahn, W. C., Sharp, P. A. et al. (2003) Lentivirus-delivered stable gene silencing by RNAi in primary cells. *RNA*, **9**, 493–501.
170. Laktionov, P. P., Dazard, J. E., Vives, E., Rykova, E. Y., Piette, J., Vlassov, V. V. and Lebleu, B. (1999) Characterisation of membrane oligonucleotide-binding proteins and oligonucleotide uptake in keratinocytes. *Nucleic Acids Res*, **27**, 2315–2324.
171. de Diesbach, P., Berens, C., N'Kuli, F., Monsigny, M., Sonveaux, E., Wattiez, R. and Courtoy, P. J. (2000) Identification, purification and partial characterisation of an oligonucleotide receptor in membranes of HepG2 cells. *Nucleic Acids Res*, **28**, 868–874.
172. Lingor, P., Michel, U., Scholl, U., Bahr, M. and Kugler, S. (2004) Transfection of "naked" siRNA results in endosomal uptake and metabolic impairment in cultured neurons. *Biochem Biophys Res Commun*, **315**, 1126–1133.
173. Overhoff, M., Wunsche, W. and Sczakiel, G. (2004) Quantitative detection of siRNA and single-stranded oligonucleotides: relationship between uptake and biological activity of siRNA. *Nucleic Acids Res*, **32**, e170.
174. Davidson, T. J., Harel, S., Arboleda, V. A., Prunell, G. F., Shelanski, M. L., Greene, L. A. and Troy, C. M. (2004) Highly efficient small interfering RNA delivery to primary mammalian neurons induces MicroRNA-like effects before mRNA degradation. *J Neurosci*, **24**, 10040–10046.
175. Muratovska, A. and Eccles, M. R. (2004) Conjugate for efficient delivery of short interfering RNA (siRNA) into mammalian cells. *FEBS Lett*, **558**, 63–68.
176. Moschos, S. A., Jones, S. W., Perry, M. M., Williams, A. E., Erjefalt, J. S., Turner, J. J., Barnes, P. J., Sproat, B. S., Gait, M. J. and Lindsay, M. A. (2007) Lung delivery studies using siRNA conjugated to TAT(48-60) and penetratin reveal peptide induced reduction in gene expression and induction of innate immunity. *Bioconjug Chem*, **18**, 1450–1459.
177. Kim, W. J., Christensen, L. V., Jo, S., Yockman, J. W., Jeong, J. H., Kim, Y. H. and Kim, S. W. (2006) Cholesteryl oligoarginine delivering vascular endothelial growth factor siRNA effectively inhibits tumor growth in colon adenocarcinoma. *Mol Ther*, **14**, 343–350.
178. Chiu, Y. L., Ali, A., Chu, C. Y., Cao, H. and Rana, T. M. (2004) Visualizing a correlation between siRNA localization, cellular uptake, and RNAi in living cells. *Chem Biol*, **11**, 1165–1175.
179. Abes, S., Moulton, H., Turner, J., Clair, P., Richard, J. P., Iversen, P., Gait, M. J. and Lebleu, B. (2007) Peptide-based delivery of nucleic acids: design, mechanism of uptake and applications to splice-correcting oligonucleotides. *Biochem Soc Trans*, **35**, 53–55.
180. Bolcato-Bellemin, A. L., Bonnet, M. E., Creusat, G., Erbacher, P. and Behr, J. P. (2007) Sticky overhangs enhance siRNA-mediated gene silencing. *Proc Natl Acad Sci U S A*, **104**, 16050–16055.
181. Lee, S. Y., Huh, M. S., Lee, S., Lee, S. J., Chung, H., Park, J. H., Oh, Y. K., Choi, K., Kim, K. and Kwon, I. C. (2009) Stability and cellular uptake of polymerized siRNA (poly-siRNA)/polyethylenimine (PEI) complexes for efficient gene silencing. *J Control Release*, **141**, 339–346.
182. Ikeda, Y. and Taira, K. (2006) Ligand-targeted delivery of therapeutic siRNA. *Pharm Res*, **23**, 1631–1640.
183. Oishi, M., Nagasaki, Y., Nishiyama, N., Itaka, K., Takagi, M., Shimamoto, A., Furuichi, Y. and Kataoka, K. (2007) Enhanced growth inhibition of hepatic multicellular tumor spheroids by lactosylated poly(ethylene glycol)-siRNA conjugate formulated in PEGylated polyplexes. *ChemMedChem*, **2**, 1290–1297.
184. Medarova, Z., Pham, W., Farrar, C., Petkova, V. and Moore, A. (2007) In vivo imaging of siRNA delivery and silencing in tumors. *Nat Med*, **13**, 372–377.
185. Derfus, A. M., Chen, A. A., Min, D. H., Ruoslahti, E. and Bhatia, S. N. (2007) Targeted quantum dot conjugates for siRNA delivery. *Bioconjug Chem*, **18**, 1391–1396.
186. Cesarone, G., Edupuganti, O. P., Chen, C. P. and Wickstrom, E. (2007) Insulin receptor

substrate 1 knockdown in human MCF7 ER+ breast cancer cells by nuclease-resistant IRS1 siRNA conjugated to a disulfide-bridged D-peptide analogue of insulin-like growth factor 1. *Bioconjug Chem*, **18**, 1831–1840.

187. Xia, C. F., Zhang, Y., Boado, R. J. and Pardridge, W. M. (2007) Intravenous siRNA of brain cancer with receptor targeting and avidin-biotin technology. *Pharm Res*, **24**, 2309–2316.

188. Oishi, M., Nagasaki, Y., Itaka, K., Nishiyama, N. and Kataoka, K. (2005) Lactosylated poly(ethylene glycol)-siRNA conjugate through acid-labile beta-thiopropionate linkage to construct pH-sensitive polyion complex micelles achieving enhanced gene silencing in hepatoma cells. *J Am Chem Soc*, **127**, 1624–1625.

189. Chu, T. C., Twu, K. Y., Ellington, A. D. and Levy, M. (2006) Aptamer mediated siRNA delivery. *Nucleic Acids Res*, **34**, e73.

190. McNamara, J. O., II, Andrechek, E. R., Wang, Y., Viles, K. D., Rempel, R. E., Gilboa, E., Sullenger, B. A. and Giangrande, P. H. (2006) Cell type-specific delivery of siRNAs with aptamer-siRNA chimeras. *Nat Biotechnol*, **24**, 1005–1015.

191. McCaffrey, A. P., Meuse, L., Pham, T. T., Conklin, D. S., Hannon, G. J. and Kay, M. A. (2002) RNA interference in adult mice. *Nature*, **418**, 38–39.

192. Bartlett, D. W. and Davis, M. E. (2008) Impact of tumor-specific targeting and dosing schedule on tumor growth inhibition after intravenous administration of siRNA-containing nanoparticles. *Biotechnol Bioeng*, **99**, 975–985.

193. Urban-Klein, B., Werth, S., Abuharbeid, S., Czubayko, F. and Aigner, A. (2005) RNAi-mediated gene-targeting through systemic application of polyethylenimine (PEI)-complexed siRNA *in vivo*. *Gene Ther*, **12**, 461–466.

194. Schiffelers, R. M., Ansari, A., Xu, J., Zhou, Q., Tang, Q., Storm, G., Molema, G., Lu, P. Y., Scaria, P. V. and Woodle, M. C. (2004) Cancer siRNA therapy by tumor selective delivery with ligand-targeted sterically stabilized nanoparticle. *Nucleic Acids Res*, **32**, e149.

195. Song, E., Zhu, P., Lee, S. K., Chowdhury, D., Kussman, S., Dykxhoorn, D. M., Feng, Y., Palliser, D., Weiner, D. B., Shankar, P. et al. (2005) Antibody mediated *in vivo* delivery of small interfering RNAs via cell-surface receptors. *Nat Biotechnol*, **23**, 709–717.

196. Heidel, J. D., Yu, Z., Liu, J. Y., Rele, S. M., Liang, Y., Zeidan, R. K., Kornbrust, D. J. and Davis, M. E. (2007) Administration in non-human primates of escalating intravenous doses of targeted nanoparticles containing ribonucleotide reductase subunit M2 siRNA. *Proc Natl Acad Sci U S A*, **104**, 5715–5721.

197. Kim, S. H., Mok, H., Jeong, J. H., Kim, S. W. and Park, T. G. (2006) Comparative evaluation of target-specific GFP gene silencing efficiencies for antisense ODN, synthetic siRNA, and siRNA plasmid complexed with PEI-PEG-FOL conjugate. *Bioconjug Chem*, **17**, 241–244.

198. Bartlett, D. W. and Davis, M. E. (2007) Effect of siRNA nuclease stability on the *in vitro* and *in vivo* kinetics of siRNA-mediated gene silencing. *Biotechnol Bioeng*, **97**, 909–921.

199. Shen, J., Samul, R., Silva, R. L., Akiyama, H., Liu, H., Saishin, Y., Hackett, S. F., Zinnen, S., Kossen, K., Fosnaugh, K. et al. (2006) Suppression of ocular neovascularization with siRNA targeting VEGF receptor 1. *Gene Ther*, **13**, 225–234.

200. Wolfrum, C., Shi, S., Jayaprakash, K. N., Jayaraman, M., Wang, G., Pandey, R. K., Rajeev, K. G., Nakayama, T., Charrise, K., Ndungo, E. M. et al. (2007) Mechanisms and optimization of *in vivo* delivery of lipophilic siRNAs. *Nat Biotechnol*, **25**, 1149–1157.

201. Bumcrot, D., Manoharan, M., Koteliansky, V. and Sah, D. W. (2006) RNAi therapeutics: a potential new class of pharmaceutical drugs. *Nat Chem Biol*, **2**, 711–719.

202. Corey, D. R. (2007) Chemical modification: the key to clinical application of RNA interference? *J Clin Invest*, **117**, 3615–3622.

Part II

Viral Small RNAs

Chapter 6

Viral Small RNA Cloning and Sequencing

Valérie Gausson and Maria-Carla Saleh

Abstract

At the current rate of technological progress, high-throughput sequencing of nucleic acids has become a commodity. These techniques are perfectly suitable for viral small RNAs sequencing and contribute to the understanding of many aspects of virus biology in the context of host–pathogen interaction. However, the generation of high quality data is still an issue and the preparation of small RNAs libraries that accurately reflect the viral siRNAs in the sample remains a challenge. In this chapter we describe how to clone and sequence libraries of viral small RNAs from infected insect samples (mosquito, drosophilidae, insect-derived cell lines).

Key words: Small RNAs, Deep sequencing, Illumina, Library preparation, RNA ligation

1. Introduction

Virus-derived small RNAs (vsiRNAs) were first identified in plants as a result of the antiviral RNAi host response (1). The dicing of the viral RNA into siRNAs was later shown in *Drosophila* (2) and the nematode *C. elegans* (3). These vsiRNAs reflect the activation of the antiviral immune system in the infected species.

Because vsiRNAs arise from replicative intermediates or structured genome sequences of viruses, their characterization greatly contributes to the understanding of many aspects of virus biology in the context of host–pathogen interaction. The abundance of vsiRNAs reflects the levels of viral replication and gene expression. The polarity of the vsiRNAs (they can correspond to the (+) strand or the (–) strand of the virus) reflects the source of dsRNA. vsiRNAs can be derived from the dicing of dsRNA in replication intermediates, in structured regions, or in convergent viral transcripts. Their mapping on the reference viral genome reflects

Ronald P. van Rij (ed.), *Antiviral RNAi: Concepts, Methods, and Applications*, Methods in Molecular Biology, vol. 721, DOI 10.1007/978-1-61779-037-9_6, © Springer Science+Business Media, LLC 2011

structured and exposed regions and allows the identification of new viral variants with deletion/insertions compared to the reference. Analysis of the vsiRNAs size is indicative of the RNAi pathway involved in their biogenesis. In *Drosophila*, the main vsiRNAs are 21 nts and a product of Dcr2 (2), whereas in plants they are 21 nts when produced from DCL-4 and 22 nts when produced by DCL-2 (4). In addition, in *Drosophila* vsiRNAs may exhibit 3′ O-methylation which is the signature of their loading in the Ago2 RISC complex (5, 6). Finally, the *de novo* assembly of vsiRNAs can be used for the discovery of new virus species (7).

Several techniques have been developed for massive sequencing of nucleic acids and they are perfectly suitable for small RNAs sequencing, including vsiRNAs. These techniques include pyrosequencing (454–Roche), sequencing-by-synthesis (Solexa – Illumina), and sequencing by ligation (SOLID – Applied Biosystems). Next generation sequencing involves direct sequencing of single DNA or even RNA molecules (Helicos BioScience) (8).

To successfully apply these techniques, one faces two steps: the preparation of small RNAs libraries that accurately reflect the vsiRNAs in the sample and the management of a considerable amount of sequencing data. Here, we describe the method we use to clone and sequence small RNAs using Illumina technology. Our protocol is adapted from Pffefer (9). The main steps are summarized in Fig. 1. Briefly, total RNA from infected samples is purified, and the small RNA fraction is recovered and ligated at the 5′ and 3′ ends to RNA adapters. The ligated RNA is then reverse-transcribed and specifically amplified with DNA primers complementary to the primers spotted on the flowcell provided by Illumina. Once the sequence data have been collected, they can be analyzed using the bioinformatic pipeline described in Chap. 7.

2. Material

2.1. Total RNAs Isolation from Infected Samples (Mosquito, Drosophilidae, Insect-Derived Cell Lines)

1. TRIzol reagent (Invitrogen, Carlsbad, CA).
2. Chloroform.
3. Isopropyl alcohol.
4. Nuclease-free 75% ethanol conserved at –20°C.
5. Nuclease-free water.
6. Deionized formamide.
7. RNaseZap RNase decontamination solution (Ambion, Austin, TX).
8. Pellet pestle (Sigma, St. Louis, MO).

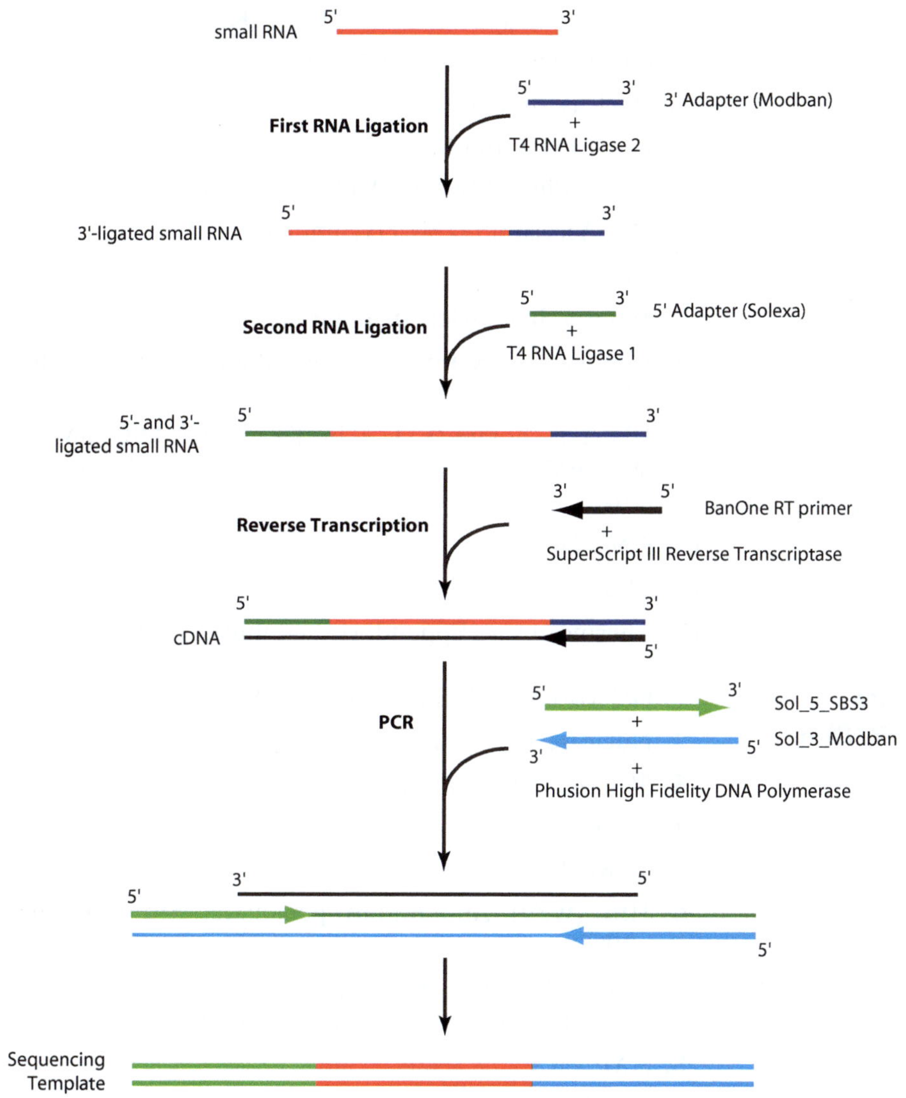

Fig. 1. Schematic representation of the small RNA library preparation (see text for details).

2.2. Preparation of Radiolabeled Size Markers

1. 100 µM of 19-nt oligoribonucleotide: 5′-rCrGrUrArCrGrCr-GrGrGrUrUrUrArArArCrGrA-3′. The RNA oligonucleotides (see item 1–3) are used as size markers and contain the PmeI GTTT/AAAC restriction site (underlined) (see Note 1).
2. 100 µM of 24-nt oligoribonucleotide: 5′-rCrGrUrArCrGr-CrGrGrArArUrArGrUrUrUrArArArCrUrGrU-3′.
3. 100 µM of 33-nt oligoribonucleotide: 5′-rCrGrUrArCrG-rCrGrGrArArUrArGrUrUrUrArArArCrUrGrUrArGrUrGr-CrUrGrArU-3′.
4. [γ-^{32}P] ATP (6,000 Ci/mmol, 10 mCi/mL).

5. T4 Polynucleotide kinase (T4 PNK) (10 U/μL).
6. Polynucleotide kinase (PNK) buffer (10×).
7. Nuclease-free water.
8. 1 M EDTA, pH 8.
9. Gel loading II buffer (Ambion, Austin, TX).
10. 30% Acrylamide/bisacrylamide (37.5:1).
11. Urea.
12. Tetramethylethylenediamine (TEMED).
13. 10% (w/v) Ammonium persulfate. Store in aliquots at −20°C.
14. 10× Tris/borate/EDTA buffer (TBE): 890 mM Tris base, 890 mM boric acid, 20 mM EDTA, pH 8.0, diluted in water.
15. 5 M NaCl.
16. Absolute ethanol.
17. Thermomixer (Eppendorf, Hamburg, Germany).

2.3. Isolation of ~19–24nt RNAs

1. Gel loading II buffer (Ambion, Austin, TX).
2. Nuclease-free water.
3. 30% Acrylamide/bisacrylamide (37.5:1).
4. Urea.
5. TEMED.
6. 10% (w/v) Ammonium persulfate. Store in aliquots at −20°C.
7. 10× TBE buffer: 890 mM Tris base, 890 mM boric acid, 20 mM EDTA, pH 8.0, diluted in water.
8. 5 M NaCl.
9. Absolute ethanol.
10. Thermomixer (Eppendorf, Hamburg, Germany).

2.4. Ligation of the 3′ Adapter

1. 10× ATP-free ligation buffer: 100 mM $MgCl_2$, 100 mM dithiothreitol (DTT), 500 mM Tris–HCl, pH 7.6, 1 mg/mL acetylated BSA (Ambion, Austin, TX).
2. DMSO.
3. 5 nM of Modban adapter 5′-rAppCTGTAGGCACCATCAAT /3ddC/-3′ (IDT, San Jose, CA) (miRNA cloning linker 1, reference 60910274). This oligonucleotide is provided lyophilized and is ready for use in cloning. Resuspend at a concentration of 100 μM in water and store at −80°C (see Note 2).
4. T4 RNA ligase 2 truncated (200 U/μL) (New England Bioloabs, Ipswich, MA).

5. Gel loading II Buffer (Ambion, Austin, TX).
6. 30% Acrylamide/bisacrylamide (37.5:1).
7. Urea.
8. TEMED.
9. 10% (w/v) Ammonium persulfate. Store in aliquots at –20°C.
10. 10× TBE buffer: 890 mM Tris base, 890 mM boric acid, 20 mM EDTA, pH 8.0, diluted in water.
11. 5 M NaCl.
12. Absolute ethanol.
13. Nuclease-free water.
14. Thermomixer (Eppendorf, Hamburg, Germany).

2.5. Preparation of Radiolabeled Decade Marker

1. Decade marker (100 ng/μL) (Ambion, Austin, TX).
2. Nuclease-free water.
3. Kinase reaction buffer (10×) (Ambion, Austin, TX).
4. [γ-^{32}P] ATP (6,000 Ci/mmol, 10 mCi/mL).
5. T4 polynucleotide kinase (10 U/μL) (Ambion, Austin, TX).
6. Cleavage reagent (10×) (Ambion, Austin, TX).
7. Gel loading II Buffer (Ambion, Austin, TX).

2.6. Ligation of the 5′ Adapter

1. 10× ATP-containing ligation buffer: 2 mM ATP, 100 mM $MgCl_2$, 100 mM DTT, 500 mM Tris–HCl, pH 7.6, 1 mg/mL acetylated BSA (Ambion, Austin, TX).
2. DMSO.
3. 100 μM of Solexa adapter, HPLC purified: 5′-rArCrArCrUr CrUrUrUrCrCrCrUrArCrArCrGrArCrGrC rUrCrUrUrCrCrGrArUrC-3′.
4. T4 RNA Ligase 1 (20 U/μL) (New England Biolabs, Ipswich, MA).
5. Gel loading II buffer (Ambion, Austin, TX).
6. 30% Acrylamide/bisacrylamide (37.5:1).
7. Urea.
8. TEMED.
9. 10% (w/v) ammonium persulfate. Store in aliquots at –20°C.
10. 10× TBE buffer: 890 mM Tris base, 890 mM boric acid, 20 mM EDTA, pH 8.0, diluted in water.
11. 5 M NaCl.
12. Absolute ethanol.
13. Nuclease-free water.
14. Thermomixer (Eppendorf, Hamburg, Germany).

2.7. Reverse Transcription

1. 100 μM RT primer BanOne, HPLC purified: 5′-ATTGATGGTGCCTACAG-3′.
2. 5× First-Strand Buffer: 250 mM Tris–HCl, pH 8.3, 375 mM KCl, 15 mM $MgCl_2$ (Invitrogen, Carlsbad, CA).
3. 20 mM (each) dNTPs Mix (dATP, dCTP, dGTP, dTTP).
4. 100 mM DTT.
5. SuperScript III Reverse Transcriptase (200 U/μL) (Invitrogen, Carlsbad, CA).
6. Thermocycler.

2.8. PCR Amplification of the Library

1. 5× Phusion HF buffer (Finnzymes, Espoo, Finland).
2. 20 mM (each) dNTPs Mix (dATP, dCTP, dGTP, dTTP).
3. 100 μM Primer Sol_5_SBS3, standard desalted: 5′-AATGATACGGCGACCACCGAACACTCTTTCCCTACACGACG-3′.
4. 100 μM Primer Sol_3_Modban, HPLC purified: 5′-CAA-GCAGAAGACGGCATACGATTGATGGTGCCTACAG-3′.
5. Phusion high-fidelity DNA polymerase (2 U/μL) (Finnzymes, Espoo, Finland).
6. Nuclease-free water.
7. NuSieve GTG agarose (Cambrex, East Rutherford, NJ).
8. 50× Tris/acetate/EDTA (TAE) buffer: 2 M Tris base, 1 M acetic acid, 50 mM EDTA, pH 8.3, diluted in water.
9. Ethidium bromide.
10. Tracklt 25 bp DNA ladder (0.5 μg/μL) (Invitrogen, Carlsbad, CA).
11. Thermocycler.

2.9. PmeI Digestion of Size Markers

1. 5 M NaCl.
2. Phenol/chloroform/isoamylic alcohol, pH 8.
3. 3 M NaAc, pH 5.5.
4. Absolute ethanol.
5. Nuclease-free water.
6. PmeI (10 U/μL) (New England Bioloabs, Ipswich, MA).
7. NEBuffer 4 (10×): 500 mM KAc, 200 mM Tris–acetate, 100 mM MgAc, 10 mM DTT, pH 7.9 (New England Bioloabs, Ipswich, MA).
8. BSA (10 mg/mL).
9. Nuclease-free water.
10. 5× Nondenaturing loading dye: 0.025% bromophenol blue, 0.025% xylene cyanol, 0.125% orange G, 50% Glycerol, 10 mM Tris–HCl, pH 7.5, 50 mM EDTA.

2.10. Gel Purification of Libraries

1. NuSieve GTG agarose (Cambrex, East Rutherford, NJ).
2. Ethidium bromide.
3. Tracklt 25 bp DNA ladder (0.5 μg/μL) (Invitrogen, Carlsbad, CA).
4. 5 M NaCl.
5. Phenol (warmed to 70°C before use).
6. Phenol/chloroform/isoamylic alcohol (25:24:1), pH 8.
7. Chloroform.
8. 70% Ethanol.
9. Nuclease-free water.

3. Methods

For all steps, use nonsticky low-binding tubes to prevent material loss.

3.1. Total RNAs Isolation

TRIzol Reagent combines acidic phenol and guanidinium thiocyanate to lyse cells, inactivate RNases, and remove lipids. Manipulate this toxic solution with gloves and in a chemical fume hood. Remember to work in an RNase-free area (cleaned with RNase Zap) and keep samples on ice as much as possible.

1. Harvest five to ten insects in a clear polypropylene tube and euthanize them by flash freezing (submerge the tube in liquid nitrogen or in a dry ice/ethanol mixture). If you are working with cells, pellet 10^7 cells by gentle centrifugation (1,300 × *g* 5 min at RT) and wash them in 1× PBS.
2. Add 200 μL TRIzol Reagent and process the sample with a Kontes pellet pestle until a homogenized tissue sample is obtained. Complete with 800 μL TRIzol Reagent and incubate 5 min at room temperature.
3. Proceed to phase separation by adding 200 μL chloroform, vortex for 1 min, and centrifuge at 12,000 × *g* for 15 min at 4°C. In order to improve RNA purity, it is preferable to sacrifice a little amount of the aqueous phase close to the interface.
4. To precipitate RNA, transfer the aqueous phase to a new prechilled tube and add 400 μL isopropyl alcohol before centrifugation at 12,000 × *g* for 15 min at 4°C.
5. Remove the supernatant and wash the RNA pellet with 500 μL chilled 75% ethanol. Mix by vortexing and centrifuge at 7,500 × *g* for 5 min at 4°C.
6. Air-dry the RNA pellet by leaving the sample open under the extractor hood and resuspend it by pipetting up and down in

deionized formamide so that final RNA concentration is 1 μg/μL. RNA can be stored at this step in nuclease-free water at −20 or −70°C. This is necessary if beta-elimination and/or 2S ribosomal RNA depletion is required.

3.2. Preparation of Radiolabeled Size Markers

3.2.1. Radiolabeling of RNAs Oligonucleotides

1. For each marker, mix in a clear polypropylene low-binding tube:
   (a) 1 μL of 10 μM marker.
   (b) 5 μL of [γ-^{32}P] ATP (6,000 Ci/mmol, 10 mCi/mL).
   (c) 0.2 μL of 10 U/μL T4 PNK.
   (d) 2 μL of 10× PNK buffer.
   (e) Up to 20 μL with nuclease-free water.
2. Incubate 30 min in water bath at 37°C, in radioactive room.
3. Stop the reaction by adding 20 μL of 30 mM EDTA, pH 8.
4. Add 1 volume of Gel loading II Buffer, incubate 2 min at 95°C, and quickly chill on ice.

3.2.2. Gel Purification of Radiolabeled Size Markers

1. Prepare a 20×20 cm gel of 15% acrylamide/bisacrylamide (37.5:1), 7 M urea, 0.5× TBE, with spacers of 0.8 mm and a 15-well comb.
2. Prerun the gel in 0.5× TBE for 30 min at 30 W. Check that your generator can deliver this power.
3. Wash carefully the wells with a needle and a syringe.
4. Load samples and load markers in outer wells. Be aware of leaving empty wells between samples.
5. Run your gel for 45 min at 15 W (750 V). Remember that the 19 nt size marker migrates with the bromophenol blue.
6. After migration, remove the upper glass plate with a spatula, cover the gel with saran wrap, place an autoradiography ruler for orientation, and expose on film for 2 min.
7. Pierce the film at the position corresponding to the markers.
8. Place the film on top of the gel using the autoradiography ruler to position.
9. Cut the size markers from the gel with a clean scalpel, using the pierced film to indicate their position.
10. Place the gel slice in a clear polypropylene low-binding tube with 300 μL of 0.3 M NaCl.
11. Elute overnight under agitation at 17°C.
12. Next day, briefly spin down the tubes and recover the supernatant avoiding polyacrylamide debris.
13. Precipitate with 3 volumes of 100% ethanol for 1 h at −20°C.
14. Centrifuge at 16,000 ×*g* for 10 min.

15. Dry out excess ethanol.
16. Dissolve in 20 μL of nuclease-free water.

3.3. Isolation of ~19–24nt RNAs

1. Prepare a size marker mix with 3 μL of each radiolabeled marker and 100 μL of gel loading blue II.
2. In a polypropylene low-binding tube, mix:
   (a) 30 μL RNA prepared at Subheading 3.1 (30 μg).
   (b) 3 μL of size marker mix from Subheading 3.3, step 1.
   (c) 33 μL gel loading blue II.
3. Denature for 30 s at 95°C and quick chill on ice.
4. Prepare a 20×20 cm gel of 15% acrylamide/bisacrylamide (37.5:1), 7 M urea, 0.5× TBE, with spacers of 1.5 mm and a 15-well comb.
5. Prerun the gel on 0.5× TBE for 30 min at 30 W.
6. Wash the wells carefully with a needle and syringe.
7. Dilute 6 μL of size marker mix prepared in Subheading 3.3, step 1 in a final volume of 40 μL and load 20 μL in the first and last lane of the gel.
8. Load samples. Be sure to leave empty wells between samples.
9. Run the gel at 30 W until the bromophenol blue reaches two third of the gel.
10. After migration, remove the upper glass plate with a spatula, cover the gel with saran wrap, and place a autoradiography ruler for orientation and expose on phosphorimager screen for 1 h (see Note 3).
11. Replace the print from the phosphorimager in top of the gel using the autoradiography ruler to position.
12. Cut the samples corresponding to the 19–24 nt position using the size markers as a guide.
13. Cut out the markers as well from the first and last gel lanes.
14. Put the gel slice in a polypropylene low-binding tube with 300 μL of 0.3 M NaCl.
15. Elute overnight under agitation at 17°C.
16. Next day, briefly spin down the tubes and recover the supernatant.
17. Precipitate with 3 volumes of 100% ethanol for 1 h at -20°C.
18. Centrifuge at 16,000×*g* for 10 min.
19. Dry out excess ethanol.

3.4. Ligation of the 3′ Adapter

1. Dissolve small RNAs and eluted markers mix in:
   (a) 10 μL RNase-free water.
   (b) 2 μL 10× ATP-free ligation buffer.
   (c) 6 μL 50% DMSO (diluted in water).
   (d) 1 μL 50 μM of Modban adapter.
2. Denature for 30 s at 95°C and quickly chill on ice.
3. Add 1 μL (200 U/μL) of truncated T4 RNA ligase 2.
4. Incubate 1 h at 37°C.
5. Stop the reaction by adding 20 μL of gel loading II buffer.
6. Prepare a 20×20 cm gel of 15% acrylamide/bisacrylamide (37.5/1), 7 M urea, 0.5× TBE, with spacers of 0.8 mm and a 15-well comb.
7. Prerun the gel on 0.5× TBE for 30 min at 30 W.
8. Wash the wells carefully with a needle and syringe.
9. Load samples. Be sure to leave empty wells between samples. Load nonligated and ligated markers at the most external wells.
10. Run the gel at 30 W until the bromophenol blue reaches two third of the gel.
11. After migration, remove the upper glass plate with a spatula, cover the gel with saran wrap, and place a autoradiography ruler for orientation and expose on phosphorimager screen for ~2.5 h.
12. Replace the print from the phosphorimager on top of the gel using the autoradiography ruler to position.
13. Cut samples in the lanes between the 37–42 nt positions using the ligated size markers as a guide. Ligation of Modban adapter to the 19 and 24 nt markers gives rise to 37 and 42 species, respectively. Remember that nonligated products will also be detectable on gel.
14. Cut out the ligated markers as well from the first and last gel lanes.
15. Put the gel slice in a polypropylene tube with 300 μL 0.3 M NaCl.
16. Elute overnight under agitation at 17°C.
17. Next day, briefly spin down the tubes and recover the supernatant.
18. Precipitate with 3 volumes of 100% ethanol for 1 h at −20°C.
19. Centrifuge at 16,000 × *g* for 10 min.
20. Dry out excess ethanol.

3.5. Preparation of Radiolabeled Decade Marker

1. In a clear polypropylene low-binding tube, mix:
   (a) 1 μL Decade marker.
   (b) 6 μL Nuclease-free water.
   (c) 1 μL 10× Kinase reaction buffer.
   (d) 1 μL [γ-^{32}P] ATP (6,000 Ci/mmol, 10 mCi/mL).
   (e) 1 μL T4 PNK.
2. Incubate 1 h at 37°C.
3. Add 8 μL of nuclease-free water and 2 μL of 10× cleavage reagent.
4. Incubate 5 min at room temperature and stop the reaction by adding 20 μL of gel loading II buffer. Store at −20°C. The radiolabeled Decade marker can be used for 15 days according to ^{32}P half-life (14.3 days).

3.6. Ligation of the 5′ Adapter

1. Dissolve 3′-ligated small RNAs and markers in:
   (a) 10 μL RNase-free water.
   (b) 2 μL 10× ATP-containing ligation buffer.
   (c) 6 μL 50% DMSO (diluted in water).
   (d) 1 μL 50 μM 5′ Solexa adapter.
2. Denature for 30 s at 95°C and quickly chill on ice.
3. Add 1 μL of 20 U/μL T4 RNA ligase 1.
4. Incubate 1 h at 37°C.
5. Stop the reaction by adding 20 μL of gel loading II buffer.
6. Prepare a 20×20 cm gel of 15% acrylamide/bisacrylamide (37.5:1), 7 M urea, 0.5× TBE, with spacers of 0.8 mm and a 15-well comb.
7. Prerun the gel on 0.5× TBE for 30 min at 15 W.
8. Wash the wells carefully with a needle and syringe.
9. Load samples. Be sure to leave empty wells between samples.
10. Run the gel at 30 W until bromophenol blue is completely out of the gel.
11. After migration, remove the upper glass plate with a spatula, cover the gel with saran wrap, place an autoradiography ruler for orientation, and expose on phosphorimager screen for 3 h.
12. Replace the print from the phosphorimager on top of the gel using the autoradiography ruler to position.
13. Cut samples in the lanes between the 69 and 74 nt positions using the ligated size markers as a guide (ligation of 5′ Solexa adapter to the 37 and 42 nt markers gives rise to 69 and 74 nt long species, respectively, as well as nonligated products).

14. Put the gel slice in a clear polypropylene low-binding tube with 300 μL of 0.3 M NaCl and 1 μL of 100 μM RT primer BanOne.
15. Elute overnight under agitation at 17°C.
16. Next day, briefly spin down the tubes and recover the supernatant.
17. Precipitate with 3 volumes of 100% ethanol for 1 h at -20°C.
18. Centrifuge at 16,000 × *g* for 10 min.
19. Dry out excess ethanol.
20. Dissolve in 10 μL of nuclease-free water.

3.7. Reverse Transcription

1. In a PCR tube, mix:
   (a) 5 μL of ligated RNA samples.
   (b) 4 μL 5× First-strand buffer.
   (c) 2 μL 20 mM dNTPs mix.
   (d) 1 μL 100 mM DTT.
   (e) Up to 18 μL with nuclease-free water.
2. Incubate 3 min at 50°C.
3. Split each sample into two PCR tubes of 9 μL.
4. In one tube, add 1 μL of 200 U/μL superscript III reverse transcriptase. Use the other tube as a negative control for reverse transcription activity (RT control, see Note 4).
5. Incubate 1 h at 50°C, then 15 min at 70°C to inactivate the enzyme.

3.8. PCR Amplification of the Library

1. In a PCR tube, mix:
   (a) 5 μL of cDNA or RT control.
   (b) 20 μL 5× Phusion HF buffer.
   (c) 1 μL 20 mM dNTPs mix.
   (d) 1 μL 100 μM Primer Sol_5_SBS3.
   (e) 1 μL 100 μM Primer Sol_3_Modban.
   (f) 1 μL Phusion DNA polymerase (2 U).
   (g) Up to 100 μL with nuclease-free water.
2. Run in thermocyler:
   (a) 94°C for 2 min.
   (b) 5 cycles: 94°C 15 s, 54°C 30 s, 72°C 30 s.
   (c) 17 cycles: 94°C 15 s, 60°C 30 s, 72°C 30 s.
   (d) 72°C for 7 min.
   (e) Maintaining at 4°C.

3. Run 5 μL of the PCR reaction in a 3% NuSieve GTG agarose gel in 0.5× TAE.
4. Check for one band between 108 and 113 bp (see Note 5).

3.9. PmeI Digestion of Size Markers

1. In a polypropylene low-binding tube, add to the 95 μL remaining of the PCR reaction:
   (a) 6 μL 5 M NaCl.
   (b) 100 μL Phenol/chloroform/isoamylic alcohol, pH 8.
2. Vortex.
3. Centrifuge at 16,000 × *g* for 5 min at 4°C.
4. Recover the aqueous phase.
5. Add 1/10 volume 3 M NaAc, pH 4.5 and 3 volumes of 100% ethanol.
6. Precipitate at –20°C for at least 3 h.
7. Centrifuge at 16,000 × *g* for 10 min.
8. Remove supernatant.
9. Dry out excess ethanol.
10. Dissolve the pellet in PmeI digestion mixture (see Note 1):
    (a) 23.7 μL of nuclease-free water.
    (b) 3 μL of 10× NEBuffer 4.
    (c) 0.3 μL of 10 mg/mL BSA.
    (d) 3 μL of 10 U/μL PmeI enzyme.
11. Incubate 2 h at 37°C.
12. After the digestion, add to each tube:
    (a) 200 μL Nuclease-free water.
    (b) 230 μL Phenol/chloroform/isoamylic alcohol, pH 8.
13. Vortex.
14. Centrifuge at 16,000 × *g* for 10 min at 4°C.
15. Recover the aqueous phase, add 1/10 volume 3 M NaAc, pH 4.5, and precipitate overnight at –20°C in 3 volumes of 100% ethanol.
16. Centrifuge at 16,000 × *g* for 10 min at 4°C.
17. Discard the supernatant.
18. Dry and dissolve the pellet in 14.4 μL of nuclease-free water and 1.6 μL of nondenaturing loading dye.

3.10. Gel Purification of Libraries

1. Prepare a 3% NuSieve GTG agarose gel (20 × 20 cm) in 0.5× TAE with 100 ng/mL ethidium bromide. Use a comb that will accommodate total sample volume (16 μL).
2. Load the samples and a DNA ladder (for example, 25 bp DNA ladder Invitrogen).

3. Migrate at 100 V. Let the bromophenol blue progressing at least 10 cm.
4. Cut the band of interest (108–113 bp). Remember that the upper band is the insert library, whereas the lower band corresponds to amplification of self-ligated adapters.
5. Weigh the gel slice and distribute 250 mg of gel into polypropylene tubes.
6. Add 2 volumes of 0.4 M NaCl. Melt the gel slice at 70°C, mixing every 2 min.
7. Add 1 volume of warm phenol, pH 8 (see Notes 6 and 7).
8. Vortex 30 s.
9. Centrifuge 5 min at 16,000 × *g* at room temperature.
10. Recover the aqueous phase.
11. Add 1 volume of phenol/chloroform/isoamylic alcohol, pH 8.
12. Vortex 30 s.
13. Centrifuge 5 min at 16,000 × *g* at room temperature.
14. Recover the aqueous phase.
15. Add 1 volume of chloroform.
16. Vortex 30 s.
17. Centrifuge 5 min at 16,000 × *g* at room temperature.
18. Recover the aqueous phase.
19. Add 1/10 3 M NaAc, pH 4.5 and 3 volume of 100% ethanol.
20. Incubate overnight at −20°C.
21. Centrifuge 10 min 16,000 × *g* at 4°C.
22. Wash with 700 μL of 70% ethanol.
23. Centrifuge 10 min 16,000 × *g* at 4°C.
24. Remove the supernatant.
25. Dry excess ethanol.
26. Dissolve in 20 μL of nuclease-free water.

At this step, your library is ready to be sent for sequencing. Some companies or core facilities may ask you to determine the concentration of your library. Use accurate technologies such as fluorometry or microfluidics-based platforms for DNA.

4. Notes

1. PmeI is an octanucleotide-recognizing rare-cutting restriction endonuclease. Size markers oligonucleotides containing the PmeI restriction site can be eliminated from the final

DNA sequencing template by PmeI digestion. Failure to remove the size markers from the final cDNA library will result in a sequence dataset that is predominated by size marker sequences.

2. The Modban adapter is a fully activated 5′ adenylated oligonucleotide and is a substrate for T4 RNA ligase in the absence of ATP. The use of this adenylated adapter improves the cloning efficiency of small RNAs, which have a 5′-phosphate and will circularize if adapters are attached using T4 RNA ligase in presence of ATP. For more details, see refs. (10) and (11).
3. Use of the phosphorimager represents a quicker and quantitative method of visualizing radioactive gels. The phosphorimager screen needs to be exposed only for one tenth of the time required for exposure to film. Because the amount of radiolabeled size markers present in your sample is subject to successive loss at each purification step, exposing your gel to a phosphorimager screen allows a sensitive gain of time. It would take an overnight exposure to get the equivalent signal intensity on film.
4. As PCR cannot discriminate between cDNA targets synthesized by reverse transcription and genomic DNA contamination, the RT control tube detects DNA contamination in the RNA preparation.
5. Adapters can self-ligate, which generates chimeric sequences. As the relative concentration of adapter primers increases, the formation of adapters dimers is promoted. If these dimers are not removed, they will ultimately be sequenced along with the intended template, wasting the capacity of the flowcell. At this stage of the library preparation, an extra lower band corresponding to self-ligated adapters can be observed. Be careful to prevent transferring the lower band when excising the library from the gel.
6. The use of warm phenol prevents agarose from solidifying by getting cold during the purification steps.
7. Commercial DNA gel extraction and clean-up kits do not have an exact size cut-off. The cloned library and the self-ligated adapters are too close in size to be adequately discriminated using a commercial kit.

Acknowledgments

The authors would like to thank Christophe Antoniewski for insight and discussion into deep sequencing, Marco Vignuzzi for comments on the manuscript, Nicolas Vodovar for helping with figures, and Ronald van Rij for being such a great and patient editor.

This work was financially supported by the Agence Nationale de la Recherche (ANR-09-JCJC-0045-01) and the European Research Council (FP7/2007-2013 ERC 242703) to MCS.

References

1. Hamilton, A. J., and Baulcombe, D. C. (1999) A species of small antisense RNA in posttranscriptional gene silencing in plants, *Science* *286*, 950–952.
2. Li, H., Li, W. X., and Ding, S. W. (2002) Induction and suppression of RNA silencing by an animal virus, *Science* *296*, 1319–1321.
3. Lu, R., Maduro, M., Li, F., Li, H. W., Broitman-Maduro, G., Li, W. X., and Ding, S. W. (2005) Animal virus replication and RNAi-mediated antiviral silencing in *Caenorhabditis elegans*, *Nature* *436*, 1040–1043.
4. Deleris, A., Gallego-Bartolome, J., Bao, J., Kasschau, K. D., Carrington, J. C., and Voinnet, O. (2006) Hierarchical action and inhibition of plant Dicer-like proteins in antiviral defense, *Science* *313*, 68–71.
5. Horwich, M. D., Li, C., Matranga, C., Vagin, V., Farley, G., Wang, P., and Zamore, P. D. (2007) The *Drosophila* RNA methyltransferase, DmHen1, modifies germline piRNAs and single-stranded siRNAs in RISC, *Curr Biol* *17*, 1265–1272.
6. Flynt, A., Liu, N., Martin, R., and Lai, E. C. (2009) Dicing of viral replication intermediates during silencing of latent *Drosophila* viruses, *Proc Natl Acad Sci U S A* *106*, 5270–5275.
7. Wu, Q., Luo, Y., Lu, R., Lau, N., Lai, E. C., Li, W. X., and Ding, S. W. (2010) Virus discovery by deep sequencing and assembly of virus-derived small silencing RNAs, *Proc Natl Acad Sci U S A* *107*, 1606–1611.
8. Anon. (2010) Human genome at ten: The sequence explosion, *Nature* 464, 670–671.
9. Pfeffer, S. (2007) Identification of virally encoded microRNAs, *Methods Enzymol* *427*, 51–63.
10. England, T. E., Gumport, R. I., and Uhlenbeck, O. C. (1977) Dinucleoside pyrophosphate are substrates for T4-induced RNA ligase, *Proc Natl Acad Sci U S A* *74*, 4839–4842.
11. Ho, C. K., Wang, L. K., Lima, C. D., and Shuman, S. (2004) Structure and mechanism of RNA ligase, *Structure* *12*, 327–339.

Chapter 7

Visitor, An Informatic Pipeline for Analysis of Viral siRNA Sequencing Datasets

Christophe Antoniewski

Abstract

High-throughput sequencing emerged as a powerful approach to characterize siRNA populations generated by hosts in response to viral infections. Here we described an informatic pipeline *visitor* to analyze in-house large sequencing datasets generated from Illumina sequencing of *Drosophila* small RNA libraries. The *visitor* perl script is designed to treat fastq sequence datasets from the Illumina sequencing platform, using a computer running under a UNIX compliant operating system (MacOS X, Linux, etc.). *visitor* first generates a detailed report of the sequence quality of the Illumina run. Then, using the Novoalign software, the script removes reads that match with the *D. melanogaster* genome from the sequencing data set. The remaining reads are aligned to a viral reference library, which can contain one or several virus genomes. *visitor* provides a hit table of identified viral siRNAs as well as graphics eps files of viral siRNA profiles. Unmatched small RNAs are also available in a fast format for de novo assembly and new virus discovery.

Key words: Illumina deep sequencing, siRNA profiling, Antiviral siRNAs

1. Introduction

High-throughput sequencing using the Illumina system is well suited for the characterization of 21nt-long viral siRNA as it typically results in 5–10 millions of ~36 nt reads per small RNA library (see Chapter 6). With a ~1 Gigabyte sequence data file in hands, the virologist has to (a) estimate the overall quality of the sequence reads and to determine the size distribution of inserts in the small RNA library, (b) separate sequence reads of viral siRNAs from sequence reads of small RNAs from the host, and (c) count and map viral siRNAs reads to one or several viral reference genomes. These tasks may appear complicated by the volume of data to analyze. However, a number of efficient softwares have been

Ronald P. van Rij (ed.), *Antiviral RNAi: Concepts, Methods, and Applications*, Methods in Molecular Biology, vol. 721,
DOI 10.1007/978-1-61779-037-9_7, © Springer Science+Business Media, LLC 2011

developed to rapidly align a large amount of sequence reads to reference genomes, using standard desktop computers (1). Here, we describe a *visitor* perl script based on the Novoalign aligner software (2).

We use *visitor* (a) to analyze the sequence quality and insert size distribution of an Illumina sequence dataset, provided in an Illumina fastq format, (b) to remove sequence reads matching with the *D. melanogaster* genome, and (c) to identify and map vsiRNAs on a collection of viral reference genomes (see Fig. 1 for an overview of the informatic workflow).

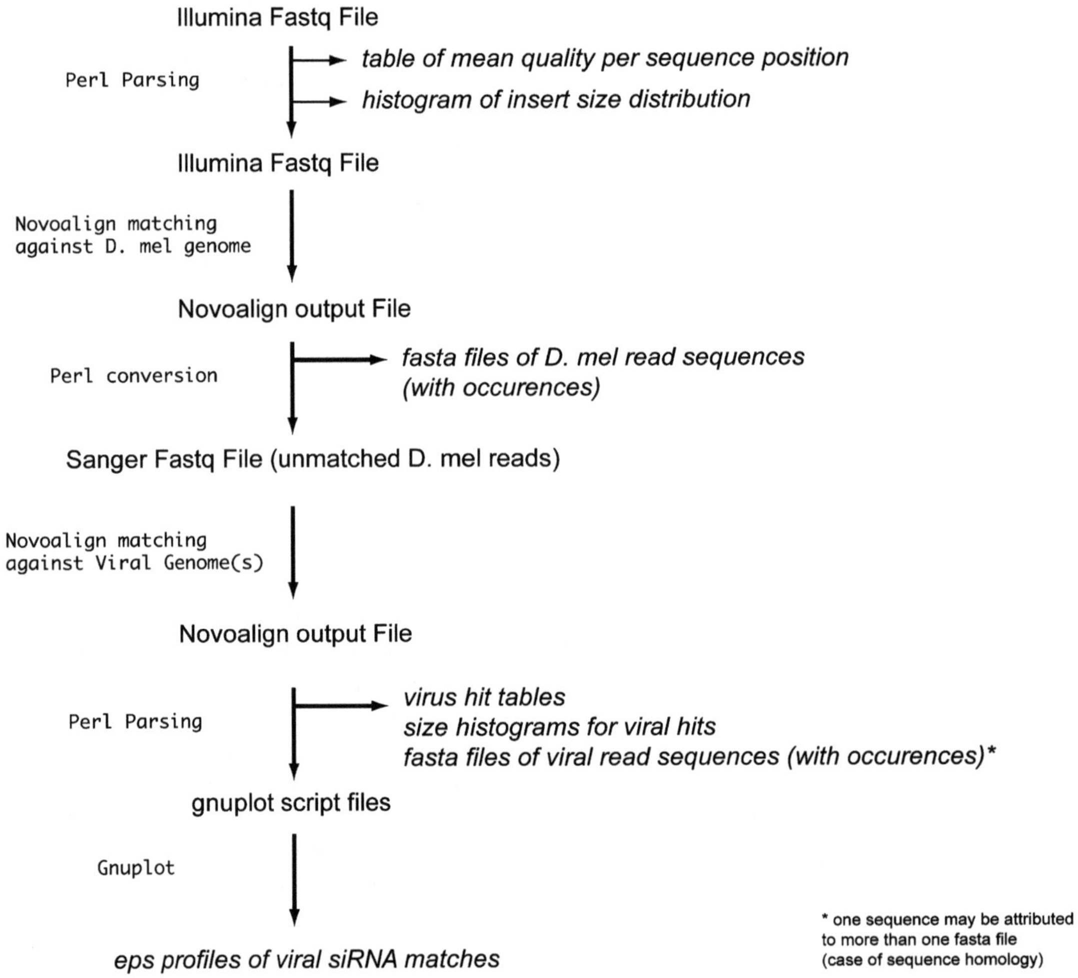

Fig. 1. Informatic workflow for analysis of viral siRNA sequence reads. *Monaco font* indicates Jobs, *Arial* indicates Inputs, and *Arial italic* indicates *Outputs*.

2. Materials

1. A computer with 8 Gb of RAM running under MacOSX (10.5 Intel or later versions) or Linux. The *visitor* script has not been tested under Window OS.
2. Minimal knowledge in UNIX management. If you do not understand the meaning of "mkdir ~/bin under terminal session," you should consider asking for help.
3. Perl installed.
4. Gnuplot installed. This is optional but highly recommended. How to install Gnuplot ?
   (a) Under Mac OS X, Xcode Developer Tools that are provided with Mac OS installation DVDs (XcodeTools.mpkg) have to be installed.
   (b) Download Gnuplot at http://sourceforge.net/projects/gnuplot/files/. If you are Leopard user, download the 4.2.5 version, if you are Snow-leopard user, the 4.4.0 should work.
   (c) Unzip the package and rename the expanded folder "gnuplot."
   (d) Move the gnuplot folder to your bin directory.
   (e) In a terminal session, navigate to your ~/bin/gnuplot directory.
   (f) Type `./configure --with-readline=builtin` and press enter key.
   (g) Type `make` and press enter key.
   (h) Type `sudo make install` and press enter key.
   (i) Type your password and press enter key.
   (j) Installation will proceed and compile the executable gnuplot file in the ~/bin/gnuplot/src directory.

3. Installation

1. If not already done, create a "bin" directory in your home directory (~/bin/).
2. Download (http://www.novocraft.com/) and install Novoalign (2.04.03 or later) and Novoindex (2.4 or later), in a ~/bin/novocraft directory.
3. Download the visitor file at http://drosophile.org/GEDlab/?page_id=254, or copy/paste the visitor script in a

text file named *visitor*. Copy the *visitor* file in your ~/bin/ directory. Make executable the *visitor* script by typing `chmod 755 ~/bin/visitor` and pressing enter.

4. Set the $PATH environmental variable so that programs in the bin directory are permanently accessible under your shell sessions:

   Using an in-line text editor (nano, vi, etc...), add these lines to your ~/.profile file:

```
PATH=$PATH:~/bin:~/bin/gnuplot/src:~/bin/novocraft
export PATH
```

For example, in terminal, type `nano ~/.profile` and press enter. Copy the two above lines in the opened window. Press Ctrl-O and press enter to write out, then press Ctrl-X to exit nano.

4. Running *VISITOR*

4.1. Prerequirements

1. If you installed gnuplot on your computer, delete the # character at line 10 of the *visitor* script, using a text editor that respects UNIX LF.
2. Create a directory ~/fasta_files, in which the reference libraries (*D. melanogaster* and viruses) will be stored.

   D. melanogaster genome: the up-to-date version can be found at ftp://ftp.flybase.net/genomes/Drosophila_melanogaster/dmel_r5.26_FB2010_03/fasta/.

   Download the dmel-all-chromosome-rX.YY.fasta.gz file (~48Mbytes), expand it, and use the resulting .fasta file.

   Viral genome(s): You can include as many viral genomes as you want in a viruses.fa file, which you save in a *fasta format*. We strongly recommend not using space characters in the fasta headers, as Novoalign will cut out headers after the first space character encountered.

   Example:

```
>my_virus_1
ATTGTGCTG
ATGGTTCCC
>my_virus_2
TTACCCGGATGCCG
TTAGGCCATGGACGTA
GCCATGATGACCATGACCA
```

Important: files should be saved with UNIX Line break characters (UNIX LF). Avoid manipulating your viruses.fa fasta file using classical text editors such as Microsoft Word which code the line break characters differently.

3. Create a working directory. For example, in terminal, type `mkdir ~/my_analysis` and press return.
4. Copy your Illumina sequence file in the working directory created above.

   This file must be in an Illumina fastq format and is most often named as s_lane_sequence.txt, where lane stands for a digit. For simplicity, we recommend to rename the file as *s[chanel number]*, for example, `s1`. It contains millions of blocks of 4 lines similar to:

```
@HWI-EAS285:1:1:1582:1499#0/1
GGATAAGGATTGGCTCTATCTCGTATGCCGTCTTCT
+HWI-EAS285:1:1:1582:1499#0/1
aa^a_]\\[_]V[^a^]R]^[VJQSXOVVJQQ]ZKZ
```

The fourth line of a block describes the quality value associated to each nucleotide of the read, whose sequence is in the second line of the block. Note that in the Illumina Casava Pipeline 1.3 and in later versions, encoding of the quality has changed. If you analyze older Illumina fastq files generated with the pipeline pre-1.3, you will be prompted to specify it when running *visitor*.

4.2. Running visitor

1. Open a terminal session.
2. Navigate to your working directory by typing `cd ~/my_analysis`
3. Type `visitor` <full path to illumina fastq file> <3′ adapter> <full path to Drosophila genome fasta file> <full path to viruse genomes fasta file> where <3′ adapter> is the nucleotide sequence of the 3′ adapter used for library preparation.

   Example:

```
visitor /Users/my_analysis/s1 CTGTAGGCACCATCA
ATCGT /Users/fasta_files/dmel-all-chromosome
-r5.26.fasta /Users/fasta_files/viruses.fa
```

At starting, you will be prompted to choose whether or not to perform the quality sequence analysis of the fastq file. This is useful when you analyze a sequencing run for the first time, but can be omitted to save time if you redo the analysis using different reference genomes.

You will also be prompted to indicate whether the Illumina fastq file was generated using the Illumina Casava pipeline 1.3 or later (most cases) or using a pipeline pre-1.3 (the Casava pipeline 1.3 was released by Illumina in march 2009).

After providing this information, *visitor* will run autonomously. Depending on the performance of your computer, the complete process may take from several minutes to a couple of hours. Hence, consider that taking two minutes to correctly type the *visitor* command line may save you a lot of time and headache.

4.3. Output Files

A successful *visitor* run generates a number of files.

1. .tab files are outputs from Novoalign jobs.
2. .fa files are fasta files of read sequences. Names of .fa files are self-explanatory. Note the format of the fasta headers in these files: `>identifier_occurence` where identifier is an arbitrary unique reference number of the sequence and occurrence is the number of time the sequence has been read in your sample.
3. minus_plot, plus_plot, .plt files are all used to generate .eps profiles. Names of all these files are also self-explanatory. These files will be generated but not used if *gnuplot* is not enabled in *visitor* (see Subheading 2, item 4). However, note that the minus_plot and plus_plot file are straight, tabulated file that can be used to generate viral profile graph using dedicated software such as R or Excel. If you are familiar with gnuplot, the .plt files are gnuplot scripts that can be modified using a text editor and run again to adapt the output format of the .eps files to your convenience (changing the legends, the axis labels, the axis ranges, etc.).
4. .eps files are virus profiles, where y-axis represents number of reads on the positive or the negative strand, and x-axis represents nucleotide coordinates in the viral genome. Note that these profiles are generated only for reads ranging from 19 to 21 nucleotides. These values can be changed in the *visitor* script at lines 306 and 307.
5. The xx_GLOBAL_REPORT.txt is a text file containing a summary of the analysis, which includes:
   (a) Mean sequence quality score for the run.
   The table represents the mean quality score for each nucleotide position in the run. It is coded on a 0 to 62 scale. Mean quality typically ranges from 20 to 40 in an acceptable run and progressively decline with the sequence position. See http://en.wikipedia.org/wiki/FASTQ_format for relationship between quality value and the sequence error probability.
   (b) Total small RNA size distribution.
   Visitor performs its own search for adapter sequences in the reads and uses the information to compute the size distribution of small RNAs in your library. This allows to monitor the number of sequences available for analysis, as well as the frequency of self-ligated 3′-5′ adapters which appears at the position 0 of the table (insert size = 0).

(c) *D. melanogaster* matching step.
This self-explanatory section describes how many reads were analyzed by Novoalign, how many reads were discarded due to poor quality value, how many reads were matched to the *D. melanogaster* genome, and how many reads were left for subsequent matching with viral genomes.

(d) Viral library matching step.
This section shows the total viral small RNA size distribution in the library as well as the small RNA size distribution of reads that matched neither the *D. melanogaster* genome nor the viral reference library. You may have a look to these sequences if you are interested in the identification of new viruses (3).

(e) Size distributions of viral hits, by virus.
This section shows the viral small RNA size distribution for each virus present in the viral reference library. Note that if a read matches with more than one virus, this read will be counted in all the corresponding distributions.

(f) GLOBAL VIRAL HITS table.
This section reports for the fraction of hit matches for each virus. Note that this is calculated as the number of matches for a given virus divided by the total number of viral matches on the viral library. If your library contains only one virus, the fraction will be equal to 1. On the other hand, if your library contains two identical sequences (>my_virus and my_virus_copy), the fractions will be 0.5 for each virus. Keep this information in mind when viral genomes with sequence similarity are included in the viral reference library.

5. Visitor Script

```
#!/usr/bin/perl -w
# GED pipeline for analysis of solexa viral sequence reads
# version 1.0 : June 8, 2010
# Usage:
# visitor <Illumina fastq file> <3' adapter sequence> <full path to
Dmel fasta library> <full path to viruses fasta library>

use File::Basename ;
$gnuplot = "absent" ;
#$gnuplot = "present"; # uncomment if gnuplot installed
```

```
$illumina_fastq_file = shift @ARGV;
$prefix_identifier = basename $illumina_fastq_file;
$prefix_identifier =~ s/\..*// ;
$adapter_sequence = shift @ARGV;
$adapter_sequence =~ tr /atgc/ATGC/ ;
$reporting = $prefix_identifier."_GLOBAL_REPORT.txt";
print "Do you want to perform the quality sequence analysis ?\n";
print "'Yes' is the default option. to skip the step (if you already
performed it), enter 'no', otherwise, type any key and press
return\n";
chomp ($sequence_analysis_subroutine = <STDIN>);

print "\nWhat is the quality format of the Solexa fastq file?\n";
print "If the Solexa file was generated with the Illumina Casava
Pipeline v1.3 or later, enter '1' (Default option)\n";
print "else enter '2':\n";
$illumina_fastq_format = "0" ;

while (($illumina_fastq_format ne "1") and ($illumina_fastq_format ne
"2")) {chomp ($illumina_fastq_format = <STDIN>);};
if ($illumina_fastq_format eq "1") {$illumina_fastq_format =
"ILMFQ";} else {$illumina_fastq_format = "SLXFQ";} ;

open (REPORTING, ">$reporting");

sub quality_control {
#### PART 1 : QUALITY CONTROL AND SIZE DISTRIBUTION OF READS ####
#generating regular expression for linker 5' seed
$linker = substr ($adapter_sequence,0,7); # 7nt first nucleotides
print "adapter seed: $linker\n";
$substA= "[TGC\\.]";
$substT= "[AGC\\.]";
$substG= "[TAC\\.]";
$substC= "[TGA\\.]";
for ($i=0 ; $i<7 ; $i++) {
    my $position = substr ($linker, $i, 1);
    if ($position eq "A") {
        $sous_motif = substr($linker, 0, $i).$substA.substr($linker, $i+1)
    }
```

```
    elsif ($position eq "T") {
        $sous_motif = substr($linker, 0, $i).$substT.substr($linker, $i+1)
    }
    elsif ($position eq "G") {
        $sous_motif = substr($linker, 0, $i).$substG.substr($linker, $i+1)
    }
        else {$sous_motif = substr($linker, 0, $i).$substC.substr($linker,
        $i+1)
    };
    push (@imperfect_motives, $sous_motif);
}
$perfect_motif = "(".substr($linker, 0, 6)."\$|"."$linker)";
$imperfect_motif = join "|", @imperfect_motives ;
$imperfect_motif= "(".$imperfect_motif.")" ;

# parsing fastq file for quality mean by position, and size distribu-
tion of trimmed reads
print "##analyzing quality score and insert size distribution##\n\n";
%score_position = ();
$scanned_lines=0;
$matched_lines=0;
open (ILLUMINA_FASQ_FILE, "$illumina_fastq_file");
while (<ILLUMINA_FASQ_FILE>) {
    $scanned_lines++;
    if (($scanned_lines % 4) == 2) { # trimming
        if (m/(.*)$perfect_motif/) {$size = length $1; $matched_
lines++; $perfect_matched++; $trimmed_size{$size}++;} # search for
perfect 7-match, or perfect 6-match at the end of the string (regular
expression engine is greedy)
        elsif (m/(.*)$imperfect_motif/) {$size = length $1; $matched_
lines++;$imperfect_matched++; $trimmed_size{$size}++;}; # search for
imperfect matches, starting from the end of the string (regular
expression engine is greedy)
    };
    if (($scanned_lines % 4) == 0) { # quality scoring by position
        if (($scanned_lines % 2000000)==0) {print $scanned_lines/4,"
scanned sequences\n";};
```

```
        $quality_string = $_ ;
        for ($i = 1; $i < 37 ; $i++) { # adapt $i max if read lengh
        is > 36 nt
            $ascii_value = substr $quality_string, $i-1, 1 ;
            $quality_value = (ord $ascii_value) -64 ;
            $score_position{$i} += $quality_value ;
        }
    }
}
close ILLUMINA_FASQ_FILE ;
print "\n";
print REPORTING "############################################\n";
print REPORTING "Mean Sequence Quality Score for the run\n" ;
print REPORTING "############################################\n\n";
print REPORTING "position\tmean score\n";
foreach $position (sort {$a<=>$b} keys %score_position) {
    $mean_quality_value = $score_position{$position}/($scanned_lines/4) ;
    print REPORTING "$position\t$mean_quality_value\n";
}
%score_position = (); # flush %score_position
print REPORTING "############################################\n";
print REPORTING "Analysis of insert sizes in the sequenced library\n" ;
print REPORTING "############################################\n\n";
print REPORTING "scanned sequences: ", $scanned_lines/4,"\n";
print REPORTING "trimmed sequences: $matched_lines\n";
print REPORTING "untrimmed sequences: ", ($scanned_lines/4)-$perfect_
matched - $imperfect_matched,"\n";
print REPORTING "perfect linker matches: $perfect_matched\n";
print REPORTING "imperfect linker matches: $imperfect_matched\n\n";
print REPORTING "# Size distribution of trimmed inserts #\n";
print REPORTING "size\tread number\n";
$sum = 0;
foreach $size (sort {$a<=>$b} keys %trimmed_size) {
    print REPORTING "$size\t$trimmed_size{$size}\n";
    $sum += $trimmed_size{$size};
}
```

```
print REPORTING "sum\t$sum\n\n";

}; # end of sub

&quality_control unless ($sequence_analysis_subroutine eq "no") ;

#### PART TWO : NOVOALIGN MATCHING AGAINST D. mel genome ####
# squashing D.mel fasta library
$Dmel_full_path = shift @ARGV ;
$Dmel_squashed_library = "Dmel.squash";
$novoindex_call = "novoindex $Dmel_squashed_library $Dmel_full_path";
system $novoindex_call ;

# Calling novoalign for fastq file matching with the squashed D. mel
library
$novoalign_output_file = $prefix_identifier."_novoalign_Dmel_matching.
tab";
$novoalign_call = "novoalign -d $Dmel_squashed_library -f $illumina_
fastq_file -F $illumina_fastq_format -a$adapter_sequence -r Random -t
30 -o N > $novoalign_output_file"; # t paramater (threshold quality)
set to 30, may be adjusted, or removed
print "starting read alignment with Dmel genome\n" ;
print "Please, be patient\n\n" ;
system $novoalign_call ;

# REGENERATION of a FASTQ *Sanger* file for next alignment with viral
genomes
$unmatched_sanger_output = $prefix_identifier."_Dmel_unmatched.sfastq" ;
open FILE, "$novoalign_output_file";
open UNMATCHED_SANGER_OUTPUT, ">$unmatched_sanger_output";
$readcount = 0;
$Dmel_readcount = 0;
$unmatched_readcount = 0;
$QC_readcount = 0;
```

```
while (<FILE>) {
    unless (m /^#/) {
        $readcount++; # this will again count all reads from the
        starting illumina fastq file.
        ($header, $sequence, $sanger_quality, $status) = (split /\t/)
        [0,2,3,4];
        if (($status eq "R") or ($status eq "U")) {
            $Dmel_readcount++;
        }
        elsif ($status eq "NM\n") {
            $unmatched_readcount++;
            print UNMATCHED_SANGER_OUTPUT "$header\n";
            print UNMATCHED_SANGER_OUTPUT "$sequence\n";
            print UNMATCHED_SANGER_OUTPUT "+\n";
            print UNMATCHED_SANGER_OUTPUT "$sanger_quality\n";
        }
        else {$QC_readcount++;};
    };
};
close FILE ;
close UNMATCHED_SANGER_OUTPUT ;
print REPORTING "##########################################\n";
print REPORTING "D. melanogaster matching step\n" ;
print REPORTING "##########################################\n\n";
print REPORTING "\n$readcount reads were analyzed by novoalign\n";
print REPORTING "$QC_readcount reads were discarded due to poor qual-
ity value\n";
print REPORTING "Dmel matched reads: ", $Dmel_readcount, "\n" ;
print REPORTING "Unmatched reads for next round: $unmatched_readcount\n" ;


#### PART THREE : NOVOALIGN MATCHING AGAINST VIRAL GENOMES - GLOBAL
MATCHING #######
# squashing virus library
print "Squashing viral library\n";
$viruses_full_path = shift @ARGV ;
$viral_squashed_library = "viruses.squash";
```

```
$novoindex_call = "novoindex $viral_squashed_library $viruses_full_
path";
system $novoindex_call ;

# Calling novoalign for sanger fastq file matching with the squashed
viruses library
$novoalign_output_file = $prefix_identifier."_novoalign_Viruses_match-
ing.tab";
$novoalign_call = "novoalign -d $viral_squashed_library -f
$unmatched_sanger_output -r Random -t 30 -o N > $novoalign_output_
file"; # t paramater (threshold quality) set to 30, may be adjusted,
or removed. A this step the sanger quality file format is automati-
cally detected by novoalign
print "starting read alignment with viruses\n" ;
print "Please, be patient\n\n" ;
system $novoalign_call ;

#parsing viral alignments for viral matches and unmatched fasta files
$fasta_unmatched_output = $prefix_identifier."_unmatched_reads.fa";
$fasta_viral_output = $prefix_identifier."_all_virus_matches.fa" ;
open FILE, "$novoalign_output_file";
open GLOBAL_VIRUS_OUTPUT, ">$fasta_viral_output";
open UNMATCHED_VIRAL_OUTPUT, ">$fasta_unmatched_output";
$viral_readcount = 0;
$unmatched_readcount = 0;
$QC_readcount = 0;
%viral_reads = ();
%unmatched_reads = ();
%histo_viral_sizes = ();
%histo_unmatched_sizes = ();
while (<FILE>) {
    unless (m /^#/) {
        ($sequence, $status) = (split /\t/)[2,4];
        if (($status eq "R") or ($status eq "U")) {
            $viral_readcount++;
            $viral_reads{$sequence}++;
            $histo_viral_sizes{length $sequence}++;
        }
```

```
        elsif ($status eq "NM\n") {
            $unmatched_readcount++;
            $unmatched_reads{$sequence}++;
            $histo_unmatched_sizes{length $sequence}++;
        }
        else {$QC_readcount++;
        };
    };
};
$i = 0;
foreach $sequence (keys %viral_reads) {
    $i++;
    print GLOBAL_VIRUS_OUTPUT ">$i","_$viral_reads{$sequence}\
    n$sequence\n";
};
foreach $sequence (keys %unmatched_reads) {
    $i++;
    print UNMATCHED_VIRAL_OUTPUT ">$i","_$unmatched_reads{$sequence}\
    n$sequence\n";
};
print REPORTING "\n#########################################\n";
print REPORTING "Viral library matching step\n" ;
print REPORTING "#########################################\n\n";
print REPORTING "\nsize distribution of viral reads\n";
print REPORTING "size (nt)\tfrequency\n";
$sum =0;
foreach $size (sort {$a<=>$b} keys %histo_viral_sizes) {
    print REPORTING "$size\t$histo_viral_sizes{$size}\n";
    $sum += $histo_viral_sizes{$size} ;
};
print REPORTING "sum\t$sum\n";
print REPORTING "\n\nsize distribution of unmatched reads\n";
print REPORTING "size (nt)\tfrequency\n";
$sum =0;
foreach $size (sort {$a<=>$b} keys %histo_unmatched_sizes) {
    print REPORTING "$size\t$histo_unmatched_sizes{$size}\n";
    $sum += $histo_unmatched_sizes{$size} ;
};
```

```
print REPORTING "sum\t$sum\n";
print REPORTING "\n";
print REPORTING "$viral_readcount viral reads identified\n";
print REPORTING "$unmatched_readcount remaining reads were not
matched\n\n";
close FILE ;
close GLOBAL_VIRUS_OUTPUT ;
close UNMATCHED_VIRAL_OUTPUT ;

#### PART FOUR : LAST NOVOALIGN MATCHING WITH VIRAL GENOMES LIBRARY,
USING THE FASTA VIRAL READS FILE ####

# Calling novoalign for fasta file matching with the squashed viruses
library
$novoalign_output_file = $prefix_identifier."_FASTA_Viruses_matching.
tab";
$novoalign_call = "novoalign -d $viral_squashed_library -f $fasta_
viral_output -r ALL -t 30 -o N > $novoalign_output_file"; # t para-
mater (threshold quality) set to 30, may be adjusted, or removed |
Note that ALl matches are now reported
print "remaking read alignment with viruses for additional analyses\n" ;
print "Please, be patient\n\n" ;
system $novoalign_call ;

# FIRST PASSAGE ON THE NOVOALIGN OUTPUT FILE TO :
# determine the number of match for each sequence in each virus
(internal repeats) for subsequent normalization
# implement virus table for fasta reporting
# implement virus histograms
%virus_table = () ; # $virus_table{$virus_name}{$target_header}=
$sequence - for fasta virus files
%repeat_table = () ; # repeat_table{virus_identifier}{read_fasta_
header}= number of matches in the *considered* viral genome (internal
repeats)
open (TAB, "$novoalign_output_file") or die ;
```

```
while (<TAB>) {
    unless (m/^#/) {
    my ($read_header, $sequence, $target_header) = (split /\t/)
[0,2,7];
    next unless (defined $target_header) ;
    $virus_table{$target_header}{$read_header}= $sequence ;
    $repeat_table{$target_header}{$read_header} += 1;
    };
}; # each sequence read is now referenced in the %repeat_table hash,
with the number of matches in the considered virus
close TAB ;

#printing fasta virus files, and histogram of read size distribution
for each virus
print REPORTING "#####################################################\n";
print REPORTING "Size distributions of viral hits, by viruses\n" ;
print REPORTING "#####################################################\n\n";
$total_viral_hits = 0;
foreach $virus (sort keys %virus_table) {
    ($temporary_virus_identifier = $virus) =~ s/>//;
    my $file = $prefix_identifier."_$temporary_virus_identifier".".fa" ;
    open (FILE, ">$file") or die ;
    %histo_virus = ();
    foreach $read (keys %{$virus_table{$virus}}) {
        print FILE "$read\n$virus_table{$virus}{$read}\n";
        ($occurence = $read ) =~ s/>\d+_// ;
        $histo_virus{length $virus_table{$virus}{$read}} +=
        $occurence ;
    };
    print REPORTING "\nread size distribution for $temporary_virus_
    identifier hits\n";
    print REPORTING "size (nt)\tfrequency\n";
    $sum = 0;
    foreach $size (sort {$a<=>$b} keys %histo_virus) {
        print REPORTING "$size\t$histo_virus{$size}\n";
        $sum += $histo_virus{$size};
    }
    print REPORTING "sum\t$sum\n";
    $viral_hit_table{$temporary_virus_identifier}=$sum ;
```

```
    $total_viral_hits += $sum;
    close FILE ;
};

# LAST PASSAGE ON THE NOVOALIGN OUTPUT FILE TO IMPLEMENT %STEP_TABLES
$step = 1 ; # this parameter may be changed for grouping reads on
blocks of nucleotides
$range_min = 19;
$range_max = 21; # will only plot 19-21nt lenght viral siRNAs. These
parameters may be also adjusted.
$range = $range_min."-".$range_max."nt";
%step_table_plus = (); # step_table_plus{$virus}{Forward offset} =
$read_occurence
%step_table_minus = (); # step_table_plus{$virus}{Reverse offset} =
$read_occurence
open (TAB, "$novoalign_output_file") or die ;
while (<TAB>) {
unless (m/^#/) {
    ($read_header, $sequence, $target_header, $target_offset,
$strand) = (split /\t/)[0,2,7,8,9];
    next unless (defined $target_header) ;
    if (($range_min <= (length $sequence)) and ((length $sequence) <=
$range_max)) {
        ($occurence = $read_header) =~ s />\d+_// ; # extract
occurence from the fasta_target_header
        if ($strand eq "R") {$target_offset = $target_offset +
(length $sequence) -1}; # reversion of the offset due to novoalign
offset assignment procedure
        $stepped_offset = $target_offset - ($target_offset % $step) ;
        $step_table_plus{$target_header}{$stepped_offset} +=
($occurence/$repeat_table{$target_header}{$read_header} ) if ($strand
eq "F");
        $step_table_minus{$target_header}{$stepped_offset} -=
($occurence/$repeat_table{$target_header}{$read_header} ) if ($strand
eq "R");
        };
    };
};
close TAB ;
```

```
$normalization_factor = 1; ######################## NORMALIZATION
FACTOR to normalize plotting if needed (for instance for comparison
of results from different libraries)
$min_data = "";
$max_data = ""; # $min_data and $max_data may be changed for gnuplot
representation (for instance, $min_data = "-400" and $max_data="+400")

# ITERATIVE VIRUS ANALYSIS AND PLOTTING, THIS PART WILL CYCLE ON
FASTA VIRUS IDENTIFIERS ($virus)

foreach $virus (sort keys %virus_table) {
    $minusplot = $prefix_identifier."_$virus"."_$range"."_minusplot";
    $minusplot =~ s/>//g;
    $plusplot = $prefix_identifier."_$virus"."_$range"."_plusplot";
    $plusplot =~ s/>//;
    $eps_output = "$virus"."_$range.eps";
    $eps_output =~ s/>//g;
    open (MINUS, ">$minusplot") or die ;
    open (PLUS, ">$plusplot") or die ;
    foreach $offset (sort {$a<=>$b} keys %{$step_table_plus{$virus}}) {
        $step_table_plus{$virus}{$offset} *= $normalization_factor ;
        print PLUS "$offset\t$step_table_plus{$virus}{$offset}\n";
        $max_offset = $offset;
    };
    foreach $offset (sort {$a<=>$b} keys %{$ step_table_
minus{$virus}}) {
        $step_table_minus{$virus}{$offset} *= $normalization_factor ;
        print MINUS "$offset\t$step_table_minus{$virus}{$offset}\n";
        if ($offset>$max_offset) {$max_offset = $offset;};
    };
    close MINUS ;
    close PLUS ;
    $script_file = $prefix_identifier."_$virus"."_$range"."_gnuplot_
plotter_script.plt";
    $script_file =~ s/>//;
    open (SCRIPT, ">$script_file") or die ;
```

```
    if ($normalization_factor == 1) {$graph_title = "Read distribu-
tion on $virus";} else {$graph_title = "$normalization_factor
Normalized Read distribution on $virus";};
    print SCRIPT "#!/usr/local/bin/gnuplot\nset output \"$eps_
output\"\nset terminal postscript colour solid\nset grid layerdefault
linetype 0 linewidth 0.500, linetype 0 linewidth 0.500\nunset key\
nset border 31 front linetype -1 linewidth 0.200\nset xzeroaxis
linetype -1 linewidth 0.500\nset yzeroaxis linetype -2 linewidth
0.500\nset ticslevel 0.3\nset title \"$graph_title\"\nset title
offset character 0, 0, 0 font \"\" norotate\nset xlabel \"Coordinates
(nt)\"\nset xlabel offset character 0, 0, 0 font \"\" textcolor lt -1
norotate\nset xrange [ * : $max_offset ] noreverse nowriteback\nset
ylabel \"Number of reads\"\nset ylabel offset character 0, 0, 0 font
\"\" textcolor lt -1 rotate by 90\nset yrange [$min_data : $max_data]
noreverse nowriteback\nGNUTERM = \"aqua\"\nplot \"$plusplot\" using
1:2 with impulses, \"$minusplot\" using 1:2 with impulses\n# EOF\n";
    close SCRIPT ;
    system "gnuplot $script_file" if ($gnuplot eq "present");
};

print REPORTING "########################################################\n";
print REPORTING "GLOBAL VIRAL HITS table\n" ;
print REPORTING "########################################################\n\n";
print REPORTING "Virus\tRead hits\tFraction\n";
foreach $virus (sort {$viral_hit_table{$b}<=>$viral_hit_table{$a}}
keys %viral_hit_table) {
    print REPORTING "$virus\t$viral_hit_table{$virus}\t", $viral_hit_
table{$virus}/$total_viral_hits, "\n";
};
####### Cleaning
close REPORTING ;
unlink "$Dmel_squashed_library", "$viral_squashed_library";
```

Acknowledgments

The author would like to thank A.L Bougé, M.C. Saleh, and N. Vodovar for helpful discussions, testing, and encouragement and Marius van den Beek for beta-testing *visitor*. This work was supported by the Institut Pasteur, the Centre National de la Recherche Scientifique and funding from the Agence Nationale pour la Recherche (AKROSS program). The *visitor* script can be downloaded at http://drosophile.org/GEDlab/?page_id=254.

References

1. Flicek, P., and Birney, E. (2009). Sense from sequence reads: methods for alignment and assembly. Nat Methods **6**, S6–S12.
2. Hercus, C. (2009). www.novocraft.com (last accessed date June, 2010).
3. Wu, Q., Luo, Y., Lu, R., Lau, N., Lai, E. C., Li, W. X., and Ding, S. W. (2010). Virus discovery by deep sequencing and assembly of virus-derived small silencing RNAs. Proc Natl Acad Sci U S A **107**, 1606–1611.

Chapter 8

Computational Prediction of Viral miRNAs

Adam Grundhoff

Abstract

While cloning and/or massive parallel sequencing of small RNAs represent powerful tools for the discovery of novel miRNAs, computational miRNA prediction represents a valuable alternative which can be performed with comparably little technical effort. This is especially true for viruses, as the number of predicted candidates generally remains low and thus within a range that may be readily confirmed by experimental means. Here, we provide a detailed protocol for the prediction of putative miRNA genes using VMir, an ab initio prediction program which we have recently designed specifically to identify pre-miRNAs in viral genomes.

Key words: microRNA, miRNA, miRNA prediction, Viral miRNA, VMir, RNAi, Noncoding RNA

1. Introduction

Mature microRNAs (miRNAs) are produced from primary precursor transcripts (called pri-miRNAs) via two endonucleolytic cleavage steps: The first step, carried out by the Drosha-containing microprocessor complex, liberates a stem-loop structure (termed pre-miRNA) from the pri-miRNA transcript. The pre-miRNA is then transported to the cytoplasm and further processed by Dicer to produce the mature miRNA duplex (see refs. (1, 2) for recent reviews on miRNA biogenesis). As neither pre- nor pri-miRNAs share any common sequence motif, it is thought that the pre-miRNA hairpin structures themselves represent the primary signal which initiates Drosha processing. Since such structures can be easily identified via minimal free energy folding, nearly all ab initio computational miRNA prediction approaches rely on the detection of genomic sequences able to form signature pre-miRNA stem loops. However, hairpins are also among the simplest secondary

Ronald P. van Rij (ed.), *Antiviral RNAi: Concepts, Methods, and Applications*, Methods in Molecular Biology, vol. 721,
DOI 10.1007/978-1-61779-037-9_8, © Springer Science+Business Media, LLC 2011

structures and are therefore predicted abundantly in all sequences. The challenge for computational miRNA prediction methods therefore is not the primary detection of the candidates per se, but to achieve efficient elimination of large numbers of false positive predictions while maintaining as many authentic pre-miRNAs as possible. As the majority of cellular miRNAs are conserved even between distantly related species, many prediction programs use evolutionary conservation as an additional filter to reduce the number of false positives. Such filters have great discriminative power, but they are of limited use in the case of viral genomes, given that viral miRNAs tend not to be conserved even between relative closely related viruses (however, some exceptions to this general rule exist (3, 4)). Therefore, in most cases it is preferable to use ab initio prediction methods to identify viral miRNAs, even though these methods produce more false positives. Fortunately, the small genome sizes of viruses will lead to a limited number of predictions which can be readily subjected to experimental confirmation. In this chapter, we provide a detailed protocol for the use of the VMir program (4–8) to perform a computational prediction of viral pre-miRNAs. We also describe how the output may be filtered depending on the method to confirm the predictions: high-throughput confirmation screens by microarray analyses, or northern blot analyses for confirmation of individual miRNAs.

2. Materials

The miRNA prediction method described here requires the VMir software package and one or more files containing the genomic sequences to be analyzed; the latter may be in raw text, FASTA or GenBank format. The latest version of the VMir software (v1.5 as of the time of this writing) can be downloaded from the following URL: http://www.hpi-hamburg.de/research/departments-and-research-groups/antiviral-defense-mechanism/software-download.html. To install the software, you will need a computer running a Windows operating system; the software will not run on Linux or MacOS systems. The VMir package also requires the Microsoft .NET Framework v2.0 or higher; in case your machine does not already have the framework, it will be installed automatically during the setup process.

Unzip/extract the downloaded software archive to a folder on your hard drive and click the Setup.exe file, then follow the instructions provided on the screen. This will install two programs: VMir Analyzer (which performs the actual analysis) and VMir Viewer (which allows viewing of results files written by VMir Analyzer). You can start both programs from the "VMir" folder located in the Programs folder of your Start Menu. You will also find shortcuts to the program's documentation in this folder.

3. Methods

The analysis is performed in two steps: First, the miRNA candidates are predicted using the VMir Analyzer, which accepts one or more input files containing the viral DNA sequences. The program predicts secondary RNA structures of overlapping sequence windows tiled across these sequences, followed by identification and scoring of putative pre-miRNA hairpins. The results of the analysis are written to an output file which can be subsequently opened in the VMir Viewer. This program allows the visualization, filtering, and export of the predicted hairpins. It can also be used to design microarrays to perform high-throughput validation of the predicted candidates.

3.1. Primary Sequence Analysis Using VMir Analyzer

1. Launch VMir Analyzer (in Windows XP, select "All Programs > VMir > VMir Analyzer" from the Start Menu). The analysis window as shown in the screenshot in Fig. 1a will appear.
2. Add one or more sequence files containing the genomic sequences you wish to analyze by clicking the "Add File" button. GenBank, FASTA, and raw text files are acceptable file formats.
3. A dialog in which you can specify several options for each of the sequence files will appear (you can also make changes later by selecting one or more files from the list and clicking the "Edit" button). Click the "Output File" button to select the name and location at which the results file should be saved (this button will be grayed out if more than one file is selected). In the "Conformation" drop-down box, select whether your input sequence is linear or circular. If you choose circular, the program will also tile windows across the junction of the head-to-tail fused input sequence. Under "Orientation(s)," you can choose "Direct," "Reverse," or "Both," depending on which of the sequence strands should be analyzed. We recommend leaving the remainder of the options at their defaults, although you may for example change the length or step size of the tiled sequence windows (see Note 1).
4. Click the "Go" button in the lower right corner of the program window. The program will analyze the sequence file and provide feedback of its progress in the task panel at the bottom half of the window. On an average computer system, you should expect the program to run for approximately 10–30 s per kb and sequence strand. Therefore, analysis of, for example, both strands of a 150-kb herpesvirus will take around 1.5 h.

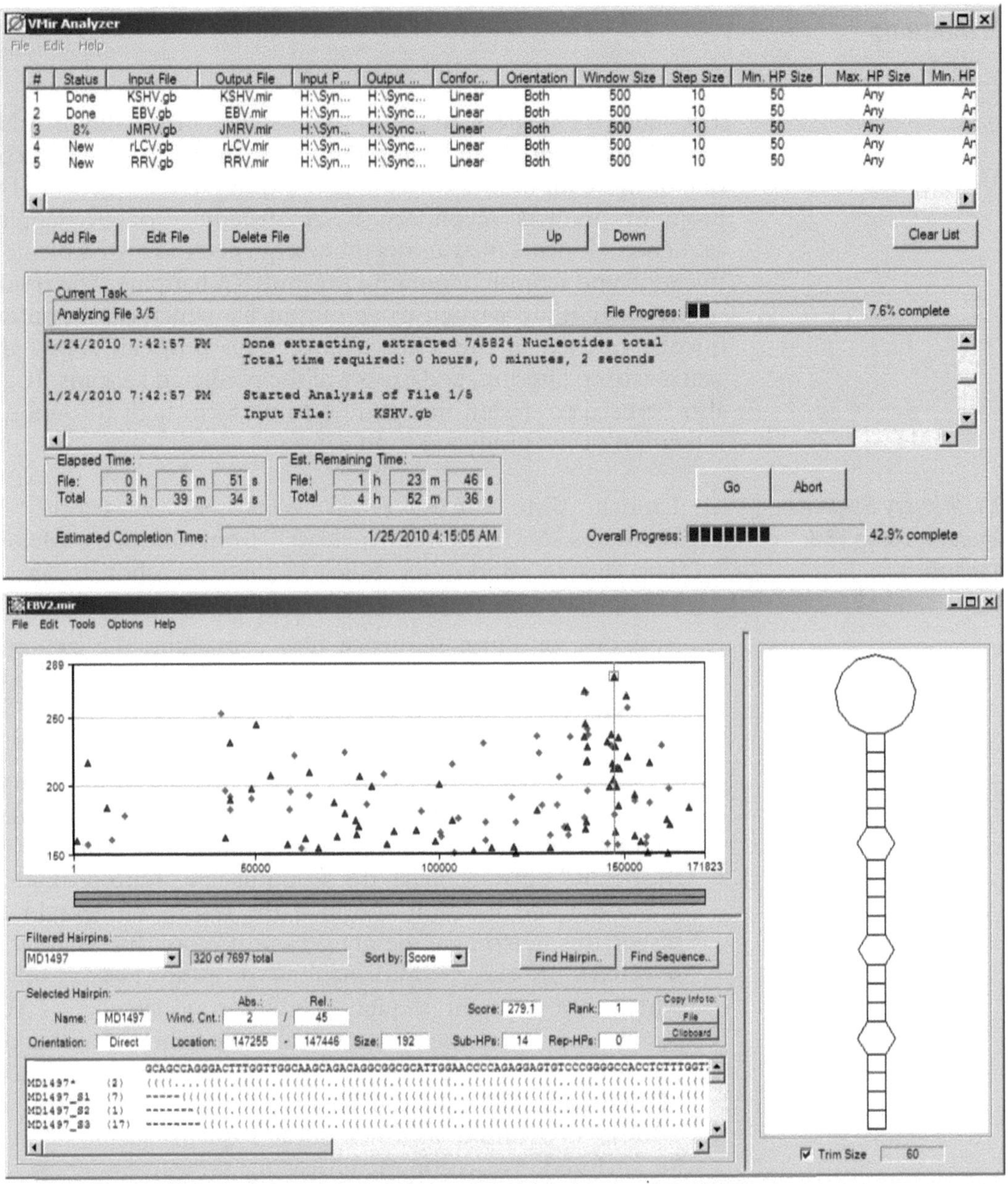

Fig. 1. VMir Screenshots. The screenshots show the main windows of the analyzer (**a**) and viewer (**b**) programs from the VMir package.

3.2. Visualization, Filtering, and Export of Results Using VMir Viewer

1. After the analysis program has finished, launch VMir Viewer (see Fig. 1b for a screenshot) by selecting "All Programs > VMir > VMir Viewer" from the Start Menu.
2. From the "File" menu, select "Open File," then browse to the location of the result file generated by VMir Analyzer.

3. The hairpins identified during the analysis will be symbolized in the chart panel by blue triangles (hairpins in direct orientation) or green diamonds (hairpins in reverse orientation) according to their location (*x*-axis) and VMir score (*y*-axis). Click on a symbol to select and view details about a hairpin in the information panel in the lower half of the program window. An image of the predicted hairpin structure will be displayed in the graphics panel to the right. By default, the program shows the complete stem-loop structure, which may be several hundred nucleotides in size. As the microprocessor complex recognizes the apical portion of the hairpin, you may wish to display only this part of the stem-loop: to do so, check the "Trim Size" box at the bottom of the picture panel and double-click on the label showing the size value to select the maximum number of nucleotides to be shown in the graphics panel.
4. When you use VMir Viewer for the first time after installation, all stem-loop structures which have been identified during the analysis step are shown. For large virus genomes (e.g., herpesviruses), this can amount to several thousand hairpins (see Fig. 2a for the primary output from an analysis of the Epstein–Barr virus (EBV) genome). The output can be filtered

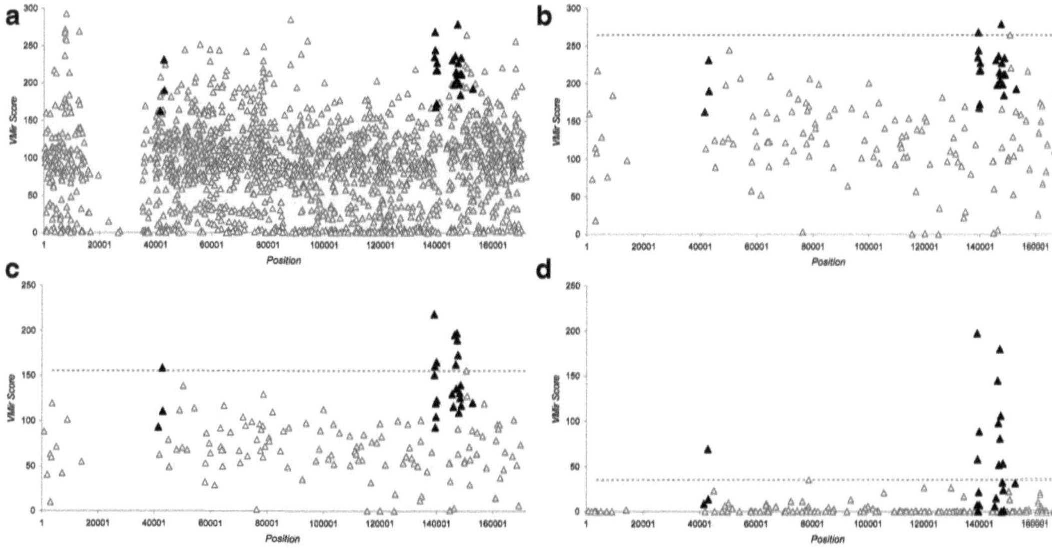

Fig. 2. VMir prediction of EBV pre-miRNAs. Depicted is the output from the prediction of pre-miRNAs encoded by Epstein–Barr virus (EBV, GenBank accession number NC_007605) under different filtering and stringency conditions. The primary analysis was performed as described in (5). Only hairpins in direct orientation are shown. *Solid black triangles* indicate the 25 authentic (i.e., experimentally confirmed) pre-miRNA stem loops encoded by EBV (3–5, 11, 12). *Open gray triangles* represent false positives. (**a**) Shows all predicted hairpins, without any filtering or stringency adjustment. In (**b**), the stem loop predictions are filtered to those which fold in 35 or more windows and which are between 50 and 220 nt in size. In (**c**) and (**d**), the stringency adjustment of the scoring algorithm was additionally set to 50 and 100%, respectively, of its maximum value. The *dashed lines* in (**b–d**) indicate the position of the highest scoring false positive prediction.

to identify those candidates with the highest probability of representing *bona fide* pre-miRNAs. The parameters of the filtering procedure are user adjustable (see Note 2): Select "Filter" from the "Edit" menu to open the filter dialog. You will be presented with a window that allows you to enter a minimal score as well as a cutoff value for the "Window Count" of the hairpins, which represents the number of overlapping sequence windows in which the hairpin was detected. We recommend to initially use values of 115 for the score and 10–35 for the window count cutoffs (see Note 3). You can also limit the output to only display the hairpins mapping to one of the sequence strands, or only hairpins with a length that falls in a specific range (see Note 4). Figure 2b shows the output from the analysis of the EBV genome after selecting a minimal window count of 35 and a hairpin length between 50 and 220 nucleotides. The filter settings are saved when you close a file and exit VMir viewer; they will be restored when you restart VMir or open another file.

5. VMir was originally designed as a low-stringency ab initio prediction method which minimizes the number of false negative predictions. Inevitably, this comes at the cost of relatively large numbers of false positives. We reasoned that the latter does not represent a problem for the analysis of viral sequences, as even large virus genomes produce a limited number of predictions which can be easily screened by high-throughput microarray methods. We realize, however, that many researchers might not want to carry out a microarray-based verification screen, but rather prefer to perform northern blotting to individually confirm the predicted candidates. In this case, it is preferable to minimize the number of false positive predictions, even if this may result in the elimination of some *bona fide* pre-miRNAs. In the current version of the software, we have therefore incorporated the possibility to increase the stringency of the predictions by imposing score penalties upon hairpins which have features that are unusual for *bona fide* pre-miRNAs (see Note 5). To use this feature, select "Adjust Score Stringency" from the "Edit" menu. A window with a slider control will open; move the slider from left to right to increase the stringency from its lowest (0%) to the highest possible value (100%). Figure 2c and d show the output of the VMir predictions with the same filter settings as in Fig. 2b, but with the stringency increased to 50% (Fig. 2c) and 100% (Fig. 2d). Note that, in contrast to the filter settings, the stringency parameter is not saved and will default to 0 every time you open a VMir file.

6. Once you have filtered the predictions according to your preferences, you can export the results by choosing the

"Export" option from the "File" menu. This will save a text file which lists the specific prediction and filtering parameters and contains detailed information about all filtered hairpins (sorted in descending order by their score). You can use this file to design probes for northern blot confirmation of the individual hairpins (see Note 6). Alternatively, if you wish to perform microarray validation experiments, VMir contains tools for the design and analysis of custom microarrays which can be accessed by selecting "Microarrays" from the "Tools" menu. See the "Designing Microarrays" and "Analyzing Microarrays" sections of the VMir documentation for detailed instructions of how to use these tools.

4. Notes

1. VMir predicts hairpins by shifting a window of defined size over the input sequence, advancing each window by a given step value. Within each window, the program performs a structure prediction by minimal free energy folding using the RNAfold algorithm (9) and identifies individual hairpins above a given size cutoff (45 nt), which are then analyzed by the scoring algorithm. By default, the window size is set to 500 nt and the step size to 10 nt. The large overlap between successive sequence windows means that each hairpin sequence will be analyzed several times, only in a slightly different sequence context at each step. This significantly increases the time of the analysis, but allows filtering of the results depending on the number of sequence windows in which a specific hairpin is detected (window count, see Note 3). Thus, if you change these settings, you will also have to adjust the filtering parameters accordingly. While the window and step size parameters may be adjusted for each file individually, there are also some global settings which may be changed by selecting "Settings" from the "Edit" menu, among them the maximum length of the hairpin segment (centered on the apical position of the terminal loop) is considered when calculating hairpin scores. See the program's documentation (contained in the installation package) for further details on these settings.
2. The optimal filter settings may differ depending on the subsequent confirmation methods you plan to use. If you want to use microarrays for high-throughput confirmation, it is preferable to perform only minimal filtering, such that the number of probes does not exceed the capacity of the microarray of your choice. The microarray design tool (see the VMir

documentation for details) calculates the number of required probes depending on the preferred array layout (i.e., the number of probe replicates and type and number of control spots) and will help you in determining the minimal filtering requirements. If no high-throughput validation methods are to be used, more stringent filtering parameters should be employed, as described in Notes 3 and 4 below.

3. We routinely use a score cutoff of 115, as more than 95% of all experimentally confirmed mammalian or viral pre-miRNA hairpins reach or exceed this value (if you adjust the stringency of the scoring algorithm, you may however also have to use a different score cutoff; see Note 4 below). The window count parameter is independent of the hairpin score, but rather provides a measure of the "robustness" with which the structures fold in the different sequence contexts generated by the sliding window. For example, if the window and step size parameters are set to their default values of 500 and 10 nt, respectively, a hairpin of 80 nt will have the opportunity to fold in up to 43 successive windows. Significantly stable structures are expected to fold in the majority of sequence contexts, and the number of windows in which a given hairpin is detected (i.e., its window count) can therefore be used as a quality criterion. We typically use window count cutoffs between 10 (least stringent) and 35 (most stringent). As the maximum number of sliding windows which contain a given hairpin sequence is dependent on the window and step size parameters, you will have to adjust the filter settings if you used other than the default values during the primary sequence analysis.

4. We usually use a size range of between 50 and 220 nucleotides. The upper limit will especially eliminate large stem-loops which are often predicted in repetitive regions.

5. The score stringency tool is based upon a statistical analysis of structural features of approximately 6,500 mammalian and viral pre-miRNA hairpins listed in release 14 of the miRNA registry (10). Hairpins which deviate significantly from this reference set in one or more categories (such as the number, size and symmetry of internal bulges, or the overall percentage of paired vs. unpaired nucleotides in the hairpin stem) are assigned a penalty which is subtracted from the hairpin score. In contrast to the original scoring system implemented by VMir Analyzer, this penalty system is much more restrictive and will eliminate false positive hairpins more efficiently (see the VMir documentation for details on how the penalty is calculated). However, as it is based on an averaged and therefore "ideal" feature set, many authentic pre-miRNAs will also be penalized and it thus may be necessary to lower the score

threshold. The effects of adjusting the stringency to 50 and 100% of its maximum value can be seen in Fig. 2c, d. In Fig. 2b, no stringency adjustment is used. If one was to perform northern blotting confirmation of the highest scoring candidates under these conditions, the first two blots would identify authentic miRNAs, but the third would already represent a false positive prediction (see the dashed line, which indicates the highest scoring false positive candidate). In contrast, 9 and 12 *bona fide* miRNAs would be identified before encountering the first false positive when adjusting the stringency as shown in Fig. 2b, c, respectively. Using a score cutoff value of 115 under conditions of maximum stringency (Fig. 2c), however, would also eliminate several authentic pre-miRNAs. The score threshold should therefore be lowered to 50 when the score stringency adjustment is set to 100%.

6. For northern blot confirmation, we routinely select probes complementary to the 35 nucleotides which flank the terminal loop on each side (it is notoriously difficult to predict which hairpin arm will give rise to mature miRNA species, and thus both arms should be tested). When no high-throughput validation screen is performed, a relatively large number of candidates may have to be tested. In these cases, it is possible to initially "multiplex" northern blots by pooling several probes prior to the labeling procedure. If a positive northern blot signal is observed, subsequent blots with individual probes can be performed. To avoid hybridization among the probes themselves, it is important that a given pool contains only one probe from a specific hairpin (i.e., avoid mixing the probes specific for the 5p and 3p arms of the same hairpin, as they may bind to each other due to the partial self-complementarity of the hairpin). Under these conditions, we have pooled up to six probes without observing a significant decrease in sensitivity.

References

1. Carthew, R. W., and Sontheimer, E. J. (2009) Origins and Mechanisms of miRNAs and siRNAs. *Cell* **136**, 642–55.
2. Kim, V. N., Han, J., and Siomi, M. C. (2009) Biogenesis of small RNAs in animals. *Nat Rev Mol Cell Biol* **10**, 126–39.
3. Cai, X., Schafer, A., Lu, S., Bilello, J. P., Desrosiers, R. C., Edwards, R., Raab-Traub, N., and Cullen, B. R. (2006) Epstein-Barr virus microRNAs are evolutionarily conserved and differentially expressed. *PLoS Pathog* **2**, e23.
4. Walz, N., Christalla, T., Tessmer, U., and Grundhoff, A. (2010) A global analysis of evolutionary conservation among known and predicted gammaherpesvirus microRNAs. *J Virol* **84**, 716–28.
5. Grundhoff, A., Sullivan, C. S., and Ganem, D. (2006) A combined computational and microarray-based approach identifies novel microRNAs encoded by human gammaherpesviruses. *RNA* **12**, 733–50.
6. Sullivan, C. S., and Grundhoff, A. (2007) Identification of viral microRNAs. *Methods Enzymol* **427**, 3–23.
7. Sullivan, C. S., Grundhoff, A. T., Tevethia, S., Pipas, J. M., and Ganem, D. (2005) SV40-encoded microRNAs regulate viral gene expression and reduce susceptibility to cytotoxic T cells. *Nature* **435**, 682–6.

8. Sullivan, C. S., Sung, C. K., Pack, C. D., Grundhoff, A., Lukacher, A. E., Benjamin, T. L., and Ganem, D. (2009) Murine Polyomavirus encodes a microRNA that cleaves early RNA transcripts but is not essential for experimental infection. *Virology* **387**, 157–67.
9. Hofacker, I. L., Fontana, W., Stadler, P. F., Bonhoeffer, S., Tacker, P., and Schuster, P. (1994) Fast folding and comparison of RNA secondary structures. *Monatshefte Chemie* **125**, 167–88.
10. Griffiths-Jones, S. (2004) The microRNA registry. *Nucleic Acids Res* **32**, D109–11.
11. Pfeffer, S., Zavolan, M., Grasser, F. A., Chien, M., Russo, J. J., Ju, J., John, B., Enright, A. J., Marks, D., Sander, C., and Tuschl, T. (2004) Identification of virus-encoded microRNAs. *Science* **304**, 734–6.
12. Zhu, J. Y., Pfuhl, T., Motsch, N., Barth, S., Nicholls, J., Grasser, F., and Meister, G. (2009) Identification of novel Epstein-Barr virus microRNA genes from nasopharyngeal carcinomas. *J Virol* **83**, 3333–41.

Chapter 9

Detection of Viral microRNAs by Northern Blot Analysis

Lydia V. McClure, Yao-Tang Lin, and Christopher S. Sullivan

Abstract

microRNAs (miRNAs) of host and viral origin have been suggested to play important roles in the viral infectious cycle. The discovery of new viral miRNAs, and understanding how viral infection alters host miRNAs, has been greatly aided by Northern blot analysis. The Northern blot method is used to detect specific RNAs that have been separated by size and immobilized onto a membrane. This method can provide specific information regarding the size of a miRNA and possible precursor structures. Thus, it represents a valuable tool in the discovery and validation of new miRNAs. Viral infection can sometimes present special challenges to utilizing Northern blot analysis. These challenges may include low miRNA expression levels, high GC content, and abundant background signal from nonspecific RNA degradation fragments triggered by the stress of lytic infection. We present a protocol for small RNA Northern blot analysis that we have successfully used to detect viral miRNAs from cells undergoing lytic infection from members of the Herpes and Polyoma virus families. Included are optimization strategies and a protocol for using radiolabeled oligonucleotides to detect larger RNAs.

Key words: Viruses, microRNA (miRNA), Urea denaturing acrylamide gel, Northern blot, Messenger RNA (mRNA), Noncoding RNA (ncRNA), Agarose formaldehyde gel, RNA-induced silencing complex (RISC), Translational inhibition

1. Introduction

Viral microRNA (miRNAs) have been identified in at least three virus families and are the subject of intense study. These small RNAs have been shown to negatively regulate expression of viral and host genes and have been implicated in processes relevant to immune evasion, tumorigenesis, and the infectious cycle. Furthermore, host miRNAs can be induced by viral infection and are likely to be utilized by viruses to optimize their replicative cycle. The discovery of viral miRNAs and understanding how infection alters cellular miRNA levels has been greatly aided by Northern blot analysis.

Ronald P. van Rij (ed.), *Antiviral RNAi: Concepts, Methods, and Applications*, Methods in Molecular Biology, vol. 721, DOI 10.1007/978-1-61779-037-9_9, © Springer Science+Business Media, LLC 2011

Northern blot analysis separates RNA based on size and allows for the identification of specific RNAs using radiolabeled probes. Alwine et al. was the first to visualize specific RNAs using this method (1). Early studies in the miRNA field relied on Northern analysis to confirm the existence and authenticity of novel short RNAs (2, 3). The Northern method was used to identify the first short-interfering RNA species in plants (4) and the first miRNAs conserved between *Caenorhabditis elegans* and mammals (4–7). The Northern blot analysis protocol described here, or similar variations, were used in the discovery of viral miRNAs from the Herpes, Polyoma, and Asco virus families (8–11). Because of its ability to give specific information regarding miRNA size and precursor structure, small RNA Northern analysis has become the gold-standard method for identifying novel small RNAs and analyzing their functions.

Several other methods are currently used to detect miRNAs and monitor their expression. Real-time PCR (RT-PCR), microarray analyses, RNase protection assays (RPA), and primer extension assays all have their specific advantages. RT-PCR, utilizing a miRNA-specific hairpin reverse transcription primer is a highly sensitive detection method for miRNAs (12). However, this method is expensive and provides little information about the authenticity of a predicted miRNA. RPA and primer extension are also sensitive methods for quantifying miRNA species, but information regarding transcript size and miRNA precursors is limited. Higher-throughput analysis of changes in relative miRNA levels is accomplished through microarray analysis or next generation sequencing. However, both have a limited dynamic linear range and typically provide no information about precursor structure. Similarly, Northern blot analysis has its own set of limitations, including limited sensitivity, inability to discern small sequence differences between orthologous miRNAs, and a requirement for a relatively large amount of starting material (total RNA) to obtain a strong signal. On the other hand, Northern blot analysis is the gold-standard method for validation of novel miRNAs as it provides information that can be used to differentiate bona fide miRNAs from background bands. Detection of some viral-relevant miRNAs can be challenging as the stress of lytic infection can trigger enormous background signal from nonspecific RNA degradation products. Furthermore, some viral miRNAs have abnormally high GC content such that complementary probes are prone to pick up background signal. We present here our protocol for Northern blot analysis and optimization strategies that are highly sensitive for viral miRNAs. We also include a protocol for using oligonucleotide probes to detect larger RNAs.

2. Materials

2.1. Total RNA Isolation

1. PIG-B reagent is a cost-effective reagent for isolating total RNA and can be stored in an amber bottle at 4°C for several months (see Note 1) (adapted from ref. (13)): guanidinium thiocyanate, Coomassie brilliant blue dye R-250, sodium citrate, citric acid, ethylenediamine (EDTA), isoamyl alcohol, Sarkosyl, saturated phenol pH 4.5, β-mercaptoethanol, 8-Hydroxyquinoline. Phenol is a hazardous poison and Guanidine Thiocyanate is a skin irritant. Wear gloves and a lab coat when preparing and using PIG-B and avoid breathing fumes by working in a chemical fume hood. Allow complete solubilization before adding each reagent by using a stir bar and a large beaker placed on a stir plate in a chemical fume hood. This recipe will prepare 1 L of PIG-B reagent and each reagent should be added in the order listed. To 250 mL of distilled water add 236 g of guanidinium thiocyanate (final concentration 2 M), 2.09 g citric acid (final 15.4 mM), 2.96 g sodium citrate (final 11.5 mM), 20 mL 0.25 M EDTA at pH 8.0 (final 5 mM), 2.5 g Sarkosyl (final concentration 0.25%), and bring to 490 mL with distilled water. Allow the solution to dissolve completely, which may take an hour or longer. When the previous reagents are in solution add 480 mL saturated phenol pH 4.5 (final concentration 48%), 21 mL Isoamyl alcohol (final concentration 2.1%), 5 mL β-mercaptoethanol (final concentration 0.5%), and 1 g 8-hydroxyquinoline (final concentration 0.1%). Adjust the pH to 4.5 with 1 N citric acid pH 2 and add 25 mg Coomassie blue. PIG-B preparation will take approximately 2 h.
2. RNaseZap RNase decontamination solution (Ambion, Austin, TX).
3. 13 mL Sarstedt tube, 95×16.8 mm and cap (Sarstedt, Toronto, ON).
4. Chloroform.
5. Isopropanol.
6. 75% Ethanol.
7. Nuclease-free Tris-EDTA (pH 7) buffer (TE, Ambion, Austin, TX).
8. RNA coprecipitant: GlycoBlue (Ambion, Austin, TX).

2.2. Denaturing Small RNA Acrylamide Analysis

2.2.1. Preparing the Denaturing Acrylamide Gel

1. Large format vertical gel system: BioRad Protean II xi (Bio-Rad, Hercules, CA).
2. Large gel casting stand (Bio-Rad, Hercules, CA).
3. Glass plates: 16×20 and 18.3×20 cm (Bio-Rad, Hercules, CA).
4. Glass plate separators: 2 mm (Bio-Rad, Hercules, CA).
5. Gel comb: 1.5 mm and 25-well (Bio-Rad, Hercules, CA).

6. RNaseZap RNase decontamination solution (Ambion, Austin, TX).
7. 75% Ethanol.
8. 50 mL Conical tubes.
9. Urea.
10. 30% Acrylamide:bisacrylamide solution (29:1).
11. Concentrated running buffer: 5× Tris–borate-EDTA (TBE) solution. Mix 108 g Tris, 55 g boric acid and 5.845 g of EDTA to 2 L with distilled water. Add reagents in this order and allow complete solubilization using a magnetic stir bar before adding the next reagent. To reduce precipitation of components out of solution, the buffer can be stored at 37°C.
12. 10% Ammonium persulfate (APS): Prepare a fresh solution in water. Use fresh or store in aliquots at –20°C for up to several months.
13. *N,N,N,N′*-Tetramethyl-ethylenediamine (TEMED).
14. 18-Gauge needle and 30-mL syringe (BD, Franklin Lakes, NJ) for flushing the wells of the gel.

2.2.2. Running the Denaturing Acrylamide Gel

1. Concentrated running buffer: 5× TE buffer, see Subheading 2.2.1, item 11.
2. Nuclease-free Tris-EDTA (pH 7) buffer (TE, Ambion, Austin, TX).
3. 2× RNA loading buffer: NorthernMax-Gly Sample Loading Dye (Ambion, Austin, TX).
4. 1% Ethidium bromide.
5. RNaseZap RNase decontamination solution (Ambion, Austin, TX).
6. Saran wrap.

2.3. Electrophoretic Transfer of Acrylamide Gel

1. Transfer system: BioRad Transblot cell (Bio-Rad, Hercules, CA).
2. Positively charged nylon membrane (Hybond-N+, Amersham, Piscataway, NJ).
3. Concentrated running buffer: 5× TE buffer, see Subheading 2.2.1, item 11.
4. Filter paper: Whatman gel blotting paper.
5. Gel holder cassette (Bio-Rad, Hercules, CA).
6. Fiber pads (Bio-Rad, Hercules, CA).
7. Transfer cooling system: Super cooling coil (Bio-Rad, Hercules, CA).

8. Power supply: BioRad PowerPac Basic (Bio-Rad, Hercules, CA).
9. Membrane UV crosslinker and hybridization oven: HL-2000 HybriLinker (UVP, Upland, CA).

2.4. Northern Blot Analysis for miRNAs

1. Expresshyb hybridization buffer solution (Clontech, Mountain View, CA). Store at 37°C. Alternatively, store Expresshyb at room temperature and warm to 37°C to completely dissolve before use (see Note 2). We store a magnetic stir bar in the Expresshyb solution and stir on a magnetic stir plate for ~5 min before using.
2. Small hybridization vessels: 20 mL disposable scintillation vials.
3. Large hybridization vessels: 100 mL Kimble hybridization roller bottles.
4. Membrane UV crosslinker and hybridization oven: HL-2000 HybriLinker (UVP, Upland, CA).
5. T4 polynucleotide kinase (PNK).
6. Radioactive γ-^{32}P adenosine 5′-triphosphate. Store at −80°C, and thaw at least 1 h prior to using. 1 mCi stored at 5 mCi/mL in 10 mM Tricine, pH 7.6.
7. Nuclease-free Tris-EDTA (pH 7) buffer (TE, Ambion, Austin, TX).
8. Probe purification columns, Illustra Microspin G-25 Sephadex Columns.
9. Wash buffer: 2× SSC/0.1% SDS. 20× SSC buffer stock: 175.3 g of NaCl, 88.2 g of Sodium Citrate to 1 L with distilled water (dH_2O). Dissolve a 10% SDS solution in water, autoclave, and store for long periods at room temperature.
10. Filter paper: Whatman gel blotting paper.
11. Thermal plastic sealer: Hot iron impulse sealer.
12. Heat sealable plastic bags: Seal-o-meal.
13. Film exposure cassette: Kodak Bio-max exposure cassette.
14. Film: Kodak Bio-Max MR film, 8 × 10 in.
15. High-energy signal intensifying screen: Kodak Bio-Max TranScreen HE.

2.5. Northern Analysis for mRNAs and Large ncRNAs

2.5.1. 1% Agarose Formaldehyde Gel Preparation

1. D2 Wide agarose gel electrophoresis system (ThermoScientific, Waltham, MA).
2. RNaseZap RNase decontamination solution (Ambion, Austin, TX).
3. GenePure LE Agarose (ISC BioExpress, Kaysville, UT).
4. 10× MOPS buffer: Add 41.8 g MOPS to 700 mL of dH_2O and adjust to pH 7.0 with 2 N NaOH. Add 20 mL of 1 M

sodium acetate and 20 mL of 10 mM EDTA, pH 8.0. Adjust the volume to 1 L with dH_2O.

5. 37% Formaldehyde.

2.5.2. Sample Preparation and Gel Running

1. RNA sample loading buffer, made fresh each time (1–10 μg of RNA resuspended in a final volume with 20 μL of loading buffer): 2 μL of 10× MOPS buffer, 3.5 μL of 37% formaldehyde, 10 μL of formamide, 1 μL of 0.1% ethidium bromide (diluted 1/10 from 1% stock solution, 3.5 μL of loading dye (0.25% bromophenol blue, 0.25% xylene cyanol)). Formamide is highly corrosive and should be worked with only in a chemical fume hood.
2. Concentrated 10× MOPS running buffer, see Subheading 2.5.1, item 4.
3. RNA running buffer: 162 mL of 37% formaldehyde, 100 mL of 10× MOPS, and 738 mL of dH_2O.

2.5.3. Transfer to Nylon Membrane

1. Concentrated 20× SSC buffer, *see* Subheading 2.4, item 9.
2. Whatman 0.45 μm nylon transfer membrane.
3. Nytran SuPerCharge TurboBlotter Kit (11 × 14 cm) (Whatman, Florham park, NJ).

2.5.4. Hybridizing the Blot

1. Methylene blue (0.02% in 0.3 M sodium acetate (pH 5.5)).

2.6. Stripping and Reprobing Blots

1. Stripping buffer: 0.1% SDS. Stock solution of 10% SDS is dissolved in water, autoclaved, and stored for long periods at room temperature.

3. Methods

3.1. Total RNA Isolation

1. It is important to minimize RNase contamination by cleaning all areas where samples will be handled with RNaseZap. If possible, equipment should be reserved only for RNA work.
2. RNA can be isolated from freshly harvested cells or from samples stored at −80°C in PIG-B in 13 mL Sarstedt tubes. Add 0.2 mL of chloroform for every 1 mL of PIG-B reagent. Typically, we use 10 mL of PIG-B and 2 mL of chloroform for a 75 cm^2 flask of confluent cells. Cap the tube and invert it for 15–30 s. Let the tube sit on ice for 5 min before centrifugation at 5,000 × *g* for 30 min at 4°C (see Note 3).
3. Transfer the top clear aqueous phase to a new 13 mL Sarstedt tube and add 5 mL of isoproponal (see Note 4). Be careful not to disturb the interface between the phases. Invert the

mixture repeatedly and let sit for 5–10 min at room temperature. Centrifuge the samples at 5,000 × *g* for 15 min at 4°C (see Notes 5 and 6).

4. The RNA is contained within the white/yellow pellet and may not be visible when preparing small amounts of RNA. Remove the supernatant and add 1 mL of ice cold 75% ethanol. Vortex to break-up the pellet and transfer the 75% ethanol containing the fragments of the pellet to a clean microfuge tube using a p1000 pipettor. Centrifuge in a table-top micro centrifuge at maximum speed (16,000 × *g*) for 10 min at 4°C (see Notes 7–9).
5. Carefully remove the supernatant using a thin-bore gel loading pipette tip and an aspirator or a pipettor. It is helpful to spin the pellet down a second time to remove all residual ethanol. Allow 5–10 min to air dry the sample at room temperature in the microfuge tube until ethanol is no longer visible. Dissolve the pellet in 20 μL of nuclease-free TE (pH 7) buffer. Adjust the resuspension volume based on your pellet size. 30 μL of TE is an appropriate volume for a 25 cm^2 flask. The samples can be stored in TE for several months at −80°C with limited degradation (see Notes 10–13). For longer-term storage, leave the RNA pellet in the 75% ethanol solution until just before using.

3.2. Denaturing Small RNA Acrylamide Analysis

3.2.1. Preparing the Denaturing Acrylamide Gel

1. These instructions are for use with the BioRad Protean II xi large format vertical gel system. Before preparing the gel make sure the glass plates (16 × 20 and 18.3 × 20 cm), glass plate separators (2 mm), comb (1.5 mm, 25-well), and other equipment to be used are scrubbed clean with RNaseZap, then rinsed with dH_2O, and 70% ethanol.
2. Prepare a 2 mm thick urea denaturing 15% acrylamide gel. Warm a 50 mL conical tube containing 24 g of urea, 25 mL of 30% acrylamide:bisacrylamide (29:1) solution, 5 mL of 5× TBE buffer, and distilled water to 50 mL at 55°C in a water bath. Vortex the solution several times during heating to help dissolve the solution. This process may take 10–20 min.
3. Once dissolved, cool the tube in an ice water bath to room temperature and then add 500 μL of fresh 10% APS and 20 μL of TEMED. Invert and quickly vortex. Add to the preassembled gel casting apparatus and fill to the top of the assembled glass plates using a 25 mL serological pipette. Insert a 25-well comb (at an angle to avoid air bubble formation) and allow the gel to polymerize at least 1 h (see Notes 14 and 15).
4. Once polymerized, use an 18-gauge needle and a 30-mL syringe to flush the acrylamide from the wells with 0.5× TBE. Clamp the gel into the running apparatus (see Note 16).

3.2.2. Running the Denaturing Acrylamide Gel

1. Assemble the running apparatus and add ~400 mL of 0.5× TBE running buffer to the upper buffer chamber of the gel apparatus. Make sure buffer does not leak from the apparatus before filling the gel box with ~3 L of 0.5× TBE running buffer. Shake the apparatus up and down to remove air bubbles from the bottom of the glass plates and anode. If air bubbles remain, use a flame to bend a glass Pasteur pipette into a loop. Use this to flush the bottom of the gel and anode with running buffer to remove any residual air bubbles.
2. Each RNA sample should be prepared with 15 μg of total RNA diluted with TE buffer. Volumes of more concentrated samples are brought up to the same volume of the least concentrated sample with TE. Add an equal volume of 2× RNA loading buffer and load into the well (see Note 17).
3. Run the gel at 300–500 V until the bromophenol blue band is 1/3 to 1/2 way down the gel. If you are planning to use scintillation vials as hybridization vessels, make sure the bromophenol blue band runs a maximum of only 8 cm from the wells.
4. Separate the glass plates, and remove extraneous portions of the gel. Mark the orientation of the gel by removing a small portion of one corner of the gel. Peel one corner of the gel off the glass plate to help move the gel into a pyrex dish containing ~100 mL of distilled water. To this, add 10 μL of 1% ethidium bromide solution. Incubate for 7 min while gently rocking. Destain by adding ~100 mL of fresh distilled water and gently rocking for 7 min. Replace with fresh distilled water. Image the stained gel on a UV box (prewashed with RNaseZap and covered in saran wrap) and photo document. Before proceeding, ensure RNA samples are not degraded and are equally loaded (Fig. 1a) (see Note 18).

3.3. Electrophoretic Transfer of Acrylamide Gel

1. These instructions are for use with the BioRad Transblot cell transfer apparatus. Cut the HybondN+ membrane slightly larger than the gel size. Set up the transfer in 0.5× TBE buffer with the gel facing the cathode (black) and the membrane towards the anode (red). Sandwich the membrane and gel with Whatman paper cut to the size of the transfer apparatus. Use a disposable seriological pipette to remove any air bubbles by rolling the pipette back and forth across the Whatman paper with gentle pressure. Sandwich the Whatman papers with fiber pads, again remove air bubbles, and slide into the gel holder cassette. Add the closed cassette, cooling coil, and a stir bar to the transfer box and fill with 0.5× TBE buffer. Place the transfer box on a stir plate and attach the cooling coil with latex rubber tubing to a sink faucet. Allow a steady, low level of cold tap water to run through the cooling coil

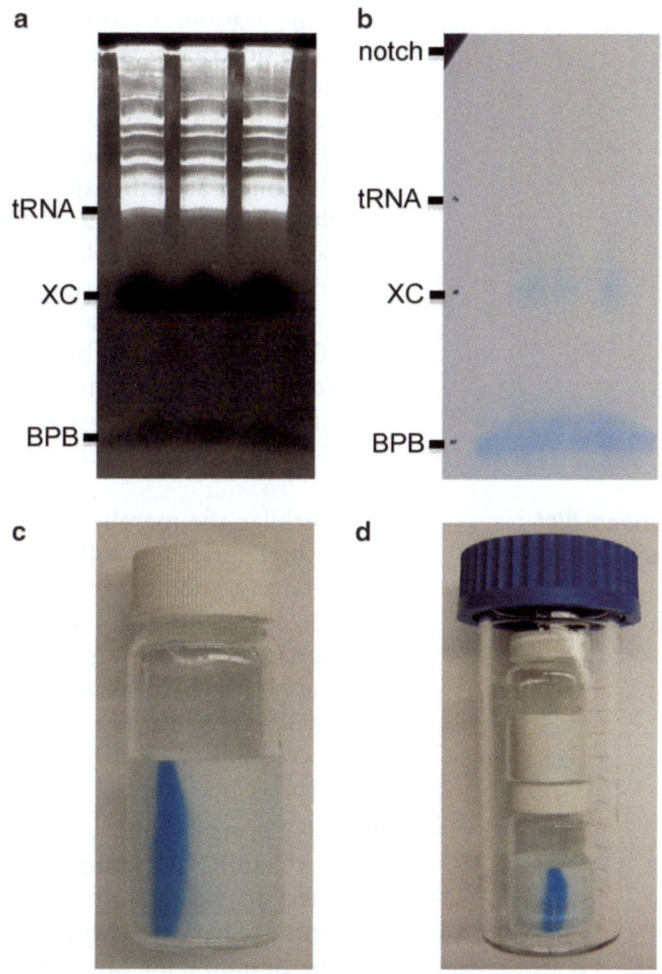

Fig. 1. Northern analysis for small RNAs. (**a**) Picture of ethidium bromide stained RNA from a denaturing 15% urea-acrylamide gel, pretransfer. Bromophenol blue (BPB), xylene cyanol (XC), and tRNA are denoted. BPB migrates at ~10 nucleotides, XC migrates at ~30 nucleotides, and tRNA migrates at ~70 nucleotides. The distinct bands seen above ~70 nucleotides indicates limited or no degradation of the RNA sample and approximately equal loading. (**b**) The nylon membrane is marked with a #2 pencil after RNA transfer at both the BPB and XC bands and at the bottom of the tRNA bands visible on the membrane under a handheld UV lamp. A notch is used to orient the membrane. (**c**) Higher-throughput analysis of specific miRNAs using radiolabeled DNA probes can be achieved using 20 mL scintillation vials. Membranes with four or fewer lanes are hybridized in scintillation vials. (**d**) Large hybridization tubes can be used for larger membranes and they also hold stacked scintillation vials rotating in a hybridization oven.

during transfer. Make sure the stir bar is steadily spinning and not blocked by the transfer cassette. Transfer at 30 V for 45 min, then 35 V for 15 min, and the remaining for 40 V for 10 min (in this order to reduce chances of heat-induced air bubble formation) (see Notes 19 and 20).

2. After the transfer is complete, remove the membrane and mark the side that was facing the gel in the lower right hand corner with a #2 pencil (see Note 21). Hang the membrane to dry for 5 min at room temperature to remove excess buffer. UV crosslink with the RNA side of the membrane facing the UV source at 1,200 μJ/m². Using a handheld long wavelength UV lamp, use a #2 pencil to mark the tRNA bands at the bottom of the visible ethidium bromide stained bands near the top portion of the gel (Fig. 1a). tRNA migrates at ~70 nucleotides. Also mark both the xylene cyanol (migrates at ~30 nucleotides) and bromophenol blue (migrates at ~10 nucleotides) dye locations (see Note 22). Dry the membrane overnight sandwiched between clean, dry Whatman paper held lightly together with a binder clip (see Note 23).

3.4. Northern Blot Analysis for miRNAs

1. Prehybridize the membrane in a glass roller bottle tube with 8 mL of Expresshyb solution for at least 40 min at 55°C. Alternatively, membranes with four or fewer lanes can be hybridized in scintillation vials containing 2 mL of Expresshyb solution. The same prehybridization and radiolabeling conditions apply for hybridization in the scintillation vials. Insert the membranes into the vessels so that the side of the membrane crosslinked with RNA is facing the inside of the tube (Fig. 1c, d) (see Notes 23 and 24).
2. An 18–22 nucleotide radiolabeled DNA probe is used to detect the miRNA of interest. Normally, the probe should be designed to be a perfect complement to the microRNA sequence. (There are exceptions to this where shorter probes may be desirable for high GC content miRNAs). Dilute the probe to 10 μM in nuclease-free water. Radiolabel 1 μL of 10 μM probe in a 10 μL kinase reaction (1 μL 10× PNK buffer, 6.75 μL of nuclease-free water, 1 μL of T4 PNK enzyme and 0.25 μL of radioactive γ-^{32}P Adenosine 5′-triphosphate (0.00125 mCi)).
3. Incubate for 30 min at 37°C.
4. Add 40 μL of TE buffer to increase the total volume to 50 μL before purifying using a Sephadex G25 spin column (see Note 25).
5. Pipette the purified radioactive probe into the Expresshyb solution avoiding direct contact with the membrane. For a typical probe, we incubate overnight at 38.5°C rotating in a hybridization oven, however, the optimal hybridization temperature may have to be empirically determined (see Notes 26 and 27, Fig. 2).
6. After approximately 12 h, wash the membrane with prewarmed (to 55°C) 2× SSC/0.1% SDS buffer for 20 min.

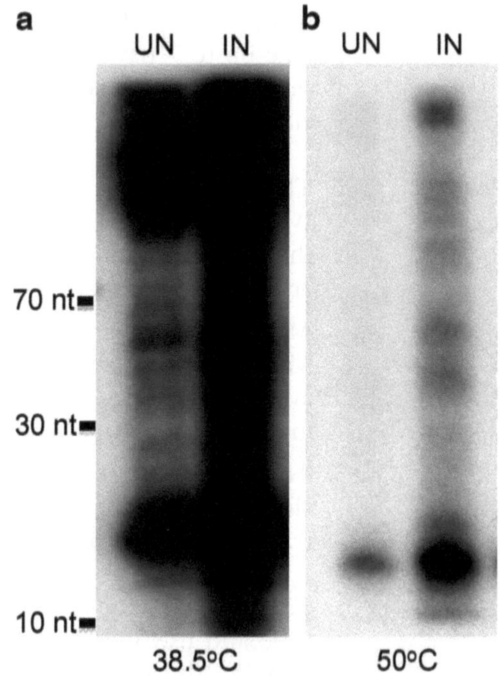

Fig. 2. Optimizing hybridization temperature based on probe GC content for microRNA Northern blot analysis. (**a**) Northern blot analysis of KSHV miR-K12-12-3p (68% GC content) shows high background signal at a 38.5°C hybridization temperature. The *left lane* shows expression of KSHV miR-K12-12-3p during latent infection (UN) and the *right lane* shows expression of miR-K12-12-3p during lytic infection (IN). Extensive RNA degradation is observed during lytic infection and there is an increase in expression of miR-K12-12-3p. (**b**) Northern blotting analysis of KSHV miR-K12-12-3p using the same radiolabeled probe at a higher hybridization temperature of 50°C shows better resolution of the microRNA with lower background signal.

Repeat wash step an additional 3 times. The wash steps are completed while rotating in the hybridization oven. The roller bottles should be washed with ≥30 mL of wash buffer each time, whereas the scintillation vials are washed with ~10 mL of wash buffer. The discarded waste is radioactive and care should be taken for proper disposal.

7. Mount the damp membrane on Whatman paper that is prewetted with 2× SSC. Once hybridized, it is important to keep the blot damp at all times to allow for successful future stripping. Heat-seal the mounted membrane inside a seal-o-meal plastic bag and use scissors to remove the excess plastic outside of the heat-seal to allow proper fit inside the exposure cassette. Place in a high-energy intensifying screen and expose to film at −80°C overnight (see Note 28).
8. Typically an overnight exposure is appropriate to detect most miRNAs. A very highly expressed miRNA will have detectable

bands after a 1-h exposure time. Less abundant miRNAs may take up to 7 days of exposure for bands to be detectable. In our experience, exposing for more than 7 days fails to appreciably increase the signal that is detected.

3.5. Northern Analysis for mRNAs and Large ncRNAs

3.5.1. 1% Agarose Formaldehyde Gel Preparation

1. These instructions are for use with the Thermo Scientific D2 Wide agarose gel electrophoresis system. Before preparing the gel, clean and dry the buffer chamber, gel tray, combs, and other equipment to be used with RNaseZap.
2. Mix 1 g of agarose and 73 mL of nuclease-free water. Microwave for 1–2 min to melt the agarose being careful to avoid boiling over. Allow the agarose solution to cool, reaching equilibrium in a 60°C water bath (see Notes 29 and 30).
3. Add 10 mL of 10× MOPS buffer and 16.2 mL of 37% formaldehyde. Mix gently and pour the gel into the assembled gel tray within the fume hood.

3.5.2. Sample Preparation and Gel Running

1. Transfer 1–10 μg of total RNA into a clean microcentrifuge tube and use a speed vacuum concentrator to dry the RNA sample. Alternatively, for more concentrated samples with a volume smaller than 20 μL, omit the concentration step.
2. Resuspend the RNA samples in 20 μL of RNA sample buffer and heat at 60°C in a water bath for 10 min. Chill the sample on ice.
3. Add ~700 mL RNA running buffer to the buffer chamber and load the RNA samples into individual wells.
4. Run the gel at ~75 V for ~3 h (see Note 31).
5. Image the agarose gel on a UV box and photo document to ensure RNA samples are not degraded and are equally loaded.

3.5.3. Transfer to Nylon Membrane

1. Remove the gel from the electrophoresis box and rinse in a glass pyrex dish containing ~100 mL of nuclease-free water by gentle agitation for 10 min. Repeat 4 times.
2. Discard the water and soak the gel in 1× SSC at room temperature for 20 min with gentle agitation. Soak the nylon transfer membrane in 1× SSC for 10–15 min while the gel is soaking (see Note 32).
3. Set up the TurboBlotter Rapid downward transfer system as described by the manufacturer. Place the stack tray on a flat surface. It is essential that the transfer is level. Arrange the transfer with 20 sheets of dry GB004 thick blotting paper in the stack tray. Add four sheets of dry GB002 thin blotting paper to the thick blotting paper stack. Add one sheet of GB002 blotting paper wet with 20× SSC. Briefly rinse the nylon transfer membrane with 20× SSC and place it on the stack.

Lastly, rinse the agarose gel with 20× SSC and arrange it on the membrane. Make sure there are no air bubbles between the gel and the membrane by carefully smoothing back and forth across the gel with gentle pressure and rolling with a seriological pipette. Wet the gel with 20× SSC and place three sheets of GB002 blotting paper wet with 20× SSC on top of the stack. Attach the buffer tray to the bottom tray and fill with 20× SSC buffer. Use a wick presoaked in transfer buffer to connect the gel stack with the buffer tray. Lay the wick across the stack to start transferring.

4. Transfer overnight at room temperature.

3.5.4. Hybridizing the Blot

1. Disassemble the transfer apparatus and rinse the blot in 2× SSC for 5 min.
2. Use a binder clip to hang the membrane for 5 min to remove excess buffer.
3. UV crosslink the membrane twice at 1,200 mJ/cm^2.
4. Mark the upper left hand corner of the membrane with a #2 pencil and label the location of the bromophenol blue and xylene cyanol bands (see Note 31).
5. Soak membrane in methylene blue solution for approximately 5 min until ribosomal RNA becomes visible. Mark locations of the RNA ladder and ribosomal RNAs with a pencil. Photograph stained blot for future reference as a loading and transfer control.
6. Prehybridization, hybridization, and exposure to film is the same as for small RNAs (Fig. 3) (see Note 33).

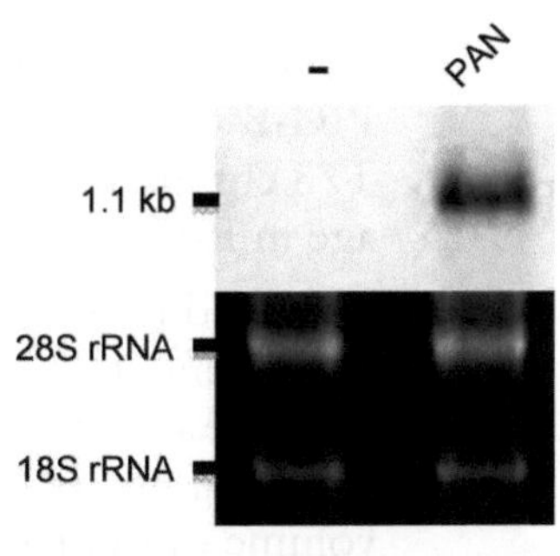

Fig. 3. Northern blot analysis of large RNAs using radiolabeled oligonucleotides as probes. Transfection of a pcDNA3.1 vector (*left lane*) or a vector expressing the KSHV lytic specific ncRNA PAN (*right lane*) in 293 T cells shows specific signal of the ncRNA from total RNA samples. 28S and 18S rRNA ethidium bromide bands from the 1% agarose formaldehyde gel are shown as loading controls.

3.6. Stripping and Reprobing Blots

1. Each membrane may be stripped and reprobed up to 8 times before signal is no longer visible or the background signal becomes too high.
2. Boil 100 mL of 0.1% SDS stripping buffer in a 500-mL beaker by microwaving. Pour onto the membrane in a glass pyrex dish rocking for 15 min.
3. Repeat the membrane wash step 4 times with boiling hot 0.1% SDS buffer. Once the hot stripping buffer cools to room temperature it can be changed. Use a Geiger counter or a phosphorimager to check the blot for removal of radiation before repeating the probe process with a new radioactive probe.

4. Notes

1. The homemade PIG-B total RNA isolation reagent is an affordable alternative to commercial reagents. Trizol (Invitrogen, Carlsbad, CA) and RNA-Bee (AMS Biotechnology, Abingdon, UK) are commercial reagents that work well and can substitute for PIG-B. All of these reagents separate total RNA from protein and most DNA. A sample lysed in PIG-B is separated into an aqueous and an organic phase by the addition of chloroform. DNA and protein are trapped in the organic phase and interface. Adding isopropanol precipitates the RNA from the aqueous phase and the 75% ethanol wash removes contaminating salts.
2. For ready use, we store our stock solution of ExpressHyb at 37°C and leave a stir bar in the ExpressHyb buffer to easily mix the solution before prehybridization.
3. We recommend that 1 mL of PIG-B reagent be used per 1×10^7 cells. An increase in PIG-B volume may be required if more cells are used. For example, we typically use 10 mL of PIG-B for harvesting a confluent monolayer of cells on a T75 cm^2 flask. The samples may be frozen for long-term storage in PIG-B at −80°C.
4. To expedite processing of multiple samples at the same time, it is useful to pre-add 5 mL of isopropanol to each clean 13 mL Sarstedt tube prior to transferring the aqueous phase. This assumes an initial volume of 10 mL of PIG-B; adjust the volume of isopropanol according to initial PIG-B levels.
5. Two phases should be visible after mixing with chloroform and incubating for 10 min, the top colorless aqueous phase and the bottom blue organic phase. If you do not see a distinct interphase, add more chloroform to the solution and

repeat the purification. A major source of contamination and RNA degradation occurs if the interphase of the organic phase is transferred with the aqueous phase into the clean 13 mL cap tube.

6. Use an alcohol resistant marker to mark the outside of the tube where the pellet is expected to form after the spin. This is useful for small volume samples and low concentration RNA preparations.
7. Carefully transfer the ethanol containing the pellet into the clean microfuge tube by vortexing and transferring with a p1000 pipettor with an aerosol barrier tip. The white or opaque pellet will be insoluble in the 75% ethanol solution and is usually visible. If low levels of RNA are to be precipitated, add GlycoBlue as a coprecipitant to increase the size and visibility of the RNA.
8. Samples are stored in nuclease-free TE buffer (pH 7) at −80°C and are stable for a few months. For long-term storage samples are left in 75% ethanol and kept at −80°C.
9. An extra wash with room temperature 75% ethanol is recommended to remove residual chloroform and to improve the solubility and purity of the RNA.
10. It is important to allow the appropriate amount of time for the RNA to dry before resuspending in TE buffer. Any visible ethanol should be removed and the sample should not be overdried. Leaving residual ethanol or overdrying the sample will result in a pellet that does not dissolve completely in TE buffer.
11. To dissolve the RNA, resuspend in nuclease-free TE buffer, and vortex for a few seconds. Let the sample sit at room temperature for 10 min. Repeat the vortex step and, if necessary, incubate at 42°C for a few minutes. The resuspended RNA solution should be clear when completely dissolved.
12. The appropriate volume needed to resuspend an RNA pellet varies among cell types and cell density. Initially, add less resuspension volume and check the concentration using a spectrophotometer. Increase the final volume accordingly with resuspension buffer until the RNA concentration is between 0.7 and 2.0 μg/μL.
13. Complete RNA solubilization is fostered by using the appropriate volume of RNA resuspension buffer. It is important to ensure the RNA concentration reading is accurate. If the RNA solution is too viscous, add more resuspension volume. This may be evident by difficulty in pipetting. An overly high concentration of RNA will affect both the RNA solubility and the accuracy of the spectrophotometer reading. As a result, the volume calculated to load into the wells of the gel will be inaccurate.

14. Cooling the gel solution before pouring the gel prevents premature polymerization during pipetting. A warm gel will polymerize quickly.
15. The degree of acrylamide gel polymerization will affect the resolution of miRNAs. When the acrylamide gel is polymerized longer we typically observe better resolution. Therefore, when possible, pour the gel a day before needed and store at room temperature.
16. It is critical to flush the wells with running buffer before loading the RNA samples. The remaining acrylamide is clear but will appear oily when flushed from the wells. Acrylamide that is not flushed from the wells prevents the RNA from sinking to the bottom of the well and can cause aberrant RNA migration while running the gel.
17. If extra wells are available, only load the middle wells of the gel. RNA that is loaded in the wells nearest to the edge of the gel tends to migrate aberrantly compared to the middle wells.
18. Intact RNA will show distinct bands by ethidium bromide staining corresponding to transfer RNA (~70 nucleotides), 5S ribosomal RNA, and other small noncoding RNAs (Fig. 1a). A smear of RNA in the wells indicates RNA degradation that may have occurred during RNA isolation or subsequent steps.
19. The transfer time is based on an individual gel transfer. If transferring two gels at the same time, transfer at 30 V for 45 min, then 35 V for 15 min, and 40 V for 15–18 min (in this order). The gradual increase is meant to minimize air bubble formation from the heat of transfer.
20. It is imperative to remove all air bubbles due to the transfer setup for a complete and intact RNA transfer from the gel onto the membrane.
21. Pencil marks pick up radiation and this signal can be used to orient the blot and mark the migration of RNA ladders.
22. The three pencil-labeled marks (for both dyes and tRNAs) show approximately where 10, 30, and 70 nucleotides migrate. Several other options for RNA markers are available, including the Decade markers (Ambion, Austin, TX) that are prepared by enzyme cleavage of a long RNA template and radiolabeled before loading onto the gel.
23. One can probe the membrane on the same day as transferring the gel by drying it in a hybridization oven. This can be done by placing the membrane in a hybridization roller bottle with the cap off. Bake in the hybridization oven at 80°C for 45 min prior to prehybridization.

24. Hybridization in scintillation vials has several advantages. The smaller hybridization tube gives a greater concentration of radiolabeled probe, higher throughput, and uses less reagents. Use of the scintillation vials is limited to membranes with 4 or fewer lanes and a gel that has only 8 cm of separation between the dye front and the wells.
25. The Sephadex G-25 spin column must be prepared prior to radiolabeled probe purification. The column is vortexed to resuspend the sephadex suspension. Remove the bottom closure and loosen the cap. Spin the column for 1 min at 700×*g* to pack the sephadex resin. Discard the liquid waste. Apply the labeling reaction containing the probe to the top of the angled resin. Do not disturb the resin when adding the liquid. The appropriate loading volume for radiolabeled oligonucleotides on the G25 column is 50 μL. Elute the radiolabeled probe by spinning the column for 2 min at 700×*g*. Most of the unincorporated radioisotopes are trapped in the column while the labeled probe is in the flow through. The column is solid radioactive waste and should be disposed of appropriately.
26. The appropriate hybridization temperature for Northern analysis should be changed based on the GC content of the probe. If possible, repetitive nucleotide strings within the DNA probe should be avoided.
27. There are three major factors that affect probing stringency: temperature, probe composition, and wash buffer conditions. If the stringency needs to be changed to achieve better results, we find it easiest to begin by changing the hybridization temperature. Usually 38.5°C works well for most 19–23 nucleotide probes with 30–50% GC content. If higher GC content probes are used, increase the hybridization temperature. For example, raising the temperature to 45°C improves the resolution for 60–65% GC content probes. In addition, higher temperature washing conditions and using more than four wash steps can achieve higher stringency. Alternatively, we have had success by decreasing the length of the probe by up to five nucleotides (length of 17 nucleotides) for those miRNAs that are particularly GC rich (70%). The best strategy will ultimately be determined empirically for each miRNA.
28. The high-energy intensifying screen is used to increase the sensitivity of the radioactive signal detection. Close the assembled intensifying screen, film, and Northern blot in a film cassette and wrap in aluminum foil to eliminate light leakage into the cassette. Place at −80°C. After removing from the −80°C freezer, allow 30 min at room temperature for the cassette and intensifying screen to thaw before developing the film in a dark-room.

29. For the gel composition, different percentages of agarose should be used based on the size of the RNA of interest.
30. We observed that some brands of agarose are not suitable for RNA Northern blot analysis due to poor transfer efficiency.
31. The bromophenol blue (~300 bp on 1% agarose gels) and xylene cyanol (~4,000 bp on 1% agarose gels) will give you an idea of how the RNA electrophoresis is progressing. You can also use a UV box to check RNA electrophoresis progression due to ethidium bromide in the RNA sample buffer.
32. Rinse the gel in buffer to remove formaldehyde to increase transfer efficiency.
33. We find that Northern blot analysis with end labeled probes is less sensitive than with probes labeled with other methods, such as random primed fill-in reactions or in vitro transcription. We use end-labeling to detect abundant RNAs. Low abundance RNAs will likely not be detected with this method. However, this method has several advantages for some applications. It is easy to prepare the probe and to strip from the membranes. It also allows for more precision and flexibility in the part of the mRNA the probe hybridizes to.

Acknowledgments

Work in the Sullivan Lab involving the protocols presented in this chapter is supported by NIH grant R01AI077746-01 and a fellowship from the UT Austin Institute for Cellular and Molecular Biology.

References

1. Alwine, J. C., Kemp, D. J., and Stark, G. R. (1977) Method for detection of specific RNAs in agarose gels by transfer to diazobenzyloxymethyl-paper and hybridization with DNA probes, *Proc Natl Acad Sci U S A* **74**, 5350–5354.
2. Lee, R. C., and Ambros, V. (2001) An extensive class of small RNAs in Caenorhabditis elegans, *Science* **294**, 862–864.
3. Lee, R. C., Feinbaum, R. L., and Ambros, V. (1993) The C. elegans heterochronic gene lin-4 encodes small RNAs with antisense complementarity to lin-14, *Cell* **75**, 843–854.
4. Hamilton, A. J., and Baulcombe, D. C. (1999) A species of small antisense RNA in posttranscriptional gene silencing in plants, *Science* **286**, 950–952.
5. Lagos-Quintana, M., Rauhut, R., Lendeckel, W., and Tuschl, T. (2001) Identification of novel genes coding for small expressed RNAs, *Science* **294**, 853–858.
6. Lau, N. C., Lim, L. P., Weinstein, E. G., and Bartel, D. P. (2001) An abundant class of tiny RNAs with probable regulatory roles in Caenorhabditis elegans, *Science* **294**, 858–862.
7. Pasquinelli, A. E., Reinhart, B. J., Slack, F., Martindale, M. Q., Kuroda, M. I., Maller, B., Hayward, D. C., Ball, E. E., Degnan, B., Muller, P., Spring, J., Srinivasan, A., Fishman, M., Finnerty, J., Corbo, J., Levine, M., Leahy, P., Davidson, E., and Ruvkun, G. (2000) Conservation of the sequence and temporal expression of let-7 heterochronic regulatory RNA, *Nature* **408**, 86–89.

8. Hussain, M., Taft, R. J., and Asgari, S. (2008) An insect virus-encoded microRNA regulates viral replication, *J Virol* **82**, 9164–9170.
9. Pfeffer, S., Sewer, A., Lagos-Quintana, M., Sheridan, R., Sander, C., Grasser, F. A., van Dyk, L. F., Ho, C. K., Shuman, S., Chien, M., Russo, J. J., Ju, J., Randall, G., Lindenbach, B. D., Rice, C. M., Simon, V., Ho, D. D., Zavolan, M., and Tuschl, T. (2005) Identification of microRNAs of the herpesvirus family, *Nat Methods* **2**, 269–276.
10. Pfeffer, S., Zavolan, M., Grasser, F. A., Chien, M., Russo, J. J., Ju, J., John, B., Enright, A. J., Marks, D., Sander, C., and Tuschl, T. (2004) Identification of virus-encoded microRNAs, *Science* **304**, 734–736.
11. Sullivan, C. S., Grundhoff, A. T., Tevethia, S., Pipas, J. M., and Ganem, D. (2005) SV40-encoded microRNAs regulate viral gene expression and reduce susceptibility to cytotoxic T cells, *Nature* **435**, 682–686.
12. Chen, C., Ridzon, D. A., Broomer, A. J., Zhou, Z., Lee, D. H., Nguyen, J. T., Barbisin, M., Xu, N. L., Mahuvakar, V. R., Andersen, M. R., Lao, K. Q., Livak, K. J., and Guegler, K. J. (2005) Real-time quantification of microRNAs by stem-loop RT-PCR, *Nucleic Acids Res* **33**, e179.
13. Weber, K., Bolander, M. E., and Sarkar, G. (1998) PIG-B: a homemade monophasic cocktail for the extraction of RNA, *Mol Biotechnol* **9**, 73–77.

Chapter 10

Detection of Viral microRNA with S1 Nuclease Protection Assay

Matthias John and Sébastien Pfeffer

Abstract

Mammalian host cells and their viral pathogens express and make use of short noncoding RNA molecules to control the infectious cycle. In order to understand their physiological role, it is necessary to develop tools for detection and quantification of these molecules. Here, we present a simple, specific, and very sensitive protocol using short radioactive DNA oligonucleotides for hybridization to homologous RNA target in a nuclease protection assay. The S1 nuclease from *Aspergillus oryzae* degrades single-stranded oligonucleotides composed of either deoxynucleotides or ribonucleotides. In contrast, double-stranded DNA, double-stranded RNA, or DNA–RNA hybrids are resistant to digestion. Subsequent analysis of the protected DNA oligonucleotide with denaturing gel electrophoresis results in radioactive signals strictly proportional to the abundance of short RNA in a given sample. The protocol works equally well for in vitro cell culture assays and for tissue samples obtained from in vivo experiments.

Key words: Nuclease protection, S1 nuclease, microRNA

1. Introduction

Micro (mi) RNAs compose a growing family of small noncoding regulatory RNA molecules that can be found in virtually all eukaryotes, from the unicellular alga *Chlamydomonas* to humans (1). Recently, miRNAs were also found in some viruses infecting mammals, such as herpesviruses and polyomaviruses (2, 3). Being practically indistinguishable from their cellular counterpart in terms of physico-chemical properties, the approaches used to study the miRNAs of viral origin mainly parallel the ones used to study cellular miRNAs. A particular challenge one could face when analyzing these tiny RNAs is their detection in tissue samples from infected animals. For example, in the case of mouse cytomegalovirus, only one in a hundred cells will be infected in

Ronald P. van Rij (ed.), *Antiviral RNAi: Concepts, Methods, and Applications*, Methods in Molecular Biology, vol. 721, DOI 10.1007/978-1-61779-037-9_10, © Springer Science+Business Media, LLC 2011

the mouse liver, which means that viral miRNAs will be strongly diluted in total RNA extracted from this organ. It might thus become challenging to detect a given miRNA with sufficient sensitivity and specificity. In this case, northern blotting is clearly not sensitive enough, although it is very specific. RT-PCR could be another approach that is more sensitive, but it can sometimes be prone to specificity issues. We found that nuclease protection assay is of sufficient specificity, and high sensitivity not only for the detection of low abundant viral miRNA in vivo (4), but also for siRNAs (5) and endogenous miRNAs (6).

2. Materials

2.1. Preparation of a Radioactive Single-Stranded DNA Probe

1. Single-stranded oligodeoxyribonucleotide (Fig. 1): dissolved in nuclease-free water at a concentration of 20 μM (~130 ng/μl), store at –20°C (see Note 1).
2. γ-[^{32}P] ATP: specific activity >6,000 Ci/mmol. Use appropriate measures for protection from this radiation hazard.
3. 10× Kinase buffer: 700 mM Tris–HCl, pH 7.6, 100 mM $MgCl_2$, 50 mM dithiothreitol (New England Biolabs); store at –20°C (see Note 2).
4. T4 Polynucleotide kinase 10 U/μl (New England Biolabs), store at –20°C.
5. Microspin column 1.5 ml packed with G-25 sephadex™ DNA grade (GE Healthcare).
6. Scintillation cocktail and liquid scintillation counter.

2.2. Hybridization and S1 Nuclease Digestion of Probe and Viral miRNA

1. 5× S1 Hybridization buffer: 1.5 M NaCl, 5 mM ethylenediaminetetraacetic acid (EDTA), 190 mM HEPES, pH 7.0, store at room temperature.
2. Sterile tRNA solution prepared from *Escherichia coli*, 10 μg/μl in nuclease-free water (Roche).
3. Triton X-100 5% in nuclease-free water, store at room temperature.

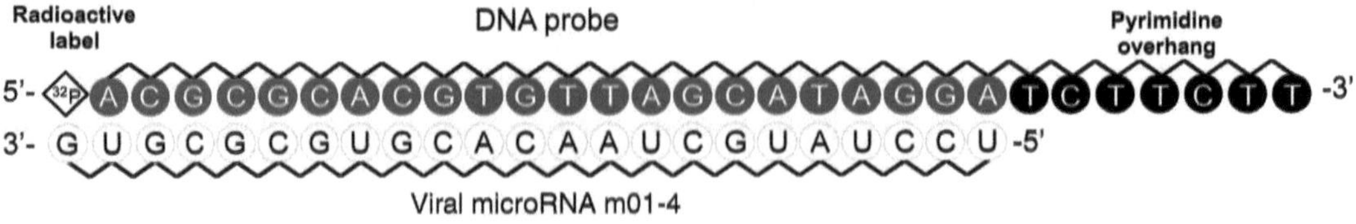

Fig. 1. Schematic illustration of a hybridized complex with a radioactive DNA probe and complementary mouse Cytomegalovirus microRNA m01-4.

4. 10× S1 Nuclease digestion buffer: 3 M NaCl, 20 mM Zn-acetate, 600 mM Na-acetate, pH 4.5; store at room temperature.
5. S1 Nuclease: 140 U/μl, store at −20°C (GE Healthcare).
6. S1 nuclease stop solution: sterile tRNA prepared from *E. coli*, dissolved at 3 μg/μl in 350 mM EDTA; store at −20°C in small aliquots.
7. Technical grade 95% ethanol.
8. Dry heating blocks for 1.5 ml reaction tubes set at 92 and 42°C.

2.3. Analysis of the Digested Sample

1. 40% Acrylamide/7.5 M urea solution (19:1 acryl:bis-acrylamide). Use appropriate measures for protection from this toxic hazard.
2. 10× Tris/Borate/EDTA (TBE) buffer: 890 mM Tris base, 890 mM boric acid, 25 mM EDTA disodium salt dihydrate. Dissolve 108 g of Tris base, 55 g of boric acid, and 9.3 g of EDTA in 1,000 ml water.
3. *N,N,N,N'*-Tetramethyl-ethylenediamine (TEMED).
4. 10% Ammonium persulfate (APS): dissolve 1 g APS in 10 ml sterile water, store at −20°C.
5. Vertical gel system with 16×20 cm glass plates and 0.4 mm sequencing spacers.
6. Formamide buffer: 25 mM EDTA, 70 μM bromophenol blue sodium salt, 90 μM xylene cyanol FF dye in formamide. Mix 9.4 ml of formamide, 0.5 ml of 500 mM EDTA, 50 μl of 10 mg/ml bromophenol blue sodium salt, and 50 μl of 10 mg/ml xylene cyanol FF. The stock solution of the dyes are prepared in nuclease-free water.
7. Chromatography paper (0.34 mm).
8. Vacuum gel dryer.
9. Phosphorimager and storage screen (GE Healthcare).

3. Methods

3.1. Preparation of a Radioactive Single-Stranded DNA Probe

1. Label the oligonucleotide. Mix the following components.
   (a) 0.75 μl 20 μM Oligonucleotide.
   (b) 5 μl γ-[^{32}P] ATP.
   (c) 2.5 μl 10× PNK kinase buffer.
   (d) 6.5 μl Nuclease-free water.
   (e) 1 μl PNK 10 U/μl

   Incubate at 37°C for 30 min.

2. Vortex a G-25 microspin column for a few seconds, pierce a hole in the cover with a small needle and break the bottom neck of the column (see Note 3).
3. Place the column on top of an empty 1.5 ml reaction tube and centrifuge the G-25 column for 1 min at 700 × *g*. Discard the reaction tube and the collected liquid.
4. Place the G-25 column on a new 1.5-ml tube.
5. Transfer the complete labeling reaction on top of the dry G-25 resin.
6. Spin the G-25 column for 1 min at 700 × *g*.
7. Discard the column. Use the collected flow-through directly in a hybridization reaction, or store at −20°C for later use.
8. Take a 1 µl aliquot, add 5 ml scintillation cocktail and determine radioactive counts in a liquid scintillation system (see Notes 4 and 5).

3.2. Hybridization and S1 Nuclease Digestion of Probe and Viral miRNA

1. For one hybridization reaction, combine in a 1.5 ml reaction tube:
   (a) 10 µl Sample (see Notes 6–8).
   (b) 5 µl 10 µg/µl tRNA solution.
   (c) 10 µl 5× Hybridization buffer.
   (d) 1 µl 5% Triton X-100.
   (e) 23 µl Sterile water.
   (f) 1 µl of Radioactive DNA oligonucleotide probe.
2. Vortex the tube thoroughly for a few seconds and briefly centrifuge to collect all liquid at the bottom of the tube.
3. Secure the cover with a safety clip.
4. Place the tube in a heating block and incubate at 92°C for 2 min.
5. Transfer the tube to a heavy-duty Plexiglas rack with cover and wait 2 h until the tube has reached room temperature (see Note 9).
6. Combine 50 µl of the hybridization reaction, 50 µl of 10× S1 Nuclease digestion buffer, 0.75 µl of S1 Nuclease, and 400 µl of sterile water (see Note 10).
7. Vortex the digestion solution.
8. Incubate tubes at 42°C for 30 min.
9. Add 10 µl of S1 nuclease stop solution and vortex briefly.
10. Add 900 µl of 95% ethanol and precipitate protected fragments overnight at −20°C.

3.3. Analysis of the Digested Sample

1. Prepare a vertical 12% acrylamide/urea gel by mixing (see Note 11):
   (a) 6 ml 40% acrylamide/7.5 M urea solution.
   (b) 2 ml 10× TBE running buffer.
   (c) 12 ml Sterile water.
   (d) 60 µl 10% APS.
   (e) 20 µl TEMED.
2. Take precipitated S1 nuclease digest samples from freezer (see Notes 7 and 8).
3. Centrifuge at 16,100 × *g* for 10 min at 4°C.
4. Carefully remove and discard radioactive supernatant.
5. Spin again for a few seconds to remove remaining droplets of liquid.
6. Resuspend RNA pellets thoroughly by pipetting and vortexing in 12 µl of formamide buffer.
7. Centrifuge tubes for a few seconds to collect liquid at the bottom of the tube.
8. Denature samples by incubating the tubes for 2 min at 92°C.
9. Place tubes immediately on ice.
10. Rinse gel pockets before loading the samples.
11. Load 4 µl of resuspended sample per lane.
12. Run for 1 h at 500 V.
13. Transfer gel to chromatography paper and dry the gel for 1 h at 85°C.
14. Expose to phosphorimager screen for 4 h or overnight.
15. Scan the screen in phosphorimager (see Note 12 and Fig. 2). Figure 3 shows the results of an actual experiment.

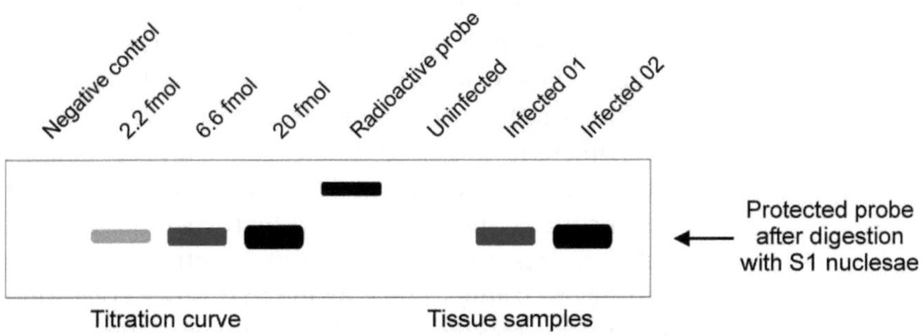

Fig. 2. Graphic representation of a S1 nuclease protection experiment after gel separation of digested probes and exposure to phosphorimaging.

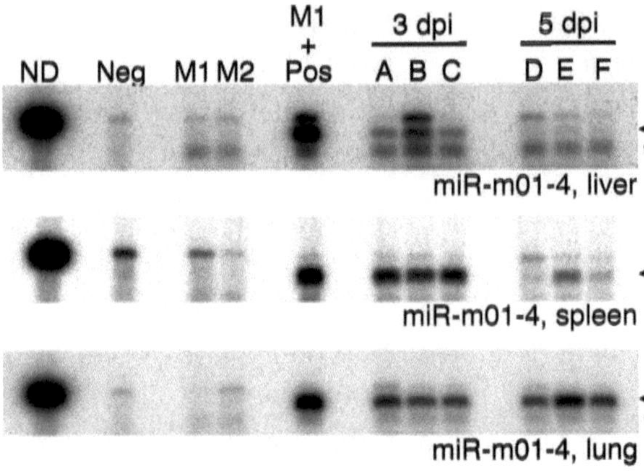

Fig. 3. Detection of viral microRNA-m01-4 in liver, spleen, and lung tissue samples derived from infected animals. *ND* nondigested radioactive DNA probe. *Neg* negative control with S1 nuclease digest of radioactive probe in presence of tRNA alone; *M1, M2* S1 nuclease digest of radioactive probe in presence of RNA derived from uninfected mice. *M1+Pos* RNA from mCMV-infected cell culture serves as positive control and is added to extracts from uninfected mouse tissue; *3 dpi* detection of viral microRNA in tissue 3 days after infection; *5 dpi* detection of viral microRNA in tissue 5 days after infection. The *arrow* indicates the correct size of protected DNA fragment after digestion in the presence of viral microRNA-m01-4 and separation on gel. Residual full-length DNA probe is detectable in several samples and appears above the protected fragment. Below the specific fragment, some nonspecific signal of short fragments is seen in liver samples from uninfected and infected mice (indicated by an *asterisk*). Adapted from (4) with permission of the American Society for Microbiology. © 2007, American Society for Microbiology.

4. Notes

1. The DNA oligonucleotide has to be of HPLC-grade purity. The first 20–22 nucleotides are fully complementary to viral miRNA target sequence and length. After hybridization, a DNA–RNA duplex with a blunt end at the site of the radioactive label is formed. Sometimes the 3′-end of a miRNA is prone to degradation by exonucleases during the isolation process and their terminal nucleotide is removed. As a consequence, DNA probes covering the whole miRNA sequence forms a duplex with the shortened target. This results in an unprotected, unpaired nucleotide carrying the radioactive label at the 5′ overhang. This free nucleotide of the hybridized probe is quickly removed by the single-strand specific S1 nuclease, resulting in loss of radioactive label. This effect is minimized using a DNA probe that is not covering the whole miRNA target sequence. A single or double nucleotide overhang of miRNA at its 3′-end results in higher signals without compromising the specificity of the probe (Fig. 1). The DNA

oligonucleotide has a noncomplementary 4–7 nucleotides pyrimidine overhang as an optimal substrate for S1 nuclease. The removal of the overhang pyrimidines is clearly visible during gel analysis as the resulting fragment migrates faster. Traces of undigested probe in the same lane are an indicator for suboptimal digestion caused by sample impurities, low nuclease concentration, or inadequate digestion times.

2. Avoid repeated freeze–thaw cycles for kinase buffer, enzyme, and radioactive γ-[^{32}P] ATP. The kinase buffer contains DDT to prevent oxidation of the enzyme. Vortexing of buffer and enzyme promotes oxidation and results in lower yield of radioactive label. The γ-[^{32}P] ATP is subjected to both radioactive decay and chemical degradation by radicals. In our hands, the labeling of the oligonucleotide works best if the γ-[^{32}P] ATP is used within 1 week of its shipment date.
3. Careful preparation of the spin column is critical. Low labeling yields are observed for probes that get stuck in a badly prepared sephadex resin.
4. Typical yield for a good DNA probe is 1,000,000–2,000,000 cpm/μl.
5. Labeled probe can be stored at −20°C and used for 3 working days. However, best performance is obtained for radioactive probes prepared the same day as hybridization.
6. Procedures to isolate small RNA are described elsewhere in this book. In Chap. 9, McClure et al. identified microRNA with a Northern Blotting protocol. In Chap. 11, Akusjarvi et al. presented a method to isolate small RNA after immunoprecipitation of RNA-induced silencing complex (RISC). Both described isolation methods results in high-quality RNA samples that are suitable for our S1 Nuclease procedure. In Dolken et al. (4), we pulverized tissue samples and used a method based on a combination of phenol and solid-phase extraction in minicolumns. This enables the separation of long and short RNA. Commercial suppliers offer this extraction method as ready-to-use kits with detailed manuals. The final eluate is enriched for short RNA. With these short RNA enriched preparations instead of total RNA, we observed better signals, lower background, and enhanced sensitivity for certain miRNAs in S1 Nuclease protection assays. Although this procedure seems rather expensive, the gain of time and improved data quality make it an interesting option. For the preparation of RNA samples, we lyse and process 20 mg of animal tissue or 500,000 cells in culture. The RNA yields are sufficient for 3–6 S1 nuclease protection assays, depending on miRNA abundance in the sample. We recommend doing the first hybridization for new miRNA targets with 4 μg of

enriched short RNA per reaction. Depending on the obtained signal, the RNA input can be reduced or increased in subsequent assays.

7. For quantification experiments, perform the S1 nuclease protection assay with a titration of a synthetic miRNA oligonucleotide of at least four different concentrations. This allows the generation of a standard curve and subsequent determination of miRNA concentration in cell culture or tissue samples.
8. Always prepare negative controls to determine background. First, digest the hybridization buffer with only tRNA and radioactive probe to control for assay performance. Second, include miRNA-free cell culture or tissue samples to see the background of your reaction. In case that there is too much background, adjust hybridization and digestion conditions (see Notes 9 and 10; Fig. 2).
9. After the heat denaturation of the hybridization reaction, tubes can be placed in a tube incubator to adjust temperatures between 37 and 50°C. Make sure that all tubes are air tight to avoid evaporation. Adjusting the hybridization conditions helps to reduce background and to increase probe specificity. It is recommended to extend the hybridization time to 4 h or even overnight if the chosen temperatures are higher than 42°C. This is necessary to reach the equilibrium between unbound probe and probe hybridized to miRNA. Hybridization temperatures higher than 50°C are not recommended, because the short DNA–miRNA hybrids may have melting temperatures between 60 and 80°C. Thus, the hybridization process may remain incomplete if the chosen temperature is too high, keeping DNA probe and miRNA target in a dissociated state.
10. Optimize the S1 nuclease digest by lowering or increasing the enzyme concentration between 50 and 200 U/reaction. An alternate approach to optimize the S1 nuclease digest is by changing the incubation temperature. The S1 nuclease is active at temperatures between 30 and 60°C. Increasing the temperature to 50°C, for example, can help to optimize the signal to background ratio of protected probe. An increase of band intensity for digested, but protected DNA fragment during gel analysis is a good indicator for effective improvement. The same is true for disappearance of any undigested probe or nonspecific signal appearing on the gel. In cases where the intensity of digested probe is fading, too much enzyme might be used or the temperature during digestion is too high. Usually, an optimization step is only needed for a new probe sequence or for S1 nuclease enzyme from a new supplier.

Once the standard conditions are defined, the protocol is quite robust and does not need frequent adjustments.

11. The acrylamide concentration and the run time of gel electrophoresis has to be chosen carefully such that a clear difference in size between undigested probe and digested probe is visible after analysis. The conditions in this chapters were successfully used for the detection of mouse cytomegalovirus miRNAs, endogenous mouse liver miRNAs and worked well for detection of siRNA in animal tissue (4–6).

12. An actual S1 nuclease protection assay is shown in Fig. 3. The negative control and two tissue samples of uninfected animals do not show any sign of viral miRNA. In contrast, a positive control and the tissue RNA preparations of six virus-infected animals give a clear signal for the viral miRNA of interest. The DNA oligonucleotide used here is not completely digested and runs above the miRNA-specific fragment. This is no problem for a qualitative or semiquantitative analysis. However, the digestion of the probe must be complete to permit quantitative analysis. Incomplete digestion of free probe is an indication for nonequilibrium condition. In this case, the observed signal intensity of a protected radioactive fragment does not represent the true amount of miRNA target in the hybridization and digestion reaction. Further optimization of digestion conditions is therefore needed to overcome the problem (see Notes 9 and 10). On the other hand, the appearance of nonspecific fragments is not problematic if they run with a different mobility from the specific protected fragment. In this example, a nonspecific band runs below the band corresponding to viral microRNA-m01-4, and it does, therefore, not interfere with detection of the miRNA.

Acknowledgments

We thank our many collaborators and the laboratory colleagues at IBMC and Roche Kulmbach GmbH who contributed to the studies described in this chapter. S.P. is supported by CNRS, Agence Nationale pour la Recherche and Institut National du Cancer

References

1. Griffiths-Jones, S., Saini, H. K., van Dongen, S., and Enright, A. J. (2008) miRBase: tools for microRNA genomics *Nucleic Acids Res.* ***36***, D154–D158.
2. Cullen, B. R. (2006) Viruses and microRNAs *Nat Genet.* ***38*** *Suppl 1*, S25–S30.
3. Pfeffer, S. (2008) Viral miRNAs: tiny but mighty helpers for large and small DNA viruses *Future Virol.* *3*, 291–298.
4. Dolken, L., Perot, J., Cognat, V., Alioua, A., John, M., Soutschek, J., Ruzsics, Z., Koszinowski, U., Voinnet, O., and Pfeffer, S.

(2007) Mouse cytomegalovirus microRNAs dominate the cellular small RNA profile during lytic infection and show features of post-transcriptional regulation *J Virol.* **81**, 13771–13782.

5. Soutschek, J., Akinc, A., Bramlage, B., Charisse, K., Constien, R., Donoghue, M., Elbashir, S., Geick, A., Hadwiger, P., Harborth, J., John, M., Kesavan, V., Lavine, G., Pandey, R. K., Racie, T., Rajeev, K. G., Rohl, I., Toudjarska, I., Wang, G., Wuschko, S., Bumcrot, D., Koteliansky, V., Limmer, S., Manoharan, M., and Vornlocher, H. P. (2004) Therapeutic silencing of an endogenous gene by systemic administration of modified siRNAs *Nature* **432**, 173–178.
6. John, M., Constien, R., Akinc, A., Goldberg, M., Moon, Y. A., Spranger, M., Hadwiger, P., Soutschek, J., Vornlocher, H. P., Manoharan, M., Stoffel, M., Langer, R., Anderson, D. G., Horton, J. D., Koteliansky, V., and Bumcrot, D. (2007) Effective RNAi-mediated gene silencing without interruption of the endogenous microRNA pathway *Nature* **449**, 745–747.

Chapter 11

Characterization of RISC-Associated Adenoviral Small RNAs

Ning Xu and Göran Akusjärvi

Abstract

RNA interference (RNAi) plays novel roles in both host antiviral defense and viral replication. It has been shown that some viruses can exploit the RNAi machinery for their own benefit by encoding for their own viral small RNAs. Here we present a collection of methods to study adenoviral small RNAs, specifically a method for immunopurification of RNA-induced silencing complex (RISC) and a biochemical assay for the activity of purified RISC associated with adenoviral small RNAs.

Key words: Adenovirus, RNA interference, RISC, Ago2, VA RNA, Viral small RNA, 293-Ago2 Cell line, S15 Extract, Immunoprecipitation

1. Introduction

One of the first biological functions established for RNA interference (RNAi) was as an antiviral defense mechanism in plants. In the early 1990s, plant scientists noticed that when they overexpressed certain plant genes from recombinant viral vectors, the corresponding mRNA was degraded rather than overexpressed, a phenomenon that was termed virus-induced gene silencing (VIGS) (1, 2). Today, we know that the RNAi response is equivalent to an immune system in plants and insects, whereas it is still debated whether RNAi naturally limits viral infections in vertebrates. Although our knowledge about the role of RNAi in virus–host interactions is still limited, emerging evidence suggests that RNAi is important for both virus replication and host antiviral defense (3). Several viruses, such as members from the herpesvirus family, simian polyomaviruses and human adenovirus, utilize this novel host cell gene regulatory mechanism to encode for

Ronald P. van Rij (ed.), *Antiviral RNAi: Concepts, Methods, and Applications*, Methods in Molecular Biology, vol. 721,
DOI 10.1007/978-1-61779-037-9_11, © Springer Science+Business Media, LLC 2011

their own subsets of small RNAs. These virus-encoded small RNAs have, in some cases, been shown to modulate viral and cellular gene expression and play important roles in both latent and productive infections (reviewed in (3)).

We have previously shown that human adenovirus (Ad) infection suppresses the activity of the two key enzyme systems in the RNAi pathway, Dicer and RNA-induced silencing complex (RISC). Human adenovirus type 5 (Ad5) encodes for two abundant 160-nucleotide-long noncoding hairpin RNAs, the virus-associated RNA (VA RNA) I and II. Both VA RNAs are processed to small RNAs (so-called mivaRNAs) that are efficiently incorporated into active RISC complex (4–6). We have established methods to isolate and functionally and structurally characterize RISC complexes from adenovirus-infected cells. For these experiments, we have established a 293-Ago2 cell line, which stably overexpresses a FLAG/HA-epitope tagged version of the Argonaute-2 protein (Ago2, the core protein of the RISC complex). By immunopurification of RISC complexes formed in Ad5 infected 293-Ago2 cells, the activity of viral small RNAs associated with RISC could be characterized in vitro (see Subheading 3.6). In addition, the small RNAs can be extracted from the immunopurified RISC and subjected to cloning and sequence analysis (see Chapter 6) or characterization by Northern blot analysis (see Chapter 9). By using these strategies, we not only identified a novel set of adenoviral small RNAs, but also characterized some of their biological activities. The in vitro RISC activity assays described here can also be used for the analysis of the biological activity of exogenous-introduced siRNAs or endogenous miRNAs by suitable design of substrate RNAs.

2. Materials

2.1. Establishment of the 293-Ago2 Cell Line

1. Cell culture medium: Dulbecco's modified Eagles medium (DMEM) supplemented with 10% newborn calf serum (NCS), 100 U/mL penicillin, and 100 μg/mL streptomycin.
2. Solution of trypsin (0.25%) in 1 mM ethylenediamine tetraacetic acid (EDTA).
3. G418 sulfate. Dissolve at 50 mg/mL in 100 mM *N*-2-hydroxyethylpiperrazine-*N*′-2-ethanesulfonate (HEPES), pH 7.3. Stable for 1 year at 2–8°C.
4. Fugene 6 reagent (Roche).
5. Plasmid pIRESneo-FLAG/HA Ago2 (Addgene, Cambridge, MA).

2.2. Infection of Adherent Cells with Adenovirus

1. Cell culture materials (see Subheading 2.1, items 1–4).
2. CsCl-purified adenovirus stock (wild-type or mutant) with a titer preferably higher than 10^{10} fluorescent focus-forming unit (FFU)/mL (7).

2.3. Preparation of S15 Extracts

1. Cell-lifters.
2. 1 mL Syringe with 23-gauge needle.
3. Phosphate-buffered saline (PBS): 20 mM potassium phosphate, pH 7.4, 130 mM NaCl. Prepare PBS in autoclaved double-distilled H_2O (ddH_2O), store at +4°C.
4. Buffer A (hypotonic): 10 mM HEPES, pH 7.9, 10 mM KCl, 1.5 mM $MgCl_2$, 1 mM dithiotreitol (DTT). Make buffer A without DTT in ddH_2O, and filter through a 0.2 μm membrane. Add the DTT from a 1 M stock solution (stored at −20°C) just before use. Keep on ice.
5. Buffer A+50% glycerol: Add 500 μL of 87% glycerol to 370 μL Buffer A. Although this practice results in a dilution of salts, it does not affect the performance of the extracts.

2.4. Immunoprecipitation of RISC

1. Anti-FLAG M2 affinity agarose gel (Sigma-Aldrich).
2. NET-1 buffer: 50 mM Tris–HCl, pH 7.5, 150 mM NaCl, 2.5% Tween 20.
3. Buffer A without DTT: 10 mM HEPES, pH 7.9, 10 mM KCl, 1.5 mM $MgCl_2$.
4. Buffer A+50% glycerol (see Subheading 2.3, item 5).
5. A roller mixer at 4°C.

2.5. Preparation of Radioactively Labeled Substrate RNA for the RISC Assay

1. Template DNA dissolved in RNase-free water (>40 ng/μL) (see Note 1).
2. 10× Nucleotide mix: 5 mM ATP, 5 mM UTP, 1 mM GTP, 1 mM CTP (see Note 2).
3. 10 mM m^7GpppG-cap nucleotide (Amersham Biosciences, 27-4635): Dissolve 5 A_{250} units of lyophilized m^7GpppG in 24 μL autoclaved ddH_2O, store at −20°C.
4. 300 mM DTT: in ddH_2O, store at −20°C.
5. T7 RNA polymerase (50 U/μL) and 10× reaction buffer provided by the supplier.
6. Porcine RNase Inhibitor (Amersham Biosciences): 39 U/μL.
7. α-^{32}P-CTP (Perkin-Elmer): 800 Ci/mmol, 20 μCi/μL.
8. RQ1 DNase (Promega): 1 U/μL.
9. Phenol: Phenol–chloroform–isoamylalcohol (25:24:1).

10. Sample loading buffer: 80% formamide (p.a. grade), 50 mM Tris–HCl, pH 7.9, 10 mM EDTA, 0.025% bromophenol blue, 0.025% xylene cyanol.
11. 10× TBE stock solution: 108 g Tris base, 55 g boric acid, 7.44 g EDTA, add deionized H_2O to 1 L and autoclave. Use 1× TBE in gel and 0.5× TBE as the running buffer.
12. Tetramethylethylenediamine (TEMED).
13. 10% APS: for 10 mL dissolve 1 g ammonium persulfate in 10 mL ddH_2O. Store in 1 mL aliquots at −20°C. Keep a working solution at 4°C no longer than 1 month.
14. Preparative 4% acrylamide 8 M urea gel (acrylamide:bis-acrylamide = 29:1) in 1× TBE buffer. Add 25 μL TEMED and 250 μL 10% APS to every 50 mL gel just before casting the gel. Use a wide comb (20 mm) so that the entire sample can be run in one well in a midi gel.
15. RNA elution buffer: 0.75 M NH_4-acetate, 10 mM Mg-acetate, 0.1 mM EDTA, 0.1% (w/v) sodium dodecyl sulfate (SDS). Store aliquots at −20°C and prewarm buffer at 37°C prior to use.
16. Ethanol: 99.9 and 70% (diluted from 99.9% with autoclaved ddH_2O).

2.6. In Vitro RISC Assay

1. RISC 4× A buffer: 60 mM HEPES-HCl, pH 7.4, 1 mM $MgCl_2$, 180 mM KCl, 2 mM DTT.
2. 100 mM ATP. Store in aliquots at −20°C.
3. 20 mM GTP. Store in aliquots at −20°C.
4. Creatine phosphate: Dissolve at 0.5 M in autoclaved ddH_2O. Store in 100 μL aliquots at −20°C.
5. Creatine phosphokinase: Dissolve at 1.2 mg/mL in autoclaved ddH_2O. It is important to prepare a fresh solution immediately before use.
6. Porcine RNase inhibitor (Amersham Biosciences): 39 U/μL.
7. 0.5 M EDTA.
8. Proteinase K buffer: 100 mM Tris–HCl, pH 7.5, 150 mM NaCl, 12.5 mM EDTA, 1% SDS. Prepare buffer in autoclaved ddH_2O and store at room temperature.
9. 20 mg/mL Proteinase K: Dissolve 20 mg lyophilized proteinase K in 1 mL of buffer: 50 mM Tris–HCl, pH 7.5, 10 mM $CaCl_2$. Store in small (e.g., 30 μL) aliquots at −20°C. Do not freeze–thaw more than five times.
10. 10 mg/mL Yeast tRNA. Dissolve in autoclaved ddH_2O and store in aliquots at −20°C.
11. Sample loading buffer (see Subheading 2.5, item 10).

12. Phenol: Phenol–chloroform–isoamylalcohol (25:24:1).
13. Chloroform–isoamylalcohol (24:1).
14. 99.9% Ethanol.
15. 8% Acrylamide 8 M urea gel (acrylamide:bis-acrylamide = 19:1) in 1× TBE. Add 25 μL TEMED and 250 μL 10% APS to every 50 mL gel solution just before casting the gel.

3. Methods

The method section summarizes the following methods:

1. Establishment of the 293-Ago2 cell line.
2. Infection of adherent cells with Adenovirus.
3. Preparation of S15 extracts from Ad-infected cells.
4. Immunopurification of RISC.
5. Strategies to synthesize radioactively labeled substrate RNA for RISC assay.
6. In vitro RISC activity assay.

3.1. Establishment of the 293-Ago2 Cell Line

The protocol describes establishment of an adherent cell line, which constitutively expresses a FLAG/HA-tagged Ago2 protein. We have used this protocol to construct 293, HeLa, and C33A cells stably overexpressing the FLAG/HA-tagged Ago2 protein.

1. Grow 293 cells on 6-cm culture plates in cell culture medium at 37°C in 5% CO_2.
2. Transfect cells with 1 μg plasmid pIRESneo-FLAG/HA Ago2 (Addgene, Cambridge, MA) using the Fugene 6 reagent as described in the protocol supplied by the manufacturer. This plasmid contains a neomycin (G418) resistance marker and codes for an amino-terminal FLAG/HA-tagged human Ago2 protein (8).
3. At 48 h posttransfection, split cells into 10-cm culture plates containing the same medium (step 1) supplemented with 400 μg/mL of G418 (see Note 3). Dilute cells such that they are not more than 25% confluent on the 10-cm culture plate at the start of the G418 selection. Change to new selection medium every 3–4 days.
4. When all 293 cells in the untransfected control plate are dead, isolate single surviving colonies of 293-Ago2 cells growing in the transfected plates by using a Gilson pipette with a sterile tip on an inverted light microscope. Transfer the clone to a 24-well plate containing prewarmed cell culture medium with G418 for further expansion of the clone.

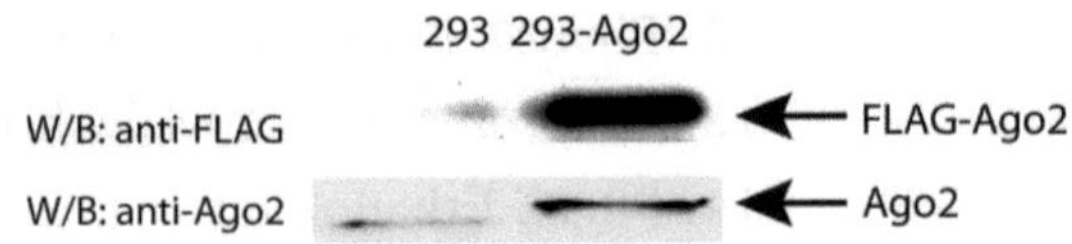

Fig. 1. Western blot analysis (W/B) of Ago2 expression in the selected 293-Ago2 cell clone and the parental 293 cell line. The membrane was detected by anti-FLAG antibody (*upper panel*) and anti-Ago2 antibody 8C7 (*lower panel*).

5. Identify FLAG/HA Ago2 expressing stable clones by Western blot analysis using the FLAG M2 monoclonal antibody.
6. Determine the level of FLAG/HA-Ago2 overexpression in individual stable clones by Western blotting using an antibody detecting both the endogenously expressed Ago2 protein and the FLAG/HA-tagged Ago2 protein (see Note 4).
7. Select a clone with a high and stable Ago2 protein expression for subsequent experiments. Avoid stable cells with an artificially too high level of Ago2 overexpression (see Note 5). In our experiment, we selected a clone showing an approximately sevenfold Ago2 overexpression (Fig. 1).
8. Grow 293-Ago2 cell line under the same condition as non-transfected 293 cells (step 1). The cells are reselected every 2–3 months by adding 400 μg/mL G418 into the culture medium. Keeping track of passage numbers and periodically freezing down cells are recommended. Work with early passage cells is recommended.

3.2. Infection of Adherent Cells with Adenovirus

1. Split 293-Ago2 cells the day before infection and seed in five 10-cm plates such that the monolayer is between 60 and 80% confluent at the time of infection. We usually infect four 10-cm 293-Ago2 plates and use an extra plate of cells for counting the number of cells on the day of infection (see Note 6).
2. Count number of cells in the control plate and calculate the amount of virus needed to obtain a multiplicity of infection of 100 FFU/cell.
3. Thaw virus stock on ice. Mix stock briefly by tapping the tube (do not vortex). Dilute virus in 3 mL of DMEM containing 2% NCS/10 cm plate. Quick-freeze the remaining virus stock in a dry ice-ethanol bath.
4. Remove culture medium.
5. Add 3 mL virus inoculum per plate.
6. Incubate for 60 min in a CO_2 incubator (37°C, 5% CO_2).
7. Remove and discard virus inoculum using an established and safe procedure.
8. To each plate, add 10 mL of DMEM containing 10% NCS.

9. Incubate cells in a CO_2 incubator (37°C, 5% CO_2).
10. Harvest at 20–24 h postinfection.

3.3. S15 Extract Preparation

The protocol for preparing cytoplasmic extracts from small volumes of cells is adapted from a protocol for preparing extracts from *Drosophila* S2 cells (9, 10). Recovering sufficient amounts of extracts requires careful handling and preferably a packed cell volume (PCV) of at least 50 µL. The protocol below is designed for four 10-cm plates of 293-Ago2 cells collected at about 20–24 h postinfection.

1. Carefully remove the growth medium by gentle aspiration. If cells are loosely attached to the plate, it is better to resuspend all cells in the growth medium with a cell-lifter and collect cells by centrifugation 5 min at 2,000 × *g* in a 50-mL plastic Falcon tube.
2. Add 1 mL ice-cold PBS per plate and resuspend cells with a cell-lifter or by gently pipetting up and down with a Gilson pipette. Transfer to four 1.5-mL Eppendorf tubes and collect cells by centrifugation at 3,000 × *g* for 2 min.
3. Remove the supernatant by gentle aspiration.
4. Resuspend the pellets in the four tubes in a total volume of 1.5 mL ice-cold PBS and combine them into a clean 2 mL Eppendorf tube, centrifugate at 3,000 × *g* for 2 min, and remove supernatant.
5. Estimate the PCV (see Note 7). Adjust the volume to 6× PCV with ice-cold Buffer A.
6. Allow cells to swell on ice for 15 min.
7. Disrupt cells by passing the solution through a 23-gauge syringe needle. Press the needle opening to the surface of the tube. Approximately 30–40 strokes are required. Be careful not to introduce bubbles. To check the efficiency of disruption, place 5 µL of the solution on a plate, add 2 µL 0.4% trypan blue, and check under the microscope. Nuclei stain blue, intact cells do not stain. Continue until more than 90% of the cells are disrupted.
8. Spin down the nuclei at 7,000 × *g* for 5 min at 4°C.
9. Transfer the supernatant (cytoplasmic extract) to a new tube. Add 0.1 volume of Buffer A + 50% glycerol.
10. Clear the supernatant by centrifugation at 15,000 × *g* for 60 min at 4°C (see Note 8).
11. Aliquot the S15 extract into Eppendorf tubes. Quick-freeze the extracts in liquid nitrogen, and store at −70°C. The extracts can be thawed three times without losing activity if they are refrozen in liquid nitrogen. A useful scheme for

aliquoting is to prepare two tubes with 30 μL extract to be used for the RISC assay (see Subheading 3.6), one tube with 5 μL extract to be used for determining the protein concentration and one tube containing the rest of the extract that will be used for immunopurification of RISC (see Subheading 3.4).

12. Assay the protein concentration of the S15 extract. The protein concentration is typically 6–8 μg/μL.

3.4. Immunopurification of RISC

1. Calculate the amounts of anti-FLAG M2 affinity gel needed for the immunoprecipitation reaction. We use 40 μL gel slurry (equal to 20 μL packed resin, see Note 9) for S15 extract prepared from four 10-cm plates.
2. Mix the bottle containing anti-FLAG M2 agarose bed slurry thoroughly by vortexing. Immediately transfer the required amount of gel suspension into a fresh 1.5-mL Eppendorf tubes (see Note 10).
3. Centrifuge the resin for approximately 5 s at 10,600 × *g*. Remove the supernatant carefully; avoid discarding any resin (see Note 11).
4. Wash the beads twice with ice-cold NET-1 buffer at a volume equal to 20 times the packed resin volume (for example, 400 μL NET-1 buffer for 20 μL packed resin), repeat step 3.
5. Add the S15 extract containing the same amount of protein (see Subheading 3.4, steps 12 and 13) to the washed resin. Mix samples gently on a roller mixer at 4°C for 2 h.
6. Centrifuge the tubes for 5 s at 10,600 × *g*. Discard the supernatant, or transfer to a fresh tube for an optional control experiment (see Note 12).
7. Wash the resin three times with ice-cold NET-1 buffer and once with buffer A (without DTT) as described in step 3. After the last wash, remove all washing buffer carefully (see Notes 11 and 12).
8. Resuspend the washed resin in 20 μL buffer A + 50% glycerol, quick-freeze in liquid nitrogen, and store at −70°C for use in RISC assay (see Subheading 3.6) or isolation of RISC-associated RNA (see Note 13).

3.5. Preparation of Radioactively Labeled Substrate RNA for the RISC Assay

In our assay system, we use capped transcripts uniformly labeled with α-^{32}P-CTP. It is essential to use a capped transcript when S15 extracts are used, since the 3′ fragment produced after RISC cleavage often is rapidly degraded. However, in assays using immunopurified RISC, both the 5′ and 3′ fragments are stable and can usually be detected on the gel (Fig. 2). Using suitable PCR primers template, DNA can be produced that allows transcription of substrate RNAs that match essentially any

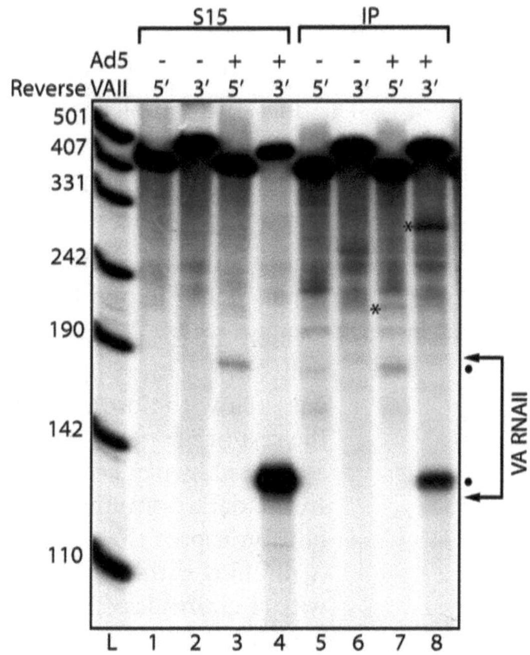

Fig. 2. In vitro RNA-induced silencing complex (RISC) activity assay. S15 cytoplasmic extracts and immunopurified RISC (IP) from uninfected 293-Ago2 cells or cells infected with Ad5 (+) were assayed for RISC activity against the synthetic transcripts with target regions complementary to the 5′ or 3′ half of VA RNAII (Reverse VAII 5′ or 3′) (4, 11). *Arrows* indicate the span of the virus-associated RNA (VA RNA) target regions in respective transcript. The positions of the 5′ end cleavage products generated by Dicer cleavage at the terminal stem in VA RNAII are indicated by a dot. Bands labeled with *asterisk* indicate the 3′ end cleavage product of the substrate RNA, which usually is degraded in S15 cytoplasmic extracts, but is often seen in reactions using immunopurified RISC. *L* DNA size ladder.

sequence of interest. In our assays, target RNAs for the RISC reaction are typically synthetic RNAs containing sequences complementary to the 5′ or 3′ halves of the adenovirus VA RNAs ((11); see Note 1).

1. Combine in an Eppendorf tube at room temperature (see Note 14):
   (a) 1.25 μL 10× T7 RNA polymerase buffer.
   (b) 4.75 μL Template DNA (200–500 ng in RNase-free water).
   (c) 1.25 μL 10× Nucleotide mix.
   (d) 1.25 μL 10 mM mGpppG-cap nucleotide.
   (e) 0.5 μL 300 mM DTT.
   (f) 2.5 μL α-^{32}P-CTP (50 μCi).
   (g) 0.5 μL Porcine RNase inhibitor.
   (h) 0.5 μL T7 RNA polymerase.

2. Incubate for 2 h at 37°C.
3. Add 1 μL RQ1 DNase, incubate for 30 min at 37°C.
4. Add 13 μL sample loading buffer. Store at −20°C or continue directly.
5. Heat samples at 100°C for 5 min, chill on ice, and load on a preparative 4% acrylamide 8 M urea gel (see Note 15).
6. Run gel in 0.5× TBE at 50 W until the bromophenol blue dye reaches the middle of the gel (see Note 16).
7. Detach the gel from the electrophoresis equipment and carefully remove the upper glass plate. Discard the lower part of gel [below the bromophenol blue dye (see Note 17)]. Wrap the rest of the gel in a plastic film (Saran wrap) and cover with a clean glass plate. Move the whole set-up to the darkroom. In the darkroom, remove the clean glass plate, place an X-ray film on top of the Saran wrap, and mark the position of the X-ray film with a waterproof pen on the Saran wrap. Cover with a clean glass plate and expose for 5 min at room temperature. Develop the X-ray film. For your own safety, work behind a Plexiglas shield.
8. Copy the position of the band corresponding to your full-length transcript from the autoradiogram back to the gel by cutting out the band on the X-ray film, which opens a window in the film. Align the film with gel using the marker pen positions and label the position of the product band on the Saran wrap through the window on the film. Cut out the full-length band on the gel with a disposable scalpel or razor blade, cut the gel slice into smaller pieces (volume 100–150 μL), and put them in an Eppendorf tube containing 500 μL RNA elution buffer.
9. Elute the transcript on a shaker at +37°C for 2 h.
10. Remove insoluble gel debris by spinning the tube at 16,000 × *g* for 5 min at room temperature and transfer the supernatant to a fresh tube. Check with a hand monitor that roughly 80% of the radioactivity is in the supernatant. If necessary, the remaining gel pieces can be eluted by adding a second 500 μL aliquot of RNA elution buffer and repeating step 9.
11. Add an equal volume of Phenol to the supernatant, vortex vigorously for 1 min, and spin at 16,000 × *g* at room temperature for 5 min. Carefully transfer the upper phase to a fresh tube and add 2.5–3 volumes of 99.9% ethanol. Precipitate the RNA at overnight −20°C.
12. Spin tube at 16,000 × *g* for 1 h at 4°C and remove the ethanol. Wash the pellet with 500 μL 70% ethanol, spin at 16,000 × *g* for 15 min, remove ethanol, and air-dry the pellet for 5–10 min at room temperature.
13. Dissolve in 30–50 μL RNase-free water. Store at −70°C.

3.6. In Vitro RISC Activity Assay

1. Prepare a reaction master-mix on ice (see Note 18). One RISC reaction should contain the following components:
   (a) 2.5 μL RISC 4× A buffer (see Note 19).
   (b) 0.1 μL 100 mM ATP.
   (c) 0.1 μL 20 mM GTP.
   (d) 0.5 μL 500 mM Creatine phosphate.
   (e) 0.25 μL Creatine kinase (freshly dissolved).
   (f) 0.05 μL Porcine RNase inhibitor.
   (g) 50,000 dpm Substrate RNA (see Note 20).
   (h) Add RNase-free water to a final volume of 5 μL.
2. Thaw S15 extract (see Subheading 3.3) and/or immunopurified RISC complexes on beads (see Subheading 3.4, step 8). Place on ice.
3. Add 5 μL of master-mix into each of reaction tube. Place the tubes at room temperature.
4. Resuspend the RISC-bound beads in buffer A by tapping the tube. Immediately transfer 5 μL of beads or 5 μL S15 extract to the reaction tube containing the master-mix. Mix by gently pipetting up and down 20 times. Be careful not to introduce any bubbles. Do not vortex (see Note 21).
5. Incubate in a cabinet incubator at 30°C for 2 h (see Note 22).
6. Add 1 μL 0.5 M EDTA to terminate reaction. Chill on ice.
7. Prepare Proteinase K mix. For one reaction, mix the following components:
   (a) 400 μL Proteinase K buffer
   (b) 2 μL 20 mg/mL Proteinase K
   (c) 1 μL 10 mg/mL tRNA.

   Add 400 μL Proteinase K mix to each reaction and incubate at 65°C for 45 min
8. Centrifuge at 16,000×*g* for 5 min at room temperature to pellet any insoluble material.
9. Transfer the supernatant to a new Eppendorf tube, extract with 400 μL phenol/chloroform/isoamyl alcohol (25:24:1), and transfer upper phase (380 μL) into a new tube (see Subheading 3.5, step 11).
10. Repeat extraction with 380 μL chloroform/isoamyl alcohol, transfer upper phase (360 μL) into a new tube (see Note 23).
11. Add 3 volumes of cold 99.9% ethanol, precipitate at −20°C overnight.
12. Pellet RNA by centrifugation at 16,000×*g* for 1 h at 4°C.

13. Remove supernatant, air-dry the pellet at room temperature for 1–5 min, and dissolve in 10–20 μL of sample loading buffer.
14. Heat samples at 100°C for 5 min, chill on ice, and collect all material at the bottom of the tube by a short centrifugation.
15. Load samples on an 8% acrylamide 8 M urea gel (see Note 24).
16. Run the gel until desired separation is obtained and dry the gel on a gel dryer.
17. Expose the gel to an X-ray film and/or expose it on a PhosphorImager screen for subsequent quantification of the results (see Fig. 2 for an example).

4. Notes

1. To generate target RNAs for mivaRNAs, the VA RNA genes were separated near the apical loop into two halves and cloned into pGL4-luciferase reporter plasmids in the reverse orientation (11). We used a forward primer with a 5′-extension encoding a minimal T7 promoter (ATATATTAATACGACT CACTATA**G**; bold nucleotide = transcription start site) and a reverse primer to generate DNA template by PCR amplification. The template DNA is purified (QIAquick PCR purification kit, QIAGEN) and used for the in vitro transcription reaction. Alternative templates can be designed to virtually any sequence by PCR amplification using one primer containing the 5′ extension encoding a minimal T7 promoter sequence. It is important that the DNA is RNase-free and without inhibitory contaminants. The length of the targets used in our study are 400–500 nucleotides and the cleavage products are 100–300 nucleotides, which can be well-separated in the 8% acrylamide 8 M urea gel. The percentage of the gel can be adjusted according to the length of RNA that you need to separate.
2. The nucleotide mix contains a lower concentration of GTP to favor incorporation of the synthetic m^7GpppG-cap nucleotide. The concentration of CTP is also lower, since α-^{32}P-CTP is used to label the transcript. The nucleotide mix stock solution can be stored in small aliquots at −20°C. Avoid repeated freeze–thawing of the stock.
3. G418 is an aminoglycoside, which blocks mammalian protein synthesis by interfering with ribosomal function. Expression of the bacterial APH gene in mammalian cells can detoxificate G418, which results in the resistance to G418 (12). The optimal killing concentration for G418 should be titrated prior to

the transfection and varies with the cell type. In an initial experiment, we treat the cells with several concentrations ranging from 100 μg/mL to 1 mg/mL. The concentration that kills the cells after 1 week of treatment is used as the selection concentration. We use a selection concentration of 400 μg/mL G418 for 293 cells.

4. To detect both the endogenously expressed and the FLAG/HA-tagged Ago2 proteins, we used antibody 8C7, which was kindly provided by G. Dreyfuss. Today, several anti-Ago2 antibodies are commercially available.
5. We avoid using clones with very high Ago2 overexpression (for example, more than ten-times overexpression compared to the 293 cell line), since the artificially high level of Ago2 protein could saturate other components of the RNAi machinery and potentially interfere with other processes in the cell.
6. The protocol can be scaled up or down, but the total cell volume at the time of harvest should be at least 50 μL to simplify the technical handling of the cell pellet during the passage through a syringe with a needle (see Subheading 3.3).
7. A simple trick to estimate the PCV is to resuspend the pellet in a small volume of Buffer A (for example 100 μL) and measure the total volume of the solution using a 200 μL pipette. The PCV is total volume minus 100 μL. Accurate measurement of PCV is important to obtain reproducible extracts.
8. Removes large cellular structures from the extract.
9. 40 μL Anti-FLAG M2 affinity gel slurry is equal to approximately 20 μL packed gel volume. Smaller amounts of resin can be used, since around 10 μL packed gel can bind >1 μg FLAG-tagged protein. However, using too small volume of anti-FLAG M2 affinity resin will cause difficulties with quantitative manipulations in the following steps.
10. For resin transfer, leave the pipette in the vial containing the agarose slurry longer than 30 s to make sure that the correct volume is taken, since the gel slurry is very viscous. Round-bottom tubes (for example, 2 mL Eppendorf tube) are not suited in this step, since it is difficult to observe the small volume of resin pellet in these tubes. If multiple immunoprecipitation reactions are simultaneously performed, we wash the resin needed for all samples together and divide it into each reaction tubes after the washing step.
11. In order to avoid losing resin, we remove the supernatant with a narrow-ended pipette tip, which we make by using a forceps to pinch the opening of the tip. During the wash steps, we usually leave a little bit supernatant above the pellet. However, try to take away all of the buffer after the last wash step.

12. The efficiency of immunoprecipitation (IP) can be estimated by comparing the amount of Ago2 protein in S15 extract before IP, after IP, and on the beads by Western blotting using the anti-FLAG antibody. The proteins bound to beads can be eluted as below:
   (a) Add 20 μL 2× SDS-PAGE Sample buffer (125 mM Tris–HCl, pH 6.8, 4% SDS, 20% (v/v) glycerol, 0.004% bromophenol blue) to each tube containing the washed beads (see Subheading 3.4, step 7).
   (b) Boil samples at 100°C for 3 min.
   (c) Centrifuge at 10,600 × *g* for 5 s at room temperature and load the supernatant, which contains the eluted protein, on the SDS-PAGE gel.
13. The immunopurified RISC-associated RNA (see Subheading 3.4, step 7) can be extracted from the beads and subjected to sequence analysis (see Chapter 6) or detected by Northern blot (see Chapter 9) (11).
   (a) Following Subheading 3.4, step 7, add 600 μL IsoB/NP40 buffer (10 mM Tris–HCl, pH 7.9, 0.15 M NaCl, 1.5 mM $MgCl_2$, 0.5% ethylphenyl-polyethylene glycol (NP40)).
   (b) Add 130 μL 5× RPS buffer (0.5 M Tris, 2.5% SDS, 50 mM EDTA).
   (c) Add 600 μL phenol/chloroform/isoamyl alcohol (25:24:1) to the beads, vortex vigorously at room temperature for 10 min.
   (d) Centrifuge at 16,000 × *g* for 5 min at room temperature. Repeat Phenol extraction once.
   (e) Carefully transfer the upper phase to a fresh tube containing 600 μL chloroform/isoamyl alcohol (24:1). Vortex for 1 min and centrifuge again.
   (f) Precipitate the supernatant overnight at −20°C with 600 μL isopropanol, 30 μL sodium acetate (3 M), and 10 μg Glycogen (Applied Biosciences).
   (g) Pellet RNA by centrifugation at 16,000 × *g* for 1 h at 4°C. Remove supernatant, air-dry the pellet at room temperature for 1–5 min, dissolve in RNase-free water or gel loading buffer.
   (h) Proceed with small RNA cloning and/or Northern blot.
14. Do not combine the in vitro transcription reaction on ice, because spermidine in the enzyme buffer will precipitate at low temperature, which is detrimental for the activity of the spermidine-dependent T7 RNA polymerase. Also, thaw the

10× T7 RNA polymerase buffer at room temperature and be careful to dissolve the precipitate in the buffer before use.

15. It is not practical to run radioactive-labeled DNA or RNA size marker beside the in vitro-transcribed RNA product in the preparative 4% acrylamide gel. In order to obtain the similar level of radioactive intensity for both size marker and the RNA product during exposure, very high amount of size marker needs to be used. If necessary, the size of the transcription product can be verified by separating 0.5 μL of the reaction mix alongside a size marker in a 4% acrylamide 8 M urea mini gel.
16. The 4% polyacrylamide gel is suitable for separating transcripts ranging in size from 200 to 500 nucleotides. For transcripts smaller than 200 nucleotides, higher percentage polyacrylamide gels should be used (6–10%). The gel can be run for longer times if a better separation of the full-length transcript is needed. However, be aware that the free radioactive nucleotides may run out of the gel into the lower buffer chamber and contaminate the electrophoresis equipment.
17. On a 4% denaturing polyacrylamide gel, the migration of the bromophenol blue dye runs with approximately 35 nucleotides (13). Therefore, the part of gel below the bromophenol blue marker contains the unincorporated radioactive nucleotides (about 80% of the input radioactivity). Discarding this part of gel will decrease the radioactivity of the whole gel and facilitate the following steps.
18. For the RISC reaction to be successful, it is important that the reaction master-mix is made fresh immediately before mixing the reaction. The crucial component appears to be the creatine kinase, an enzyme that is needed to provide the RISC reaction with ATP.
19. The final concentrations of components in the RISC reaction mixture, including the salts from the cell extracts, are: 15 mM HEPES-KOH, pH 7.4, 50 mM KCl, 1 mM $MgCl_2$, 1 mM ATP, 0.2 mM GTP, 10 μg/mL RNasin, 30 μg/mL creatine kinase, 25 mM creatine phosphate, 0.5 mM DTT, 2.5% glycerol.
20. A purified transcript can be used for RISC assay up to 2 weeks after synthesis, if stored at −70°C. However, use of fresh transcripts usually results in higher cleavage efficiency and lower background.
21. A negative control can be done by adding 1 μL 0.5 M EDTA at $t=0$, and incubate at 30°C for the same time as other samples.
22. Since the reactions are performed in a small volume (10 μL), it is important that the Eppendorf tubes are incubated in a

cabinet incubator rather than in a water bath to prevent evaporation and condensation of water under the lid.

23. Do not try to recover all of the supernatant during the phenol extraction, because contaminates in the interphase frequently result in a smearing of bands on the gel.
24. The length of the expected cleavage products determines the percentage of the gel. 8% Gel are suitable to resolve products ranging from 100 to 500 nucleotides; longer fragments should be run on a 4 or 6% gel.

Acknowledgments

This work was supported by the Swedish Cancer Society and the Uppsala RNA Research Centre (URRC).

References

1. Kumagai, M. H., Donson, J., della-Cioppa, G., Harvey, D., Hanley, K., and Grill, L. K. (1995) Cytoplasmic inhibition of carotenoid biosynthesis with virus-derived RNA. *Proc Natl Acad Sci U S A.* ***92***, 1679–1683.
2. Ruiz, M. T., Voinnet, O., and Baulcombe, D. C. (1998) Initiation and maintenance of virus-induced gene silencing. *Plant Cell.* ***10***, 937–946.
3. Cullen, B. R. (2009) Viral and cellular messenger RNA targets of viral microRNAs. *Nature.* ***457***, 421–425.
4. Andersson, M. G., Haasnoot, P. C., Xu, N., Berenjian, S., Berkhout, B., and Akusjärvi, G. (2005) Suppression of RNA interference by adenovirus virus-associated RNA. *J Virol.* ***79***, 9556–9565.
5. Aparicio, O., Razquin, N., Zaratiegui, M., Narvaiza, I., and Fortes, P. (2006) Adenovirus virus-associated RNA is processed to functional interfering RNAs involved in virus production. *J Virol.* ***80***, 1376–1384.
6. Sano, M., Kato, Y., and Taira, K. (2006) Sequence-specific interference by small RNAs derived from adenovirus VAI RNA. *FEBS Lett.* ***580***, 1553–1564.
7. Wold, W. S. M., and Tollefson, A. E. (2007) Adenovirus methods and protocols. *Methods in Molecular Medicine* Vol. 131. Humana, Totowa, NJ.
8. Meister, G., Landthaler, M., Patkaniowska, A., Dorsett, Y., Teng, G., and Tuschl, T. (2004) Human Argonaute2 mediates RNA cleavage targeted by miRNAs and siRNAs. *Mol Cell.* ***15***, 185–197.
9. Zamore, P. D., Tuschl, T., Sharp, P. A., and Bartel, D. P. (2000) RNAi: double-stranded RNA directs the ATP-dependent cleavage of mRNA at 21 to 23 nucleotide intervals. *Cell.* ***101***, 25–33.
10. Tuschl, T., Zamore, P. D., Lehmann, R., Bartel, D. P., and Sharp, P. A. (1999) Targeted mRNA degradation by double-stranded RNA in vitro. *Genes Dev.* ***13***, 3191–3197.
11. Xu, N., Segerman, B., Zhou, X., and Akusjärvi, G. (2007) Adenovirus virus-associated RNAII-derived small RNAs are efficiently incorporated into the RNA-induced silencing complex and associate with polyribosomes. *J Virol.* ***81***, 10540–10549.
12. Ausubel, F. M., Brent, R., Kingston, R. E., Moore, D. D., Seidman, J. G., Smith, J. A., and Struhl, K. (Ed.) (1995) *Current Protocols in Molecular Biology.* Wiley, New York.
13. Fritsch, E. F., Sambrook, J., Maniatis, T. (Ed.) (1989) *Molecular Cloning.* Cold Spring Harbor Laboratory Press, Cold Spring Harbor, NY.

Part III

RNAi-Based Antiviral Defense

Chapter 12

Identification of Viral Suppressors of RNAi by a Reporter Assay in *Drosophila* S2 Cell Culture

Koen W.R. van Cleef, Joël T. van Mierlo, Marius van den Beek, and Ronald P. van Rij

Abstract

The RNA interference (RNAi) pathway plays an important role in antiviral immunity in insects. To counteract the RNAi-mediated immune response of their hosts, several insect viruses, such as Flock house virus, Drosophila C virus, and Cricket paralysis virus, encode potent viral suppressors of RNAi (VSRs). Because of the importance of RNAi in antiviral defense in insects, other insect viruses are likely to encode VSRs as well. In this chapter, we describe a detailed protocol for an RNAi reporter assay in *Drosophila* S2 cells for the identification of VSR activity.

Key words: Insect virus, Antiviral immunity, RNAi, Viral suppressor of RNAi (VSR), Reporter assay

1. Introduction

During prolonged coevolution of virus and host, viruses have developed various sophisticated strategies to evade the immune defenses of their hosts. In insects, RNA interference (RNAi) is an important antiviral defense mechanism (reviewed in Chapter 1 and refs. (1, 2)). The RNAi machinery is triggered by viral double-stranded RNA (dsRNA), which is cleaved by Dicer-2 into viral small interfering RNAs (v-siRNAs). The v-siRNAs are incorporated into an RNA-induced silencing complex (RISC) where they guide the recognition and cleavage of complementary viral RNAs by Argonaute-2 (Ago-2) and thereby restrict viral replication. To interfere with the antiviral RNAi defense system, several insect viruses encode potent viral suppressors of RNAi (VSRs). These VSRs include Flock house virus (FHV) B2 (3), Drosophila C virus (DCV) 1A (4), and Cricket paralysis virus (CrPV) 1A (4,

Ronald P. van Rij (ed.), *Antiviral RNAi: Concepts, Methods, and Applications*, Methods in Molecular Biology, vol. 721,
DOI 10.1007/978-1-61779-037-9_12, © Springer Science+Business Media, LLC 2011

5). The VSRs target different steps in the RNAi pathway. For example, both FHV B2 and DCV 1A block Dicer-mediated cleavage of long dsRNA into siRNAs by binding long dsRNA molecules, whereas FHV B2 also sequesters siRNAs to prevent their incorporation into RISC (4, 6–9). In addition to these dsRNA-binding activities, several VSRs interact with components of the RNAi machinery directly. For instance, FHV B2 interacts with Dicer in order to suppress siRNA biogenesis (10), whereas CrPV 1A inhibits RISC activity via an interaction with Ago-2 (11).

Given the importance of RNAi as an antiviral defense mechanism in insects, many more insect viruses are likely to encode VSRs. This chapter provides a detailed protocol that can be used to routinely screen potential VSRs for their ability to suppress RNAi in the *Drosophila* Schneider 2 (S2) cell line. In brief, S2 cells are first cotransfected with a plasmid that expresses the potential VSR and copper-inducible expression plasmids for the firefly and *Renilla* luciferases. Two days after transfection, the cells are treated with dsRNA to silence expression of the firefly luciferase reporter (dsRNA feeding). The *Renilla* luciferase reporter is not silenced and functions as an internal control which can be used to normalize the data. Several hours after dsRNA treatment, expression of the luciferase reporters is induced with $CuSO_4$. The cells are lysed the next day and luciferase activity is quantified by dual-luciferase reporter (DLR) assays. The data are presented as firefly/*Renilla* ratios and, therefore, increased ratios indicate RNAi suppression by a potential VSR. Once a VSR has been identified, additional reporter assays and biochemical experiments can be performed to determine which step in the RNAi pathway is targeted by the VSR.

A flow chart of the RNAi reporter assay and two variants thereof are shown in Fig. 1. In experiments to identify VSR activity, we routinely induce RNAi by adding dsRNA to the culture supernatant (dsRNA feeding, see Note 1). In Fig. 2, a representative example is presented of an experiment that demonstrates the VSR activities of both DCV 1A and CrPV 1A, using the standard RNAi reporter assay and the two variants. In one of the variants, siRNAs are cotransfected with the plasmids (siRNA cotransfection). In contrast to dsRNA, siRNAs do not require processing by Dicer-2. Successful suppression of RNAi in this variant therefore implies that the VSR interferes with steps downstream of Dicer-2 cleavage. In the other variant, dsRNA is cotransfected with the plasmids, rather than added to the culture supernatant (dsRNA cotransfection). Notably, whereas DCV 1A inhibits Dicer-2 cleavage of dsRNA (4), the protein is unable to suppress RNAi in this experimental setup. Presumably, the cotransfected dsRNA is processed into siRNAs before the VSR is expressed at sufficient levels to suppress Dicer-2 cleavage. The main protocol (see Subheadings 2 and 3) describes the RNAi reporter assay using dsRNA feeding to induce RNAi; details on the variants are described in Subheading 4.

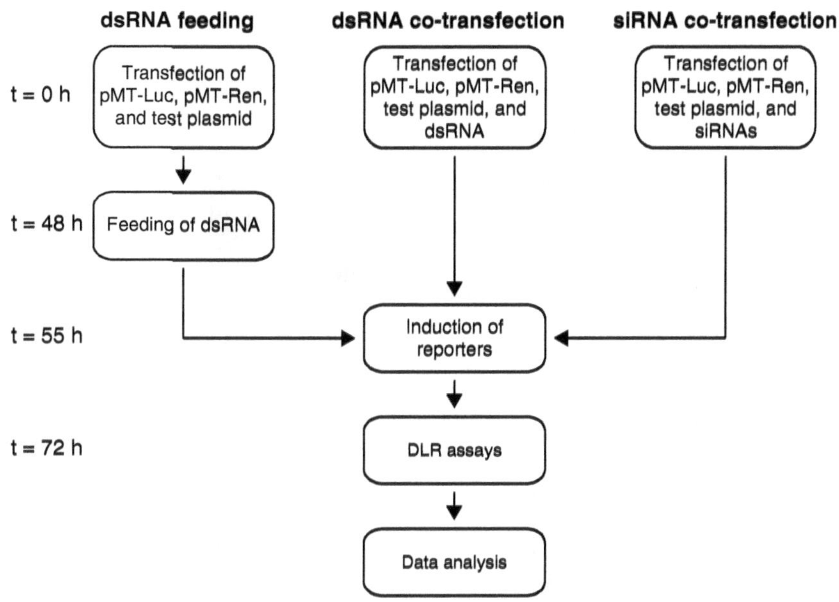

Fig. 1. Flow chart of the RNAi reporter assay. See text for details.

2. Materials

2.1. Cell Culture

1. *Drosophila* S2R+ cells (Drosophila Genomics Resource Center, https://dgrc.cgb.indiana.edu) (see Note 1).
2. Culture medium: Schneider's *Drosophila* Medium (Invitrogen, Carlsbad, CA) supplemented with 10% heat-inactivated Fetal Calf Serum (FCS) (Biochrom, Berlin) and 1% Penicillin/Streptomycin (Invitrogen). Store at 4°C.
3. Cell scrapers (Corning, Corning, NY).
4. 25 cm^2 Cell culture flasks (Corning).
5. 96-Well cell culture plates (Corning).
6. Hemocytometer.

2.2. Generation of Templates for In Vitro Transcription

1. Phusion High-Fidelity DNA polymerase (Finnzymes, Espoo), including 5× Phusion HF Buffer (see Note 2).
2. 10 mM dNTP mix (Roche, Mannheim).
3. Plasmid templates for polymerase chain reaction (PCR) containing the GL3 firefly luciferase and green fluorescent protein (GFP) target sequences.
4. The following primers (10 μM):

   T7-Luc-F: 5′-TAATACGACTCACTATAGGGAGATATGAAGAGATACGCCCTGGTT-3′

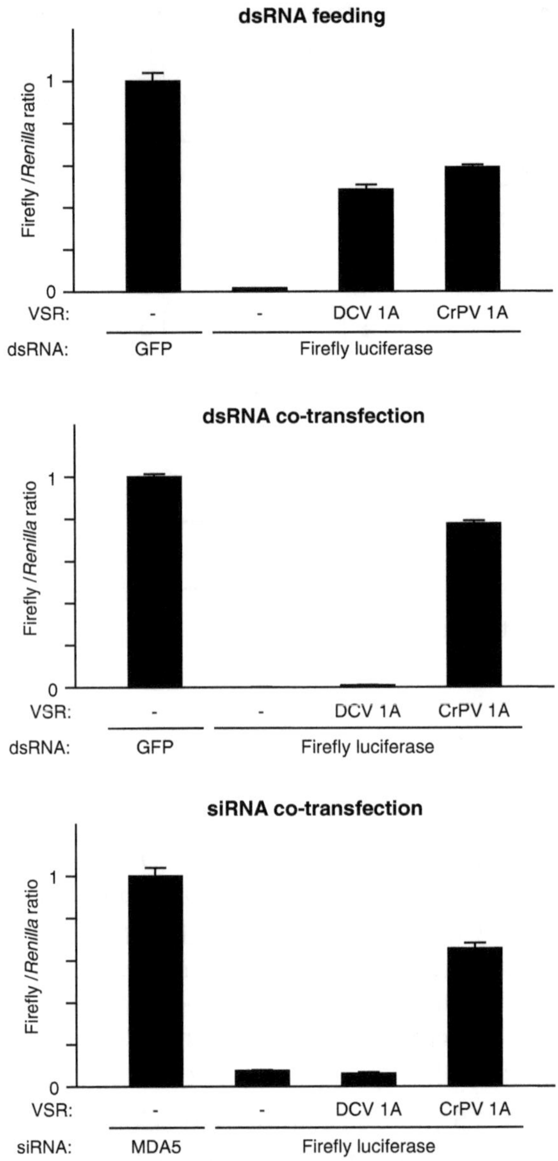

Fig. 2. Example of an experiment that demonstrates the VSR activities of DCV 1A and CrPV 1A. S2R+ cells were transfected with plasmids pMT-Luc, pMT-Ren, and either a plasmid that expresses one of the indicated VSRs or the corresponding empty plasmid. Firefly luciferase-specific or nonspecific (GFP) dsRNA was introduced into the cells by feeding at 48 h after plasmid transfection (*upper panel*) or by cotransfection with the plasmids (*middle panel*). Firefly luciferase-specific or nonspecific (MDA5) siRNAs were introduced into the cells by cotransfection with the plasmids (*lower panel*). To induce expression of the firefly and *Renilla* luciferase reporters, the culture supernatant was supplemented with $CuSO_4$ at 55 h after transfection. At 72 h after transfection, DLR assays were performed to determine the activity of the reporters. The firefly/*Renilla* ratios were calculated and the data were normalized to the nonspecific dsRNA or siRNA controls. Error bars represent the standard deviations of three independent samples. DCV 1A interferes with RNAi at the level of Dicer and only suppresses RNAi if the luciferase reporter is silenced by dsRNA feeding after plasmid transfection. Notably, whereas DCV 1A inhibits Dicer-2 cleavage of dsRNA (4), the protein is unable to suppress RNAi after dsRNA transfection. Presumably, the cotransfected dsRNA is processed into siRNAs before the VSR is expressed at sufficient levels to suppress Dicer-2 cleavage. CrPV 1A, which interferes with RNAi at the level of RISC, is active in all variants of the RNAi reporter assay.

T7-Luc-R: 5′-TAATACGACTCACTATAGGGAGATAAAA CCGGGAGGTAGATGAGA-3′

T7-GFP-F: 5′-TAATACGACTCACTATAGGGAGAAGCTG ACCCTGAAGTTCATCTG-3′

T7-GFP-R: 5′-TAATACGACTCACTATAGGGAGAGGTG TTCTGCTGGTAGTGGTC-3′

5. Thermal cycler.

2.3. In Vitro Transcription and dsRNA Formation

1. T7-promoter-flanked firefly luciferase and GFP PCR products (see Subheading 3.2).
2. RiboMAX Large Scale RNA Production System-T7 (Promega, Madison, WI), containing T7 Transcription 5× Buffer, rNTPs (25 mM each), and T7 Enzyme Mix (see Note 3).
3. GenElute Mammalian Total RNA Miniprep Kit (Sigma-Aldrich, St. Louis, MO) (see Note 4).
4. Heating block.
5. Spectrophotometer.

2.4. Transfection

1. S2R+ cells cultured in a 96-well plate, seeded 1 day prior to transfection at a density of 5×10^4 cells per well (see Subheading 3.1). The cells should be 40–80% confluent on the day of transfection.
2. Effectene Transfection Reagent (Qiagen, Hilden), including Buffer EC and Enhancer.
3. Plasmids pMT-Luc and pMT-Ren, which express the firefly (pMT-Luc) and *Renilla* (pMT-Ren) luciferases from the copper-inducible metallothionein promoter (4) (see Note 5).
4. Plasmids that express the potential VSRs which are to be tested for VSR activity as well as the corresponding empty plasmid (see Note 6).
5. A plasmid that expresses a known VSR (see Note 7).

2.5. dsRNA Feeding

1. Transfected S2R+ cells in a 96-well plate (see Subheading 3.4).
2. Firefly luciferase and GFP dsRNA (see Subheading 3.3).
3. Culture medium (see Subheading 2.1, item 2).

2.6. Induction of the Reporters

1. Transfected S2R+ cells in a 96-well plate that are fed with dsRNA (see Subheading 3.5).
2. 50 mM $CuSO_4$: Dissolve 2.5 g of $CuSO_4.5H_2O$ in 200 ml of H_2O. Filter sterilize, aliquot, and store at −20°C.
3. Culture medium (see Subheading 2.1, item 2).

2.7. DLR Assays

1. Transfected S2R+ cells in a 96-well plate that are fed with dsRNA and of which the reporters have been induced (see Subheading 3.6).
2. Phosphate-buffered saline (PBS): Dissolve 80 g of NaCl, 2 g of KCl, 11.5 g of $Na_2HPO_4.7H_2O$, and 2 g of KH_2PO_4 in 1 l of H_2O to prepare a 10× stock solution. Adjust the pH to 7.3 with HCl, sterilize by autoclaving, and store at room temperature. To prepare a working solution, add nine volumes of H_2O to one volume of 10× PBS stock solution.
3. Dual-Luciferase Reporter Assay System (Promega), containing Passive Lysis Buffer (PLB), Luciferase Assay Reagent II (LAR II), and Stop & Glo Reagent (see Note 8).
4. Rocking platform or orbital shaker.
5. Luminometer tubes.
6. Luminometer.

3. Methods

3.1. Cell Culture

1. Culture S2R+ cells in 25 cm^2 flasks at 25°C without CO_2. The cells should be passaged when approaching confluency.
2. To split the cells, scrape the cells with a cell scraper in their culture supernatant and resuspend them by gently pipetting up and down several times.
3. Transfer the desired amount of cells to new cell culture flasks and/or plates in an appropriate volume of fresh culture medium. Split the cells 1:5 into 25 cm^2 flasks for routine maintenance. For RNAi reporter assays, count the cells using a hemocytometer and seed 5×10^4 cells in 100 μl of culture medium per well in a 96-well plate. Since the RNAi reporter assays are performed in triplicate, seed three wells for each potential VSR that is to be tested as well as for all the controls (see Subheading 3.4).

3.2. Generation of Templates for In Vitro Transcription

1. Perform PCR reactions to generate T7-promoter-flanked templates for in vitro transcription (see Subheading 3.3). Primers T7-Luc-F and T7-Luc-R are used for the amplification of firefly luciferase; primers T7-GFP-F and T7-GFP-R are used for the amplification of GFP. Both primer sets introduce a T7 promoter sequence at both the 5′ and 3′ end of the amplified fragment. Prepare 50-μl PCR reactions containing:
   (a) 1 pg to10 ng of plasmid template DNA.
   (b) 10 μl of 5× Phusion HF Buffer.
   (c) 1 μl of 10 mM dNTP mix.
   (d) 2.5 μl of 10 μM forward primer.

(e) 2.5 µl of 10 µM reverse primer.

(f) 0.5 µl of 2 U/µl Phusion DNA polymerase.

(g) H_2O to 50 µl.

Mix by gently pipetting up and down.

2. Place the PCR reactions in a thermal cycler. To amplify the target sequences, start with an initial denaturation at 98°C for 30 s, followed by 30 cycles of denaturation at 98°C for 10 s, annealing at 61°C for 20 s, and extension at 72°C for 20 s. Finish with a final extension of 10 min at 72°C.

3. Analyze 5 µl of the PCR product by standard agarose gel electrophoresis and ethidium bromide staining. The T7-promoter-flanked firefly luciferase and GFP PCR fragments should have lengths of 489 and 483 bp, respectively (see Note 9). The PCR products can be used directly in subsequent in vitro transcription reactions without purification.

3.3. In Vitro Transcription and dsRNA Formation

1. Firefly luciferase and GFP dsRNA is generated by in vitro transcription reactions on their corresponding T7-promoter-flanked PCR fragments (see Subheading 3.2). Since the PCR fragments contain a T7 promoter sequence at both the 5′ and 3′ end, both strands of the dsRNA duplex are generated in a single reaction. Set up 20-µl in vitro transcription reactions containing:

(a) 5 µl of T7-promoter-flanked PCR product.

(b) 4 µl of T7 Transcription 5× Buffer.

(c) 6 µl of rNTPs (25 mM each).

(d) 2 µl of T7 Enzyme Mix.

(e) 3 µl of H_2O.

Mix by gently pipetting up and down.

2. Incubate at 37°C for 2–4 h.

3. Place the reaction mixture in a heating block preheated to 80°C and incubate for 10 min.

4. Switch off the heating block, but do not take out the reaction mixture.

5. Let the reaction mixture slowly cool to room temperature in the heating block to allow dsRNA formation.

6. Clean up the in vitro-transcribed dsRNA using the GenElute Mammalian Total RNA Miniprep Kit according to the manufacturer's instructions (see Note 10).

7. Verify the integrity of the dsRNA by standard agarose gel electrophoresis and ethidium bromide staining, and determine its concentration and purity using a spectrophotometer.

8. Aliquot the purified dsRNA preparations and store at −80°C (see Note 11).

3.4. Transfection

1. Transfections are done in triplicate for the following experimental conditions.
   (a) Nonspecific dsRNA control (no VSR; GFP dsRNA).
   (b) Specific dsRNA control (no VSR; firefly luciferase dsRNA).
   (c) Positive VSR control (known VSR; firefly luciferase dsRNA).
   (d) Test samples (each of the potential VSRs; firefly luciferase dsRNA).

   The nonspecific and specific dsRNA controls are included to monitor the efficiency and specificity of the dsRNA-induced silencing of the firefly luciferase reporter, whereas the positive VSR control is incorporated to determine whether the sensitivity of the assay allows detection of VSR activity.
2. For transfection of S2R+ cells in a well of a 96-well plate, combine the following amounts of plasmid DNA (see Notes 12 and 13):
   (a) 50 ng of one of the following plasmids.
      (i) The empty plasmid (nonspecific and specific dsRNA control).
      (ii) The plasmid that expresses a known VSR (positive VSR control).
      (iii) A plasmid that expresses a protein which is to be tested as a VSR (test sample).
   (b) 12.5 ng of pMT-Luc.
   (c) 3 ng of pMT-Ren.
3. Adjust the volume to a total of 30 μl with Buffer EC.
4. Add 0.8 μl of Enhancer to condensate the DNA. Mix by vortexing for 1 s.
5. Incubate at room temperature for 2–5 min.
6. Briefly spin down.
7. Dilute the Effectene Transfection Reagent 8× in Buffer EC and add 2.5 μl of the dilution to the DNA-Enhancer mixture to create condensed Effectene-DNA complexes. Mix by vortexing for 10 s.
8. Incubate at room temperature for 5–10 min.
9. Add the transfection mixture to the culture supernatant of the S2R+ cells. There is no need to refresh the medium prior to transfection.
10. Incubate the cells at 25°C for 48 h.

3.5. dsRNA Feeding

1. Dilute the purified dsRNA preparations in culture medium to a final concentration of 20 ng/μl.

2. At 48 h after plasmid transfection, add 10 μl of the diluted dsRNA targeting either firefly luciferase (specific dsRNA control, positive VSR control, and test samples) or GFP (nonspecific dsRNA control) to the culture supernatant of the cells. There is no need to refresh the medium prior to dsRNA feeding (see Note 14).
3. Incubate the cells at 25°C for 7 h.

3.6. Induction of the Reporters

1. Dilute the 50 mM $CuSO_4$ stock solution 10× in culture medium and induce expression of the firefly and *Renilla* luciferase reporters by adding 16 μl of the $CuSO_4$ dilution to the culture supernatant of the cells at 7 h after dsRNA feeding.
2. Incubate the cells at 25°C for 17 h.

3.7. DLR Assays

1. At 17 h after induction of the reporters, completely remove the culture supernatant from the cells.
2. Add 100 μl of PBS to the cells.
3. Gently swirl the culture plate.
4. Completely remove the PBS from the cells.
5. Apply 100 μl of 1× PLB to the cells.
6. Place the culture plate on a rocking platform or orbital shaker and shake gently at room temperature for 15 min to ensure complete lysis of the cells (see Note 15).
7. Predispense 25 μl of LAR II into the number of luminometer tubes required to complete the desired number of DLR assays.
8. Program a luminometer to perform a premeasurement delay of 2 s followed by a measurement period of 10 s (see Note 16).
9. Transfer 10 μl of the cell lysate into a luminometer tube predispensed with LAR II and mix by pipetting up and down several times (see Note 17). It is not necessary to clear the lysate of residual cell debris first.
10. Place the tube in the luminometer and measure the firefly luciferase reporter activity.
11. Remove the tube from the luminometer.
12. Add 25 μl of Stop & Glo Reagent and mix by vortexing briefly.
13. Place the tube in the luminometer and measure the *Renilla* luciferase reporter activity.
14. Discard the tube and proceed with the next DLR assay (see Note 18).

3.8. Presentation of the Data

1. To normalize the data from the DLR assays, calculate the firefly/*Renilla* ratio for each sample.
2. Determine the mean firefly/*Renilla* ratio as well as the standard deviation of the triplicates for each experimental condition.
3. Present the data in diagrams similar to those shown in Fig. 2 (see Note 19). The controls provide important information regarding the quality of the experiment. First, successful silencing of the firefly luciferase reporter should be evident from a lower firefly/*Renilla* ratio of the specific dsRNA control than of the nonspecific dsRNA control (see Notes 20 and 21). Second, when compared to the specific dsRNA control, the positive VSR control must present a higher firefly/*Renilla* ratio (see Notes 22 and 23).

4. Notes

1. To silence the firefly luciferase reporter, we add dsRNA to the culture supernatant of S2R+ cells (dsRNA feeding, or soaking). The dsRNA is taken up by the cells and processed by the RNAi machinery (12). The S2 cell line is highly heterogeneous in morphology, growth rate, and other characteristics. Be aware that, due to variable passage history and culture conditions, not all sublines of the S2 cell line possess the ability to efficiently take up dsRNA from the culture supernatant. When using an S2 cell line other than S2R+, make sure that the cells are capable to do so. Different S2 cell lines might require optimization of the protocol. When using S2 cells that do not take up dsRNA from the culture supernatant, you can consider transfection of the dsRNA into the cells.
2. Phusion DNA polymerase works very well in our hands, but any thermostable DNA polymerase (such as Taq) can be used to amplify the templates for in vitro transcription. Keep in mind that the indicated cycling conditions are optimized for amplification with Phusion DNA polymerase. Use of other polymerases may require optimization of the PCR reaction.
3. The RiboMAX Large Scale RNA Production System-T7 is specifically designed to produce large amounts of in vitro-transcribed RNA. However, other T7 RNA polymerase-based in vitro transcription methods can also be used.
4. The GenElute Mammalian Total RNA Miniprep Kit is designed to isolate total RNA from mammalian cells and tissues, but the kit can also be used to clean up RNA. Although the kit is not optimized for dsRNA, we obtain good results using the RNA clean-up procedure of the kit. When using

other commercial kits or methods to clean up dsRNA, verify its functionality.

5. Plasmids pMT-Luc and pMT-Ren are derived from vector pMT/V5-His B (Invitrogen).
6. We generally express our proteins of interest from vector pAc5.1/V5-His (Invitrogen). The *Drosophila* actin 5 (Ac5) promoter in this vector allows high-level, constitutive expression in S2 cells. In addition, the vector contains a C-terminal V5 epitope and a polyhistidine (6×His) tag which can be used to confirm expression of the protein.
7. As a positive control, you can include a known VSR in your experiment. All established VSRs can be used as a positive control when the firefly luciferase reporter is silenced by dsRNA feeding 2 days after plasmid transfection. However, when the reporter is silenced by cotransfection of dsRNA or siRNAs with the plasmids, it is important to use a VSR that interferes with RNAi at steps downstream of Dicer (for example, CrPV 1A, see Fig. 2).
8. Instructions for preparation and storage of 1× PLB, LAR II, and the Stop & Glo Reagent are described in the manufacturer's technical manual. It is important that all reagents and samples are at ambient temperature when performing the DLR assays, since the activity of the luciferase reporters is temperature-sensitive.
9. The firefly luciferase and GFP PCRs should generate single T7-promoter-flanked fragments with the indicated sizes. If no products or nonspecific products are observed on the agarose gel, optimization of the PCR reaction may be required. Alternatively, it might be necessary to purify the correct fragment from gel before continuing with in vitro transcription.
10. It is not necessary to remove the DNA template by digestion with DNase.
11. Avoid multiple freeze–thaw cycles and keep the RNA on ice whenever it is thawed for use.
12. Since the experiments are performed in triplicate, it is convenient to prepare a master mix for each experimental condition.
13. Instead of inducing RNAi by dsRNA feeding at 48 h after plasmid transfection, the firefly luciferase reporter can be silenced by cotransfection of dsRNA or siRNAs with the plasmids. For dsRNA cotransfection, add 10 ng of either firefly luciferase (specific dsRNA control, positive VSR control, and test samples) or GFP (nonspecific dsRNA control) dsRNA to the mixture of plasmids during the transfection procedure. For siRNA cotransfection, add 2 μl of a 1 μM stock solution to the mixture of plasmids during the transfection procedure. We purchase our firefly luciferase-specific and nonspecific

control (MDA5) siRNAs from Dharmacon (Lafayette, CO). When performing the dsRNA or siRNA cotransfection variants of the assay, omit the dsRNA feeding step (see Subheading 3.5).

14. Some researchers use FCS-free culture medium during dsRNA feeding. In our experiments, we do not observe any difference in the efficiency of dsRNA-mediated silencing when feeding is performed in either the presence or absence of FCS.
15. The cell lysates can be stored at −20°C for up to 1 month if you wish to continue with the DLR assays later. For long-term storage, the lysates should be stored at −80°C. Prevent multiple freeze–thaw cycles, since this can cause gradual loss of reporter activity.
16. Single-sample, multiple-sample, and plate-reading luminometers can be used to perform the DLR assays. It is recommended that multiple-sample and plate-reading luminometers are equipped with reagent injectors. This is not required for single-sample luminometers.
17. It is important not to mix by vortexing, but by pipetting up and down. Vortexing can create a microfilm of the luminescent solution along the sides of the tube which can escape mixing with the Stop & Glo Reagent in subsequent steps.
18. It is more convenient to first measure the firefly luciferase activities in all the samples, before measuring the *Renilla* luciferase activities.
19. As an alternative, you can present the data as fold silencing relative to the nonspecific dsRNA control. To calculate the fold silencing for a specific experimental condition, divide the mean firefly/*Renilla* ratio of the nonspecific dsRNA control by that of the experimental condition.
20. If you do not observe silencing in the specific dsRNA control, make sure that the dsRNA preparations are of sufficient quality (see Subheading 3.3). It is important to work under RNase-free conditions to prevent degradation of the RNA.
21. Low absolute firefly and *Renilla* luciferase counts may indicate a low transfection efficiency and may require optimization of the assay.
22. If you do not observe suppression of RNAi by either the positive VSR control or the test samples, confirm their expression (for example, by Western blot analysis).
23. While interpreting data from the RNAi reporter assay, it is important to realize that virtually any dsRNA-binding protein can suppress RNAi when overexpressed (13). Where possible, confirm the activity of an identified VSR in cells infected with the virus carrying the VSR.

Acknowledgments

The authors would like to thank Walter Bronkhorst for helpful discussions. This work was supported by a fellowship from the Nijmegen Centre for Molecular Life Sciences, by a VIDI fellowship from the Netherlands Organization for Scientific Research (project number 864.08.003), and by Horizon Breakthrough fellowships from the Netherlands Genomics Initiative (project numbers 93519018 and 93518020).

References

1. Ding, S.W., and Voinnet, O. (2007) Antiviral immunity directed by small RNAs *Cell* **130,** 413–26.
2. van Rij, R.P., and Berezikov, E. (2009) Small RNAs and the control of transposons and viruses in Drosophila *Trends Microbiol* **17,** 163–71.
3. Li, H., Li, W.X., and Ding, S.W. (2002) Induction and suppression of RNA silencing by an animal virus *Science* **296,** 1319–21.
4. van Rij, R.P., Saleh, M.C., Berry, B., Foo, C., Houk, A., Antoniewski, C., and Andino, R. (2006) The RNA silencing endonuclease Argonaute 2 mediates specific antiviral immunity in Drosophila melanogaster *Genes Dev* **20,** 2985–95.
5. Wang, X.H., Aliyari, R., Li, W.X., Li, H.W., Kim, K., Carthew, R., Atkinson, P., and Ding, S.W. (2006) RNA interference directs innate immunity against viruses in adult Drosophila *Science* **312,** 452–4.
6. Aliyari, R., Wu, Q., Li, H.W., Wang, X.H., Li, F., Green, L.D., Han, C.S., Li, W.X., and Ding, S.W. (2008) Mechanism of induction and suppression of antiviral immunity directed by virus-derived small RNAs in Drosophila *Cell Host Microbe* **4,** 387–97.
7. Chao, J.A., Lee, J.H., Chapados, B.R., Debler, E.W., Schneemann, A., and Williamson, J.R. (2005) Dual modes of RNA-silencing suppression by Flock House virus protein B2 *Nat Struct Mol Biol* **12,** 952–7.
8. Lingel, A., Simon, B., Izaurralde, E., and Sattler, M. (2005) The structure of the flock house virus B2 protein, a viral suppressor of RNA interference, shows a novel mode of double-stranded RNA recognition *EMBO Rep* **6,** 1149–55.
9. Lu, R., Maduro, M., Li, F., Li, H.W., Broitman-Maduro, G., Li, W.X., and Ding, S.W. (2005) Animal virus replication and RNAi-mediated antiviral silencing in Caenorhabditis elegans *Nature* **436,** 1040–3.
10. Singh, G., Popli, S., Hari, Y., Malhotra, P., Mukherjee, S., and Bhatnagar, R.K. (2009) Suppression of RNA silencing by Flock house virus B2 protein is mediated through its interaction with the PAZ domain of Dicer *FASEB J* **23,** 1845–57.
11. Nayak, A., Berry, B., Tassetto, M., Kunitomi, M., Acevedo, A., Deng, C., Krutchinsky, A., Gross, J., Antoniewski, C., and Andino, R. (2010) Cricket paralysis virus antagonizes Argonaute 2 to modulate antiviral defense in Drosophila *Nat Struct Mol Biol* **17,** 547–54.
12. Saleh, M.C., van Rij, R.P., Hekele, A., Gillis, A., Foley, E., O'Farrell, P.H., and Andino, R. (2006) The endocytic pathway mediates cell entry of dsRNA to induce RNAi silencing *Nat Cell Biol* **8,** 793–802.
13. Lichner, Z., Silhavy, D., and Burgyán, J. (2003) Double-stranded RNA-binding proteins could suppress RNA interference-mediated antiviral defences *J Gen Virol* **84,** 975–80.

Chapter 13

Dicer Assay in Drosophila S2 Cell Extract

Baojun Yang and Hongwei Li

Abstract

Double-stranded RNA (dsRNA) is the trigger of RNA interference (RNAi)-mediated gene regulation. Dicer processes dsRNAs into short interfering RNAs (siRNAs), which are incorporated into the effector RNA-induced silencing complex (RISC) and direct degradation of homologous target mRNAs. In plants and invertebrates, the RNAi machinery also acts as an antiviral mechanism through production of viral siRNAs by Dicer and silencing of replicating viruses. Viral suppressors of RNAi (VSRs) are encoded by some viruses and serve as a strategy to counteract the RNAi-based antiviral immunity. In this chapter, we describe a Dicer activity assay in extracts prepared from *Drosophila melanogaster* S2 cells. We also introduce a simple procedure to study VSR activity in the *in vitro* Dicer assay.

Key words: RNAi, *Drosophila*, Dicer, dsRNA, siRNA, Virus, Suppressor

1. Introduction

RNAi is a small RNA-mediated gene regulatory mechanism that is evolutionally conserved in most eukaryotic organisms (1). There are two major classes of small RNAs, short interfering RNAs (siRNAs) and microRNAs (miRNAs) (2). These 21- to 25-nucleotide (nt) small RNAs are produced by an RNaseIII family enzyme called Dicer and regulate gene expression through the effector protein Argonaute (Ago) in a sequence-specific manner (2). In *Drosophila*, there are distinct RNAi components which are required for production and function of siRNAs and miRNAs (3, 4). Dicer-2 serves as the central engine to sense double-stranded RNA (dsRNA) triggers and to process them into siRNAs (5). The siRNAs are subsequently loaded onto the RNA-induced silencing complex (RISC) and guide the RISC to recognize cognate mRNAs for destruction (6, 7). The target mRNAs is cleaved by Argonaute 2 (Ago2) protein at the central position of a siRNA/mRNA pairing

Ronald P. van Rij (ed.), *Antiviral RNAi: Concepts, Methods, and Applications*, Methods in Molecular Biology, vol. 721,
DOI 10.1007/978-1-61779-037-9_13, © Springer Science+Business Media, LLC 2011

(4, 6, 8). For the biogenesis of miRNAs, Pasha and Drosha form the microprocessor that converts primary miRNAs into precursor miRNAs (pre-miRNAs) (9). A complex of Dicer-1 and Loquacious is required for the processing of stem-loop structures of pre-miRNAs into mature miRNAs (10–12). AGO1 is involved in mature miRNA production and also has impacts on miRNA-directed regulation of target mRNA (4).

In the siRNA pathway, the dsRNA triggers are generally derived from viral infection, direct introduction of exogenous dsRNAs, transcription of an inverted repeat transgene, processing of fold-back structures of RNAs, mobilization of transposable elements, or activities of endogenous RNA-dependent RNA polymerase, especially in plants (13). In plants and invertebrates infected with viruses, dsRNAs generated from replicating viruses can be processed by RNAi machinery, leading to accumulation of viral-specific siRNAs and inhibition of viral replication (13–19). Virus-specific RNAi is induced during viral infection and plays a role in antiviral immunity (15, 20–22). In *Drosophila*, evidence from extensive studies using cultured cells and whole flies strongly supports the concept of antiviral RNAi (14, 15, 21, 23). On the other hand, viruses also evolve strategies to counteract the RNAi-based immunity. Nonstructural or structural proteins encoded by a variety of viruses have been identified as viral suppressors of RNAi (VSRs) (13, 15, 21, 22, 24, 25). Mechanistically, most characterized VSRs suppress RNAi by sequestration of dsRNAs or siRNAs. VSRs bind to dsRNAs or siRNAs in a sequence-independent way and protect dsRNAs from being processed by Dicer or prevent incorporation of siRNAs into the RISC (22, 26–28).

In the model system *Drosophila melanogaster*, *in vitro* biochemical analysis has been widely used as a tool to study mechanisms underlying RNAi and virus–RNAi interactions. Using crude or purified extracts prepared from *Drosophila* S2 cells, major components of the RNAi machinery, such as Dicer-2, Ago-2, and R2D2, were identified and their functions have been illustrated (4–8, 29, 30). In the analysis of antiviral RNAi, VSRs such as flock house virus B2 (FHVB2) and *Drosophila* C virus 1A proteins inhibit dsRNA processing in S2 extracts (15, 21, 27). In this chapter, we describe a detailed method for Dicer activity assay using extracts prepared from *Drosophila* S2 cells, and we introduce a simple procedure to study VSR activity in an *in vitro* Dicer assay.

2. Materials

2.1. Preparation of dsRNAs

2.1.1. Preparation of DNA Templates for dsRNAs

1. Taq DNA polymerase (5 units/μl) and 10× reaction buffer (Promega). Store at −20°C.
2. 10 mM dNTPs: dATP, dGTP, dCTP, and dTTP at a concentration of 10 mM, store at −20°C.

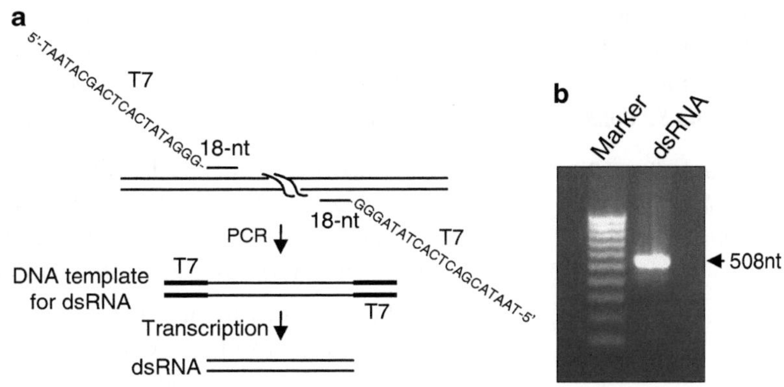

Fig. 1. Preparation of dsRNA. (**a**) Schematic presentation of dsRNA production. (**b**) The 508-nt dsRNA of eGFP synthesized by in vitro transcription is visualized on 1% agarose gel by ethidium bromide staining. Marker, 1 Kb DNA ladder.

3. Primers for amplification of DNA templates for eGFP dsRNA: The primers were designed by adding a T7 promoter sequence to the 5′-end of 18-nt eGFP-specific oligonucleotides, giving forward primer 5′-**TAATACGACTCACTATAGGG**CGAGGAGCTGTTCACCGG-3′ and reverse primer 5′-**TAATACGACTCACTATAGGG**TCCTCGATGTTGTGGCGG-3′. The T7 promoter sequence is indicated in bold (Fig. 1a).
4. Phenol/chloroform (1:1): mix an equal volume of buffer-saturated phenol and chloroform, store at 4°C.
5. 3 M sodium acetate (pH 5.2).
6. Isopropanol.
7. 70% Ethanol.

2.1.2. Synthesis of dsRNAs by In Vitro Transcription and Labeling of dsRNAs with [α-^{32}P] UTP

1. MEGAScript Kit (Ambion), store at −20°C. The kit contains all reagents required, including buffer, enzyme, nucleotides, and stop solution (5 M Ammonium Acetate, 100 mM EDTA).
2. [α-^{32}P] UTP (6,000 Ci/mmol, 10 mCi/ml) (MP Biomedicals, LLC), store at −20°C.
3. DEPC-H_2O: add 1 ml of diethyl pyrocarbonate (DEPC) to 1 l H_2O, and mix thoroughly. Incubate overnight at room temperature and autoclave.
4. Phenol/chloroform, see Subheading 2.1.1, item 4.
5. Isopropanol.
6. 70% Ethanol.

2.2. In Vitro Dicer Activity Assay Using Extracts from Drosophila S2 Cells

2.2.1. Preparation of Dicer Extracts from Drosophila S2 Cells

1. Schneider's *Drosophila* Medium supplemented with 10% fetal bovine serum and 100 units/ml of penicillin and 100 μg/ml of streptomycin (Invitrogen). Store at 4°C.
2. *Drosophila* Schneider 2 (S2) cells (Invitrogen).
3. 10× Phosphate-buffered saline (PBS): dissolve 80 g NaCl, 11.5 g $Na_2HPO_4 \cdot 7H_2O$, 2 g KCl, and 2 g KH_2PO_4 in 800 ml of H_2O, adjust pH to 7.4, and add H_2O up to 1 l. Sterilize by autoclaving and store at room temperature.
4. 50× Stock solution of protease inhibitor: dissolve one complete protease inhibitor cocktail tablet (Roche) in 1 ml H_2O to make a 50× stock solution. Store aliquots at –20°C.
5. Hypotonic buffer: 10 mM Hepes-KOH, pH 7.0, 2 mM $MgCl_2$, 0.1% β-mercaptoethanol, 1× Protease inhibitor. To make 10 ml of the buffer, mix 0.1 ml of 1.0 M Hepes-KOH, pH 7.0, 0.2 ml of 100 mM $MgCl_2$, 10 μl of β-mercaptoethanol, 0.2 ml of 50× Stock solution of protease inhibitor, and 9.5 ml of H_2O together. Prepare fresh every time.
6. Dounce tissue homogenizers with pestle B (2 ml, Kontes).
7. 80% Glycerol: autoclave and store at room temperature.

2.2.2. In Vitro Dicer Activity Assay

1. 10× Buffer K: 200 mM Hepes-KOH, pH 7.0, 20 mM DTT, 20 mM $MgCl_2$. Make 10 ml of the 10× buffer by mixing 2 ml of 1 M Hepes, pH 7.0, 0.2 ml of 1 M DTT, 0.2 ml of 1 M $MgCl_2$, and 7.6 ml of H_2O together. Prepare fresh every time.
2. 0.5 M Creatine Phosphate (disodium salt, Roche): dissolve 80 mg creatine phosphate in 500 μl H_2O, store aliquots at –20°C.
3. 1 mg/ml Creatine Phosphate Kinase (Roche): dissolve creatine phosphate kinase in storage buffer (50 mM Tris–HCl, pH 7.0, 1 mM DTT, 0.1 mM EDTA, 1 μM ATP, 50% Glycerol), and store at –20°C.
4. RNasin Ribonuclease Inhibitor, 40 units/μl (Promega).
5. 2× PK buffer: 1.0% SDS, 10 mM Tris–HCl, pH 7.5, 1 mM EDTA, and 1 mg/ml Protease K. Prepare fresh every time from individual stock solutions (20% SDS, 1 M Tris–HCl, pH 7.5, 500 mM EDTA, pH 8.0, 20 mg/ml Protease K).
6. Glycogen (20 μg/μl, Invitrogen). Store at –20°C.
7. TE buffer (10 mM Tris–HCl, pH 8.0, 1 mM EDTA).
8. Sample buffer: deionized formamide/TE (1:1), prepare fresh every time.

2.2.3. Preparation of Size Marker for siRNAs

1. Chemically synthesized 21-nt siRNA duplex (Dharmacon) or 22-nt DNA oligonucleotide.
2. [γ-^{32}P] ATP (6,000 Ci/mmol, 10 mCi/ml). Store at –20°C.
3. T4 DNA polynucleotide kinase (PNK, 10 units/μl) and 10× PNK reaction buffer (Promega).

2.2.4. Denaturing Urea Polyacrylamide Gel Electrophoresis (Urea-PAGE)

1. Acrylamide solution 40% (Acrylamide: Bis-Acrylamide 19:1), store at 4°C.
2. 5× TBE (electrophoresis running buffer): dissolve 54.0 g of Tris base and 27.5 g of boric acid in 800 ml of H_2O, add 20 ml of 0.5 M EDTA (pH 8.0), and adjust the volume to 1 l. Sterilize by autoclaving.
3. 10% Ammonium persulfate. Store at 4°C.
4. TEMED, store at 4°C.
5. 6× gel loading buffer: 0.25% xylene cyanol FF, 0.25% bromophenol blue, and 30% glycerol. Prepare with DEPC-H_2O, autoclave and store aliquots at –20°C.
6. Dual adjustable SLAB gel unit (CBS Scientific): glass plates (W16 cm×L20 cm), spacer (1-mm thick), and comb (20-well).
7. X-ray film (BioMax MS Film, Kodak) or a screen of PhosphorImager (Molecular Dynamics).

2.3. Inhibition of Dicer Activity by Viral Proteins

1. pGEX-FHVB2: FHVB2 gene was cloned into pGEX-4T2 (GH Healthcare) for expression of GST-tagged recombinant B2 protein in *Escherichia coli* (27).
2. 1× PBS buffer: diluted from 10× PBS as described in Subheading 2.2.
3. Glutathione Sepharose 4 Fast Flow (GE Healthcare): 50%, equilibrated with 1× PBS.
4. Elution buffer: 50 mM Tris–HCl, 10 mM reduced glutathione, pH 8.0.
5. Ampicillin stock solution (100 mg/ml). Filter sterilize and store at –20°C.
6. LB medium (Luria–Bertani Medium): dissolve 10 g of tryptone, 5 g yeast extract, and 10 g NaCl to 900 ml of H_2O, adjust pH to 7.0, and add H_2O up to 1 l. Sterilize by autoclaving and keep at 4°C.
7. LB/amp: LB medium supplemented with 100 μg/ml Ampicilin.
8. 100 mM IPTG. Filter sterilize and store at –20°C.

3. Methods

3.1. Preparation of dsRNAs

Currently, *in vitro* transcription by T7 RNA polymerase is the most commonly used method to synthesize long dsRNAs. Usually, dsRNAs with the size of 300- to 1,000-bp exhibit a potent RNAi effect in transfected S2 cells and can be efficiently produced by T7 RNA polymerase (7, 31, 32). We used PCR fragments flanked by T7 promoter sequence at both ends as a DNA template for transcription.

3.1.1. Preparation of DNA Templates for dsRNAs

1. Set up PCR reaction for dsRNA template in a total volume of 100 μl. Assemble the following components:
   (a) 10 μl 10× Reaction buffer.
   (b) 2 μl 10 mM dNTPs.
   (c) 5 μl 10 μM forward primer.
   (d) 5 μl 10 μM reverse primer.
   (e) 1 μl eGFP PCR fragment or plasmid containing the eGFP gene (0.1 μg/μl).
   (f) 1 μl Taq DNA polymerase (5 units/μl).
   (g) 76 μl H_2O.

   Split the reaction mix into two 200-μl PCR tubes and run the following program on a PCR thermocycler: 94°C for 5 min; followed by 94°C for 45 s, 55°C for 45 s, 72°C for 1 min for 30 cycles; 72°C for 10 min.
2. Run 3 μl of PCR products on 1% agarose gel to check specificity of amplification.
3. Combine PCR products from the two tubes and add an equal volume of phenol/chloroform (1:1). Vortex for 30 s, and centrifuge at 16,000 ×*g* for 10 min at room temperature, and carefully transfer the aqueous phase to a new tube.
4. Add 1/10 volume of 3 M sodium acetate and an equal volume of isopropanol, mix thoroughly, and keep at room temperature for 10 min.
5. Centrifuge at 16,000 ×*g* for 10 min at room temperature. Discard the supernatant and rinse the DNA pellet with 800 μl of 70% ethanol. Air-dry the pellet for 10–20 min at room temperature.
6. Dissolve the pellet in 30 μl H_2O, and measure the OD260 to calculate the concentration of DNA. Store at −20°C.

3.1.2. Synthesis of dsRNAs by In Vitro Transcription and Labeling of dsRNAs with [α-^{32}P] UTP

Ambion's MEGAScript T7 Kit provides an efficient tool to produce dsRNAs with high yield and quality (Fig. 1b). According to manufacturer's manual, dsRNAs are synthesized by *in vitro* transcription and uniformly labeled with [α-^{32}P] UTP as below.

1. Assemble the *in vitro* transcription reaction on ice (see Note 1):
   (a) 2 μl of 10× Reaction buffer.
   (b) 2 μl of ATP (75 mM).
   (c) 2 μl of CTP (75 mM).
   (d) 2 μl of GTP (75 mM).
   (e) 0.5 μl of UTP (75 mM).
   (f) 2 μl of template DNA (0.5 μg/μl).
   (g) 2 μl of Enzyme mix.
   (h) 4 μl of [α-^{32}P] UTP.
   (i) 3.5 μl of DEPC-H_2O, bringing the final volume up to 20 μl.
2. Incubate at 37°C for 2–4 h (see Note 2).
3. To remove the template DNAs, add 1 μl of DNaseI (2 units/μl) into the reaction, and incubate at 37°C for 15 min.
4. Add 115 μl of H_2O and 15 μl of stop solution, mix well.
5. Add 150 μl of phenol/chloroform (1:1), and vortex for 30 s.
6. Centrifuge at 16,000 ×*g* for 10 min at room temperature, and carefully transfer the aqueous phase to a new tube.
7. Add an equal volume of isopropanol, mix thoroughly, and incubate at room temperature for 10 min.
8. Centrifuge at 16,000 ×*g* for 10 min at room temperature. Discard the supernatant and rinse the dsRNA pellet with 800 μl of 70% ethanol. Air-dry the pellet for 10–20 min at room temperature.
9. Dissolve the pellet in 100 μl H_2O and measure OD260 to calculate concentration of dsRNAs (see Note 3). Store the labeled dsRNAs at −20°C.

3.2. In Vitro Dicer Activity Assay Using Extracts from Drosophila S2 Cells

Processing of dsRNA into siRNAs by Dicer extract was first established by Hannon and coworkers (5, 7). Crude and purified extracts prepared from *Drosophila* S2 cell have been widely used in biochemical analysis of RNAi machinery. In our study, we adopted the same method with some modifications.

3.2.1. Preparation of Dicer Extracts from Drosophila S2 Cells

1. Passage *Drosophila* S2 cells into two 100-mm cell culture dishes containing 10 ml of fresh complete Schneider's *Drosophila* medium at a concentration of 1×10^6 cells/ml, and incubate at 28°C.
2. At 3 days after passage, collect S2 cell culture into two 15-ml centrifuge tubes.
3. Centrifuge at 1,000 ×*g* for 3 min at room temperature to pellet the cells, discard the supernatants (medium).

4. Gently resuspend the cell pellets in 10 ml 1× PBS. Centrifuge at 1,000 × *g* for 3 min to pellet the cells.
5. Repeat step 4 once.
6. Resuspend the cell pellets in 4 ml of ice-cold Hypotonic buffer containing 60 mM KCl, and combine cell suspensions into one ice-cold 15-ml centrifuge tube. Centrifuge at 1,000 × *g* for 5 min at 4°C, remove the supernatant (see Note 4).
7. Estimate the volume of the cell pellet, and add 50–80% volume of ice-cold Hypotonic buffer (no KCl). Gently resuspend the cells.
8. Transfer the cell suspension into an ice-cold Dounce tissue homogenizer, and keep on ice for 10 min.
9. Slowly dounce 30 times on ice, and transfer the lysate to a 1.5-ml microcentrifuge tube.
10. Centrifuge at 20,000 × *g* at 4°C for 20 min (see Note 5).
11. Carefully transfer the supernatant into a 1.5-ml microcentrifuge tube, and do not touch the pellets.
12. Add 80% glycerol to a final concentration of 10%. Store the S2 extract at −80°C.

3.2.2. In Vitro Dicer Activity Assay

1. Assembly the Dicer assay reaction on ice in a volume of 10 μl:
   (a) 1 μl 10× Buffer K.
   (b) 1 μl 10 mM ATP.
   (c) 0.5 μl 0.5 M Creatine phosphate.
   (d) 0.3 μl 1 mg/ml Creatine phosphate kinase (see Note 6).
   (e) 0.2 μl Ribonuclease inhibitor (40 units/μl).
   (f) 1 μl ^{32}P-labeled dsRNA (50 nM).
   (g) 6 μl S2 extract (*see* Subheading 3.2.1).
2. Incubate at 30°C for 2 h.
3. Add 90 μl of 2× PK buffer to the 10 μl Dicer assay reaction, and incubate at 65°C for 10 min.
4. Add 100 μl TE buffer to bring the volume up to 200 μl, and then mix with an equal volume of phenol/chloroform (1:1). Vortex for 10 s.
5. Centrifuge at 16,000 × *g* for 10 min at room temperature, carefully transfer the aqueous phase to a new microcentrifuge tube, and add 1 μl of 20 μg/μl Glycogen.
6. Add 3× volume of ice-cold ethanol, keep at −20°C for at least 2 h.
7. Centrifuge at 20,000 × *g* for 10 min at 4°C, discard supernatants.

8. Air-dry the RNA pellet at room temperature for 10–20 min (see Note 7).
9. Dissolve RNAs in 20 μl sample buffer. Heat at 95°C for 5 min and chill on ice for 1 min.
10. Centrifuge the tube for a few seconds to collect the sample at the bottom, and add 2 μl of 6× gel loading buffer. The sample is ready for loading on Urea-PAGE.

3.2.3. Preparation of Size Marker for siRNAs

Both 21-nt siRNA and 22-nt DNA oligonucleotide can be used as the siRNA size marker in the Urea-PAGE after labeled with [γ-^{32}P] ATP (see Note 8).

1. Set up the reaction in a total volume of 20 μl:
   (a) 2 μl of 10× PNK buffer.
   (b) 2 μl of siRNA or DNA oligonucleotide (10 μM).
   (c) 1 μl of T4 PNK (10 units/μl).
   (d) 4 μl of [γ-^{32}P] ATP.
   (e) 11 μl of H_2O.
2. Incubate at 37°C for 45 min; add 180 μl of sample buffer and 40 μl of 6× loading buffer. Store at −20°C (see Note 9).

3.2.4. Denaturing Urea Polyacrylamide Gel Electrophoresis (Urea-PAGE)

1. Prepare 15% denaturing urea polyacrylamide gel in a total volume of 45 ml: mix 4.5 ml of 5× TBE, 9 ml of H_2O, and 18.9 g of urea in a 100 ml beaker, heat in a microwave oven for 10–15 s to completely dissolve urea. Then add 16.9 ml of 40% acrylamide solution while continuously stirring with a small magnetic bar. Adjust the volume up to 45 ml with H_2O, add 240 μl of 10% Ammonium persulfate and 10 μl of TEMED.
2. Immediately pour gel with a 25 ml serological pipette, place the comb, and keep at room temperature for about 30 min until polymerized.
3. Prerun the gel in 0.5× TBE for 30 min at 200 V.
4. Rinse the well with 0.5× TBE using a 30 ml syringe, and load the heat-denatured RNA sample prepared from Dicer assay (see Subheading 3.2.2). The ^{32}P-labeled siRNA or DNA oligonucleotide is loaded as a size marker (see Note 10).
5. Run the gel at 200 V until bromophenol blue reaches about 2/3 of gel. siRNAs should migrate between the bromophenol blue and xylene cyanol FF.
6. Carefully remove the gel from the glass plate, drain off the extra buffer from the gel, and wrap it in Saran wrap.
7. Expose the gel to a sheet of X-ray film or a PhosphorImager screen (Fig. 2). Exposure time will depend on the strength of radioactive signal on the gel (see Note 11).

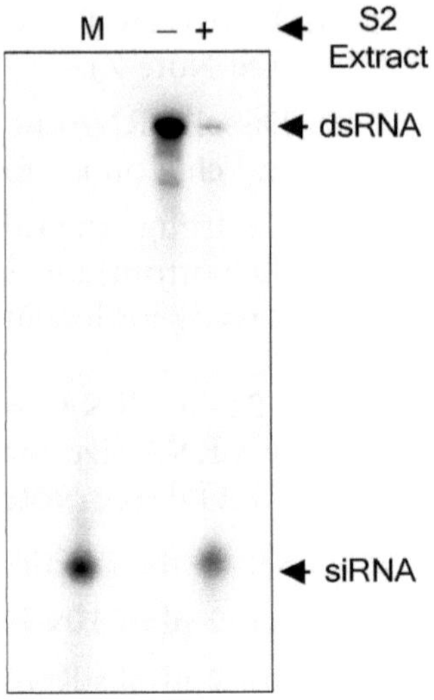

Fig. 2. Processing of dsRNAs into siRNAs in in vitro Dicer assay. The ^{32}P-labeled 508-bp dsRNA of eGFP was incubated with Dicer extracts prepared from *Drosophila* S2 cells. M, a ^{32}P-labeled 21-nt siRNA, is used as size marker.

3.3. Inhibition of Dicer Activity by Viral Proteins

In *Drosophila* S2 cells, FHV induces antiviral RNAi, evidenced by accumulation of abundant viral siRNAs (33). In addition, B2 protein of FHV is identified as a viral suppressor of RNAi. The suppression mechanism has been investigated recently (21, 27, 33). B2 has the affinity to bind long dsRNAs and siRNAs, and the binding activities are essential for suppression of FHV-induced antiviral RNAi in *Drosophila* cells. We also demonstrated that recombinant FHVB2 protein purified from *E. coli* is able to inhibit dsRNA processing by Dicer in S2 extracts in a dose-dependent manner (Fig. 3).

3.3.1. Expression and Purification of GST-Tagged FHVB2 Protein from E. coli

1. Inoculate a single colony of *E. coli* BL21 cells containing a recombinant plasmid pGEX-FHVB 2–10 ml of LB/amp medium. Incubate overnight at 37°C with vigorous shaking.
2. Inoculate the 10 ml culture to prewarmed 240 ml of LB, and grow at 37°C with shaking for 0.5–1 h until the OD600 reaches 0.5. Add 2.5 ml of 100 mM IPTG (a final concentration of 1.0 mM), and incubate for additional 4–5 h.
3. Pellet the bacteria by centrifuging the culture at 4,000 × *g* for 10 min at 4°C, and resuspend the cell pellet in 5 ml of ice-cold 1× PBS. Disrupt the cells by sonication on ice.
4. Add another 4.5 ml of 1× PBS to the cell lysate, and 0.5 ml of 20% Triton X-100 (a final concentration of 1%). Mix gently on a rotary shaker at 4°C for 30 min to aid in solublization of the fusion protein.
5. Centrifuge at 4,000 × *g* for 10 min at 4°C, and transfer the supernatant to a 15-ml centrifuge tube. Add 1 ml of the 50%

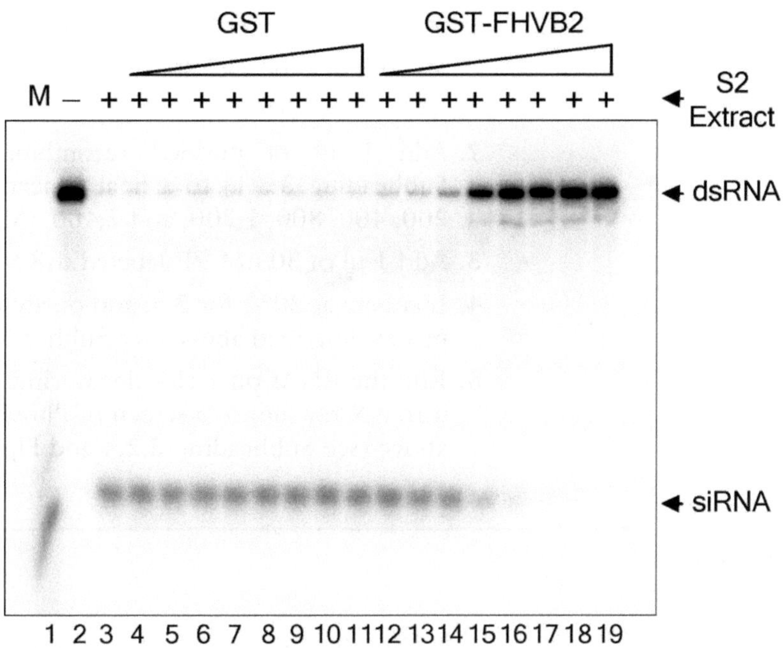

Fig. 3. FHVB2, a viral suppressor of RNAi, inhibits dsRNA processing in S2 extracts. ^{32}P-labeled dsRNA of eGFP was incubated with S2 extracts in presence of purified GST (*lane 4–11*) or GST-tagged flock house virus B2 (GST-FHVB2) proteins (*lane 12–19*), which were used at increasing concentrations (25, 50, 100, 200, 400, 800, 1,200, 2,400 nM). *M*, a labeled 21-nt siRNA, is used as size marker.

slurry of Glutathione Sepharose 4 Fast Flow, incubate on a rotary shaker at 4°C for 30–60 min.

6. Centrifuge at 500×*g* for 5 min, and carefully discard the supernatant. Wash the Sepharose medium with 10 ml of ice-cold 1× PBS twice.
7. Gently resuspend the Sepharose medium in 0.5 ml of elution buffer, incubate on a rotary shaker at 4°C for 10 min.
8. Centrifuge at 500×*g* for 5 min and carefully transfer the supernatant (the first eluted protein) into a fresh centrifuge tube.
9. Repeat steps 7 and 8 for a second elution. The concentration of GST-tagged protein is determined by standard Bradford assays and pool those eluted proteins together if both have significant amounts of proteins (see Note 12).

3.3.2. Inhibition of dsRNA Processing by Viral Proteins in S2 Extracts

1. Prepare a master mix for a reaction volume of 10 μl by combining the following components:
   (a) 1 μl 10× Buffer K.
   (b) 1 μl 10 mM ATP.
   (c) 0.5 μl 0.5 M Creatine phosphate.

(d) 0.3 μl 1 mg/ml Creatine phosphate kinase.

(e) 0.2 μl Ribonuclease inhibitor (40 units/μl).

(f) 5 μl S2 extract.

2. Add 1 μl of purified recombinant viral proteins (see Subheading 3.3.1) to a final concentration of 25, 50, 100, 200, 400, 800, 1,200, and 2,400 nM.
3. Add 1 μl of 50 nM ^{32}P-labeled dsRNA.
4. Incubate at 30°C for 2 h, and purify RNAs from the reaction mix as described above (see Subheading 3.2.2).
5. Run the RNAs on a 15% denaturing Urea-PAGE and expose it to a X-ray film or a screen of PhosphorImager as described above (see Subheading 3.2.4 and Fig. 3).

4. Notes

1. Before the template DNA is used in *in vitro* transcription for production of uniformly labeled dsRNA, it should be first tested in a small volume of transcription reaction without addition of [α-^{32}P] UTP. The dsRNA product needs to be analyzed by electrophoresis to ensure that the dsRNA is synthesized with high specificity and correct size.
2. Alternatively, dsRNAs can also be synthesized and labeled in the following reaction. Although it is not as efficient as the Ambion's MEGAScript T7 Kit, it can generate dsRNAs with sufficient quality and quantity for *in vitro* Dicer assay.

(a) 4 μl 1 M Hepes-KOH, pH 7.5.

(b) 2 μl 50 mM Spermidine.

(c) 6 μl 100 mM $MgCl_2$.

(d) 5 μl 100 mM DTT.

(e) 2 μl of ATP, CTP, GTP (75 mM each).

(f) 0.5 μl 75 mM UTP.

(g) 4 μl template DNA (0.5 μg/μl).

(h) 1 μl Ribonuclease Inhibitor (40 units/μl).

(i) 2 μl T7 RNA polymerase (20 units/μl).

(j) 4 μl [α-^{32}P] UTP.

(k) 15.5 μl of DEPC-H_2O, bringing the final volume up to 50 μl.

Incubate at 37°C for 2–4 h, and purify dsRNA product as described in Subheading 3.1.2.

3. At the step of dsRNA purification, some NTPs can be coprecipitated with dsRNA, leading to inaccurate measurement of dsRNA concentration. To solve this problem, Qiagen's RNAeasy columns could be used for further purification of dsRNAs.
4. Starting from this step, the sample should be kept or manipulated on ice as long as possible. Addition of KCl up to 60 mM in the hypotonic buffer is to prevent the cells from swelling too much and bursting when the cells are rinsed.
5. The S2 cell lysate can be centrifuged at 100,000 ×*g* at 4°C for 30 min (Optima TLX Benchtop Ultracentrifuge, Beckman Coulter). The supernatant could give a better result in *in vitro* dsRNA processing.
6. Processing of dsRNAs by Dicer-2 is an ATP-dependent event (5, 10, 29, 34). In the Dicer activity assay, ATP is supplied in the reaction up to a concentration of 1 mM. The presence of the phosphate-creatine kinase system could further facilitate siRNA production. However, efficient dsRNA processing can be still achieved without addition of the creatine phosphate and creatine phosphate kinase.
7. During purification of RNAs from Dicer assay reaction, it is easy to lose the RNA because of its trace amount. After extraction of Dicer reaction with phenol/chloroform, 1 μl of glycogen (20 μg/μl) is added into the aqueous phase to facilitate precipitation of RNAs by ethanol. Alternatively, 1 μl of sheared salmon sperm DNA (nuclease-free, 10 μg/μl) can also be used as carriers and performs as efficiently as glycogen. We usually use the Geiger counter to monitor the presence of RNAs in the desired fraction during the process. To minimize the loss of RNAs, washing the RNA pellet is not recommended. The RNA pellet can be kept at −20 or −80°C if the Urea-PAGE is not performed immediately.
8. Without any specific modification, chemically synthesized siRNA duplex, single-stranded siRNA, and DNA oligonucleotides usually have a 5′-OH group. By labeling with [γ-^{32}P] ATP, the 21-nt siRNA and 22-nt DNA oligonucleotides can be used as size markers in the Urea-PAGE for analysis of siRNAs that are processed from dsRNAs by the Dicer extract. In the gel, the 21-nt siRNA (single-stranded) migrates slightly slower compared to the 22-nt DNA oligonucleotide.
9. If labeled siRNA duplex is used as the size marker, it should be denatured by incubation at 95°C for 5 min followed by the ice-bath for 2 min before loading. Even after heat-treatment, the siRNA duplex could be still detected and migrates more slowly than the single-stranded siRNA. If DNA oligonucleotide or single-strand siRNA is used, heat-treatment is not necessary.

10. Fresh labeled size marker usually gives a much stronger signal than siRNAs processed from labeled dsRNAs by the Dicer extract, resulting in an overexposed image of siRNA marker. Before loading, labeled siRNA size marker should be measured by the Geiger counter, and appropriate dilution should be made to give the amount of loaded marker with a reading of ~200–300 cpm. Also, it is recommended to leave one or two lanes blank between the sample and the size marker.
11. The signal from siRNAs processed from the fresh labeled dsRNA by the extract is usually strong so that exposure of the film or PhosphorImager screen could be carried out in one or a few hours. The weak signal due to poor labeling or old [α-^{32}P] UTP requires a longer exposure time, which may result in diffuse bands of the siRNA. In that case, the gel has to be dried before film exposure.
12. To monitor the quantity and quality of the purified recombinant viral protein, it is necessary to analyze the protein on SDS-PAGE before use for Dicer inhibition assay.

Acknowledgments

This work was supported partially by Hawaii Community Foundation (the Victoris S. and Bradley L. Geist Foundation, 20071370) and the grant (P20RR018727) from the Centers of Biomedical Research Excellence, NIH.

References

1. Hannon, G.J. (2002) RNA interference. *Nature*, **418**, 244–251.
2. Tomari, Y. and Zamore, P.D. (2005) Perspective: machines for RNAi. *Genes Dev*, **19**, 517–529.
3. Lee, Y.S., Nakahara, K., Pham, J.W., Kim, K., He, Z., Sontheimer, E.J. and Carthew, R.W. (2004) Distinct roles for Drosophila Dicer-1 and Dicer-2 in the siRNA/miRNA silencing pathways. *Cell*, **117**, 69–81.
4. Okamura, K., Ishizuka, A., Siomi, H. and Siomi, M.C. (2004) Distinct roles for Argonaute proteins in small RNA-directed RNA cleavage pathways. *Genes Dev*, **18**, 1655–1666.
5. Bernstein, E., Caudy, A.A., Hammond, S.M. and Hannon, G.J. (2001) Role for a bidentate ribonuclease in the initiation step of RNA interference. *Nature*, **409**, 363–366.
6. Hammond, S.M., Boettcher, S., Caudy, A.A., Kobayashi, R. and Hannon, G.J. (2001) Argonaute2, a link between genetic and biochemical analyses of RNAi. *Science*, **293**, 1146–1150.
7. Hammond, S.M., Bernstein, E., Beach, D. and Hannon, G.J. (2000) An RNA-directed nuclease mediates post-transcriptional gene silencing in Drosophila cells. *Nature*, **404**, 293–296.
8. Liu, J., Carmell, M.A., Rivas, F.V., Marsden, C.G., Thomson, J.M., Song, J.J., Hammond, S.M., Joshua-Tor, L. and Hannon, G.J. (2004) Argonaute2 is the catalytic engine of mammalian RNAi. *Science*, **305**, 1437–1441.
9. Denli, A.M., Tops, B.B., Plasterk, R.H., Ketting, R.F. and Hannon, G.J. (2004) Processing of primary microRNAs by the Microprocessor complex. *Nature*, **432**, 231–235.
10. Jiang, F., Ye, X., Liu, X., Fincher, L., McKearin, D. and Liu, Q. (2005) Dicer-1 and R3D1-L

catalyze microRNA maturation in Drosophila. *Genes Dev*, **19**, 1674–1679.

11. Liu, X., Park, J.K., Jiang, F., Liu, Y., McKearin, D. and Liu, Q. (2007) Dicer-1, but not Loquacious, is critical for assembly of miRNA-induced silencing complexes. *RNA*, **13**, 2324–2329.
12. Saito, K., Ishizuka, A., Siomi, H. and Siomi, M.C. (2005) Processing of pre-microRNAs by the Dicer-1-Loquacious complex in Drosophila cells. *PLoS Biol*, **3**, e235.
13. Li, H.W. and Ding, S.W. (2005) Antiviral silencing in animals. *FEBS Lett*, **579**, 5965–5973.
14. Flynt, A., Liu, N., Martin, R. and Lai, E.C. (2009) Dicing of viral replication intermediates during silencing of latent Drosophila viruses. *Proc Natl Acad Sci U S A*, **106**, 5270–5275.
15. van Rij, R.P., Saleh, M.C., Berry, B., Foo, C., Houk, A., Antoniewski, C. and Andino, R. (2006) The RNA silencing endonuclease Argonaute 2 mediates specific antiviral immunity in Drosophila melanogaster. *Genes Dev*, **20**, 2985–2995.
16. Wang, M.B. and Metzlaff, M. (2005) RNA silencing and antiviral defense in plants. *Curr Opin Plant Biol*, **8**, 216–222.
17. Ding, S.W. and Voinnet, O. (2007) Antiviral immunity directed by small RNAs. *Cell*, **130**, 413–426.
18. Campbell, C.L., Keene, K.M., Brackney, D.E., Olson, K.E., Blair, C.D., Wilusz, J. and Foy, B.D. (2008) Aedes aegypti uses RNA interference in defense against Sindbis virus infection. *BMC Microbiol*, **8**, 47.
19. Campbell, C.L., Black, W.C., Hess, A.M. and Foy, B.D. (2008) Comparative genomics of small RNA regulatory pathway components in vector mosquitoes. *BMC Genomics*, **9**, 425.
20. Aliyari, R. and Ding, S.W. (2009) RNA-based viral immunity initiated by the Dicer family of host immune receptors. *Immunol Rev*, **227**, 176–188.
21. Aliyari, R., Wu, Q., Li, H.W., Wang, X.H., Li, F., Green, L.D., Han, C.S., Li, W.X. and Ding, S.W. (2008) Mechanism of induction and suppression of antiviral immunity directed by virus-derived small RNAs in Drosophila. *Cell Host Microbe*, **4**, 387–397.
22. Li, W.X., Li, H., Lu, R., Li, F., Dus, M., Atkinson, P., Brydon, E.W., Johnson, K.L., Garcia-Sastre, A., Ball, L.A. *et al.* (2004) Interferon antagonist proteins of influenza and vaccinia viruses are suppressors of RNA silencing. *Proc Natl Acad Sci U S A*, **101**, 1350–1355.
23. Wang, X.H., Aliyari, R., Li, W.X., Li, H.W., Kim, K., Carthew, R., Atkinson, P. and Ding, S.W. (2006) RNA interference directs innate immunity against viruses in adult Drosophila. *Science*, **312**, 452–454.
24. Li, W.X. and Ding, S.W. (2001) Viral suppressors of RNA silencing. *Curr Opin Biotechnol*, **12**, 150–154.
25. Li, F. and Ding, S.W. (2006) Virus counterdefense: diverse strategies for evading the RNA-silencing immunity. *Annu Rev Microbiol*, **60**, 503–531.
26. Lichner, Z., Silhavy, D. and Burgyan, J. (2003) Double-stranded RNA-binding proteins could suppress RNA interference-mediated antiviral defences. *J Gen Virol*, **84**, 975–980.
27. Lu, R., Maduro, M., Li, F., Li, H.W., Broitman-Maduro, G., Li, W.X. and Ding, S.W. (2005) Animal virus replication and RNAi-mediated antiviral silencing in Caenorhabditis elegans. *Nature*, **436**, 1040–1043.
28. Singh, G., Popli, S., Hari, Y., Malhotra, P., Mukherjee, S. and Bhatnagar, R.K. (2009) Suppression of RNA silencing by Flock house virus B2 protein is mediated through its interaction with the PAZ domain of Dicer. *FASEB J*, **23**, 1845–1857.
29. Liu, Q., Rand, T.A., Kalidas, S., Du, F., Kim, H.E., Smith, D.P. and Wang, X. (2003) R2D2, a bridge between the initiation and effector steps of the Drosophila RNAi pathway. *Science*, **301**, 1921–1925.
30. Liu, X., Jiang, F., Kalidas, S., Smith, D. and Liu, Q. (2006) Dicer-2 and R2D2 coordinately bind siRNA to promote assembly of the siRISC complexes. *RNA*, **12**, 1514–1520.
31. Echeverri, C.J. and Perrimon, N. (2006) High-throughput RNAi screening in cultured cells: a user's guide. *Nat Rev Genet*, **7**, 373–384.
32. March, J.C. and Bentley, W.E. (2007) Methods for gene silencing with RNAi. *Methods Mol Biol*, **388**, 427–434.
33. Li, H., Li, W.X. and Ding, S.W. (2002) Induction and suppression of RNA silencing by an animal virus. *Science*, **296**, 1319–1321.
34. Nykänen, A., Haley, B. & Zamore, P.D. (2001) ATP requirements and small interfering RNA structure in the RNA interference pathway. *Cell*, **107**, 309–321.

Chapter 14

Slicer Activity in *Drosophila melanogaster* S2 Extract

Arabinda Nayak and Raul Andino

Abstract

Reconstitution of RNA-inducing silencing complex (RISC) in vitro is a powerful biochemical technique to analyze crucial steps in RNA interference (RNAi) pathways. RISC contains an RNase enzyme, Argonaute, which is guided by small interfering RNA (siRNA) to recognize and silence its targets. *Drosophila* S2 cell extract is a good source of enzymes and factors that faithfully recapitulate essential steps in RISC assembly and function. In this chapter, we will describe how to prepare enzymatically active cell-free S2 extract to analyze the Slicer activity of RISC as well as the effects of viral RNAi suppressors which block this process.

Key words: *Drosophila*, Cell-free extracts, RNAi, Argonaute, siRISC, mRNA cleavage

1. Introduction

RNA interference (RNAi) is a mechanism for double-stranded RNA (dsRNA)-induced, sequence-specific silencing of cognate genes (1–3). Cellular events like replication of RNA viruses, convergent transcription of cellular genes, and transcription of parasitic genetic elements such as transposons generate dsRNAs, which are processed into discrete 21–25 nt RNA duplexes called small interfering RNAs (siRNAs) (2, 4). Artificial introduction of dsRNA into *Drosophila* S2 cells by transfection triggers a similar response (5). siRNA duplexes act as signature molecules that guide the formation of RNA-inducing silencing complex (RISC). In this chapter we discuss about RISC initiated by siRNA hence referred as siRISC. The loaded siRISC then selects its mRNA target by sequence homology and induces target mRNA degradation. RISC competent for target degradation has been purified from *Drosophila* S2 cells (6) and the minimal components required for such cleavage have been defined as a 232 kDa

Ronald P. van Rij (ed.), *Antiviral RNAi: Concepts, Methods, and Applications*, Methods in Molecular Biology, vol. 721,
DOI 10.1007/978-1-61779-037-9_14,

complex (7). Although several additional protein components like Tudor-SN (Tudor-Staphylococcal Nuclease), VIG (Vasa Intronic Gene), and dFXR (*Drosophila* Homologue of Human Fragile X mental Retardation protein) have been shown to copurify with siRISC (8, 9), Argonaute 2 is the sole nuclease required for mRNA target cleavage (10). Indeed, messenger RNA cleavage can be reconstituted in vitro using recombinant Argonaute 2 and single-stranded siRNA, suggesting that siRNA and Argonaute 2 are the core of the RISC machinery (11, 12).

Recent biochemical and structural analyses of key components of the RNAi pathways shed light on the molecular mechanisms that dictate the specific function of each RNAi component (13–17). In *Drosophila*, RNase III motifs of Dicer 2 catalyze the cleavage of long dsRNA to produce 21–25 nt siRNAs, a process called *Dicing* (4, 18–20). siRNAs are a hallmark of the *Dicing* process and have a characteristic 3′ hydroxyl with dinucleotide overhang and a 5′ monophosphate (19). In the subsequent step, *Assembly*, each siRNA duplex molecule is channeled through a sequential modification process with the aid of several protein factors, including R2D2, Dicer 2, and Argonaut 2, to form a catalytically competent RISC called holo-RISC (12, 21–24). R2D2 is an RNA-binding adaptor protein that senses the thermodynamically stable end of the double-stranded siRNA, and with the aid of Dicer 2, which recognizes the unstable end, facilitates siRNA loading (21, 22, 25, 26). Both Dicer 2 and Argonaute 2 contain PAZ (Piwi/Argonaute/Zwile) domains that have been shown to recognize and bind to 3′ 2-nt overhangs of siRNA (6, 18, 27, 28, 33). The overlapping PAZ domain in Dicer 2 and Argonaute 2 facilitates the transfer of the siRNA from Dicer 2 to Argonaute 2, providing fidelity to this transition process (29). This is supported by the finding that Dicer 2 and Argonaute 2 interact with each other in *Drosophila* cells (6, 22). In the final stage of *Assembly* which leads to activation of RISC, Argonaute 2 initiates duplex unwinding at the less stable end of the siRNA (25, 30, 31). Subsequently, the passenger strand is cleaved and the guide strand retained (10, 12, 32). This process is facilitated by C3PO (*component 3* promoter of RISC), a Mg^{2+}-dependent endoribonuclease (33). Argonaute 2 complexed with guide strand is a fully competent multiturnover RNase enzyme capable of several rounds of target RNA degradation (34). Argonaute 2 possesses RNase H activity (Slicer) and catalyzes a Mg^{2+}-dependent mRNA cleavage at position between 10 and 11 nt from the 5′ end of the guide strand siRNA (14, 20, 34, 35). This process is called *Slicing*. In *Drosophila* cell-free extract, both *Dicing* and *Assembly* are ATP-dependant process, whereas *Slicing* is an ATP-independent processes (7, 23).

We have shown in *Drosophila* that RNAi serves as an adaptive antiviral defense mechanism against invading insects *Dicistroviruses* (36). As a countermeasure, insect viruses encode RNAi suppressors to perturb RISC function and pathogenesis. Our current studies reveal that *Drosophila C Virus* (DCV) and *Cricket Paralysis Virus* (CrPV) encode RNAi suppressors (1A) with different biochemical properties and hence inhibit different steps in the RNAi pathway (36, 37). RNAi suppressor DCV-1A is an RNA-binding protein that inhibits the *Dicing* steps in RISC *Assembly*; on the other hand, CrPV-1A inhibits *Slicing* activities of holo-RISC. In the absence of specific pharmacological inhibitors of RNAi components, viral suppressors provide exciting opportunities to dissect and understand intricate molecular mechanism of the RISC assembly and function in vivo and in vitro. In this chapter, we discuss the Slicer activity of RISC in cell-free S2 extract and some insights into the effects of RNAi suppressors on RISC catalysis. This RISC cleavage assay can be used as a tool to understand the molecular requirements of various viral suppressors to exert their function on RISC in vitro. Additionally, it could be used for candidate siRNA target validation for applications that include the study of loss of function phenotype in cell culture for reverse genetics approach as well as for therapeutic intervention in animals.

2. Materials

2.1. Cell Culture and Extract Preparation

1. *Drosophila* S2 cells (Invitrogen).
2. Schneider's *Drosophila* medium (Invitrogen).
3. Penicillin streptomycin glutamine solution (100×) (Invitrogen).
4. Fetal bovine serum (FBS), heat inactivated (Invitrogen).
5. *Drosophila* culture medium: Schneider's *Drosophila* medium, 10% FBS, 1× Penicillin Streptomycin Glutamine solution.
6. Tissue culture treated, 150 cm^2 cell culture flask (Corning).
7. Trypan blue (Invitrogen).
8. Hemocytometer.
9. Dulbecco's phosphate buffer saline (D-PBS), no calcium or magnesium salt: 0.2 g/L KH_2PO_4, 2.16 g/L $Na_2HPO_4 7H_2O$, 0.2 g/L KCl, 8.0 g/L NaCl.
10. Protease inhibitor tablet, EDTA Free (Roche).
11. S2 Lysis buffer: 30 mM HEPES-KOH (pH 7.4), 100 mM KOAc, 2 mM $MgCl_2$, add one EDTA free, protease inhibitor tablet (Roche) per 10 mL of buffer.
12. Protein assay kit (Bio-Rad).
13. Stereomicroscope (Nikon).
14. Sterile 1.5 mL eppendorf tube, sterile 15 and 50 mL graduated conical tubes.

2.2. DNA Template Preparation for In Vitro Transcription

1. DNA template, e.g., Luciferase template (pGL3 Firefly luciferase reporter vector, Promega).
2. Forward and Reverse primer: T7-Luc Forward: (gcg TAATACGACTCACCTATAGGGTGGAGAG CAACTGCATAAGG), T7-Luc Reverse: AGAATCTCACG CAGGCAGTTC.
3. Thermostable DNA polymerase enzyme *Pfu* and buffer (Stratagene).
4. Deoxyribonucleotide triphosphate (dNTP) mixture containing 10 mM each dATP, dGTP, dCTP, and dTTP (Fermentas).
5. GFX PCR DNA and Gel Band purification kit (GE Healthcare).
6. Nuclease-free water (Ambion).
7. Spectrophotometer/Nanodrop (Thermo Scientific).
8. Thermal Cycler.

2.3. In Vitro Transcription of RNA and Gel Purification

1. DNA template with bacteriophage T7 promoter.
2. 5× Transcription buffer: 400 mM HEPES-KOH (pH 7.5), 120 mM $MgCl_2$, 10 mM spermidine, 200 mM DTT.
3. NTP set, 100 mM solutions 4× 0.25 mL (Fermentas).
4. RNasin Ribonuclease Inhibitor (Promega).
5. TURBO DNase (Ambion).
6. Phenol:chloroform:isoamyl alcohol (25:24:1) (Invitrogen).
7. Ammonium acetate stop solution: 5 M NH_4OAc, 100 mM EDTA (Ambion).
8. RNA storage solution: 1 mM sodium citrate, pH 6.4 (Ambion).
9. 2× Gel-loading buffer II: 95% Formamide, 18 mM EDTA, and 0.025% SDS, Xylene Cyanol, and Bromophenol Blue (Ambion).
10. 6% Acrylamide denaturing gel: 1× TBE (Tris-Borate-ethylenediaminetetraacetic acid), 6 M urea, 6% acrylamide.
11. 1.5% TBE agarose gel and 10× TBE running buffer: 108 g Tris base, 55 g boric acid, 40 mL 0.5 M EDTA (pH 8.0) per liter of solution.
12. RNA elution buffer: 200 mM Tris–Cl, pH 7.5, 25 mM EDTA (pH 8.0), 600 mM sodium chloride, 2% w/v sodium dodecyl sulfate.
13. T7 RNA polymerase enzyme.
14. Absolute ethanol (200 proof).
15. Fluor-coated thin layer chromatography (TLC) plate (Ambion).

2.4. Radiolabel GTP Cap Labeling

1. Gel-purified RNA substrate.
2. Guanylyltransferase (GTR), 10× capping reaction buffer, *S*-adenosyl methionine (SAM) (10 mM) (Ambion).
3. [α-^{32}P] GTP (3,000 Ci/mmol) (Perkin Elmer).
4. MicroSpin G-50 column (GE HealthCare).
5. Glycogen (20 mg/mL, Roche).
6. SAM (New England Biolab): SAM bought from NEB is 32 mM and dilute it to 10 mM with SAM dilution buffer. SAM decomposes quickly when not stored at −20°C and is best for one time use.
7. SAM dilution buffer: 10% ethanol, 5 mM H_2SO_4.
8. Ultrafree-MC centrifugal filter units, pore size 0.1 μM (Millipore).

2.5. Slicer Assay

1. Amino acid mixture, complete (1 mM) (Promega).
2. Creatine phosphate (Fluka).
3. Creatine phosphokinase (CPK) (Calbiochem).
4. 40× RISC cleavage buffer (38): Prepare creatine phosphate in water, aliquot, and store at −70°C. Creatine kinase can be prepared in 30 mM HEPES pH 7.4, 50% glycerol and can be stored at −70°C for at least 6 months to 1 year. Store 1 mM amino acid mix at −70°C. Store rest of the components at −20°C and prepare 40× buffer fresh each time.
   (a) 50 μL Nuclease-free water
   (b) 20 μL 500 mM Creatine monophosphate
   (c) 20 μL 1 mM Amino acid mix
   (d) 2 μL 1 M DTT
   (e) 2 μL 20 U/μL RNasin
   (f) 4 μL 100 mM ATP
   (g) 1 μL 100 mM GTP
   (h) 6 μL 2 U/μL CPK
   (i) 16 μL of 1 M Potassium acetate
5. Radiolabeled capped luciferase mRNA substrate, as prepared in Subheading 3.4.
6. Luciferase siRNA (Luc siRNA): Passenger strand: 5′CUUACGCUGAGUACUUCGAdTdT, Guide strand: 5′ UCGAAGUACUCAGCGUAAGdTdT (Dharmacon).
7. GAPDH siRNA negative control (Ambion).
8. Proteinase K solution (10 mg/mL).

3. Methods

3.1. Drosophila S2 Cell Extract Preparation

1. Culture S2 cells at 25°C in *Drosophila* culture medium to a cell density of 2.0–3.0 × 10^7 cells/mL (see Note 1). Five T150 cm^2 flasks with 20 mL media per flask will yield sufficient cells for a good extraction.
2. Check cell number and viability by Trypan blue staining using a Hemocytometer (cells should have >90% viability). Harvest cells by gently tapping the flask and pipetting with a 10 mL pipet with the conditioned media (the media in which cells are grown in the flask). Transfer the cells to a prechilled 50 mL graduated conical tubes (25 mL cell suspension per 50 mL conical tube).
3. Centrifuge cells for 10 min at 1,000 × *g* at 4°C (see Note 2) using a prechilled centrifuge.
4. Pool all cells into a single 50 mL conical tube by resuspending each pellet with 5 mL of cold D-PBS and centrifuge for 5 min at 1,000 × *g* at 4°C.
5. Discard supernatant, resuspend the pellet in 5 mL of D-PBS, and transfer cell suspension to a prechilled 15 mL graduated conical tube and centrifuge at 1,000 × *g* for 5 min.
6. Discard the supernatant and measure the packed cell volume (PCV) (expect a cell volume of ~1 mL for a 100 mL culture).
7. Resuspend the cell pellet in 1 PCV of S2 Lysis buffer containing protease inhibitors.
8. Aliquot S2 cell suspension (100 μL/aliqout) into prechilled 1.5 mL eppendorf tubes, transfer to −80°C until further use.
9. To prepare the extract for your experiments, remove aliquots from −80°C and thaw on ice for 60 min.
10. Mix the tube content by tapping the tube, centrifuge for 20 min at 13,000 × *g*, and carefully collect the supernatant. Estimate the protein concentration of the extract using the protein assay kit. Supernatant will have both *Dicing* and *Slicing* activities (see Note 3).

3.2. Preparation of Radiolabeled GTP Capped mRNA

3.2.1. Designing Template DNA for Target RNA Synthesis

1. Design forward primers against the gene of interest preceded by a T7 promoter sequence. We add three nucleotides (ggc or gcg) upstream of the T7 promoter sequence (Fig. 1a) for efficient RNA polymerase binding and RNA production. T7 promoter sequence is followed by 20–22 nucleotide sequences corresponding to the gene of interest. Design a 20–22 nucleotides reverse primer complementary to the gene to produce a PCR fragment ranging from 200 to 600 bp in length.

a `T7-Luc Forward:` `gcg` *TAATACGACTCACTATAGGG* **TGGAGAGCAACTGCATAAGG**

`T7-Luc Reverse:` **AGAATCTCACGCAGGCAGTTC**

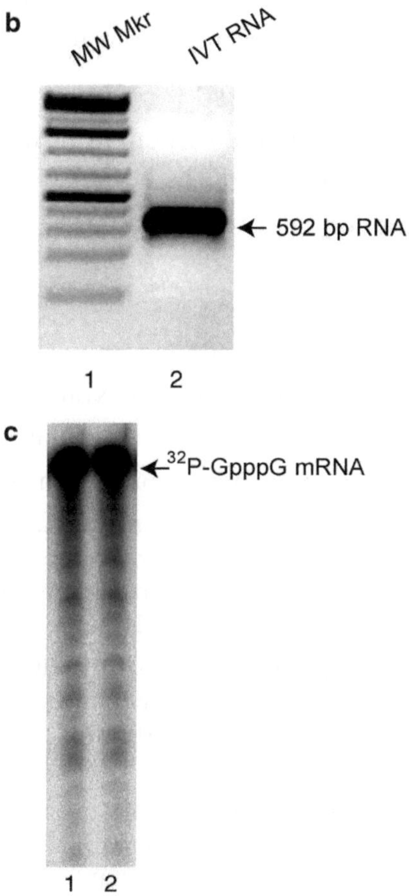

Fig. 1. Radiolabeled capped mRNA synthesis. (**a**) Primers for polymerase chain reaction (PCR) amplification of a DNA template for in vitro transcription. Forward primer (T7-Luc Forward) containing firefly luciferase (Luc) sequence (*bold capital*) flanked by a T7 promoter sequence (*italics*) at the 5′ end was designed. The reverse primer (T7-Luc Reverse) is complementary to Luc, but does not contain a T7 promoter. Primer pairs were used to amplify a 592 base pair DNA product from pGL3 template. The amplified PCR product with T7 promoter sequence was gel-purified and used as a template for the transcription reaction using recombinant T7 RNA polymerase. (**b**) RNA synthesized in vitro was gel-purified. A molecular weight marker (*lane 1*) was run along gel-purified RNA (*lane 2*) on a 1.5% tris-borate-ethylenediaminetetraacetic acid (TBE) agarose gel electrophoreis. (**c**) Gel-purified RNA was used as a substrate for radiolabeled GTP capping. Integrity of the capped transcripts was verified by autoradiography using 6% denaturing urea-polyacrylamide gel electrophoresis (*lane 1* and *2*).

2. Amplify the desired DNA fragment in 50 μL PCR reactions. Set up multiple parallel PCR reactions to have sufficient starting material for multiple in vitro transcription reactions (for example, one 8-tubes strip of PCR reactions, corresponding to 400 μL of PCR reaction mix).

3. To remove any nonspecific amplification products and other contaminating materials, purify the PCR DNA fragment using low-melting agarose and GFX PCR DNA Gel Band purification kit. Dissolve purified DNA in nuclease-free water.
4. Quantify using a Nanodrop or Spectrophotometer and dilute to a final concentration of 0.1 μg/μL.

3.3. In Vitro Transcription of Target mRNA

1. Set up a 120 μL RNA transcription reaction, by mixing:
   (a) 35 μL DNA template (0.1 μg/μL)
   (b) 24 μL 5× Transcription buffer
   (c) 36 μL 25 mM each NTP mix
   (d) 19 μL RNase-free water
   (e) 4.5 μL T7 polymerase (15 μM/μL)
   (f) 1.5 μL (20 U/μL) RNase inhibitor (see Note 4)
   Incubate the reaction at 37°C from 6 h to overnight
   Add 1 μL TURBO DNase and incubate for 25 min at 37°C
2. Add 20 μL ammonium acetate stop solution and 60 μL nuclease-free water to 121 μL in vitro transcription reaction. Mix with equal volume (200 μL) of phenol:chloroform:isoamyl alcohol (25:24:1) and centrifuge at 13,000 × *g* for 10 min.
3. Carefully collect aqueous supernatant (approximately 160 μL) so as not to contaminate with organic solvent and precipitate with 2.5 volumes of absolute ethanol (200 proof). Place the tube at −20°C for 6 h to overnight.
4. Centrifuge for 15 min at 13,000 × *g* at 4°C. Remove the supernatant, wash the pellet with 1 mL of 70% ethanol, and spin for 5 min at 4°C. Discard the supernatant, air-dry the pellet for a brief period (be careful not to overdry the pellet), and dissolve in 100 μL of RNA storage solution.
5. Quantify RNA concentration using a spectrophotometer or a Nanodrop. The typical yield is about 0.5 mg of 600 bases RNA per in vitro transcription. Check the integrity and quality of RNA on a 1.5% TBE agarose gel (see Note 4).
6. Gel purification of the in vitro-transcribed RNA: Mix RNA sample with equal volume of 2× gel-loading buffer II, heat for 5 min at 75°C, and immediately transfer the denatured sample to ice for 2 min. Run the sample using a 6% denaturing urea-polyacrylamide gel.
7. Visualize the RNA by UV shadowing: carefully remove the gel from glass plates and cover it with a plastic wrap. Place the plastic-wrapped gel on top of a fluor-coated TLC plate and visualize the RNA band in a dark room by exposing gel with a hand-held UV light source (254 nm). RNA band will appear as a purple band, excise the band and cut the gel band into small pieces.

8. Elute RNA by incubating gel pieces in RNA elution buffer overnight at 37°C with constant shaking.
9. Extract the eluate with phenol as described in steps 2–4. Resuspend the RNA pellet in nuclease-free water, quantify the sample, and dilute to a final concentration of 10 pmol/μL. Store in small aliquots of RNA sample and store at −70°C for future use. Check the quality of the RNA on a 1.5% TBE agarose gel (Fig. 1b).

3.4. GTP Cap Labeling of Target mRNA

1. Set up a 30 μL reaction by mixing
   (a) 2.0 μL Gel-purified RNA (10 pmol/μL)
   (b) 3.0 μL 10× Capping reaction buffer
   (c) 2.0 μL SAM (10 mM stock)
   (d) 1.0 μL RNasin (20 U/μL)
   (e) 2.0 μL GTR
   (f) 10 μL [α-^{32}P] GTP (3,000 Ci/mmol) (see Note 5)
   (g) 10 μL Nuclease-free water

   Incubate the reaction for 60 min at 37°C
2. Remove unincorporated nucleotides using MicroSpin G-50 column. Mix the flow-through with an equal volume of 2× gel-loading buffer II, heat for 5 min at 75°C, and run the sample through a 6% urea-polyacrylamide gel as described in Subheading 3.3, step 6.
3. Carefully remove the gel from glass plates and cover the gel with a plastic wrap. Expose the gel on X-ray film for 1–5 min and detect the radiolabeled band by autoradiography (Fig. 1c). Excise the gel band, break it into small pieces, and elute radiolabeled capped RNA in RNA elution buffer overnight as described Subheading 3.3, step 8.
4. Extract the eluate with phenol:chloroform:isoamyl alcohol (25:24:1) and precipitate with 2.5 volumes of absolute ethanol and 1 μL of 20 mg/mL glycogen. Incubate the sample overnight at −20°C. Pellet the sample at 13,000 × *g* for 10 min at 4°C, wash with 70% ethanol, spin for 5 min at 4°C, and dissolve the pellet in nuclease-free water.
5. Measure the specific activity of the transcript via scintillation counting, dilute the sample to an activity of 30,000 cpm/μL, and store in small aliquots at −20°C. For an alternative method for GTP cap labeling, see Note 6.

3.5. In Vitro Slicer Assay

In S2 extracts, RISC *Assembly* and mRNA *Slicing* can be achieved by incubating siRNA duplex and cognate target mRNA. Since cap-radiolabeled mRNA is stable in S2 extract, a 5′ cleaved product induced by the siRNA duplex can be detected by denaturing polyacrylamide gel electrophoresis. The RNAi suppressor CrPV-1A blocks mRNA cleavage by interfering with the function of Argonaute 2; DCV-1A, on the other hand, has no observable effects on Slicer activity (37). Thus, CrPV-1A functions as an inhibitor of mRNA degradation mediated by Argonaute 2 enzyme and serves as a negative regulator of the RISC cleavage activity. Recombinant CrPV-1A and DCV-1A purified from *Escherichia coli* retain their functionality and can be supplemented in the RISC assay (37).

1. Assemble 35 μL Slicer assay reactions on ice by mixing in the following order:
   (a) 14 μL S2 extract (17 mg/mL, 40% of reaction volume)
   (b) 10.2 μL 40× cleavage buffer
   (c) 1.0 μL siRNA (100 nM)
   (d) 7.8 μL S2 lysis buffer

   Incubate at 25°C for 30 min. This incubation period allows the RISC to fully assemble in S2 extract. Subsequently add:
   (e) 1.0 μL recombinant RNAi suppressor (1A) protein (350 nM)
   Mix and incubate at 25°C for another 10 min. Then add:
   (f) 1.0 μL Capped mRNA (30,000 cpm/~10 nM 600 bases mRNA).

   Incubate at 25°C for additional 2 h 30 min.

   Follow Note 7 and 8 while performing the reaction.
2. Add 130 μL of RNA elution buffer and 35 μL proteinase K solution (10 mg/mL) to the Slicer reaction and incubate at 65°C for 30 min.
3. Extract with phenol:chloroform:isoamyl (25:24:1) alcohol and add 1 μL glycogen (20 mg/mL) and 480 μL of absolute ethanol to the aqueous phase (~160 μL). Precipitate RNA as described in Subheading 3.4, step 4.
4. Dissolve the pellet in 25 μL 2× gel-loading buffer II, denature the sample, and resolve the 5′ cleaved RNA product in a 6% urea-polyacrylamide gel. Analyze the 5′ cleaved product by phosphorimaging (Fig. 2).

4. Notes

1. While culturing S2 cells, do not split to a density below 0.5×10^6/mL. S2 cells tend to clump at low density and grow poorly.
2. While making S2 extract, care should be taken to maintain cold temperature throughout the procedure to prevent

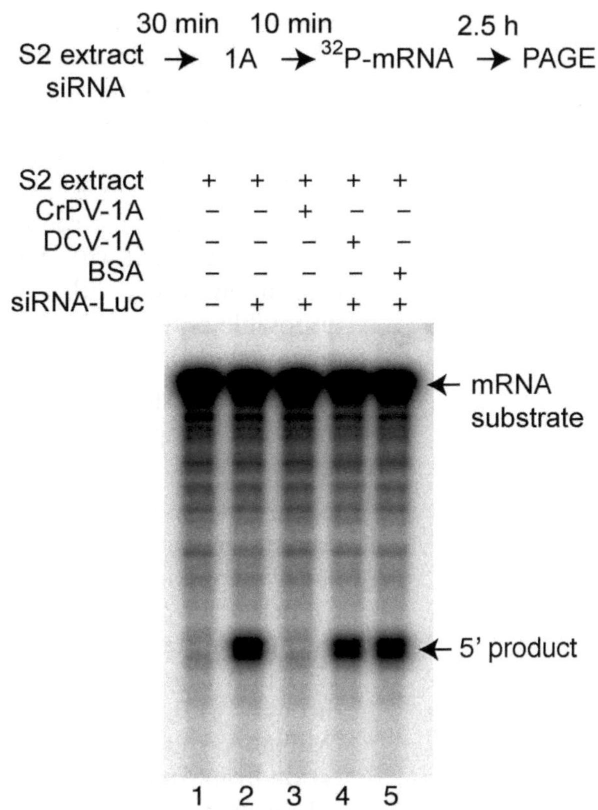

Fig. 2. Slicer assay in S2 extracts. S2 extracts were programmed with small interfering RNA (siRNA) duplex for 30 min followed by addition of suppressor proteins (1A) for 10 min. RNA-inducing silencing complex (RISC) cleavage reaction was initiated by incubating mRNA target (Fig. 1c) for additional 2 h 30 min. A 100 nucleotides 5′ cleaved product was analyzed using 6% denaturing polyacrylamide gel electrophoresis (PAGE). *Lane 1* no protein no siRNA control; *lane 2* no protein control; *lane 3* 350 nM CrPV-1A; *lane 4* 350 nM DCV-1A; *lane 5* 350 nM BSA control. As expected, supplementation of CrPV-1A in S2 extracts inhibited Argonaute 2-mediated mRNA target cleavage; DCV-1A, on the other hand, did not have any effects.

inactivation of the RISC components. Significant Slicer activity can be lost exposing the extract to 30–37°C for 30 min (33). Once prepared, S2 extract can be stored at −80°C up to 6 months without significant loss of Slicer activity.

3. High-speed centrifugation (100,000 × *g*) cell-free extract of untreated S2 cells (S100) can be prepared and used for the RISC cleavage assay. However, treatment of S2 cells with dsRNA by soaking or transfection results in sedimentation of holo-RISC complex with high molecular weight ribosomal fractions rendering S100 extract inefficient in RISC cleavage assay (5).
4. Follow aseptic technique when working with RNA. Use latex or vinyl gloves while handling reagents and RNA samples to

prevent RNase contamination from the skin surface or other sources. Ribonucleases (RNases) are very stable enzymes and care should be taken to avoid introducing RNases into the RNA sample during or after the isolation procedure. Keep isolated RNA on ice all the time. Purified RNA may be stored at −20 or −70°C in RNA storage solution (Ambion) up to 1 year. For analyzing the RNA sample, electrophoresis tanks should be cleaned with detergent solution (e.g., 0.5% SDS), thoroughly rinsed with RNase-free water. The use of sterile, disposable polypropylene tubes throughout the procedure is recommended.

5. Always use fresh [α-^{32}P] GTP and gel-purified RNA for synthesis of the capped mRNA target
6. Previously, our lab used Vaccinia GTR enzyme from Ambion to cap mRNA. GTR from Ambion has recently been discontinued. We recommend obtaining an expression construct to purify recombinant GTR from bacteria and follow the protocol below for capped mRNA synthesis. Prepare 10× capping reaction buffer [500 mM Tris–HCl (pH 8.0), 60 mM KCl, 12.5 mM $MgCl_2$, 12.5 mM DTT, 0.5 mg/mL BSA], pass through a 0.1 μM ultrafree-MC centrifugal filter unit (Millipore), aliquot and store at −20°C (Personal communication from Ambion). Set a 20 μL reaction containing 2.0 μL gel-purified RNA (10 pmol/μL), 2.0 μL 10× capping reaction buffer, 1.0 μL SAM (10 mM stock), 1.0 μL RNasin (20 U/μL), 4.0 μL GTR (7.5 μM/μL), 10 μL [α-^{32}P] GTP (3,000 Ci/mmol). Incubate the reaction for 1 h at 37°C and process the sample as described in the Subheading 3.4.
7. Presence of 5′ phosphate in siRNA duplex is essential for productive RISC formation and mRNA cleavage. Synthetic siRNA with 5′OH undergoes rapid phosphorylation in S2 extracts (7) and can be used for analyzing the slicer activity.
8. RNAi is an ATP-dependent process and requires an ATP regenerating system in cell-free extract for efficient target cleavage (7, 23, 34). Supplementation of exogenous CPK enzyme and creatine phosphate substrate in the reaction facilitates transfer of phosphates from substrate to ADP to form ATP. While performing the *Slicing* assay, always use a fresh aliquot of creatine phosphate and creatine kinase.

Acknowledgments

We thank Phillip Zamore, Benjamin Haley, Carla Saleh, and Ronald van Rij for their suggestions in preparing S2 extract. We thank Michel Tassetto from the Andino laboratory for testing this protocol.

Thanks to Ashley Acevedo, Cecily Burrill, and Mark Borja for critical reading and valuable input while preparing the manuscript. Arabinda Nayak is a research fellow supported by N.I.H. grants (AI40085 and AI064738) to Raul Andino.

References

1. Kennerdell, J. R., and Carthew, R. W. (1998) Use of dsRNA-mediated genetic interference to demonstrate that frizzled and frizzled 2 act in the wingless pathway, *Cell* **95**, 1017–1026.
2. van Rij, R. P., and Berezikov, E. (2009) Small RNAs and the control of transposons and viruses in *Drosophila*, *Trends Microbiol* **17**, 163–171.
3. Yang, D., Lu, H., and Erickson, J. W. (2000) Evidence that processed small dsRNAs may mediate sequence-specific mRNA degradation during RNAi in *Drosophila* embryos, *Curr Biol* **10**, 1191–1200.
4. Zamore, P. D., Tuschl, T., Sharp, P. A., and Bartel, D. P. (2000) RNAi: double-stranded RNA directs the ATP-dependent cleavage of mRNA at 21 to 23 nucleotide intervals, *Cell* **101**, 25–33.
5. Hammond, S. M., Bernstein, E., Beach, D., and Hannon, G. J. (2000) An RNA-directed nuclease mediates post-transcriptional gene silencing in *Drosophila* cells, *Nature* **404**, 293–296.
6. Hammond, S. M., Boettcher, S., Caudy, A. A., Kobayashi, R., and Hannon, G. J. (2001) Argonaute2, a link between genetic and biochemical analyses of RNAi, *Science* **293**, 1146–1150.
7. Nykanen, A., Haley, B., and Zamore, P. D. (2001) ATP requirements and small interfering RNA structure in the RNA interference pathway, *Cell* **107**, 309–321.
8. Caudy, A. A., Myers, M., Hannon, G. J., and Hammond, S. M. (2002) Fragile X-related protein and VIG associate with the RNA interference machinery, *Genes Dev* **16**, 2491–2496.
9. Caudy, A. A., Ketting, R. F., Hammond, S. M., Denli, A. M., Bathoorn, A. M., Tops, B. B., Silva, J. M., Myers, M. M., Hannon, G. J., and Plasterk, R. H. (2003) A micrococcal nuclease homologue in RNAi effector complexes, *Nature* **425**, 411–414.
10. Rand, T. A., Petersen, S., Du, F., and Wang, X. (2005) Argonaute2 cleaves the anti-guide strand of siRNA during RISC activation, *Cell* **123**, 621–629.
11. Rivas, F. V., Tolia, N. H., Song, J. J., Aragon, J. P., Liu, J., Hannon, G. J., and Joshua-Tor, L. (2005) Purified Argonaute2 and an siRNA form recombinant human RISC, *Nat Struct Mol Biol* **12**, 340–349.
12. Miyoshi, K., Tsukumo, H., Nagami, T., Siomi, H., and Siomi, M. C. (2005) Slicer function of *Drosophila* Argonautes and its involvement in RISC formation, *Genes Dev* **19**, 2837–2848.
13. Lingel, A., Simon, B., Izaurralde, E., and Sattler, M. (2003) Structure and nucleic-acid binding of the *Drosophila* Argonaute 2 PAZ domain, *Nature* **426**, 465–469.
14. Parker, J. S., Roe, S. M., and Barford, D. (2004) Crystal structure of a PIWI protein suggests mechanisms for siRNA recognition and slicer activity, *EMBO J* **23**, 4727–4737.
15. Collins, R. E., and Cheng, X. (2005) Structural domains in RNAi, *FEBS Lett* **579**, 5841–5849.
16. Patel, D. J., Ma, J. B., Yuan, Y. R., Ye, K., Pei, Y., Kuryavyi, V., Malinina, L., Meister, G., and Tuschl, T. (2006) Structural biology of RNA silencing and its functional implications, *Cold Spring Harb Symp Quant Biol* **71**, 81–93.
17. Jinek, M., and Doudna, J. A. (2009) A three-dimensional view of the molecular machinery of RNA interference, *Nature* **457**, 405–412.
18. Tomoyasu, Y., Miller, S. C., Tomita, S., Schoppmeier, M., Grossmann, D., and Bucher, G. (2008) Exploring systemic RNA interference in insects: a genome-wide survey for RNAi genes in *Tribolium*, *Genome Biol* **9**, R10.
19. Bernstein, E., Caudy, A. A., Hammond, S. M., and Hannon, G. J. (2001) Role for a bidentate ribonuclease in the initiation step of RNA interference, *Nature* **409**, 363–366.
20. Elbashir, S. M., Lendeckel, W., and Tuschl, T. (2001) RNA interference is mediated by 21- and 22-nucleotide RNAs, *Genes Dev* **15**, 188–200.
21. Liu, Q., Rand, T. A., Kalidas, S., Du, F., Kim, H. E., Smith, D. P., and Wang, X. (2003) R2D2, a bridge between the initiation and effector steps of the *Drosophila* RNAi pathway, *Science* **301**, 1921–1925.
22. Pham, J. W., Pellino, J. L., Lee, Y. S., Carthew, R. W., and Sontheimer, E. J. (2004) A Dicer-2-dependent 80 s complex cleaves targeted

mRNAs during RNAi in *Drosophila*, *Cell* **117**, 83–94.

23. Pham, J. W., and Sontheimer, E. J. (2005) Molecular requirements for RNA-induced silencing complex assembly in the *Drosophila* RNA interference pathway, *J Biol Chem* **280**, 39278–39283.
24. Tomari, Y., Du, T., Haley, B., Schwarz, D. S., Bennett, R., Cook, H. A., Koppetsch, B. S., Theurkauf, W. E., and Zamore, P. D. (2004) RISC assembly defects in the *Drosophila* RNAi mutant armitage, *Cell* **116**, 831–841.
25. Tomari, Y., Matranga, C., Haley, B., Martinez, N., and Zamore, P. D. (2004) A protein sensor for siRNA asymmetry, *Science* **306**, 1377–1380.
26. Liu, X., Jiang, F., Kalidas, S., Smith, D., and Liu, Q. (2006) Dicer-2 and R2D2 coordinately bind siRNA to promote assembly of the siRISC complexes, *RNA* **12**, 1514–1520.
27. Song, J. J., Liu, J., Tolia, N. H., Schneiderman, J., Smith, S. K., Martienssen, R. A., Hannon, G. J., and Joshua-Tor, L. (2003) The crystal structure of the Argonaute2 PAZ domain reveals an RNA binding motif in RNAi effector complexes, *Nat Struct Biol* **10**, 1026–1032.
28. Yan, K. S., Yan, S., Farooq, A., Han, A., Zeng, L., and Zhou, M. M. (2003) Structure and conserved RNA binding of the PAZ domain, *Nature* **426**, 468–474.
29. Siomi, H., and Siomi, M. C. (2009) On the road to reading the RNA-interference code, *Nature* **457**, 396–404.
30. Khvorova, A., Reynolds, A., and Jayasena, S. D. (2003) Functional siRNAs and miRNAs exhibit strand bias, *Cell* **115**, 209–216.
31. Okamura, K., Ishizuka, A., Siomi, H., and Siomi, M. C. (2004) Distinct roles for Argonaute proteins in small RNA-directed RNA cleavage pathways, *Genes Dev* **18**, 1655–1666.
32. Matranga, C., Tomari, Y., Shin, C., Bartel, D. P., and Zamore, P. D. (2005) Passenger-strand cleavage facilitates assembly of siRNA into Ago2-containing RNAi enzyme complexes, *Cell* **123**, 607–620.
33. Liu, Y., Ye, X., Jiang, F., Liang, C., Chen, D., Peng, J., Kinch, L. N., Grishin, N. V., and Liu, Q. (2009) C3PO, an endoribonuclease that promotes RNAi by facilitating RISC activation, *Science* **325**, 750–753.
34. Haley, B., and Zamore, P. D. (2004) Kinetic analysis of the RNAi enzyme complex, *Nat Struct Mol Biol* **11**, 599–606.
35. Schwarz, D. S., Tomari, Y., and Zamore, P. D. (2004) The RNA-induced silencing complex is a Mg^{2+}-dependent endonuclease, *Curr Biol* **14**, 787–791.
36. van Rij, R. P., Saleh, M. C., Berry, B., Foo, C., Houk, A., Antoniewski, C., and Andino, R. (2006) The RNA silencing endonuclease Argonaute 2 mediates specific antiviral immunity in *Drosophila melanogaster*, *Genes Dev* **20**, 2985–2995.
37. Nayak, A., Berry, B., Tassetto, M., Kunitomi, M., Acevedo, A., Deng, C., Kruchinsky, A., Gross, J., Antoniewski, C., and Andino, R. (2010) *Cricket paralysis virus* (CrPV) antagonizes Argonaute 2 to modulate antiviral defense in *Drosophila*, *Nat Struct Mol Biol* **17**(5), 547–554.
38. Haley, B., Tang, G., and Zamore, P. D. (2003) In vitro analysis of RNA interference in *Drosophila melanogaster*, *Methods* **30**, 330–336.

Chapter 15

Gel Mobility Shift Assays for RNA Binding Viral RNAi Suppressors

Tibor Csorba and József Burgyán

Abstract

The host–virus interaction is a continuous coevolutionary race involving both host defence strategies and virus escape mechanisms. RNA silencing is one of the main processes employed by eukaryotic organisms to fight viruses. However, viruses encode suppressor proteins to counteract this antiviral mechanism. Virtually all plant viruses encode at least one suppressor. In spite of being highly diverse at the protein level, a large group of these proteins inhibit RNA silencing very similarly, by sequestration of double-stranded RNA or small-interfering RNA molecules, the central players of the pathway. The RNA binding capacity of virus suppressor proteins can be studied by the electrophoretic mobility shift assay method. Also known as gel retardation assay, gel mobility assay, gel shift assay or band shift assay, EMSA is an in vitro technique used to characterize protein:DNA or protein:RNA interactions. The method had been developed based on the observation that protein: nucleic acid complexes migrate slower through a non-denaturing polyacrylamide gel than the free nucleic acid fragments.

Here, we provide a detailed protocol for the analysis of crucifer-infecting *Tobacco mosaic tobamovirus* (cr-TMV) silencing suppressor protein p122 RNA binding capacity.

Key words: Antiviral silencing, Viral suppressor proteins, siRNA, RNA binding, Electro mobility shift assay, Band shift assay, cr-TMV p122

1. Introduction

The electrophoretic mobility shift assay (EMSA) assay was developed based on the method described earlier (1, 11). EMSA is performed by incubating purified native protein or a mixture of proteins with a labelled RNA in a binding reaction. After the formation of protein: RNA complexes, they are separated based on their differences in electrophoretic mobility in a native acrylamide gel. The speed at which the RNA or the protein: RNA complexes move through the gel is determined by their size, charge

Ronald P. van Rij (ed.), *Antiviral RNAi: Concepts, Methods, and Applications*, Methods in Molecular Biology, vol. 721, DOI 10.1007/978-1-61779-037-9_15,

and shape. As the protein: RNA complex is less mobile compared to the free RNA, the signal is "shifted" up on the gel (see Fig. 1a–c). The protein–RNA interaction depends on the RNA binding affinity of the protein. The formed complexes are stabilized by the low ionic strength buffer used and by the caging effect of the acrylamide gel matrix during the course of electrophoretic separation. In the correct experimental conditions (see Note 1) the ratio of free to bound RNA is identical to the ratio formed in the binding reaction before entering in the gel. Knowing the concentration of

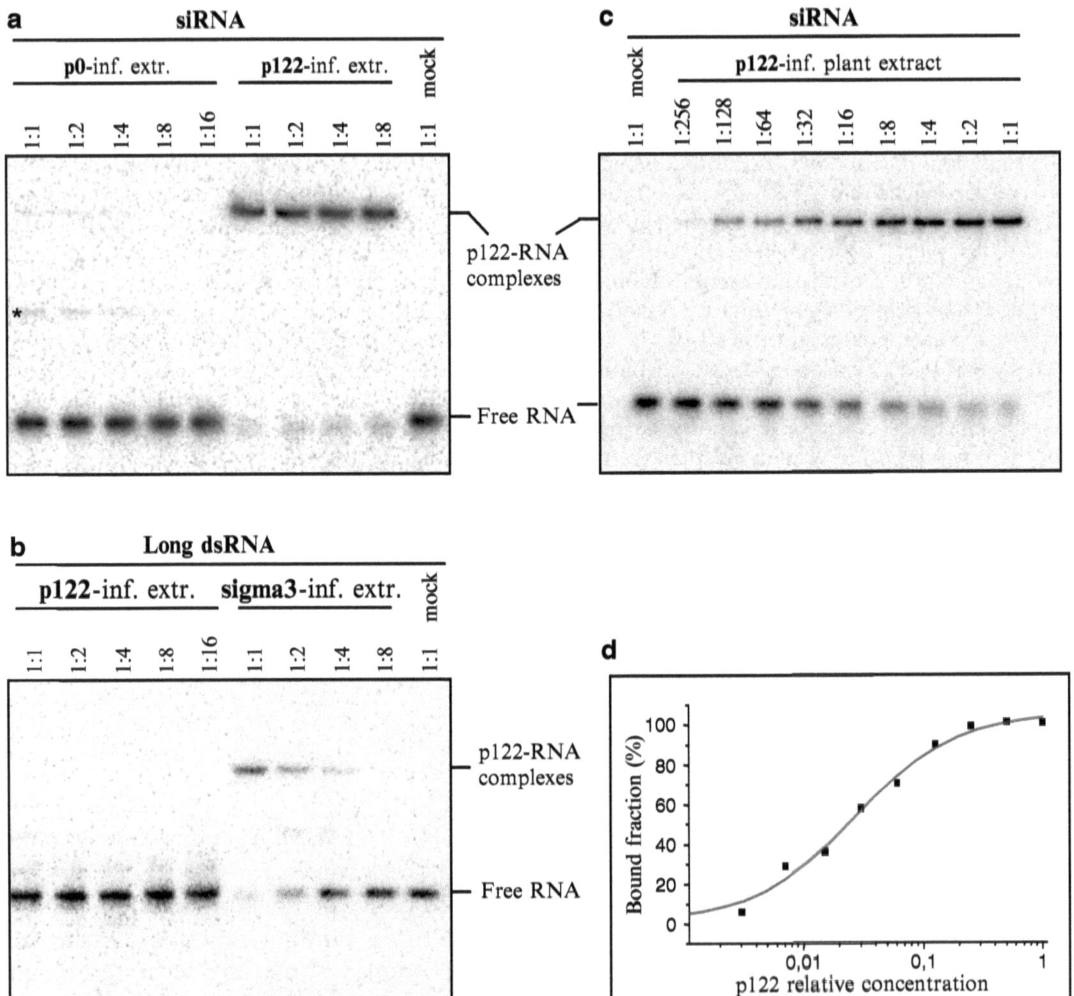

Fig. 1. Tobamoviral (cr-TMV) p122 protein but not polerovirus (BWYV) P0 protein binds 21 nt "bona fide" siRNA. ^{32}P-labelled siRNA molecules were incubated with a dilution series of P0 or p122-agroinfiltrated *Nicotiana benthamiana* plant extracts as indicated. 1 μg of total protein (1:1) or twofold dilutions (1:2, 1:4, etc.) was used in the band shift reactions. Empty-vector infiltrated plant extract (mock) was the negative control (**a**). p122 does not bind long dsRNA molecules (50 bp). dsRNA-binding Reovirus sigma3 protein infiltrated plant extract was used as positive control (**b**). The relative binding affinity of p122 protein was determined from a dilution series p122-siRNA complexes (**c**). Signals were quantified with ImageQuant software and a sigmoid curve was fitted with Origin software. Relative K_d = 0.026 μg total protein (**d**).

the protein of interest the dissociation constant (K_d) is estimated as the concentration of the protein required to give 50% saturation. If a mixture of proteins or a crude preparation is used (and the concentration of protein of interest is therefore not known) the relative dissociation constants can be estimated for different types of nucleic acids, such as siRNA, long dsRNA, single-stranded (ss) RNA, ssDNA, dsDNA, as a function of protein concentration and the relative affinities can be compared.

The method described in this chapter has been used to demonstrate the RNA binding affinity of cr-TMV p122 (2) and other virus silencing suppressor proteins (3–6).

2. Materials

2.1. siRNA Preparation

1. Synthetic 2′-ACE protected siRNAs (Dharmacon). Deprotection buffer: 100 mM acetic acid, adjusted to pH 3.8 with TEMED.
2. DNA oligonucleotides.
3. [γ-32P] ATP (3,000 Ci/mmol, 10 mCi/mL).
4. T4 polynucleotide kinase (PNK) enzyme (10 unit/μL) and 10× PNK buffer: 700 mM Tris–HCl, pH 7.6, 100 mM MgCl2, 50 mM DTT.
5. Band shift buffer: 83 mM Tris–HCl, pH 7.5, 100 mM NaCl, 66 mM KCl, 0.8 mM $MgCl_2$, 10 mM DTT. Add DTT fresh from a stock solution (see Note 2).
6. Penguin Electrophoresis System from Owl Technologies (20 × 20 cm glass plates, 1.5 mm width).
7. 40% acrylamide/bis-acryamide (39:1) stock solution in 0.5× TBE buffer. For 10 mL gel mix: 3.76 mL of 40% acrylamide, 1 mL of 5× TBE buffer, 5.14 mL MQ water, 100 μL of 10% APS, and 5 μL of TEMED.
8. 0.5× TBE (Tris–borate-ethylenediaminetetraacetic acid) electrophoresis buffer: 0.045 M Tris base, 0.045 M boric acid, 1 mM EDTA, pH 8.0.
9. 6× loading dye: 0.25% bromophenol blue, 0.25% xylene-cyanol, 40% glycerol.
10. 300 mM $NaCl_2$ solution.

2.2. Protein Expression and Preparation

1. *Agrobacterium tumefaciens* C58C1 bacteria transformed with pBIN61-p122 construct (for transient expression of p122 protein in *Nicotiana benthamiana* leaves, see Note 3).
2. Agroinfiltration buffer: 10 mM morpholineethanesulfonic acid (MES), 0.15 mM acetosyringone, 10 mM $MgCl_2$.

3. Protein extraction buffer: 83 mM Tris–HCl, pH 7.5, 100 mM NaCl, 66 mM KCl, 0.8 mM $MgCl_2$, 10 mM DTT (see Note 2). This buffer is identical to the band shift buffer.
4. 0.2-μm Syringe filter.
5. Bradford assay.

2.3. Binding Reaction

1. Band shift buffer: 83 mM Tris–HCl, pH 7.5, 100 mM NaCl, 66 mM KCl, 0.8 mM $MgCl_2$, 10 mM DTT (see Note 2).
2. 0.02% Poly (oxyethylene) x-sorbitane-monolaurate (Tween-20).

2.4. Native Gel Preparation and Complex Separation

1. Materials for running native acrylamide gels (see Subheading 2.1, items 6–9).
2. Gel dryer.

2.5. Signal Detection and Band Quantification

1. PhosphorImager screen and Storm 860 Molecular Imager (GE Healthcare).
2. ImageQuant software (GE Healthcare).
3. Origin software (www.origin.com).

3. Methods

3.1. siRNA Preparation

1. Design synthetic RNA oligonucleotides (19, 21, 26, and 50 nt long) as siRNA pairs that generate perfect dsRNA duplexes with or without 2 nt 3′ overhangs after annealing (see Note 4). Deprotect the 2′-ACE protected RNA molecules according to manufacturer's instructions. Briefly, dissolve the RNA pellet in 2′-deprotection buffer, divide in aliquots, incubate at 60°C for 30 min and lyophilize to dryness before use. Resuspend one aliquot of deprotected RNA in MilliQ water and store other aliquots at −80°C.
2. Phosphorylate 20 pmols of one strand of the siRNAs with γ-^{32}P ATP in using T4 polynucleotide kinase. Combine in an Eppendorf tube on ice:
   (a) 12 μL water.
   (b) 2 μL 10× PNK buffer.
   (c) 2 μL γ-^{32}P ATP.
   (d) 2 μL oligonucleotides (10 pmol/μL).
   (e) 2 μL PNK (10 unit/μL).
3. Phosphorylate complementary strand. Assemble the reaction as in Subheading 3.1, step 2, but use 2 μL of cold 10 mM ATP.
4. Add cold-phorphorylated complementary strand in fivefold molar excess to the ^{32}P labelling reaction in 1× T4 PNK buffer

in a total volume of 40 μL. Heat the mix containing both strands to 95°C for 1 min, and very slowly cool down in a water bath (about 1°C/3 min) to facilitate the strand annealing. Turn off the power supply of the water bath and let it cool down to room temperature (see Note 5). The 1:5 molar ratio ensures that the all of the labelled RNA will be converted to double-stranded form. The dsRNA will be separated from the non-annealed ssRNAs by purification from a native gel (see Subheading 3.1, step 6).

5. Store an aliquot of the labelled RNA in single-stranded form to analyze the p122 affinity for ssRNA molecules.
6. Repeat the same procedure for ssDNA and dsDNA to assess the p122 affinity for these nucleic acids.
7. Use native 1.5 mm thick 15% acrylamide gels in 0.5× TBE (see Notes 6 and 7) for the labelled nucleic acid purification. Add 8 μL of 6× loading dye to the 40 μL reaction mix and load on gel 12 μL/well.
8. Run the gels at 10 V/cm for 3–4 h, disassemble and locate the duplexes by exposing for 1–2 min to X-ray film. The procedure should be done behind a shield: remove the glass plate from one side of the gel and wrap the other glass plate with the gel in cling film. Place the film on the top of the gel. Mark the position of the film with a marker pen. Develop the film after exposure and dry it. Afterwards place back the film on the top of the gel and mark the position of labelled RNA/DNA with a needle by making punctures through the film. Then unwrap the gel and excise the RNA/DNA band.
9. Add 0.5 mL of 300 mM NaCl solution to the gel slices and incubate overnight at 4°C on a rotary shaker. Cut the gel slices into small pieces to enhance the yield of nucleic acid extraction. Precipitate the nucleic acid in 2.5 volumes of ethanol, wash the precipitates two times with 1 mL of 70% ethanol, dry the pellet and resuspend in 100 μL band shift buffer. The general yield of purification is about 50% of the input.

3.2. Protein Expression and Preparation

Protein p122 was expressed in plant leaves using *Agrobacterium* infiltration-mediated transient expression (1). The protein is expressed at high level as the RNA silencing system is blocked by its presence (see Note 8). However, if the protein of interest can be purified by the mean of other methods, its RNA binding affinity can also be studied by EMSA (see Note 9).

1. Resuspend *Agrobacterium* C58C1 harbouring pBIN-p122 in agroinfiltration buffer, and incubate at 25°C for 3 h. Infiltrate the second and third youngest leaves of a six-leaf stage *Nicotiana benthamiana* plant with a *Agrobacteria* suspension

at OD_{600} = 0.5. For negative control infiltrate leaves with pBIN-empty vector containing *Agrobacterium*.

2. Prepare the native protein extracts from infiltrated patches at 3 days after infiltration. Perform the extraction on ice using ice-cold materials: grind 0.25 g leaf tissue in 1 mL of band shift buffer. Centrifuge the crude extract twice for 15 min at 15,000 × *g* and filter through a 0.2 μm syringe filter. Store the supernatants in −80°C in single-use aliquots. Quantify the concentration of total protein by Bradford assay.

3.3. Binding Assay

1. Assemble the in vitro binding reactions on ice. In a 10 μL reaction, use labelled RNA or DNAs at a final concentration of 1 pM in 1×band shift buffer containing 0.02% of Tween-20 (2). Prepare master-mix containing band shift buffer, labelled RNA and Tween-20 and aliquot in eppendorf tubes (9 μL/tube).
2. Prepare a twofold serial dilution in band shift buffer of protein extracts from plants infiltrated with p122 or empty vector, starting from a concentration of 1 μg/μL (see Fig. 1a–c). Add 1 μL of p122-containing or control extracts to 9 μL the band shift reaction mix (see Note 10).
3. Incubate the reactions at 25°C for 30 min. Add 2 μL of loading dye, spin down quickly. Load the reactions directly onto the pre-runned and pre-chilled gel.

3.4. Native Gel Electrophoresis

1. Pre-run a 5% native acrylamide gel in 0.5× TBE for 20 min at 10 V/cm at 4°C (see Note 7).
2. Load reactions (5–10 μL) on gel and run the gel at 10 V/cm for 2 h.
3. Carefully disassemble the gel after electrophoresis: the gel should stick to one of the glass plates. Place a piece of 3 mm Whatman paper, which covers the entire gel, on top of the gel. Turn the stack of paper, gel and glass plate upside-down and carefully detach the gel from the glass plate by lifting the plate upwards. The gel surface is covered with a nylon wrap and desiccated in the gel dryer.
4. Expose the dried gel overnight to a PhosphorImager screen.

3.5. Signal Detection, Quantification and Data Analysis

1. Detect the radioisotope signal in a PhosphorImager.
2. Quantify the signal of the protein: nucleic acid complexes by ImageQuant software, normalize the data to the signal of band shift reaction without p122 protein (extracts from empty-vector infiltrated plants), where is no complex present (see Fig. 1c).
3. Represent the data in a graph with the relative dilution of p122 on the *y*-axis, and the signal from the p122: nucleic acid complexes on the *x*-axis using Origin software (see Fig. 1d).

4. Determine the inflection point value of the dilution of p122 by fitting a sigmoid curve according to manufacturer's instructions. The inflection point represents the relative dissociation constant. The relative dissociation constants of a given extract for different types of nucleic acid molecules can be directly compared to each other.

4. Notes

1. Sources of problems can be the presence of denaturing agents, high voltage, high temperature, proteases or other factors that disrupt the protein: RNA binding.
2. Dithiothreitol (DTT) should always be freshly added to solutions. Stock solutions of 1 M concentrations is stored in aliquots at –20°C.
3. Binary vectors are cloning vectors that are able to replicate in both *E. coli* and *A. tumefaciens*, and can be used for plant transformation. The gene of interest and a selection marker are located between the left and right borders of the T-DNA region of the plasmid. In a transformation event the T-DNA region is delivered from the plasmid into the plant genome where it integrates and expresses the gene of interest.
4. The central molecules of RNA silencing pathways in plants are siRNAs with a length of 21–24 nt. To assess the importance of the length of the RNA molecule for binding by p122 protein, we designed shorter or longer molecules having 3′2 nt overhangs, in addition to 21 nt siRNA. Longer single-stranded or double-stranded RNAs can be prepared by T7, T3 or SP6 in vitro transcription of sense and reverse complementary strands, annealed and gel purified. Depending on their base content, long RNAs can form intra-molecular loops, which may interfere with formation of perfect dsRNA molecules.
5. Alternatively you can set up a PCR programme in a thermal cycler to cool down the mixture at about 1°C/3 min.
6. Glass plates, spacers and combs should be properly cleaned with detergent. Clear from time to time with 0.5 M NaOH solution to remove trace of grease or other contaminations. Take great care to properly rinse the glass plates, spacers and combs with ethanol and distilled water before use.
7. Wash the wells with buffer prior to loading the samples. However, the washing has to be mild, because a harsh wash can cause small damages to the bottom of the wells, which may cause proteins or protein complexes to stick to wells.

8. This is a crude protein preparation. There could be differences in expression levels between different RNAi suppressor proteins, however these are usually expressed at high level.
9. Some proteins can be purified as recombinant proteins in *E. coli*. This technique was used efficiently to purify different silencing suppressor proteins (5, 7–10).
10. Pipette the protein extract dilutions on the wall of the eppendorf tubes on ice and spin down afterwards to start the binding reactions simultaneously.

Acknowledgments

TCS and JB are funded by the European Commission (FP6 Integrated Project SIROCCO LSHG-CT-2006-037900).

References

1. Voinnet, O., Rivas, S., Mestre, P., and Baulcombe, D. (2003) An enhanced transient expression system in plants based on suppression of gene silencing by the p19 protein of tomato bushy stunt virus, *Plant J* **33**, 949–956.
2. Csorba, T., Bovi, A., Dalmay, T., and Burgyan, J. (2007) The p122 subunit of Tobacco Mosaic Virus replicase is a potent silencing suppressor and compromises both small interfering RNA- and microRNA-mediated pathways, *J Virol* **81**, 11768–11780.
3. Merai, Z., Kerenyi, Z., Kertesz, S., Magna, M., Lakatos, L., and Silhavy, D. (2006) Double-stranded RNA binding may be a general plant RNA viral strategy to suppress RNA silencing, *J Virol* **80**, 5747–5756.
4. Merai, Z., Kerenyi, Z., Molnar, A., Barta, E., Valoczi, A., Bisztray, G., Havelda, Z., Burgyan, J., and Silhavy, D. (2005) Aureusvirus P14 is an efficient RNA silencing suppressor that binds double-stranded RNAs without size specificity, *J Virol* **79**, 7217–7226.
5. Lakatos, L., Csorba, T., Pantaleo, V., Chapman, E. J., Carrington, J. C., Liu, Y. P., Dolja, V. V., Calvino, L. F., Lopez-Moya, J. J., and Burgyan, J. (2006) Small RNA binding is a common strategy to suppress RNA silencing by several viral suppressors, *EMBO J* **25**, 2768–2780.
6. Silhavy, D., Molnar, A., Lucioli, A., Szittya, G., Hornyik, C., Tavazza, M., and Burgyan, J. (2002) A viral protein suppresses RNA silencing and binds silencing-generated, 21- to 25-nucleotide double-stranded RNAs, *EMBO J* **21**, 3070–3080.
7. Zhang, X., Yuan, Y. R., Pei, Y., Lin, S. S., Tuschl, T., Patel, D. J., and Chua, N. H. (2006) Cucumber mosaic virus-encoded 2b suppressor inhibits Arabidopsis Argonaute1 cleavage activity to counter plant defense, *Genes Dev* **20**, 3255–3268.
8. Vargason, J., Szittya, G., Burgyan, J., and Hall, T. M. (2003) Size selective recognition of siRNA by an RNA silencing suppressor, *Cell* **115**, 799–811.
9. Bortolamiol, D., Pazhouhandeh, M., Marrocco, K., Genschik, P., and Ziegler-Graff, V. (2007) The polerovirus F box protein P0 targets ARGONAUTE1 to suppress RNA silencing, *Curr Biol* **17**, 1615–1621.
10. Hemmes, H., Lakatos, L., Goldbach, R., Burgyan, J., and Prins, M. (2007) The NS3 protein of Rice hoja blanca tenuivirus suppresses RNA silencing in plant and insect hosts by efficiently binding both siRNAs and miRNAs, *RNA* **13**, 1079–1089.
11. Garner, M. M., and Revzin, A. (1981) A gel electrophoresis method for quantifying the binding of proteins to specific DNA regions: application to components of the *Escherichia coli* lactose operon regulatory system, *Nucleic Acids Res* **9**, 3047–3060.

Chapter 16

dsRNA Uptake in Adult *Drosophila*

Benjamin Obadia and Maria-Carla Saleh

Abstract

RNA interference (RNAi) is a conserved sequence-specific gene silencing mechanism that is induced by double-stranded RNA (dsRNA). The development of methods that allow internalization of dsRNA and concomitant silencing of the desired gene has not stopped since the first demonstration of RNAi in *Caenorhabditis elegans*. In this chapter, we describe how to introduce exogenous dsRNA into adult *Drosophila* in order to interfere with endogenous or viral gene expression.

Key words: Double-stranded RNA (dsRNA), dsRNA uptake, RNA interference, Viruses, Insects

1. Introduction

Double-stranded RNA (dsRNA) can trigger sequence-specific gene inactivation through RNA interference (RNAi) mechanisms in cells, as first demonstrated after injection into the worm *Caenorhabditis elegans* (1). This post-/cotranscriptional silencing effect was soon discovered in many diverse organisms ranging from plants (2) to fruit flies (3). If introduced into cells, dsRNA precursors (that can vary in length and origin) are diced into shorter dsRNAs (called small interfering RNAs or siRNAs) which serve as the guides for translational inhibition (reviewed in ref. (4)).

Drosophila biology has greatly benefited from the use of RNAi. dsRNA injection into embryos before cellularization stage has been used as a tool to study developmental aspects or gene function (3). However, many genes cannot be silenced during development without having a lethal effect, and syncytial stage in embryos does not permit the targeting of specific cells. Injection of dsRNA into the hemolymph of *Drosophila* larvae has also been attempted but triggering RNAi in most tissues, except for hemocytes, seems impossible (5). In adult *Drosophila*, injecting dsRNA

Ronald P. van Rij (ed.), *Antiviral RNAi: Concepts, Methods, and Applications*, Methods in Molecular Biology, vol. 721,
DOI 10.1007/978-1-61779-037-9_16,

can induce a specific knockdown, allowing an *in vivo* functional analysis of the targeted genes (6–8). *In vitro* evidence shows that *Drosophila* S2 cells (derived from embryonic precursors of haemocytes) can autonomously take-up long dsRNAs (~200–500 bp) from exogenous medium (9, 10), reminiscent of the environmental RNAi shown in *C. elegans* (reviewed in ref. (11)).

In an immunological context, it was established that the core components of the RNAi machinery and the siRNA pathway are essential for antiviral defense (12–14). Interestingly, an effective dsRNA uptake mechanism is essential to generate a systemic antiviral state from a locally initiated infection, showing that, upon viral infection, dsRNAs can spread throughout the whole adult *Drosophila* (15). Taken together, these data foster the following model of immunity: during lysis, infected cells release viral dsRNAs, which can be taken up from a distance by hemocytes and perhaps other adult tissues in the flies in order to generate a systemic antiviral state. A mechanism for dsRNA uptake has been shown in other insects, like the mosquito *Aedes aegypti* (16).

In this chapter, we describe how to prepare and inject dsRNA in adult *Drosophila*. We have used this protocol to initiate a strong antiviral response in the context of a viral infection (15) but in our hands this same protocol has successfully been used to silence endogenous gene expression.

2. Material

2.1. Fly Stocks

Flies are grown on a standard yeast/agar medium at 25°C on a 12-h light/dark cycle. All adult flies used for injection must be at least 5 days old (see Note 1).

2.2. Total RNA Isolation

1. TRIzol Reagent (Invitrogen, Carlsbad, CA).
2. Chloroform.
3. Isopropyl alcohol.
4. Nuclease-free 75% ethanol conserved at –20°C.
5. Nuclease-free water.
6. RNaseZap RNase decontamination solution (Ambion, Austin, TX).
7. Pellet pestle (Sigma, St. Louis, MO).

2.3. Reverse Transcription

1. SuperScript II Reverse Transcriptase (200 U/μL) (Invitrogen, Carlsbad, CA).
2. First-strand buffer (5×): 250 mM Tris–HCl, pH 8.3, 375 mM KCl, 15 mM $MgCl_2$.

3. 10 mM (each) dNTPs Mix (dATP, dCTP, dGTP, dTTP).
4. 100 mM DTT (dithiothreitol).
5. RNase OUT Ribonuclease Inhibitor (40 U/μL) (Invitrogen, Carlsbad, CA).
6. Random Hexamers (50 ng/μL).
7. Nuclease-free water.
8. Nanodrop UV spectrophotometer.

2.4. Polymerase Chain Reaction

1. Thermocycler.
2. Thermophilic DNA polymerase (*Thermus aquaticus*) (5 U/μL).
3. PCR Buffer (10×): 200 mM Tris–HCl, pH 8.4, 500 mM KCl.
4. 10 mM (each) dNTPs Mix (dATP, dCTP, dGTP, dTTP).
5. 25 mM $MgCl_2$.
6. Nuclease-free water.

2.5. dsRNA Production

1. T7 RNA polymerase buffer (5×): 400 mM Hepes-KOH, pH 7.5, 120 mM $MgCl_2$, 10 mM Spermidine (Sigma, St. Louis, MO), 200 mM DTT (Sigma, St. Louis, MO) (see Notes 2 and 3). Store in aliquots at −20°C for several months.
2. T7 RNA polymerase (20 U/μL) (Ambion, Austin, CA).
3. RNase OUT Ribonuclease Inhibitor (40 U/μL) (Invitrogen, Carlsbad, CA).
4. 100 mM (each) NTPs (ATP, CTP, GTP, UTP) (Fermentas, Burlington, ON).
5. 5 M Ammonium acetate (Ambion, Austin, TX).
6. Phenol/chloroform/isoamylic alcohol (25/24/1), pH 6.6 (Ambion, Austin, TX).
7. Isopropyl alcohol.
8. Nuclease-free 75% ethanol.
9. 1 M Tris–HCl, pH 7.0 (Ambion, Austin, TX).
10. Nuclease-free water.
11. Nanodrop UV spectrophotometer.

2.6. Microinjection Needles Fabrication

1. P-97 Flaming/Brown micropipette puller (Sutter Instrument Company, Novato, CA).
2. Borosilicate glass tubes (#3-000-203-G/X: outside diameter: 1.14 mm, inside diameter: 0.53 mm, overall length: 3.5″) (Drummond Scientific Company, Broomall, PA) or (#BF100-50-10: outside diameter: 1.00 mm, inside diameter: 0.50 mm, overall length: 10 cm) (Sutter Instrument Company, Novato, CA).

3. Mineral oil.
4. Pasteur pipette.

2.7. Injection

1. Automatic nanoinjector Nanoject II variable volume (Drummond Scientific Company, Broomall, PA). The Nanoject II is set up to inject 50.6 nL by moving positions of the dip switches up (U) or down (D) as DDUD from position 1–4, respectively.
2. Stereozoom microscope.
3. Cold-light reflector lamp.
4. CO_2-anaesthetizing system (Genesee Scientific, San Diego, CA).

3. Methods

3.1. Template Selection

In vitro dsRNA synthesis can be made from various DNA templates generated by PCR. Either genomic DNA or cDNA can be used to amplify the gene region targeted by the silencing assay. Keep in mind that only small interfering RNAs (generated from the dsRNA sequence by the RNAi machinery) that match with transcript sequences will generate a knockdown (see Note 4). For this reason, it is preferable to proceed on cDNA templates (to avoid intronic DNA contaminations) obtained from flies expressing the gene of interest. Alternatively it is possible to design primers within a single exon if genomic DNA is to be used as matrix. The complete protocol from RNA isolation to dsRNA production is described in Subheadings 3.3–3.6.

3.2. Primers Design

MIT's Primer3 program (http://frodo.wi.mit.edu/primer3/) is suitable to design oligonucleotides, for amplification of the target sequence. Optimal oligo size is 22 nt (±2) length for a 60°C annealing temperature. At the 5′-end of each oligonucleotide, add a 23 nt T7 promoter sequence (5′-TAATACGACTCACTAT AGGGAGA-3′) (the G in bold is the first base incorporated into RNA during transcription). Order primers in salt-free water at 100 μM. dsRNA products ranging from 200 to 1,500 bp have been shown to work in RNAi but ~300–600 bp are usually used for an efficient uptake and silencing. Also remember to check that the total dsRNA product sequence does not contain ≥19-mer matching other gene sequences to prevent nonspecific knockdown of gene expression (off-target effect).

3.3. Total RNA Isolation

TRIzol Reagent combines acidic phenol and guanidinium thiocyanate in order to lyse cells, inactivate RNases, and remove lipids.

Manipulate this toxic solution with gloves and in a chemical fume hood. Remember to work in an RNase-free area (cleaned with RNase Zap) and keep samples on ice as much as possible.

1. Harvest three to five flies, expressing the gene of interest, in a clear polypropylene tube and euthanize them by flash freezing in liquid nitrogen or in a dry ice/ethanol mixture (see Note 5).
2. Add 200 μL TRIzol Reagent and process the flies with a pellet pestle until a homogenized tissue sample is obtained. Complete with 800 μL TRIzol Reagent and incubate 5 min at room temperature.
3. Proceed to the phase separation by adding 200 μL chloroform, vortex for 1 min, and centrifuge at 12,000 × *g* for 15 min at 4°C (see Note 6).
4. To precipitate RNA, transfer the aqueous phase to a new prechilled tube and add 400 μL isopropyl alcohol before centrifuging at 12,000 × *g* for 15 min at 4°C (see Note 7).
5. Remove the supernatant and wash the RNA pellet with 500 μL of chilled 75% ethanol. Mix by vortexing and centrifuge at 7,500 × *g* for 5 min at 4°C.
6. Remove ethanol and air dry the RNA pellet by leaving the sample open under the extractor hood and dissolve it by pipetting up and down in 40–80 μL nuclease-free water (see Notes 8 and 9).

3.4. Reverse Transcription

1. Prepare RNA at a concentration of 1 μg/μL (see Note 10).
2. In a PCR tube, mix:
   (a) 1 μL Random Hexamers.
   (b) 1 μL 10 mM dNTPs mix.
   (c) 1 μL RNA sample (or nuclease-free water as a negative control).
   (d) 9 μL Nuclease-free water.
3. Heat 5 min at 65°C, and quick chill on ice (see Note 11).
4. Spin down and add:
   (a) 4 μL 5× First-strand buffer.
   (b) 2 μL 100 mM DTT.
   (c) 1 μL 40 U/μL RNase OUT.
5. Mix gently, spin briefly down, and incubate 2 min at 25°C before adding 1 μL of 200 U/μL SuperScript II Reverse Transcriptase.
6. Incubate at 42°C for 50 min to allow elongation and then inactivate the reaction by heating at 70°C for 15 min.
7. Aliquot and store cDNA at −20°C if necessary.

3.5. Polymerase Chain Reaction

Once cDNA from desired transcriptome is obtained, perform PCR with synthesized primers.

1. For a single PCR reaction, mix:
   (a) 2.5 μL 10× PCR buffer.
   (b) 2 μL 25 mM $MgCl_2$.
   (c) 0.5 μL 10 mM dNTPs Mix.
   (d) 16.75 μL Nuclease-free water.
   (e) 0.25 μL 5 U/μL thermophilic DNA polymerase.
2. Transfer 22 μL of PCR Mix in a PCR tube, and add:
   (a) 1 μL 10 μM Forward primer.
   (b) 1 μL 10 μM Reverse primer.
   (c) 1 μL cDNA (or nuclease-free water as a negative control).
3. Gently mix, spin down briefly, and run in thermocyler (33 cycles: 94°C 30 s, 60°C 30 s, 72°C 60 s) (see Note 12).
4. Check the result on a 1% agarose gel.

3.6. dsRNA Production

1. In a polypropylene tube, prepare a T7 Mix:
   (a) 20 μL 5× T7 RNA polymerase buffer.
   (b) 5 μL of each 100 mM NTP (ATP, CTP, GTP, UTP).
   (c) 2 μL 40 U/μL RNase OUT.
   (d) 1 μL 20 U/μL T7 RNA polymerase.
2. Per one T7 *in vitro* transcription reaction, in a 1.5 mL polypropylene tube, mix:
   (a) 1 μg PCR product.
   (b) 43 μL T7 Mix preparation.
   (c) Up to 100 μL with nuclease-free water.
3. Incubate at 37°C for 16 h or overnight (in a dry bath incubator or in a bacterial incubator to avoid condensation).
4. After *in vitro* transcription, add 340 μL nuclease-free water, 60 μL 5 M Ammonium acetate (samples can be stored at −20°C at this stage if needed), and 500 μL phenol/chloroform/isoamylic alcohol stored at 4°C.
5. Mix by vortexing 15 s, and centrifuge at 16,000×*g* (maximum speed in a microcentrifuge) for 5 min at 4°C.
6. Transfer aqueous phase in a new tube, add 1 volume (~500 μL) of isopropyl alcohol, mix by inversion, and incubate at least 1 h at −20°C.
7. Centrifuge at 16,000×*g* for 15 min at 4°C, wash the pellet with 800 μL of 75% ethanol, and centrifuge again as previously.

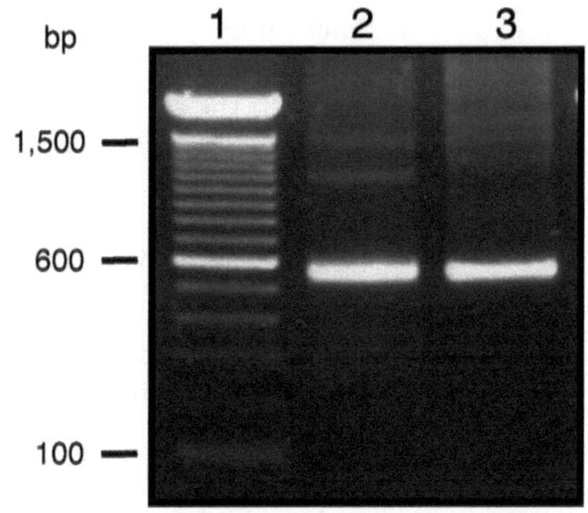

Fig. 1. Checking for annealing of dsRNA. After in vitro transcription a 1/10th dilution of the obtained dsRNA was subjected to electrophoresis in a 1% agarose gel, stained with ethidium bromide. *Lane 1*: molecular weight marker. *Lanes 2* and *3*: dsRNA after annealing. Note the presence of a sharp band at the desired size (around 500 bp).

8. Air dry the RNA pellet and dissolve it in nuclease-free water or 10 mM Tris–HCl.
9. Measure RNA concentration, and adjust it at 3 μg/μL (samples can be stored at −20°C at this stage until annealing).
10. Incubate the RNA sample for 15 min at 65°C in a dry bath incubator, then remove the heating block, or switch off the incubator, and let it cool down to room temperature for an efficient annealing.
11. Check on a 1% agarose gel (Fig. 1), and store in 20 μL aliquots at −20 or −80°C.

3.7. Preparation of Needles for Microinjection

1. In order to make <1 μm needles from borosilicate tubes, the micropipette puller machine program is set up by the following parameters: heat: 515, pull: 60, velocity: 60, time: 250, pressure: 500 (see Note 13).
2. Once needles are pulled, back-fill microinjection needles with mineral oil using a Pasteur pipette previously thinned (see Note 14). Take care to remove any air bubbles inside.
3. Needles can be stored at room temperature for several months.
4. Just before use, delicately break the tip of the needle with thin tweezers, taking care of making the tip thin enough to avoid major injury to flies, but thick enough to ensure full delivery of the solution (see Note 15). Empty almost all of the mineral oil from the needle, and fill it with the solution to inject.

3.8. Injection

1. Flies must be CO_2-anaesthetized and positioned for intrathoracical injection (see Note 16). Needles are applied between the supraalar bristles (SA1, SA2) and the presutural bristle (PS), in the intrascutal suture level (Fig. 2). Note that introduction of the needle in this region should not encounter any resistance, as if you are introducing the needle in a hole (see Note 17).
2. 50 nL of an appropriate dsRNA solution (3 μg/μL) or a buffer solution as a control (10 mM Tris–HCl, pH 7.5) are injected using the nanoinjector.
3. After injection, let flies recover in appropriate vials in horizontal position (to avoid flies to get stuck to the medium), and verify the appearance of melanization spots at the injection site (see Note 18).
4. Track the silencing of your targeted gene by a method of choice. Efficiency of silencing will be dependent on the dsRNA preparation and time-dependent on your target (mRNA stability, protein turnover).

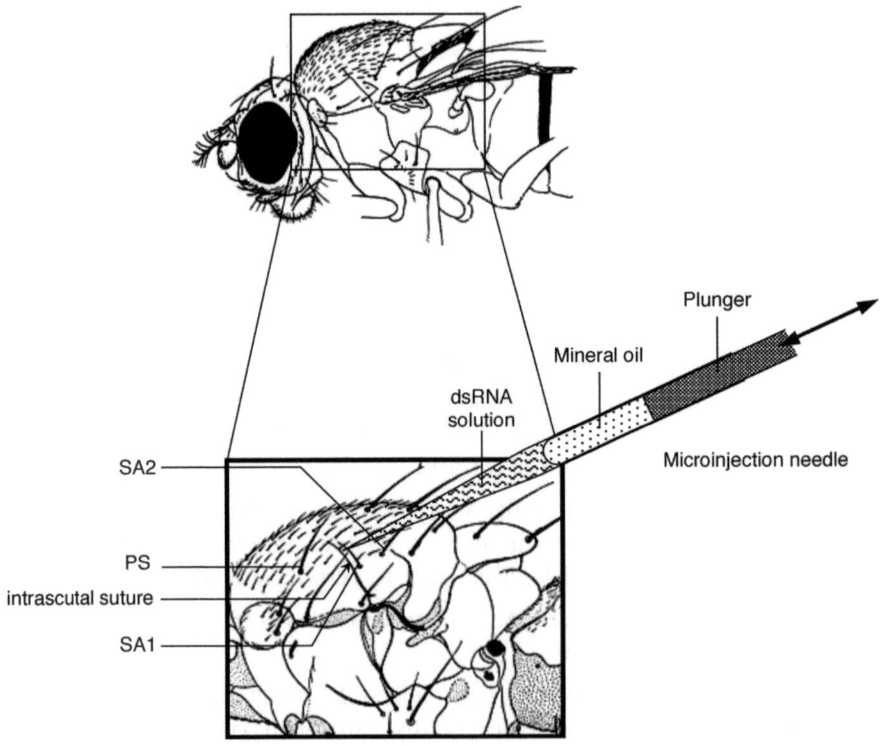

Fig. 2. Scheme from intrathoracic injection in adult *Drosophila*. Needles are gently introduced between the supraalar bristles (SA1, SA2) and the presutural bristle (PS), in the intrascutal suture level (see insight). Reproduced from Biology of *Drosophila* with permission from Cold Spring Harbor Laboratory Press (18).

4. Notes

1. Younger flies are unable to take up dsRNA as also shown in ref. (7).
2. This homemade buffer has been optimized over commercial buffers for better results yields.
3. It is important to use KOH in order to buffer HEPES and not NaOH. Spermidine stimulates T7 RNA polymerase activity, DTT works as a reducing agent to avoid disulfide bonds. DTT is not easily dissolved so it may be necessary to heat at 55°C the solution until crystals disappear. Adjust concentration of reagents in nuclease-free water. Dissolve DTT and spermidine in water.
4. Coding sequences but also 3′ untranslated regions work as excellent targets.
5. Manipulating liquid nitrogen needs precaution, you should prefer prepare a "slurry" mixture by carefully adding pure ethanol on dry ice. The fact that tubes can stand right in the mixture provides a better contact with the slurry and so a better cold convection to your samples.
6. After centrifugation, you can observe three phases: a lower red phenol/chloroform phase containing proteins and lipids, a viscous white interphase containing DNA, and an upper colorless aqueous phase containing RNA.
7. In order to improve RNA purity, it is preferable to sacrifice a little amount of the aqueous phase close to the interface.
8. Do not let the RNA sample dry completely. Once water added, you can heat the sample at 60°C for 10 min on a dry bath to help dissolution.
9. RNA can be stored in nuclease-free distilled water at −20 or −80°C. There is no consensus on RNA storage, but in water, in a Tris–EDTA buffer (10 mM Tris, pH 7.0, 1 mM EDTA), or in 1 mM sodium citrate (pH 6.4), RNA is stable at least 1 year at −80°C. RNA can also be stored in pure deionised formamide at −70°C, or in ethanol from the precipitation step at −20°C.
10. A 20 μL reaction volume can be used for 1 ng to 5 μg of total RNA.
11. The heat step allows the secondary structure denaturation.
12. As a standard running condition, you can use: (a) 94°C for 5 min, (b) 94°C for 30 s, (c) 60°C for 30 s, (d) 72°C for 1 min, (e) steps (b)–(d) 33 cycles, (f) 72°C for 5 min, (g) holding at 4°C.

13. For more technical information about the parameters, download the supplier's pipette cookbook: http://www.sutter.com/products/product_sheets/p97.html.
14. In order to make Pasteur pipette thinner, light a Bunsen burner, hold the pipette at each extremity, and heat the thinnest part of it on the flame for few seconds, turning continuously the pipette until the thin part becomes red. When it starts to melt, remove quickly from the flame and pull constantly until ductility is lost. Then, cut the thinner Pasteur pipette at an appropriate distance to fill the micropipette.

    Remind to not wear gloves with Bunsen burner.
15. In a nutshell, proceed with care, and practice a lot!
16. To improve fly survival, it is best not to leave flies more than half an hour on the CO_2-anaesthetizing pad. We suggest injecting flies in small groups, knowing that a good experimenter can inject up to a hundred flies per hour.
17. Injection can be performed intraabdominally (6), but in our hands lethality after injection is higher.
18. Melanization spots appear a few minutes after injury, but for convenience and to prevent repeated anaesthesia at the same day, you should check the day after. For an overview of melanization, see ref. (17).

Acknowledgments

The authors would like to thank Marco Vignuzzi for comments on the manuscript. This work was financially supported by the Agence Nationale de la Recherche (ANR-09-JCJC-0045-01) and the European Research Council (FP7/2007-2013 ERC 242703) to MCS. BO is a University of Paris VI and Ministère de la Recherche fellow.

References

1. Fire, A., Xu, S., Montgomery, M. K., Kostas, S. A., Driver, S. E., and Mello, C. C. (1998) Potent and specific genetic interference by double-stranded RNA in *Caenorhabditis elegans*, *Nature 391*, 806–811.
2. Waterhouse, P. M., Graham, M. W., and Wang, M. B. (1998) Virus resistance and gene silencing in plants can be induced by simultaneous expression of sense and antisense RNA, *Proc Natl Acad Sci U S A 95*, 13959–13964.
3. Kennerdell, J. R., and Carthew, R. W. (1998) Use of dsRNA-mediated genetic interference to demonstrate that frizzled and frizzled 2 act in the wingless pathway, *Cell 95*, 1017–1026.
4. Siomi, H., and Siomi, M. C. (2009) On the road to reading the RNA-interference code, *Nature 457*, 396–404.
5. Miller, S. C., Brown, S. J., and Tomoyasu, Y. (2008) Larval RNAi in *Drosophila*? *Dev Genes Evol 218*, 505–510.
6. Dzitoyeva, S., Dimitrijevic, N., and Manev, H. (2001) Intra-abdominal injection of double-stranded RNA into anesthetized adult

Drosophila triggers RNA interference in the central nervous system, *Mol Psychiatry* **6**, 665–670.

7. Goto, A., Blandin, S., Royet, J., Reichhart, J. M., and Levashina, E. A. (2003) Silencing of Toll pathway components by direct injection of double-stranded RNA into *Drosophila* adult flies, *Nucleic Acids Res* **31**, 6619–6623.
8. Dzitoyeva, S., Dimitrijevic, N., and Manev, H. (2003) Gamma-aminobutyric acid B receptor 1 mediates behavior-impairing actions of alcohol in *Drosophila*: adult RNA interference and pharmacological evidence, *Proc Natl Acad Sci U S A* **100**, 5485–5490.
9. Saleh, M. C., van Rij, R. P., Hekele, A., Gillis, A., Foley, E., O'Farrell, P. H., and Andino, R. (2006) The endocytic pathway mediates cell entry of dsRNA to induce RNAi silencing, *Nat Cell Biol* **8**, 793–802.
10. Ulvila, J., Parikka, M., Kleino, A., Sormunen, R., Ezekowitz, R. A., Kocks, C., and Ramet, M. (2006) Double-stranded RNA is internalized by scavenger receptor-mediated endocytosis in *Drosophila* S2 cells, *J Biol Chem* **281**, 14370–14375.
11. Whangbo, J. S., and Hunter, C. P. (2008) Environmental RNA interference, *Trends Genet* **24**, 297–305.
12. Wang, X. H., Aliyari, R., Li, W. X., Li, H. W., Kim, K., Carthew, R., Atkinson, P., and Ding, S. W. (2006) RNA interference directs innate immunity against viruses in adult *Drosophila*, *Science* **312**, 452–454.
13. Galiana-Arnoux, D., Dostert, C., Schneemann, A., Hoffmann, J. A., and Imler, J. L. (2006) Essential function in vivo for Dicer-2 in host defense against RNA viruses in *Drosophila*, *Nat Immunol* **7**, 590–597.
14. van Rij, R. P., Saleh, M. C., Berry, B., Foo, C., Houk, A., Antoniewski, C., and Andino, R. (2006) The RNA silencing endonuclease Argonaute 2 mediates specific antiviral immunity in *Drosophila melanogaster*, *Genes Dev* **20**, 2985–2995.
15. Saleh, M. C., Tassetto, M., van Rij, R. P., Goic, B., Gausson, V., Berry, B., Jacquier, C., Antoniewski, C., and Andino, R. (2009) Antiviral immunity in *Drosophila* requires systemic RNA interference spread, *Nature* **458**, 346–350.
16. Sanchez-Vargas, I., Scott, J. C., Poole-Smith, B. K., Franz, A. W., Barbosa-Solomieu, V., Wilusz, J., Olson, K. E., and Blair, C. D. (2009) Dengue virus type 2 infections of *Aedes aegypti* are modulated by the mosquito's RNA interference pathway, *PLoS Pathog* **5**, e1000299.
17. Tang, H. (2009) Regulation and function of the melanization reaction in *Drosophila*, *Fly (Austin)* **3**, 105–111.
18. Ferris, G. F. (1994) External morphology of the adult. *in* Demerec M. (ed.) Biology of Drosophila. Cold Spring Harbor Laboratory Press. pp 368–419.

immunity against viruses in adult Drosophila. Science 312: 1252–[illegible]

16. [illegible]

Drosophila triggers RNA interference in the central nervous system. [illegible]

[illegible]

Part IV

Antiviral RNAi Therapy

Chapter 17

Design of Small Interfering RNAs for Antiviral Applications

Diana Rothe, Erik J. Wade, and Jens Kurreck

Abstract

RNA interference (RNAi) is an evolutionarily conserved mechanism for sequence-specific target RNA degradation in animals and plants, which plays an essential role in gene regulation. In addition, it is believed to function as a defense against viruses and transposons. In recent years, RNAi has become a widely used approach for studying gene function by targeted cleavage of a specific RNA. Moreover, the technology has been developed as a new therapeutic option that has already made its way into clinical testing.

Treatment of viral infections remains a serious challenge due to the emergence of new viruses and strain variation among known virus species. RNAi holds great promise to provide a flexible approach that can rapidly be adapted to new viral target sequences. A major challenge in the development of an efficient RNAi approach still remains the design of small interfering RNAs (siRNAs) with high silencing potency. While large libraries with validated siRNAs exist for silencing of endogenously expressed genes in human or murine cells, siRNAs still have to be designed individually for new antiviral approaches. The present chapter describes strategies to design highly potent siRNAs by taking into consideration thermodynamic features of the siRNA, as well as the structural restrictions of the target RNA. Furthermore, assays for testing the siRNAs in reporter assays as well as options to improve the properties of siRNAs by the introduction of modified nucleotides will be described. Finally, experimental setups will be outlined to test the siRNAs in assays with infectious viruses.

Key words: Cell viability assay, Plaque assay, RNA interference, shRNA, Short hairpin RNA, Small interfering RNA

1. Introduction

Despite the advances in medical sciences, the threat posed by viruses has remained a serious problem, particularly outside industrialized countries. While the human immunodeficiency virus (HIV) and the hepatitis B and C viruses (HBV, HCV) continue to claim millions of lives every year, viruses like the SARS coronavirus appear as human pathogens and novel variants of

Ronald P. van Rij (ed.), *Antiviral RNAi: Concepts, Methods, and Applications*, Methods in Molecular Biology, vol. 721,
DOI 10.1007/978-1-61779-037-9_17, © Springer Science+Business Media, LLC 2011

well-known viruses, like the H1N1/09 influenza virus, cause pandemics of global dimensions. Thus, there is a pressing unmet medical need to develop new antiviral drugs. While small molecules have dominated antivirals for decades, biologics like monoclonal antibodies and nucleic acids therapies have more recently been considered promising alternatives (1).

Among the various oligonucleotide-based strategies, RNA interference (RNAi) is widely regarded as a particularly powerful technology (2–4). RNAi is a posttranscriptional gene silencing mechanism that is triggered by double-stranded RNA. For applications in mammalian cells, small interfering RNAs (siRNA) of approximately 19 base pairs with two nucleotide overhangs at the 3′ ends of both strands become incorporated into the RNA-induced silencing complex (RISC) and induce cleavage of a complementary target RNA. Within just a few years, RNAi has become a standard molecular biological method to study gene functions. Moreover, the technology has been developed into a new therapeutic approach and approximately a dozen clinical trials based on RNAi are currently underway. Among these trials are various applications of RNAi to treat infections with the respiratory syncytial virus, HBV and HIV (5).

Silencing of endogenously expressed genes has become a routine procedure and predesigned or validated siRNAs with silencing guarantee against virtually any human or murine gene are commercially available. In contrast, the design of siRNAs with high antiviral activity still remains a challenging task. Several features have been described that influence the success of an RNAi approach (6):

- Thermodynamic design of the siRNA.
- Structure of the siRNA antisense strand.
- Structure of the target RNA.

Most of the publically available tools for the design of siRNAs exclusively optimize the base composition of the siRNA. More recently, however, a design algorithm was developed for the selection of siRNA with particularly high potency and specificity, which not only focuses on the design of the siRNA, but also takes into consideration the secondary structures of the siRNA and their target site (7).

A major challenge for the long-term inhibition of viruses by RNAi is the prevention of viral escape. Thus, siRNAs need to be directed against well-conserved target sites. Unfortunately, simply directing siRNAs to protein coding regions is insufficient, since silent mutations can cause the siRNAs to lose their inhibitory potential. Target regions with an important function in the structure of the RNA may prove to be a better choice to avoid viral escape. It has, for example, been shown that an siRNA against the highly conserved cis-acting replication element (CRE) was capable of inhibiting various enteroviruses over a long period, while an siRNA targeted against protein-encoding regions of the RNA, which are not organized into functional three dimensional structures, led to rapid viral escape (8). Since even the most careful selection of a single target site might be insufficient for sustained viral silencing, use of multiple siRNAs against multiple target sites may be necessary to prevent viral escape. In the case of coxsackievirus B3 and HIV-1, resistance rapidly emerged when single molecules were used, but a combination of three or four siRNAs, respectively, targeting distinct regions of the genome, were able to prevent the emergence of resistance (9, 10).

A general requirement for the application of RNAi in vivo is the stabilization of siRNAs by the introduction of modified nucleotides (11). Based on the experience in the antisense field, numerous modified nucleotides have been assessed for their applicability to enhance the stability of siRNAs against nucleases. Phosphorothioates, nucleotides with modifications at the 2′ position (e.g., 2′-*O*-methyl-RNA and 2′-fluoro-nucleotides), as well as locked nucleic acids are among the most widely employed building blocks for RNAi applications. A fully modified siRNA with a drastically increased half-life in human serum had a significantly higher activity in a vector-based in vivo model of HBV infection as compared to the unmodified form (12). Furthermore, the introduction of modified nucleotides not only improves the stability of siRNAs, but can also reduce off-target effects (13) and improves the antiviral activity of siRNAs directed against highly structured target regions (14).

The present chapter describes a systematic method for the design of highly potent antiviral siRNAs (summarized in Fig. 1). Experimental procedures to test these siRNAs in reporter assays (GFP Reporter Assay and Dual-Luciferase Reporter Assay) as well as in assays with infectious viruses (Cell viability assay and Plaque reduction assay) will be outlined. Finally, options to improve the properties of siRNAs by introduction of modified nucleotides will be described.

2. Material

2.1. GFP Reporter Assay

1. CT- or NT-GFP Fusion TOPO TA Expression Kit (Invitrogen, Carlsbad, CA, USA). Store the competent TOP10 cells at –80°C, and all other components of the kit at –20°C.
2. Materials for isolation of viral DNA or RNA. We recommend use of a commercial isolation kit, for example, the High Pure Viral Nucleic Acid Kit (Roche Diagnostics, Mannheim, Germany).

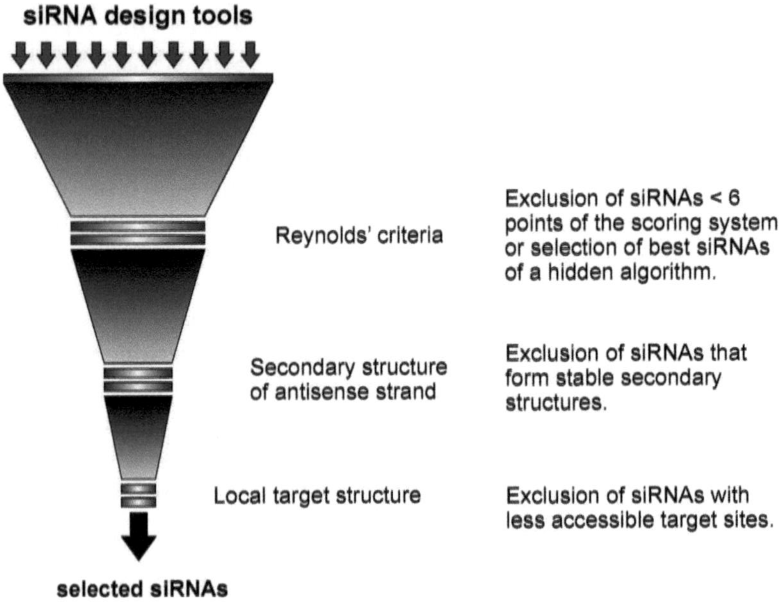

Fig. 1. Systematic selection of antiviral siRNAs. In the first step, siRNAs are scored according to a system based on the thermodynamic features of the siRNA (15). Next siRNAs with antisense strands that are predicted to form stable secondary structures are excluded. Finally, the accessibility of the target site in the viral RNA is tested to end up with a pool of siRNAs which are suitable for testing in vitro.

3. Materials for reverse transcription if the genome of the isolated virus is RNA.
4. Materials for PCR and *Taq* polymerase if a polymerase with proofreading activity is used in PCR.
5. Materials for the Gel purification of DNA. We recommend use of a commercially available gel purification kit, for example, the NucleoSpin Extract II (Macherey-Nagel, Düren, Germany).
6. LB medium: Dilute 1% (w/v) Tryptone, 1% (w/v) NaCl, and 0.5% (w/v) yeast extract in H_2O and autoclave to sterilize. LB Agar plates additionally contain 1.5% (w/v) Agar. For selection, add 100 μg/ml ampicillin. Store LB medium and LB Agar plates at 4°C, but warm to room temperature before use.
7. Materials for the isolation of plasmid DNA. For test-restrictions and sequencing, any DNA Plasmid Miniprep (Kit) is sufficient. For transfection-grade plasmids, we recommend to use a commercially available Plasmid DNA Isolation Kit, for example, the NucleoBond Xtra Midi Kit (Macherey-Nagel).
8. Primer for sequencing of pcDNA3.1/CT-GFP-TOPO: T7 (forward): 5′-TAA TAC GAC TCA CTA TAG GG-3′ and

GFP Reverse (reverse): 5′-GGG TAA GCT TTC CGT ATG TAG C-3′.

9. Primer for sequencing of pcDNA3.1/NT-GFP-TOPO: GFP Forward (forward): 5′-CGA CAC AAT CTG CCC TTT CG-3′ and BGH Reverse (reverse): 5′-TAG AAG GCA CAG TCG AGG-3′.

2.2. Dual-Luciferase Reporter Assay

1. psiCHECK-2 Vector (Promega, Madison, WI, USA). Store at −20°C.
2. The same materials which are described for GFP Reporter Assay (see Subheading 2.1) except items 1, 8, and 9.
3. Restriction enzymes Not I and Xho I. We recommend use of commercially available restriction enzymes with optimized buffer solutions.
4. T4 DNA Ligase. We recommend use of a commercial DNA ligation kit with a corresponding buffer.
5. Primer for sequencing of psiCHECK-2 vectors: Reverse primer: 5′-CTC ATT TAG ATC CTC ACA C-3′.
6. Dulbecco's Phosphate-buffered salt solution (PBS) without Ca and Mg (PAA Laboratories GmbH, Cölbe, Germany).
7. Dual-Luciferase Reporter Assay System (Promega). Store all components of the Kit at −20°C. For long-term storage, prepare aliquots and store at −70°C.
8. Luminometer.

2.3. Cell Culture and Transfection

1. Dulbecco's modified Eagle's medium (DMEM) (PAA Laboratories) supplemented with 10% fetal calf serum (FCS) (PAA), 2 mM glutamine, non-essential amino acids.
2. Dulbecco's PBS without Ca and Mg (PAA Laboratories).
3. Opti-MEM (Invitrogen) or serum-free DMEM.
4. Lipofectamine 2000 transfection reagent (Invitrogen). Store at 4°C.
5. Fluorescence microscope with GFP filter and additional Cy3 filter if the transfection efficiency of siRNAs into other cell lines has to be tested.

2.4. Lysis and Western Blot

1. Dulbecco's PBS without Ca and Mg (PAA).
2. DTT Lysis buffer: 62.5 mM Tris–HCl, pH 6.8, 2% (w/v) SDS, 50 mM DTT, 10% (v/v) glycerol, and 0.05 (w/v) bromophenol blue. Store in aliquots at −20°C.
3. Tris-buffered saline (TBS): 20 mM Tris, 150 mM NaCl, pH 7.5. Prepare a tenfold stock solution containing 200 mM Tris and 1.5 mM NaCl. Adjust to pH 7.5 with HCl and store at room temperature. For 1× TBS, dilute the stock solution 1:10.

4. TTBS: Prepare a solution containing TBS and 0.1% Tween 20. Store at room temperature.
5. Materials for standard western blot including Hybond-P PVDF membranes (Amersham/GE Healthcare, Uppsala, Sweden) and non-fat dry milk powder.
6. Ponceau solution: 0.2% Ponceau S, 3% acetic acid in water. Store at room temperature.
7. Primary antibodies: Polyclonal Anti-GFP rabbit serum (Invitrogen), store at −20°C. Use at a 1:5,000 dilution in western blot. Mouse monoclonal anti-Actin antibody (Chemicon/Millipore), store at 4°C. Use at a 1:5,000 dilution in western blot.
8. Secondary antibodies: Immuno Pure Goat Anti-Rabbit IgG (H+L), Peroxidase Conjugated (Pierce, now part of Thermo Fisher Scientific, Rockford, IL, USA). Immuno Pure Goat Anti-Mouse IgG (H+L), Peroxidase Conjugated (Pierce). Store at −20°C.
9. ECL Western blotting substrate (Pierce).
10. Chemiluminescence Imager or material for wet development inclusive X-ray films.

2.5. Cell Viability Assay

1. Materials for siRNA transfection (see Subheading 2.3, items 1–4).
2. Virus stock of interest.
3. Cell Proliferation Kit II (XTT) (Roche Diagnostics). Store both the XTT labeling reagent and the electron-coupling reagent in small aliquots at −20°C and protect from light.
4. ELISA plate reader for measurement of absorbance at a defined wavelength between 450 and 500 nm, with a reference wavelength greater than 650 nm.

2.6. Plaque Reduction Assay

1. Materials for siRNA transfection (see Subheading 2.3, items 1–4).
2. Virus stock of interest.
3. Difco Agar Noble (Becton Dickinson, Franklin Lakes, NJ, USA).
4. Eagle's MEM: Dilute 9.53 g of Minimum Essential Medium (MEM) powder (Invitrogen) and 2.2 g $NaHCO_3$ in 500 ml water. Add antibiotics, penicillin, and streptomycin. Sterilize by filtration through a 0.2-µm filter. Store at 4°C.
5. Neutral red solution: Dilute 0.02 g neutral red (Merck) in 50 ml PBS and filter to remove undissolved particles. Avoid exposing the solution to light for longer than necessary. The dilution has to be prepared freshly.

3. Methods

3.1. Bioinformatics: siRNA Design

1. The first major step for the design of highly active siRNAs consists of the selection of siRNA sequences with favorable thermodynamic features. The thermal stability of both ends, as well as base preferences of active siRNAs at certain positions, influence the silencing efficiency of an siRNA (15). For the design of siRNAs, publically available web tools are used (examples are given in Table 1). It is advisable to generate an siRNA pool with at least two different web tools. The BLAST option should be applied to avoid unspecific binding to endogenous target sequences, thereby reducing off-target effects. As an example, siRNA predesign using *RNAi Designer* from Invitrogen is described below:
   (a) Go to the *RNAi Designer* website from Invitrogen (see Table 1).
   (b) Choose *siRNA* as "Target Design Options."
   (c) Enter the complete or partial target sequence. Use highly conserved regions of the virus (see Note 1).
   (d) Deselect all options for target region selection (ORF, UTR).

Table 1
List of web-based algorithms for the design of siRNAs

Source	ULR
Dharmacon	http://www.dharmacon.com/DesignCenter
Integrated DNA Technologies	http://www.eu.idtdna.com/Scitools/Applications/RNAi/RNAi.aspx
Sonnhammer Lab	http://sirna.sbc.su.se/
Invitrogen	https://rnaidesigner.invitrogen.com/rnaiexpress/
Qiagen	www1.qiagen.com/GeneGlobe/
Sfold Algorithm	http://sfold.wadsworth.org/index.pl
Whitehead Institute	http://jura.wi.mit.edu/bioc/siRNAext/ (requires registration)
MWG-Biotech	www.eurofinsdna.com (requires registration)
Applied Biosystems	http://www5.appliedbiosystems.com/tools/siDesign/

(e) Select the BLAST database (e.g., if human viruses are to be targeted, choose *human* to avoid target sites within the human genome; for initial testing in animal models, homology to the animal's genome should be avoided as well).

(f) "GC content" should be set to a minimum *of* 25% and a maximum *of* 55%.

(g) Select *default motif pattern* (see Note 2).

(h) Submit by pressing the *RNAi Design* button. A summary of designed siRNAs is given in a new window. *Sort* the output by choosing *ranking*.

(i) Select siRNAs with 4½ to 5 stars and copy the sequences.
In general, choose the best siRNAs obtained with a design tool. If the siRNA design tool did not include a BLAST option, an NCBI BLAST should follow (see Note 3).

2. The obtained siRNA pool is subsequently filtered with respect to their thermodynamic features:

(a) Convert all sequences to obtain the antisense strand of each siRNA in a 5′ to 3′ orientation. Antisense strands are complementary to their target RNA (see Fig. 2).

(b) Remove duplicate siRNAs. Only 100% identical siRNA sequences should be excluded since a single nucleotide shift can sometimes influence the efficiency of an siRNA.

(c) Calculate the score of each siRNA based on the scoring system given in Table 2.

(d) Exclude siRNAs that have less than 6 points of Reynolds' criteria.

3. The second major step is to exclude siRNAs whose antisense strand is predicted to form secondary structures. To this end, the change in Gibbs free energy (ΔG) of the siRNA antisense strand is determined. Negative ΔG values indicate stable secondary structures, whereas positive ΔG values indicate instable secondary structures resulting in a high degree of unfolded

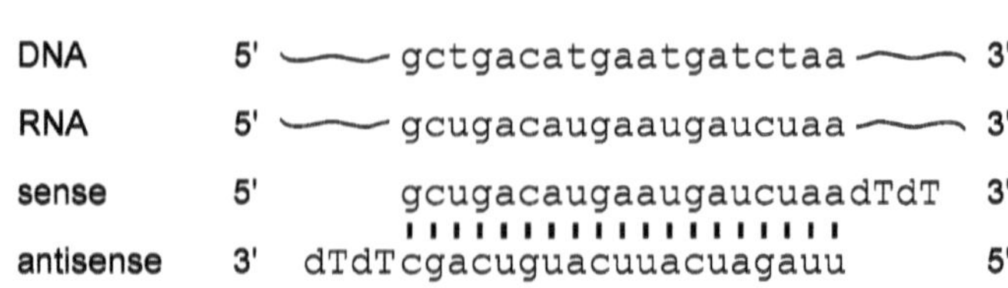

Fig. 2. Orientation of sense and antisense strands of an siRNA in relation to corresponding strands of DNA and RNA.

Table 2
Scoring system based on the thermodynamic features of the siRNA (according to Reynolds et al. (15))

Criterion	Score
GC content between 30 and 52% (in the 19mer siRNA duplex)	1 Point
A or U at positions 15–19	1 Point for each
Lack of internal repeats (T_m < 20°C)	1 Point
A at position 19	1 Point
A at position 3	1 Point
U at position 10	1 Point
G or C at position 19	−1 Point
G at position 13	−1 Point

Positions refer to the sense strand of the siRNA in 5′ to 3′ orientation

Table 3
Additional websites

Source	URL
Rensselaer bioinformatics web server	http://mfold.bioinfo.rpi.edu/
Burnet Institute mfold server	http://mfold.burnet.edu.au/
NCBI BLAST	http://blast.ncbi.nlm.nih.gov/Blast.cgi
Ensembl BLAST	http://ensembl.org/Multi/blastview/
Invitrogen	www.invitrogen.com (search for "GFP Fusion TOPO TA Expression Kit")
Promega	www.promega.com (search for "psiCHECK-2")
New England Biolabs	www.neb.com/nebecomm/tech_reference/restriction_enzymes/cleavage_olignucleotides.asp
Turner energy rules	http://www.bioinfo.rpi.edu/zukerm/rna/energy/

single-stranded RNA. The ΔG determination using the publically available *UNAfold* software is described. This software extended and replaced the *mfold* software of Zuker (16).

(a) Go to the Rensselaer bioinformatics web server (see Table 3) and use the DINAMelt Web Server.

(b) Select the application *Nucleic Acid Quikfold*.

(c) Enter all siRNA antisense strand sequences (5′ to 3′) which were obtained after first selection step (thermodynamic design of siRNAs, steps 1 and 2) separated by semicolons.

(d) Choose *RNA* (*3.0*) in the "energy rules" field. The RNA is now folded at the fixed temperature of 37°C. The hybridization temperature can be modified by using *RNA* (*2.3*). Additionally, "sequence type" has to be specified as *linear*. Other settings in this window are not necessary.

(e) *Submit* to *DINAMelt Server*. A summary of potential folding results is given in a new window (see Note 4). Exclude all siRNAs whose antisense strand is predicted to form secondary structures indicated by at least one structure with a negative ΔG value. Structures with ΔG values around zero have a folding probability of 0.5 (17) and should be further analyzed (see Note 5).

4. The accessibility of the target RNA sequence will be considered in the third major step of siRNA design. The procedure consists of two steps: First, folding of the entire target RNA molecule and, second, the determination of the change in Gibbs free energy (ΔG) of the local RNA target site. ΔG determination using the publically available *mfold3.2* software, which should be used preferentially for long RNA sequences, is subsequently described (16).

(a) Go to the "Rensselaer bioinformatics web server" (see Table 3) and use the Mfold Web Server.

(b) Select the application RNA*fold*.

(c) Enter the "name" and the "sequence" to be folded.

(d) Choose "RNA sequence is" *linear*. The folding temperature is fixed at 37°C. This temperature can be modified by using the *RNA mfold version 2.3 server*. The "upper bond on the number of computing folding" can be limited to 10 or less for long sequences as they may increase the computational load and resulting response time. Other settings in this window are not necessary.

(e) Sequences shorter than 800 bases can be folded immediately (as of March 2010). For longer RNA sequences up to 6,000 bases, select *batch* and enter your *E-mail address*.

(f) Submit your entries by pressing the *Fold RNA* button. RNA sequences longer than 6,000 bases can be folded by the alternative "Burnet Institute mfold Server" (see Table 3). The application *Submit a RNA sequence* is based on *mfold version 3.0*. Parameters in this window can be set according to the *mfold version 3.2* described above, except the folding temperature, which is not fixed at 37°C. Additionally, limit the "upper bond on the number of computing folding" to 10. Both "Rensselaer bioinformatics web server" and "Burnet Institute mfold Server" will give a summary of potential folding results.

(g) ΔG is calculated for a number of potential secondary structures (listed as "Structure 1, 2, 3 ..." beneath "View Individual Structures"). The more negative ΔG, the more stable secondary structure is going to be, which also indicate a higher chance of its presence in a biological system.

(h) Save at least ten structures using the *Download all foldings* button.

(i) Find the corresponding target structures of the siRNAs in every potential folding.

(j) Select the most promising siRNA target sites by checking the following points:

  (i) Number of identical or similar target sites. The program will give various structures and corresponding energies (ΔG). Good target accessibility in different structure predictions increases the possibility of overall target accessibility.

  (ii) Number of bonds. High number of bonds (base pairs) within the target sequence indicates a strong secondary structure and less accessibility for an siRNA.

(k) Determine the ΔG value of the local RNA target structure for the selected siRNA target sites:

  (i) Find the corresponding siRNA target site and determine all bases within the target site that are involved in forming a secondary structure. Add also bases or base pairs which are part of the close neighborhood, for example bases that form loops, even if these are not part of the siRNA target site.

  (ii) Go to *Thermodynamic Details* of chosen structure in the Output window (folding results).

  (iii) Find the corresponding bases (marked by numbers) and add all single ΔG values of involved bases. Energies of helices and loops have to be included if one base of this structure is part of the target site. This calculation results in an approximation of ΔG_{local} which is usually sufficient. If precise calculation of ΔG_{local} is necessary, please refer to the Turner energy rules (Table 3).

(iv) Calculate the change in Gibbs free energy (ΔG) of all local RNA target structures (ΔG_{local}). Exclude all siRNAs whose target structure showed strong variability based on at least ten folded structures (≥5 different target structures). Furthermore, accessible target sites are characterized by low ΔG_{local} values (18). Therefore, select siRNAs in the order of their ΔG_{local} values with a threshold of about −15 kcal/mol (19). The selected siRNAs are potentially efficient siRNAs and their silencing efficiency can further be tested in Reporter Assays (see Subheading 3.2 and Note 6).

3.2. In Vitro Testing: Reporter Assays

Reporter assays are the method of choice to verify the efficiency of the selected siRNAs. The virus target sequence is cloned upstream or downstream of a GFP or luciferase reporter and is then cotransfected with the selected siRNAs. The virus target sequence, which is a partial viral sequence, should be long enough to simulate the structural environment of siRNA target sequence (20). In contrast, very long sequences may complicate reverse transcription and/or PCR amplification of the viral target sequence and may negatively influence the transfection efficiency of eukaryotic cells (see Note 7).

3.2.1. GFP Reporter Assay

The protocol for TOPO cloning described in the following section is a shortened version of the instruction manual (see Note 8). The TOPO cloning allows the ligation of the PCR-amplified target sequence of the virus into either pcDNA3.1/CT-GFP-TOPO or pcDNA3.1/NT-GFP-TOPO. The terms CT and NT indicate that GFP will be expressed fused to, respectively, the C-terminus and N-terminus of the target sequence.

1. Isolate the virus DNA or RNA using a commercial DNA or RNA virus isolation kit.
2. Perform a reverse transcription reaction if the isolated virus contains an RNA genome.
3. Design the primer for TOPO cloning. Do not add 5′ phosphates to the PCR primers. The pcDNA3.1/CT-GFP-TOPO does not contain an ATG initiation codon. The PCR product must, therefore, be cloned in frame with GFP. If the target sequence to be amplified lacks the ATG initiation codon, then the Kozak consensus sequence (Kozak consensus sequence is: (G or A)NNATGG (N = any base)) needs to be incorporated into the forward primer. The PCR product for pcDNA3.1/NT-GFP-TOPO may also be cloned in frame with GFP, but this is not essential if a translational stop codon is introduced upstream of the virus target sequence. We observed an improved GFP expression using this modified reporter

plasmid, in which the viral target sequence is part of the untranslated region of the GFP reporter. The stop codon can be incorporated into the forward primer, for example: 5′-**TGA** NNN N...

4. Run the PCR to amplify your target cDNA sequence.
5. Gel-purify the DNA fragment of interest (partial virus cDNA target sequence).
6. Add Taq DNA polymerase buffer, 0.4 mM dATP, and 0.5 U of Taq DNA polymerase to the gel-purified DNA fragment (see Note 9).
7. Incubate 10 min at 72°C.
8. Place the vials on ice.
9. Add 1 μl of salt solution and 1 μl of TOPO vector to 4 μl of the PCR product to set up the TOPO Cloning reaction (see Note 10).
10. Mix gently and incubate for 5 min at room temperature.
11. Place the tube on ice and store the sample at −20°C until later use, or proceed directly with transformation.
12. Thaw a tube containing 100 μl chemically competent *Escherichia coli* cells. Alternatively, use 50 μl TOP10 cells provided with the GFP Fusion TOPO TA Expression Kit (see Note 11).
13. Add 2 μl of the TOPO Cloning reaction (ligation reaction) to the tube containing 100 μl of chemically competent *E. coli* cells and mix gently. Do not mix by pipetting up and down.
14. Place the tube on ice and incubate for 30 min.
15. Heat-shock the *E. coli* cells for 60 s at 42°C without shaking.
16. Transfer the tube to ice and incubate for 3 min.
17. Add 700 μl of room temperature LB medium without antibiotics.
18. Cap the tube tightly and shake the tube horizontally at 37°C for 45 min.
19. Sediment the transformed *E. coli* cells by centrifugation at 3,000 × *g* for 5 min and discard 700 μl of the supernatant.
20. Resuspend the *E. coli* cells in residual LB medium.
21. Spread all of the transformation on a prewarmed selective plate (containing the antibiotic ampicillin) and incubate overnight at 37°C.
22. Pick ten colonies.
23. Culture all colonies overnight in 3 ml LB medium containing 100 μg/ml ampicillin.

24. Isolate plasmid DNA. In general, low amount and purity are sufficient for restriction analysis and/or sequencing.
25. To verify the orientation of the insert, analyze the plasmids by restriction analysis. To this end, it is advantageous to use at least one restriction site located within the virus target sequence and one that is present in the vector (see Note 12).
26. Clones containing inserts in the correct orientation should be sequenced to confirm that the original sequence has not been mutated (see Subheading 2.1, items 8 and 9).
27. Culture the clone of choice in 200 ml LB medium containing 100 μg/ml ampicillin.
28. Isolate the plasmid DNA. Plasmid DNA for transfection into eukaryotic cells must be very clean and free from phenol and sodium chloride to prevent cell death and to avoid low transfection efficiencies.

3.2.2. Dual-Luciferase Reporter Assay

The Dual-Luciferase Reporter Assay is a method to measure the activity of both firefly and Renilla luciferases in one sample. The virus target sequence is cloned in the 3′ untranslated region (3′-UTR) of the Renilla luciferase gene. In-frame cloning is not required as the translational stop codon of Renilla luciferase is present in psiCHECK-2 vector (cloning of the virus target sequence into this vector is subsequently described). The firefly luciferase, which is also expressed by the psiCHECK-2 vector, allows normalization of the Renilla luciferase activity.

1. Isolate the virus DNA or RNA using commercially available DNA or RNA virus isolation kit.
2. Perform a Reverse Transcription reaction if the isolated virus is based on RNA.
3. As in-frame cloning with Renilla luciferase gene is not required, the design of PCR primers follows common rules. Both primers contain one restriction site that is used for cloning into psiCHECK-2 vector (see Note 13). For example, use:

Forward primer: ccg CTC GAG nnn nnn nnn nnn nnn …

Reverse primer: ata gtt ta G CGG CCG C nnn nnn nnn nnn nnn …

The enzyme recognition sites (Xho I in the forward primer and Not I in the reverse primer) are indicated by upper case letters. The lower case letters show additional nucleotides for efficient restriction of the site located at the end of the PCR fragment. The bases which are part of the viral target sequence are indicated as nnn.

4. Run a PCR to amplify your target cDNA sequence.
5. Gel-purify the DNA fragment of interest (partial virus cDNA target sequence) and prepare an analytic gel to estimate the DNA concentration.
6. Add 5 U of Not I and Xho I (or other enzyme if other restriction sites were used) per µg DNA and complement the reaction with buffer and BSA.
7. Incubate for 20 h at 37°C.
8. Inactivate the restriction enzymes by incubation at 65°C for 20 min.
9. Place the tube on ice.
10. Ligate the DNA fragment into the psiCHECK-2 vector that was previously digested with Not I and Xho I using a standard protocol.
11. Transform the ligation reaction into *E. coli* cells and analyze colonies as described in Subheading 3.2.1, steps 12–28.
12. Clones containing inserts in the correct orientation should be sequenced (see Subheading 2.2, item 5).

3.3. Cellular Assays: Transfection of Mammalian Cells

For a first functional assay of the siRNAs in a reporter assay, it is advisable to use the cell line which will also be employed for virus assays. As an example, a protocol for transient transfection of HeLa cells in 24-well plates using Lipofectamine 2000 transfection reagent will be described (see Note 14). The transfection efficiency into other cell lines can be tested by transfection of a random siRNA that is labeled with Cy3 at the 5′ end of the siRNA sense strand (siRNA can be transfected with or without a reporter plasmid). We recommend the use of Cy3, as FITC is not distinguishable from GFP in fluorescence microscopy.

1. Seed 5×10^4 HeLa cells per well in a volume of 500 µl medium without antibiotics.
2. Incubate overnight at 37°C and 5% CO_2.
3. Dilute 0.8 µg of the reporter plasmid (pcDNA3.1/NT-GFP-TOPO, pcDNA3.1/CT-GFP-TOPO or psiCHECK-2 with inserted virus target sequence) and 10 nM of siRNA (designed antiviral siRNA or a negative control siRNA with no matches either to the viral or target organism genome) in Opti-MEM or serum-free medium to a total volume of 50 µl. The molar concentrations are related to the final volume of 600 µl.
4. Premix 2 µl of Lipofectamine 2000 transfection reagent with 48 µl Opti-MEM.
5. Incubate for 5 min at room temperature.
6. Combine Lipofectamine and DNA/siRNA mixture and incubate for at least 20 min.

7. Add the final mix (100 μl) to the cells. Do not remove the culture medium.
8. Incubate overnight at 37°C and 5% CO_2.

3.4. Analysis of siRNA Silencing Efficiency

If one of the TOPO vectors described above served as a reporter in cotransfection with siRNAs (see Subheading 3.3), the first analysis of siRNA silencing efficiency can be performed using fluorescence microscopy. This requires a filter for GFP detection. After replacement of medium from the transfected culture cells with 500 μl PBS, GFP expression of the reporter plasmid is determined (see Fig. 3). Highly efficient siRNAs strongly reduce GFP expression compared to the transfection of reporter plasmid without an siRNA, while the control siRNA should have no effect. These results should be confirmed by Western Blot analysis as described in Subheading 3.4.1. In the case of psiCHECK-2 cotransfection, both Renilla and Firefly luciferases will be measured (see Subheading 3.4.2).

3.4.1. Western Blot

1. Remove the medium from the culture cells.
2. Wash the cells with 500 μl of PBS.
3. Remove PBS by aspiration.
4. Add 30 μl of DTT lysis buffer per well and transfer the lysate to an eppendorf tube.
5. Boil the lysate for 5 min at 95°C.

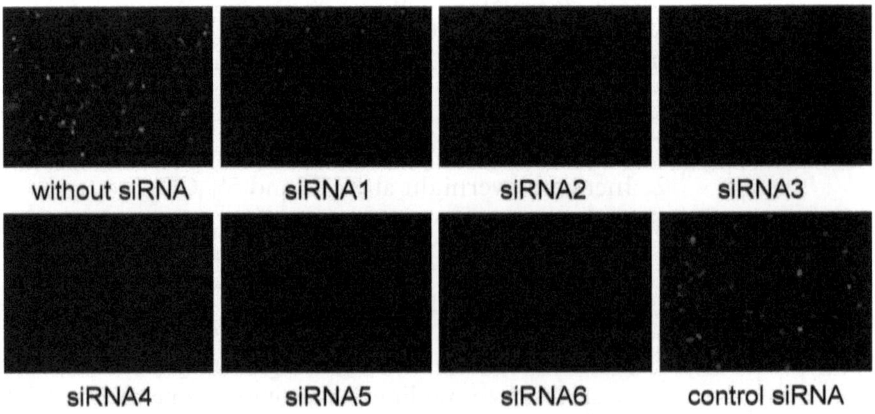

Fig. 3. Fluorescence microscopy of GFP reporter assay. The viral sequence encoding the RNA-dependent RNA polymerase ($3D^{Pol}$) of Echovirus 30 (EV-30) was cloned downstream of GFP. Cells were transfected with 0.8 μg of GFP-$3D^{Pol}$ plasmid and different siRNAs targeting $3D^{Pol}$ of EV-30. Cotransfection of 10 nM of siRNA2 to siRNA6 strongly reduced GFP expression compared to the transfection of reporter plasmid without an siRNA. The siRNA1 only leads to partial inhibition of target gene expression. In the panel labeled without siRNA, the cells were transfected with GFP reporter, but without siRNA. The control siRNA panel shows the results of a control siRNA with no matches either to the viral or target organism genome.

6. Separate the proteins on a 15% (w/v) SDS-polyacrylamide gel (see Note 15).
7. Transfer the proteins to PVDF membrane using your method of choice.
8. Control the protein transfer by Ponceau staining. Strip the membrane by washing with TBS.

   Shake the membrane during the following washing steps and antiserum incubations.
9. Rinse the membrane in TTBS.
10. Block with 5% non-fat milk powder in TTBS for 1 h at room temperature.
11. Rinse the membrane 3 times in TTBS.
12. Wash 3 times for 10 min in TTBS.
13. Incubate with GFP antiserum at a dilution of 1:5,000 in TTBS for 1 h at room temperature or overnight at 4°C.
14. Rinse the membrane 3 times in TTBS.
15. Wash 3 times for 10 min in TTBS.
16. Incubate with the secondary Goat Anti-Rabbit antibody conjugated with horseradish peroxidase, at a dilution of 1:5,000 in TTBS for 1 h at room temperature or overnight at 4°C.
17. Rinse the membrane 3 times in TTBS.
18. Wash 3 times for 10 min in TTBS.
19. Combine 500 μl of each component of the ECL Western Blotting Substrate and transfer the mixture to the membrane.
20. Incubate 1 min in the dark.
21. Detect the signals using your method of choice (Chemiluminescence Imager or development using an X-ray film).
22. Strip the membrane by washing 3 times for 10 min in TTBS.
23. Block with 5% nonfat milk powder in TTBS for 1 h or with 10% non-fat milk powder in TTBS for 10 min at room temperature.
24. To confirm equal loading of the samples, reprobe the membrane with an antibody directed against a housekeeping gene, for example, actin. Repeat all steps starting from step 9. Use the mouse monoclonal anti-Actin antibody at a dilution of 1:5,000.

3.4.2. Dual-Luciferase Measurement

1. Remove the growth medium from the culture cells.
2. Wash the cells with 500 μl of PBS.
3. Remove PBS by aspiration.

4. Add 100 μl of 1× passive lysis buffer (PLB) per well and shake the culture plate gently on an orbital shaker or on a rocking platform for 15 min at room temperature. The culture plate can be stored at −20°C for later use, or analyzed directly.
5. Transfer the lysate to a tube and sediment residual cell debris by centrifugation for 30 s at full speed in a tabletop centrifuge. Transfer the cleared lysates to a fresh tube for subsequently measurement.
6. Thaw adequate aliquots of LARII reagent (see Note 16) and prepare an adequate amount of Stop & Glo reagent (50 μl of each reagent per measurement is needed). As the Stop & Glo substrate is supplied in a 50× concentration, dilute 1 volume of 50× Stop & Glo substrate in 50 volumes Stop & Glo buffer. Protect both reagents from light and keep all components at room temperature during measurement.
7. Program your luminometer. (Default parameters: 2 s premeasurement delay and 10 s measurement period for each reporter.)
8. Pipette 50 μl of LARII reagent into a luminometer tube.
9. Add 20 μl of the sample (PLB lysate) or diluted sample and mix gently.
10. Measure the activity of Firefly luciferase.
11. Add 50 μl of Stop & Glo reagent and mix gently.
12. Measure the activity of Renilla luciferase.
13. Repeat the measurement for all samples starting from step 8.

Select efficient siRNAs for the following functional analysis in virus assay. Highly potent antiviral siRNAs reduce the GFP expression virtually to completion (Western Blot) or show at least 80% inhibition of Renilla luciferase (Dual-Luciferase Reporter Assay).

3.5. Virus Assays: Functional Analysis

A crucial test is the evaluation of the antiviral activity of designed siRNAs in assays with infectious virus. Two methods for in vitro testing are described below. The cell viability assay is an indirect method to quantify the efficiency of an siRNA, whereas the plaque reduction assay can show the direct inhibition of virus replication by an siRNA.

3.5.1. Cell Viability Assay

The XTT-based cell viability assay is used for the quantification of cell proliferation and viability. Metabolically active cells cleave the yellow tetrazolium salt to an orange formazan dye. The absorbance measured at a defined wavelength directly correlates with cell viability. We recommend the use of the XTT reagent due to its high sensitivity compared to other tetrazolium salts, such as MTT.

1. Seed 2–3 × 10^4 HeLa cells per well in a volume of 100 μl medium without antibiotics in a 96 well plate.
2. Incubate overnight at 37°C and 5% CO_2.

   Prepare the following siRNA transfection mixtures in quadruplicate.
3. Dilute siRNA (designed antiviral siRNA or a negative control siRNA with no matches either to the viral or target organism genome) in Opti-MEM or serum-free medium to a volume of 25 μl. The molar concentrations are related to the final volume of 150 μl.
4. Premix 0.5 μl of Lipofectamine 2000 transfection reagent with 24.5 μl Opti-MEM.
5. Incubate for 5 min at room temperature.
6. Combine Lipofectamine and siRNA and incubate for at least 20 min.
7. Add the final mix (50 μl per well) to the cells.
8. Incubate for 4 h at 37°C and 5% CO_2 (see Note 17).
9. Calculate the amount of virus required to reach a specific multiplicity of infection (m.o.i.) based on the titer of the virus stock and the number of seeded cells. Dilute the virus suspension to a specific m.o.i. An infection using an m.o.i. of 0.1 is subsequently described (see Note 18).

   Example: The virus titer is 2.5 × 10^7 plaque forming units (pfu)/ml. If 50 μl per well of the virus suspension is added to the cells, the titer will be 1.25 × 10^6 pfu per well. This value is further divided by the number of seeded cells resulting in an absolute m.o.i. value. Finally, the dilution factor is calculated by dividing by the desired m.o.i.
10. Prepare the dilution of virus suspension in refrigerator-cold serum-free medium without antibiotics (see Note 19).
11. Incubate at 37°C for 5 min before adding to the cells. Virus dilutions can be stored at 4°C for several minutes, if necessary.
12. Add 50 μl per well of the diluted virus suspension. Include negative control wells that remain uninfected. To this end, use serum-free medium without antibiotics and handle these cells identical to infected cells.
13. Incubate for 30 min at 37°C and 5% CO_2 (see Note 20).
14. Remove the virus suspension and add 100 μl of medium containing FCS and antibiotics.
15. Incubate at 37°C and 5% CO_2 until beginning the cell viability assay (XTT measurement).

16. Prepare an adequate amount of XTT reagent. For example, mix 100 μl electron-coupling reagent with 5 ml XTT labeling reagent.
17. Add 50 μl per well of fresh XTT reagent and incubate 4 h at 37°C and 5% CO_2. Do not remove the medium.
18. Measure the absorbance in a microtiter plate reader at a defined wavelength between 450 and 500 nm. The reference wavelength should be greater than 650 nm.

3.5.2. Plaque Reduction Assay

Generally, each plaque has its origin in the viral infection of a single cell. The reduction of virus titer shows directly the efficiency of an siRNA as the virus spread is prevented.

1. Seed 1.5–2.5 × 10^5 HeLa cells per well in a volume of 500 μl of medium without antibiotics in a 24 well plate.
2. Incubate overnight at 37°C and 5% CO_2.
3. Dilute siRNA (designed antiviral siRNA or a negative control siRNA with no matches either to the viral or target organism genome) in Opti-MEM or serum-free medium to a volume of 50 μl. The molar concentrations are related to the final volume of 600 μl.

   Prepare as many siRNA transfected wells as needed to titrate the different virus dilutions. Include positive control wells which remain untransfected, but will be infected with different virus dilutions.
4. Premix 2 μl of Lipofectamine 2000 transfection reagent with 48 μl Opti-MEM.
5. Incubated for 5 min at room temperature.
6. Combine Lipofectamine and siRNA and incubate for at least 20 min.
7. Add the final mix (100 μl per well) to the cells.
8. Incubate for at least 4 h at 37°C and 5% CO_2 (see Note 17).

   Cells should be close to 100% confluent before overlaying the infected cell monolayer.
9. Prepare serial 1:10 dilutions of the virus stock in refrigerator-cold serum-free medium without antibiotics (see Note 19). In general, an appropriate range of dilutions is 10^{-4} to 10^{-7} (depending on the titer of virus stock).
10. Incubate at 37°C for 5 min before adding to the cells. Virus dilutions can be stored at 4°C for several minutes, if necessary.
11. Add 250 μl per well of the diluted virus suspension. Include negative control wells that remain uninfected. To this end, use serum-free medium without antibiotics and handle these cells identically to infected cells.

12. Incubate for 30 min at 37°C and 5% CO_2 (see Note 20).
13. Remove the virus suspension by gentle aspiration.
14. Add 4 ml FCS to 20 ml Eagle-MEM and prewarm to 41.5°C.
15. Melt 0.28 g of Difco Agar Noble in 16 ml H_2O. Cool to 40–45°C and add the prewarmed Eagle-MEM to the Agar Noble solution.
16. Incubate for 15 min at 41.5°C. This makes a 0.7% Agar Noble solution (Eagle Overlay) which is used to overlay the infected cell monolayer.
17. Add 500 µl per well of Eagle Overlay and leave the plates for 5 min at room temperature. Take care not to dislodge any cells.
18. Incubate at 37°C and 5% CO_2. Plaques should be visible within around 3 days depending on the virus.
19. Add 500 µl of neutral red solution to each of the wells and incubate for 3–4 h at 37°C and 5% CO_2 (see Note 21).
20. Remove the stain by gentle aspiration.
21. Invert the plate and count the plaques in each well.
22. Calculate the viral titer of each well.

$$\text{pfu/ml} = \text{counted plaques} / (\text{dilution factor} \times \text{volume of diluted virus added to the well}).$$

An alternative protocol can be used to accelerate the staining (see Note 22). Efficient siRNAs are able to reduce the virus titer at least one order of magnitude. Figure 4 shows an example of such a plaque reduction assay.

3.6. Chemical Modification: Generating Stabilized siRNAs

Selected (antiviral) siRNAs which show high efficiency may be chemically modified to decrease potential off-target effects and/or to enhance serum stability. One approach to reduce off-target effects is to prevent the incorporation of the unwanted sense strand in RISC. This can, for example, be achieved by an LNA modification at the 5′ end (21), or by introduction of a 2′-*O*-methyl modification at position 1 and 2 (13) of the sense strand. However, a specific reduction of off-target effects was observed when position 2 of the antisense strand was additional 2′-*O*-methyl modified (13). In theory, chemical modifications of an siRNA will also result in increased serum stability. In practice, siRNA modifications are often accompanied by decreased silencing efficiency compared to the unmodified siRNA. LNA modifications introduced at any position except 1, 10, 12, and 14 in the antisense strand do not cause substantial loss of silencing efficiency (21). Another way to combine high silencing efficiency, increased serum stability, and reduced off-target effects is complete modification

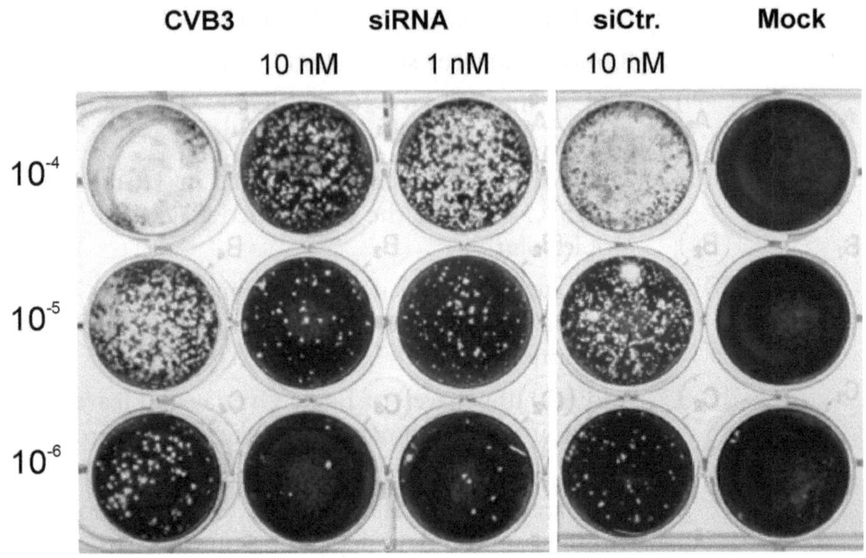

Fig. 4. Plaque reduction assay. After transfection of a virus-specific siRNA at 1 and 10 nM concentrations, cells were infected with different dilutions of Coxsackievirus B3 (CVB3). The infection with a virus dilution of 10^{-5} of 10 nM transfected cells targeting CVB3 resulted in fewer visible plaques than for the 10^{-6} dilution of the virus alone. The CVB3 titer was reduced more than one order of magnitude. Mock: uninfected cells, siCtr.: control siRNA with no matches to either the viral or target organism genomes.

with different nucleotide analogs as shown for an siRNA targeting the HBV genome (12) (Fig. 5). However, chemical modifications are dependent on the application as well as on the virus to be inhibited.

With the strategy described here, it will be possible to identify efficient siRNAs against virtually any given virus – as we have recently exemplified for echovirus 30 (22). The next step will be to transfer the antiviral RNAi approach to an in vivo model. One of the main challenges will then be efficient delivery of the siRNA to the desired target tissue. The following chapters describe virus-based approaches to deliver shRNA expression cassettes as well as different strategies for the application of chemically presynthesized siRNAs.

4. Notes

1. In order to identify conserved regions in the viral genome, it is advisable to select sequences of various strains and perform a BLAST search. Regions with high homology can be considered conserved and represent suitable siRNA target sites.
2. Tuschl's motif pattern strongly restricts the number of siRNAs for subsequent selection, which may become a problem if short sequences are targeted.

sense 5' B **GGA**cuucucuc**AA**uuuucu**TT** B 3'

antisense 3' **T**$_s$**T**ccuGAAGAGAGuuAAAAGA 5'

AGT = deoxy A, G &T
AG = 2'-O-methyl A & G
cu = 2'-fluoro C & U
B = 3',5' inverted deoxy abasic
s = phosphorothioate

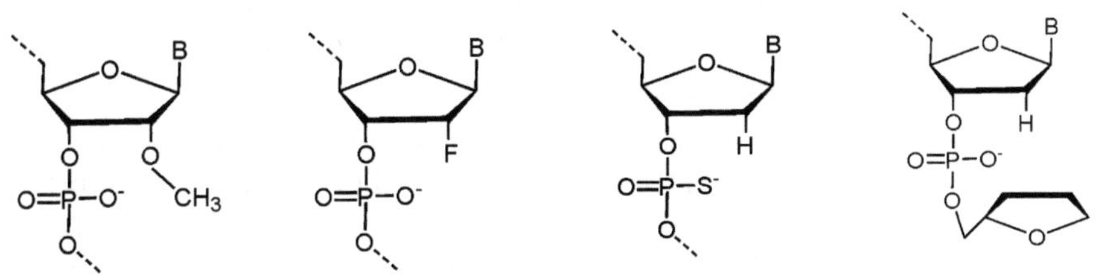

Fig. 5. Chemical modifications of the hepatitis B virus siRNA (figure modified from ref. (12)). The siRNA is fully modified (*top*) with different nucleotide analogs (*bottom*) and showed an increased half-life in human serum and a significantly higher activity in a vector-based in vivo model of HBV infection as compared to the unmodified form.

3. The NCBI BLAST tool or/and the Ensembl BLAST tool can be used for homology searches (see Table 3).
4. If the summary of potential folding is: "No folding possible" and/or the software are/is showing the term ∞ for ΔG, then the RNA sequence does not form any secondary structure.
5. If all siRNAs (antisense strand) give negative ΔG values, examine all structures again and use only siRNAs that show stem-loop structures with ≥2 nucleotide free at the 5′ terminus and ≥4 free nucleotides at the 3′ terminus for further analyses (17).
6. If no siRNAs meet the indicated criteria, first check if another partial sequence of the virus can be targeted. Alternatively, the threshold for ΔG of the local RNA target site can be adjusted, especially if many siRNAs remain whose antisense strands do not form stable secondary structures (see also Note 5). To this end, select at least four siRNAs in order of their ΔG_{local} values. The threshold of Reynolds' criteria should not be changed.
7. In general, it is best to include a partial target RNA with a length of 1–2 kb that spans the siRNA target sequence.
8. Both an extended protocol and a trouble shouting guide of TOPO cloning can be found in the complete instruction manual provided with the kit.

9. For long virus target sequences, use a polymerase with proofreading activity. Since the amplified sequences will lack 3′ A-overhangs, which are essential for the described cloning method, additional incubation with *Taq* polymerase is necessary.
10. If *E. coli* is to be transformed by electroporation, the salt solution must be diluted fourfold to a final concentration of 50 mM NaCl and 2.5 mM $MgCl_2$ to prevent arcing.
11. Optionally, 1.5 μl of 1.42 mM β-mercaptoethanol can be added to the chemically competent *E. coli* cells followed by an incubation on ice for 20 min. This procedure may increase the transformation efficiency. Store the 1.42 mM solution of β-mercaptoethanol at 4°C.
12. Vector maps and sequences of pcDNA3.1/CT-GFP-TOPO and pcDNA3.1/NT-GFP-TOPO are available on the Invitrogen website (see Table 3). Vector map and sequence of psiCHECK-2 is available on the Promega website (see Table 3).
13. Other restriction sites may also be suitable. In this case, customize the additional bases for efficient restriction and the incubation time. An extensive table is provided by New England Biolabs (see Table 3).
14. The use of other cell lines or/and other transfection reagents will require different conditions than those given here.
15. Density of SDS-polyacrylamide gel depends on molecular weight (MW) of the target sequence. If pcDNA3.1/CT-GFP-TOPO was used, add the MW of GFP (around 15 kD) to the calculated MW of the target sequence for an estimate, though some sequences may influence gel mobility, leading to measured weights other than the predicted value.
16. LARII aliquots have to be thawed at room temperature, as the reagent is heat-labile.
17. Incubation overnight is also possible if the cells are plated at a lower density.
18. An m.o.i. of 0.1 is suitable for many viruses, but in some cases may need to be adjusted. The control infection should show as many individual plaques as can be reliably counted. An effective siRNA will appreciably reduce the number of plaques, while a weak siRNA may only lead to a slight reduction. A large number of distinguishable plaques on the control plate will simplify the recognition of weak siRNAs, which may be desirable if further optimization is possible.
19. FCS within a medium may interfere with viral infection efficiency.
20. Incubation time depends on the infection cycle of a specific virus.

21. Neutral red as well as crystal violet is taken up by healthy cells, but not by dead cells. Plaques therefore appear as clear circles against the red/violet background. If the clear areas overlap too much, it is difficult to recognize individual plaques and infection time should be reduced. If plaques are too small to be easily read, try extending infection time, though it could also be an indication of a defective or weakened strain.

22. The staining can be accelerated by using crystal violet solution. To this end, fix the cells with 10% TCA for 10 min at room temperature. Then remove the agar without disturbing the cell monolayer and add 500 μl of crystal violet solution (0.5% solution in PBS) to each of the wells and incubate for 5 min at room temperature (see Note 21). Remove the stain by aspiration and wash with PBS. Invert the plate, and count the plaques of each well. Calculate the viral titer as described in Subheading 3.5.2, step 22.

References

1. Fox, J. L. (2007) Antivirals become a broader enterprise. *Nat Biotechnol* **25**, 1395–1402.
2. Grimm, D. (2009) Small silencing RNAs: state-of-the-art. *Adv Drug Deliv Rev* **61**, 672–703.
3. Kurreck, J. (2009) RNA interference: from basic research to therapeutic applications. *Angew Chem Int Ed Engl* **48**, 1378–1398.
4. Castanotto, D., and Rossi, J. J. (2009) The promises and pitfalls of RNA-interference-based therapeutics. *Nature* **457**, 426–433.
5. Haasnoot, J., Westerhout, E. M., and Berkhout, B. (2007) RNA interference against viruses: strike and counterstrike. *Nat Biotechnol* **25**, 1435–1443.
6. Kurreck, J. (2006) siRNA efficiency: structure or sequence-that is the question. *J Biomed Biotechnol* **2006**, 83757.
7. Wang, X., Wang, X., Varma, R. K., Beauchamp, L., Magdaleno, S., and Sendera, T. J. (2009) Selection of hyperfunctional siRNAs with improved potency and specificity. *Nucleic Acids Res* **37**, e152.
8. Lee, H. S., Ahn, J., Jee, Y., Seo, I. S., Jeon, E. J., Jeon, E. S., Joo, C. H., Kim, Y. K., and Lee, H. (2007) Universal and mutation-resistant anti-enteroviral activity: potency of small interfering RNA complementary to the conserved cis-acting replication element within the enterovirus coding region. *J Gen Virol* **88**, 2003–2012.
9. ter Brake, O., t Hooft, K., Liu, Y. P., Centlivre, M., von Eije, K. J., and Berkhout, B. (2008) Lentiviral vector design for multiple shRNA expression and durable HIV-1 inhibition. *Mol Ther* **16**, 557–564.
10. Merl, S., and Wessely, R. (2007) Anti-coxsackieviral efficacy of RNA interference is highly dependent on genomic target selection and emergence of escape mutants. *Oligonucleotides* **17**, 44–53.
11. Behlke, M. A. (2008) Chemical modification of siRNAs for in vivo use. *Oligonucleotides* **18**, 305–319.
12. Morrissey, D. V., Blanchard, K., Shaw, L., Jensen, K., Lockridge, J. A., Dickinson, B., McSwiggen, J. A., Vargeese, C., Bowman, K., Shaffer, C. S., Polisky, B. A., and Zinnen, S. (2005) Activity of stabilized short interfering RNA in a mouse model of hepatitis B virus replication. *Hepatology* **41**, 1349–1356.
13. Jackson, A. L., Burchard, J., Leake, D., Reynolds, A., Schelter, J., Guo, J., Johnson, J. M., Lim, L., Karpilow, J., Nichols, K., Marshall, W., Khvorova, A., and Linsley, P. S. (2006) Position-specific chemical modification of siRNAs reduces "off-target" transcript silencing. *RNA* **12**, 1197–1205.
14. Dutkiewicz, M., Grunert, H. P., Zeichhardt, H., Lena, S. W., Wengel, J., and Kurreck, J. (2008) Design of LNA-modified siRNAs against the highly structured 5′ UTR of coxsackievirus B3. *FEBS Lett* **582**, 3061–3066.
15. Reynolds, A., Leake, D., Boese, Q., Scaringe, S., Marshall, W. S., and Khvorova, A. (2004) Rational siRNA design for RNA interference. *Nat Biotechnol* **22**, 326–330.

16. Zuker, M. (2003) Mfold web server for nucleic acid folding and hybridization prediction. *Nucleic Acids Res* **31**, 3406–3415.
17. Patzel, V., Rutz, S., Dietrich, I., Koberle, C., Scheffold, A., and Kaufmann, S. H. (2005) Design of siRNAs producing unstructured guide-RNAs results in improved RNA interference efficiency. *Nat Biotechnol* **23**, 1440–1444.
18. Schubert, S., Grunweller, A., Erdmann, V. A., and Kurreck, J. (2005) Local RNA target structure influences siRNA efficacy: systematic analysis of intentionally designed binding regions. *J Mol Biol* **348**, 883–893.
19. Westerhout, E. M., Ooms, M., Vink, M., Das, A. T., and Berkhout, B. (2005) HIV-1 can escape from RNA interference by evolving an alternative structure in its RNA genome. *Nucleic Acids Res* **33**, 796–804.
20. Sun, G., and Rossi, J. J. (2009) Problems associated with reporter assays in RNAi studies. *RNA Biol* **6**, 406–411.
21. Elmen, J., Thonberg, H., Ljungberg, K., Frieden, M., Westergaard, M., Xu, Y., Wahren, B., Liang, Z., Orum, H., Koch, T., and Wahlestedt, C. (2005) Locked nucleic acid (LNA) mediated improvements in siRNA stability and functionality. *Nucleic Acids Res* **33**, 439–447.
22. Rothe, D., Werk, D., Niedrig, S., Horbelt, D., Grunert, H. P., Zeichhardt, H., Erdmann, V. A., and Kurreck, J. (2009) Antiviral activity of highly potent siRNAs against echovirus 30 and its receptor. *J Virol Methods* **157**, 211–218.

Chapter 18

RNAi-Inducing Lentiviral Vectors for Anti-HIV-1 Gene Therapy

Ying Poi Liu, Jan-Tinus Westerink, Olivier ter Brake, and Ben Berkhout

Abstract

RNA interference (RNAi)-based gene therapy for the treatment of HIV-1 infection provides a novel antiviral approach. For delivery of RNAi inducers to CD4+ T cells or CD34+ blood stem cells, lentiviral vectors are attractive because of their ability to transduce nondividing cells. In addition, lentiviral vectors allow stable transgene expression by inserting their cargo into the host cell genome. However, use of the HIV-1-based lentiviral vector also creates specific problems. The RNAi inducers can target HIV-1 sequences in the genomic RNA of the lentiviral vector. As the RNAi-inducing cassette contains palindromic sequences, the lentiviral vector RNA genome will have a perfect target sequence for the expressed RNAi inducer. Vectors encoding microRNAs face the putative problem that the vector RNA genome can be inactivated by Drosha processing. Here, we describe the design of lentiviral vectors with single or multiple RNAi-inducing antiviral cassettes. The possibility of titer reduction and some effective countermeasures are also presented.

Key words: RNAi, Lentiviral vector, siRNA, shRNA, miRNA, HIV-1, Gene therapy, Antiviral, Titer, Transduction

1. Introduction

RNA interference (RNAi) is an evolutionary conserved gene silencing mechanism in plants, insects, fungi, and nematodes that is triggered by double-stranded RNA (dsRNA) derived from viral replication intermediates or transposable elements (1–5). The dsRNA is processed by Dicer into 21–25 base pair (bp) small interfering RNAs (siRNAs) with 2-nt 3′ overhangs (6). The siRNA duplex is incorporated into the RNA-induced silencing complex (RISC). The passenger strand of the siRNA is degraded and the guide strand of the siRNA targets RISC to cleave the perfectly complementary mRNA.

Ronald P. van Rij (ed.), *Antiviral RNAi: Concepts, Methods, and Applications*, Methods in Molecular Biology, vol. 721, DOI 10.1007/978-1-61779-037-9_18, © Springer Science+Business Media, LLC 2011

In mammals, RNAi plays a role in regulation of gene expression at the posttranscriptional level via microRNAs (miRNAs). These molecules are expressed as primary miRNAs in the nucleus that are processed by the Drosha-DGCR8 complex into precursor miRNAs (pre-miRNAs). Pre-miRNAs are hairpin RNAs of approximately 70 nucleotides (nt) that are transported to the cytoplasm by Exportin-5 and further processed by the Dicer/TRBP/PACT endonuclease complex into imperfect ~22 nt miRNA duplexes (7, 8). The single-stranded mature miRNA programs RISC to complementary mRNA sequences to cause mRNA cleavage or translational repression, depending on the level of complementarity between the miRNA and the mRNA target (9, 10).

RNAi soon became a very attractive therapeutic approach to mediate gene silencing of disease-associated mRNAs and transcripts encoded by pathogenic viruses because of its high efficiency and sequence-specificity (11–13). The development of gene constructs that express siRNAs, the short hairpin RNAs (shRNAs) and artificial miRNAs (Fig. 1a), made it possible to induce stable gene knockdown in mammalian cells (14–16). This is particularly interesting for the treatment of some chronic diseases and persistent viral infections that require a long-term supply of RNAi inducers.

We and others have shown that RNAi can be employed to efficiently inhibit HIV-1 replication by targeting viral mRNAs (17–27). Like therapy with a single antiretroviral drug, the use of a single anti-HIV-1 shRNA led to the emergence of RNAi-escape variants that acquired mutations in the target sequence (25, 28, 29). Furthermore, HIV-1 can alter the local structure of its mRNA target such that it becomes inaccessible to RISC (29). To prevent viral escape, a combinatorial RNAi attack is required in which multiple viral sequences are targeted simultaneously, thus raising the genetic threshold towards resistance. Different strategies have been proposed to achieve a combinatorial RNAi attack, including multiple shRNAs (17, 18), a miRNA-like polycistron (20), or an extended shRNA (e-shRNA) (19, 21) (Fig. 1b–d). The e-shRNA design consists of multiple active shRNA-units that are stacked on top of each other.

For the delivery of RNAi inducers to HIV-1 susceptible cells, lentiviral vectors are attractive delivery systems due to their efficiency to transduce many cell types and nondividing cells, including hematopoietic stem cells (30–33). The lentiviral vector becomes stably integrated into the host cell chromosomes and allows long-term transgene expression (34–40). Furthermore, lentiviral vectors exhibit a reduced risk of insertional oncogenesis compared to retroviral vectors (41–43). For the treatment of an HIV-1-infected patient, it has been proposed to transduce CD34+ hematopoietic stem cells ex vivo with a lentiviral vector encoding anti-HIV-1 shRNAs or miRNAs (Fig. 2). The transduced cells

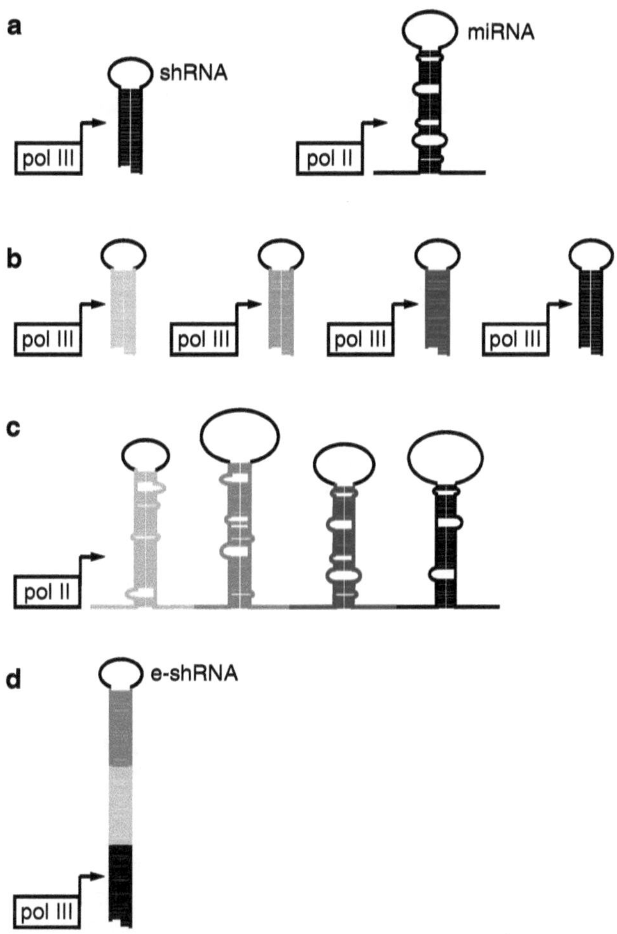

Fig. 1. Expression strategies for single or multiple RNAi inducers. (**a**) Mono-RNAi expression vector using a shRNA driven by an RNA polymerase III promoter or an artificial miRNA driven by an RNA polymerase II promoter. (**b**) Combinatorial RNAi using multiple different shRNA expression cassettes in a single vector. (**c**) Combinatorial RNAi by expression of a miRNA-like polycistron expressed from an RNA polymerase II promoter to target different mRNAs. (**d**) Combinatorial RNAi using extended-shRNAs (e-shRNAs) that encode multiple effective siRNAs in one single hairpin.

will subsequently be engrafted back into the patient. The transduced CD34+ cells will give rise to an HIV-1 resistant myeloid and lymphoid cell population. Recently, the first successful clinical trial was reported using lentiviral vectors for a gene therapy of hematopoietic stem cells (44). This treatment stabilized disease progression in adrenoleukodystrophy patients, an otherwise deadly disease. This finding confirms the potential of lentiviral-mediated gene therapy of hematopoietic stem cells for other diseases.

To obtain a combinatorial RNAi attack against HIV-1, we previously generated several gene constructs that encode multiple viral inhibitors. The most straightforward approach was the

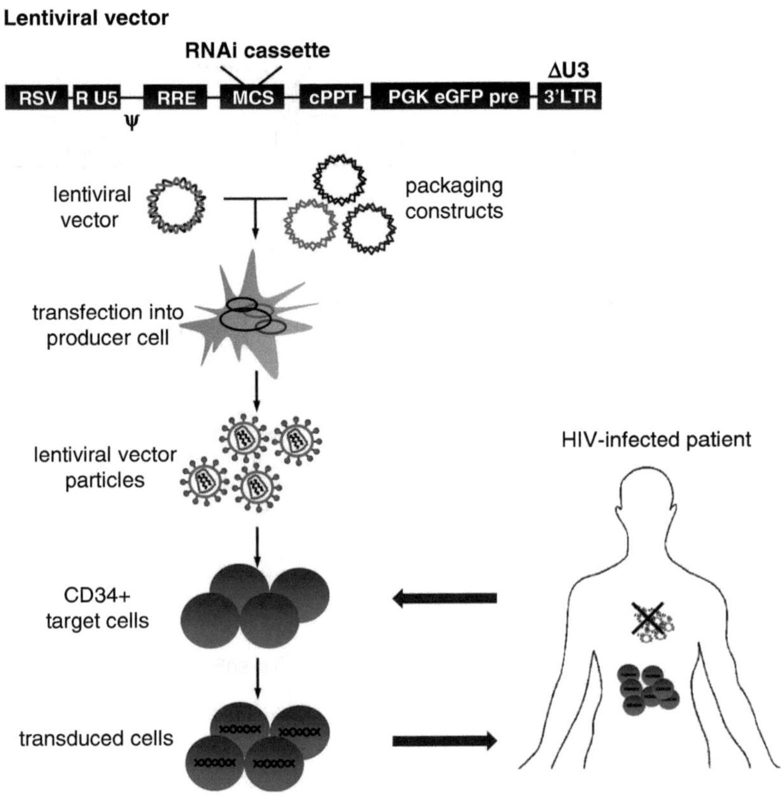

Fig. 2. Lentiviral vector production and the ex vivo gene therapy protocol. We use a third-generation self-inactivating lentiviral vector, which is shown at the *top*. The vector RNA genome is expressed from the Rous Sarcoma Virus promoter (RSV) and encodes the rev responsive element (RRE), a multiple cloning site (MCS), central polypurine tract (cPPT), and the enhanced green fluorescent protein (eGFP) reporter gene driven by the phosphoglycerate kinase promoter (PGK) with the posttranscriptional regulatory element (PRE) from hepatitis B virus. Both the cPPT and the PRE sequence have been introduced into the vector to enhance transduction efficiency. For vector production, the lentiviral vector plasmid is cotransfected with the packaging constructs (plasmids expressing Gag-Pol, Rev, and VSV-G protein) into a producer cell line. Lentiviral vector particles can be harvested from the supernatant. Subsequently, the particles can be purified and added to target cells for transduction. For treatment of an HIV-infected patient, CD34+ cells can be withdrawn to undergo an ex vivo transduction treatment with lentiviral vectors encoding RNAi inducers against HIV-1. These transduced cells can be engrafted back into the patient, where they hopefully will resist HIV-1 infection and prevent the gradual collapse of the immune system.

repeated use of the same promoter cassette to express different shRNAs (Fig. 1b) (45, 46). However, this design triggered recombination-mediated deletion of lentiviral vector sequences due to the promoter repeats (17). This problem can be avoided by the use of different promoters for shRNA expression. In this way, we successfully generated a lentiviral vector construct expressing four different shRNAs from distinct promoters (18). Furthermore, we generated an antiviral miRNA polycistron that encodes four active miRNAs (Fig. 1c) and e-shRNA constructs that express two or three siRNAs (Fig. 1d) (19–21). However, we and others showed that these RNAi-inducing cassettes can negatively affect the lentiviral vector titer, which will constrain clinical applications (47–50).

We depicted numerous possible causes for titer reduction of lentiviral vectors encoding RNAi cassettes in Fig. 3. For instance, we described that shRNA vectors that target HIV-1 sequences may also act on the identical viral sequences in the vector RNA genome (Fig. 3, left: vector targeting), thus causing a profound titer reduction via vector targeting (48). This titer reduction could be alleviated by inhibition of the RNAi pathway (50). Alternatively, this effect can be avoided by selecting antiviral shRNAs that do not match the lentiviral vector genome. Another more general problem expected for shRNA cassettes occurs in the producer cell, where the expressed shRNA can target the lentiviral vector genome via the shRNA sequences that also form an integral part of the vector genome (self-targeting). Interestingly, we did not observe such a reduction in vector titer, which is likely

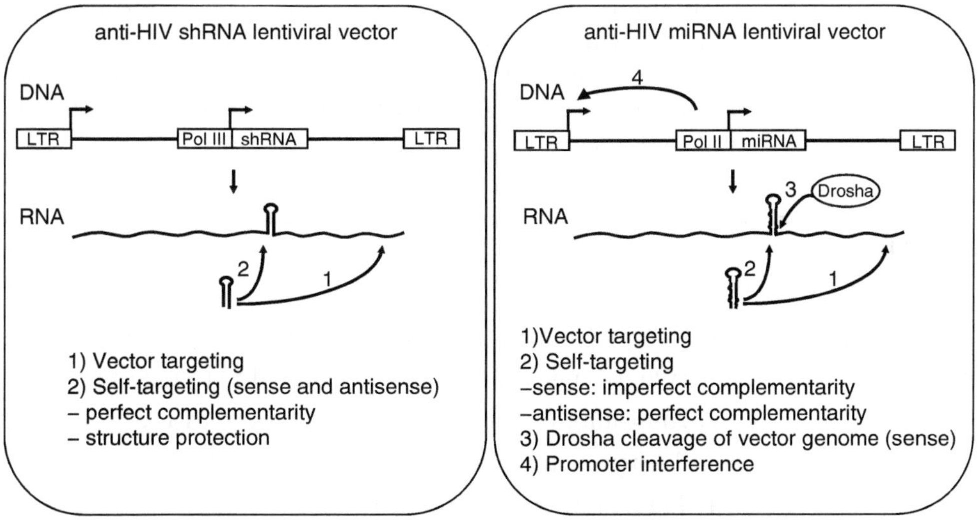

Fig. 3. Possible causes for reduced titers of lentiviral vectors encoding shRNA and miRNA. Different mechanisms that can account for titer reduction in (*left*) shRNA vectors and (*right*) miRNA vectors are illustrated. Anti-HIV-1 shRNAs could potentially target HIV-1 sequences in the vector backbone during lentiviral vector production (mechanism 1, vector targeting). Any shRNA has the potential to target its complementary sequence as part of the lentiviral vector genome (mechanism 2, self-targeting). This second mechanism causes no problem because the target is protected in a hairpin structure and thus inaccessible to RISC. Anti-HIV-1 miRNAs can also target the lentiviral vector (mechanism 1). Expressed mature miRNAs can target their own sequence as part of the lentiviral vector genome (mechanism 2). If the miRNA cassette is cloned in the sense orientation, the target as part of the lentiviral vector genome will be imperfectly complementary. When the miRNA cassette is cloned in the antisense orientation, a perfectly complementary target sequence will reside in the lentiviral vector genome. The primary miRNAs in the vector RNA genome can be processed by Drosha and thus trigger degradation of the vector genome (mechanism 3). As the lentiviral vector genome and the miRNA are transcribed from an RNA polymerase II promoter, promoter interference could possibly reduce the titer or diminish miRNA expression (mechanism 4).

due to occlusion of this target sequence by the stable stem-loop structure of the shRNA, such that it remains inaccessible to RISC (29, 51–54).

The titer reduction observed for vectors encoding miRNAs could not be repaired by inhibition of the RNAi pathway, suggesting that these vectors encounter other problems (Fig. 3, right). Our data suggest that Drosha-mediated cleavage of the miRNA sequences within the vector genome causes only a slight reduction in titer if the cassette is cloned in sense orientation. Another possible problem is that the miRNA sequence as part of the vector genome is targeted by the expressed miRNA. Depending on the orientation of the miRNA cassette, there will be a partially (sense) or fully (antisense) complementary target in the vector RNA genome. Intriguingly, we observed that the titer could be restored to almost normal levels by substitution of the promoter that drives miRNA expression by an inducible promoter (50). This result hints at promoter interference as possible mechanism for titer reduction. Thus, care should be taken in the design of RNAi cassettes in the context of a lentiviral vector. In this chapter, we describe the cloning strategies for the construction of single RNAi vectors (shRNA and miRNA). We provide practical information for the construction of vectors for a combinatorial RNAi attack, including multiple shRNAs, a miRNA-like polycistron, and e-shRNAs. Furthermore, we provide practical information and advice for the production of lentiviral vectors encoding RNAi inducers against HIV-1.

2. Materials

2.1. shRNA Vector Construction

1. pSUPER vector containing the human H1 promoter (OligoEngine, Seattle, WA, USA) (15) (see Note 1).
2. Annealing buffer: 100 mM NaCl, 50 mM HEPES, pH 7.4.
3. T4 ligase (Invitrogen, Carlsbad, CA, USA).
4. Ligase buffer (Invitrogen).
5. Restriction enzymes: BglII and HindIII.
6. T7 primer (sequence: TAATACGACTCACTATAGGG) and M13 rp primer (sequence: CAGGAAACAGCTATGACC).
7. Reddymix PCR Master mix (Abgene, Epsom, UK).
8. GT116 *Escherichia coli* cells (Δdcm, sbcCD) that are specifically designed for cloning and propagation of shRNA-expressing plasmids with hairpin structures (Invivogen, San Diego, CA, USA).
9. Gene Pulser II, cell electroporator (Bio-Rad, Hercules, CA, USA).

10. 2 mm Electroporation cuvettes (Eurogentec, San Diego, CA, USA).
11. Luria Broth (LB) medium.
12. Betaine (Sigma, St Louis, MO, USA).
13. BigDye Terminator v3.1 Cycle Sequencing Kit (Applied Biosystems Inc, Foster City, CA, USA).
14. Lentiviral vector plasmids are derived from the construct pRRLcpptpgkgfppreSsin (55), which we renamed JS1.

2.2. Extended shRNA (e-shRNA) Vector Construction

1. The same reagents as described in Subheading 2.1.

2.3 miRNA Vector Construction

1. pcDNA6.2-GW/EmGFP-miR plasmid (Invitrogen) (20).
2. Oligonucleotides F1 (CAGGTCGACGGATCCTATTTCCTTCAAATGAATG), R1 (ATGGCAGGAAGAAGCGGAGGTGCTACAGAAGCTGTC), F2 (ATGGCGGGAGGAAGCGGTTGGTACTGCTAGCTGTAGAA), R2 (GACCTCGAGTGCGGCCAGATCTAAGCTGGAGTTCTACAGCTA), and oligonucleotides for generating the antiviral miRNA (forward oligo: CTTCTGTAGCACCTCCGCTTCTTCCTGCCATGTAGTGTTTAGTTATCTAATGGCGGGAGGAAGCGGTTGGTACTGCTAGC; reverse oligo: GCTAGCAGTACCAACCGCTTCCTCCCGCCATTAGATAACTAAACACTACATGGCAGGAAGAAGCGGAGGTGCTACAGAAG).
3. Restriction enzymes: BglII, BamHI, EcoRV, NruI, and XhoI.
4. The same reagents as described in Subheading 2.1, items 2–4 and 7–14.

2.4. Lentiviral Vector Production

1. Dulbecco's Modified Eagle's Medium (DMEM) supplemented with 10% fetal bovine serum (FCS, Hyclone, Ogden, UT, USA), penicillin (100 U/ml), streptomycin (100 μg/ml), and minimal essential medium nonessential amino acids (DMEM/10%FCS).
2. Advanced Roswell Park Memorial Institute (RPMI) medium supplemented with 1% fetal bovine serum (FCS), 2 mM L-glutamine, 40 U/ml penicillin, and 40 μg/ml streptomycin.
3. Human Embryonic Kidney (HEK) 293T cell line (ATCC).
4. SupT1 T cell line (ATCC).
5. Phosphate-buffered saline solution, pH 7.4.
6. 0.05% Trypsin with EDTA solution.
7. Lentiviral vectors encoding RNAi inducers were generated by excising the shRNA, e-shRNA, or miRNA cassette from

the expression plasmid and inserting them in the multiple cloning site of the lentiviral vector JS1, as described in Subheadings 3.1–3.3.

8. Lentiviral vector packaging constructs: pSYNGP (56) for the expression of a human codon-optimized gag-pol sequence without RRE, pRSV-Rev (57) for Rev expression which interacts with the RRE on the vector genome and/or wild-type HIV-1 gag-pol transcripts, pVSV-g (57) for pseudotyping the vector with the VSV-g envelope protein.
9. RNAi pathway competitor/inhibitor constructs that were used for improvement of lentiviral vector production in case of RNAi-related production problems:
   (a) A pSUPER construct that contains five repeated H1 promoters that express shRNAs against different HIV-1 targets, p5xshRNA (17), to provide excess shRNAs.
   (b) An siRNA against human Dicer to knock down Dicer function and to saturate RISC (sequence: TCAACCA GCCACTGCTGGA).
   (c) pshDrosha, a shRNA expression construct against the Drosha enzyme to knock down the Drosha level (a kind gift from Bryan Cullen, Duke University).
10. 70 μM Nylon cell strainers (BD Falcon, Bedford, MA, USA).
11. OptiMEM Reduced Serum Medium.
12. OptiMEM supplemented with penicillin (30 U/ml), streptomycin (30 μg/ml), and $CaCl_2$ (100 μg/ml), hereafter called OptiMEM$^+$ medium.
13. Lipofectamine-2000 reagent (Invitrogen).
14. 0.45-μm Cellulose acetate filters (Whatman, Clifton, NJ, USA).
15. 100 kD MWCO Amicon Ultra centrifugal filter devices (Millipore, Billerica, MA, USA).
16. Capsid p24-ELISA. Commercial ELISA kits are available; we perform the CA-p24 ELISA as described previously (58).
17. BD FACSCanto II flow cytometer (BD Biosciences, San Jose, CA, USA).

3. Methods

In this chapter we provide practical information for the construction of single RNAi vectors including shRNA and miRNA constructs (Fig. 1a). In some cases, an intensified RNAi or a combinatorial

RNAi approach is required to obtain the desired effect, which requires a multi-siRNA/miRNA vector. We describe different cloning strategies to obtain combinatorial RNAi vectors, including multiple shRNAs, a miRNA-like polycistron, or e-shRNAs (Fig. 1b–d). These vectors are first cloned into a regular expression vector. Subsequently, the efficacy of the RNAi construct can be tested on the endogenous target or by cotransfection with a luciferase reporter construct that contains the RNAi target sequence. After determining the efficacy, the RNAi gene cassette can be transferred to the lentiviral vector. The standard procedure of producing lentiviral vector particles is described, but we also provide an optimized protocol for the production of lentiviral vector encoding antiviral (e-)shRNAs that target the vector RNA genome.

3.1. shRNA Vector Construction

Oligonucleotides encoding different shRNAs are designed as follows: Compatible sense and antisense DNA oligonucleotides are custom ordered with appropriate overhangs and BamHI and HindIII restriction sites (Fig. 4a). The annealed oligonucleotides contain the BamHI and HindIII restriction sites which can be inserted into the BglII and HindIII-digested pSUPER vector. We use the standard shRNA design that includes perfect complementary 19-nucleotide sense and antisense strands connected with the 9-nucleotide pSUPER loop design (UUCAAGAGA) (15) (Fig. 4a).

1. Dissolve the DNA oligonucleotides in sterile, nuclease-free milliQ water to a final concentration of 3 μg/μl.
2. Dissolve 1 μl of each oligonucleotide in 48 μl annealing buffer for annealing of the forward and reverse oligonucleotides.
3. Anneal the oligonucleotides by heating the solution to 94°C in a beaker of hot water for 5 min and cooling down to room temperature by placing the beaker containing the samples on the bench (see Note 2). The annealed oligonucleotides can be used immediately in a ligation reaction or stored at −20°C.
4. For digestion of the vector, digest 5 μg of pSUPER DNA with 3 μl of 10 U/μl BglII and 3 μl of 10 U/μl HindIII for 2 h at 37°C. Subsequently, heat inactivate for 20 min at 65°C. Following digestion, purify the linearized vector from a 1% agarose gel (see Note 3). Note that the insert contains the BamHI and HindIII sites, instead of BglII and HindIII, but BglII and BamHI have compatible ends.
5. Ligate 1 μl of the annealed oligonucleotides with 40 ng BglII/HindIII-digested pSUPER vector overnight at 16°C.
6. Prior to transformation, ligation mixes should be treated with BglII to reduce the background level of vector self-ligation.

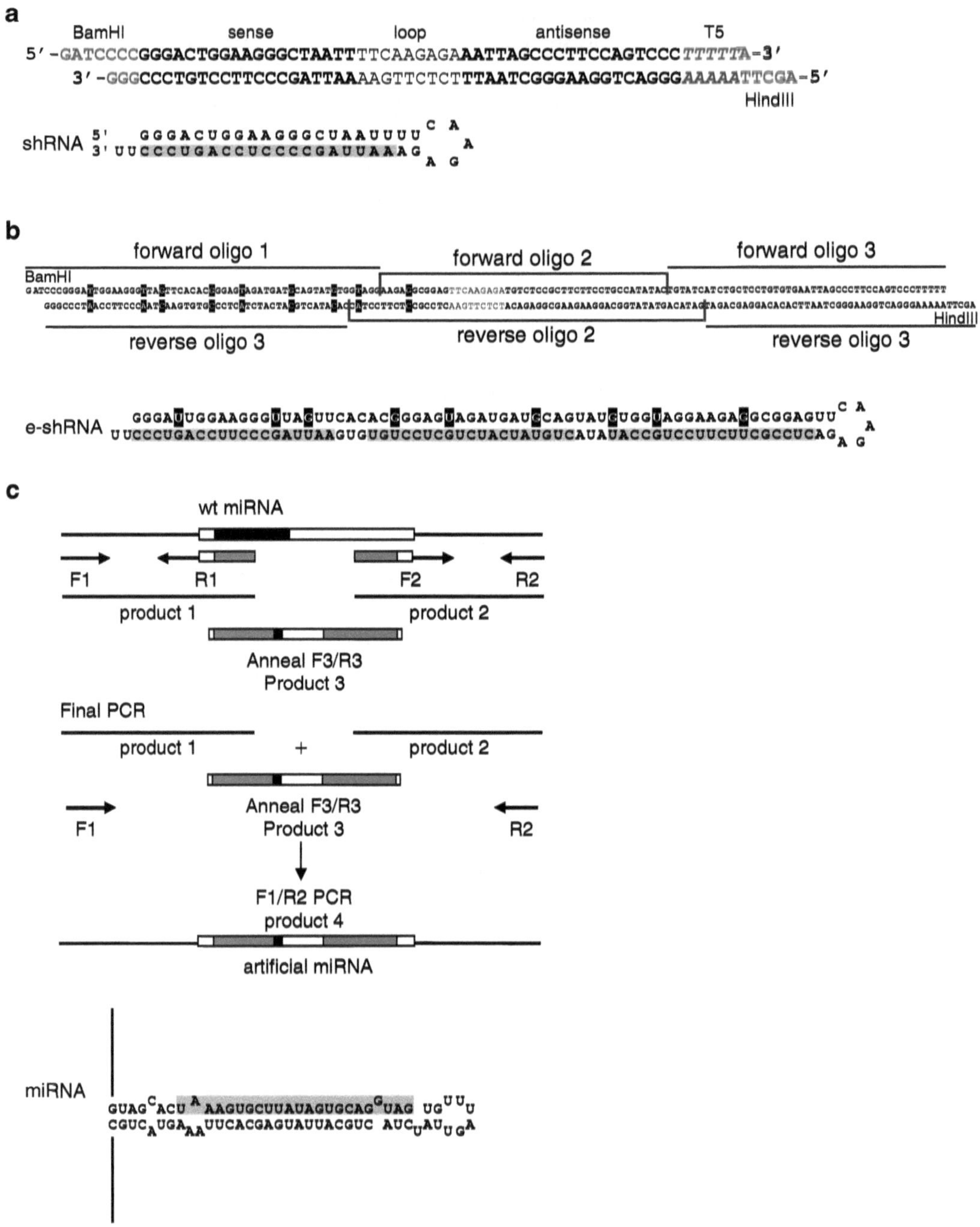

Fig. 4. Generation of shRNA, e-shRNA, or miRNA expression cassettes. (**a**) Schematic of the annealed oligonucleotides that encode the shRNA transcript. Appropriate overhangs with BamHI and HindIII sites are created to enable immediate ligation into the BglII/HindIII-digested pSUPER vector. The 19-nucleotide sense and antisense sequences are connected with a 9-nt spacer that forms a 5-nt loop in the hairpin. The transcriptional termination signal consists of five thymidines (T5). The shRNA transcript folds a small hairpin with a 2-nt UU 3′-overhang. The guide strand of the shRNA is marked in *gray*. (**b**) Construction of an e-shRNA. Multiple oligonucleotides are used that together make up the forward and reverse strands. These are dissolved in annealing buffer and annealed, thus creating appropriate overhangs for cloning in the pSUPER vector. G-U base pairs can be introduced in the hairpin stem (marked in *black*). The e-shRNA folds an extended hairpin with a 2-nt UU 3′ overhang that encodes multiple siRNAs (in this case 3, marked in *gray*). (**c**) Construction of an

The BglII site is destroyed upon successful cloning of the oligo pair into the vector. Thus, only the empty vectors will be cut by the BglII enzyme. Add 0.5 μl (10 U/μl) BglII to the ligations and incubate for 30 min at 37°C.

7. For transformation, use 1 μl of the ligation mix and 50 μl GT116 cells by electroporation (25 μF, 200 Ω, and 2.5 kV).
8. Add 1 ml of LB medium to the cuvettes and recover the cells for 45 min in an incubator at 37°C.
9. Transfer the cells to eppendorf tubes and centrifuge for 5 min at 4,000 rpm and remove ~800 μl of the supernatant, resuspend the cells, and plate two different volumes to ensure that at least one plate will form separate colonies.
10. Perform colony PCR to screen the colonies. Dissolve each colony in 50 μl of LB medium, vortex, and use 2.5 μl of this suspension as the template for PCR. Use T7 and M13rp primers for the PCR reaction. Include a negative and positive control in the PCR, e.g., the empty vector and a construct that yields a similarly sized PCR product.
11. To sequence the palindrome-containing shRNA constructs, we use the BigDye Terminator Cycle Sequencing Kit, but samples were denatured at 98°C and 1 M Betaine was added to the reaction (see Note 4).
12. Combine multiple (different or identical) shRNA expression cassettes (consisting of the promoter and the shRNA) using standard cloning techniques and insert them into the multiple cloning site of the lentiviral vector JS1.

3.2. Extended shRNA (e-shRNA) Vector Construction

Construction of the e-shRNA is similar to the shRNA construction. A notable difference is that we use multiple oligonucleotides that together will form the forward primer and multiple oligonucleotides that will form the antisense primer (Fig. 4b) (see Note 5). An example is given in Fig. 4b in which six oligonucleotides are needed to form the e-shRNA with the BamHI and HindIII restriction sites. The annealed oligonucleotides can be inserted into the BglII and HindIII-digested pSUPER vector. Destabilizing G-U base pairs can be introduced in the hairpin stem of the e-shRNA by modification of the passenger strand

Fig. 4. (continued) artificial miRNA. The wild-type (wt) miRNA sequence (in this case, miR-20) was amplified from cellular genomic DNA. Step 1: The 5′ flank of the pri-miRNA (product 1) was generated using a forward primer F1 encoding a BamHI site and a reverse primer R1 that partially encodes the HIV-1 sequence (*gray*). Step 2: The 3′ flank of the pri-miRNA (product 2) was amplified using forward primer F2 that partially encodes HIV-1 sequences and a reverse primer R2 encoding BglII and XhoI sites. Step 3: two complementary oligonucleotides F3 and R3 that encode the stem-loop part of the miRNA were annealed. Step 4: Products 1, 2, and 3 are used as a template and fused together in a PCR reaction with the outer forward F1 and reverse R2 primers, resulting in an artificial antiviral miRNA. The artificial miRNA transcript is expressed as a pri-miRNA with flanking sequences (depicted as a *line*). The mature antiviral miRNA is marked in *gray*.

(by introducing A to G mutations or C to T mutations), thus maintaining the guide strand sequence (Fig. 4b) (see Note 6).

1. Dissolve DNA oligonucleotides in sterile milliQ water to a final concentration of 3 μg/μl.
2. Add 1 μl of each oligonucleotide (3 μg/μl) in annealing buffer in a total volume of 50 μl for annealing of the forward and the reverse oligonucleotide (see Subheading 3.1, step 3).
3. Clone the annealed oligonucleotide into the pSUPER vector as described in Subheading 3.1, steps 4–10.
4. Sequence the e-shRNA constructs as described in Subheading 3.1, step 11.
5. Digest the e-shRNA cassette (e-shRNA and the H1 promoter) from the pSUPER construct and insert into the multiple cloning site of the lentiviral vector JS1 using standard cloning techniques.

3.3. miRNA Vector Construction

Construction of an artificial miRNA consists of four steps, as depicted in a schematic in Fig. 4c (20). This procedure requires a natural miRNA as a template.

1. PCR amplify the 5′ flank of the pri-miRNA with a forward primer (F1) with a BamHI site and a reverse primer (R1) encoding the HIV-1 sequence at its 3′ end using a natural miRNA as a template. Purify the PCR product (Product 1) from agarose gel. Note that primers are always designed to have at least 18 nt overlapping sequence with the template.
2. PCR amplify the 3′ flank of the pri-miRNA with a forward primer (F2) containing HIV-1 sequences and a reverse primer (R2) encoding BglII and XhoI sites and purify the PCR product (product 2) from gel.
3. Anneal two complementary oligonucleotides that create the stem-loop structure of the antiviral miRNA as described in Subheading 3.1, steps 1–3. The antiviral miRNA is designed in such that it resembles as much as possible the natural miRNA (e.g., mismatches, bulges, and loop). Bulges and mismatches in the hairpin stem are introduced by modifying the passenger strand of the miRNA. The guide strand of the miRNA is designed to have 100% complementarity with the target.
4. The annealed oligonucleotides and the two PCR products have short, 18-nt overlapping sequence segments that allow fusion by PCR with the outer forward and reverse primers used for the production of the first and second PCR product. To perform the fusion PCR, use 10 ng of each PCR product and 10 ng of the annealed oligonucleotides as the template and amplify using a standard PCR program with primer F1 and R2.

5. Purify the antiviral miRNA gene from agarose gel and ligate it into a BamHI and XhoI-digested pcDNA6.2-GW/EmGFP-miR vector.
6. To construct a miRNA polycistron, an additional miRNA hairpin can be added to an existing miRNA construct. The miRNA hairpin insert is digested with BamHI and XhoI, purified, and subsequently religated into the BglII/XhoI sites of the pcDNA6-miRNA vector. By repeating this procedure, constructs with multiple miRNAs cassettes were obtained. Note that BamHI and BglII have compatible sites; after ligation, the BglII site is destroyed (see Note 7).
7. Sequence the miRNA construct using the standard sequencing reaction and program using the BigDye Terminator Cycle Sequencing Kit. If this is unsuccessful, use the hairpin sequencing protocol described in Subheading 3.1, step 11.
8. For cloning of the miRNA expression cassette into the lentiviral vector, remove the EmGFP encoding sequence by digestion with DraI and religation of the vector (see Note 8). Digest the miRNA expression cassette (consisting of the miRNA and the CMV promoter) with NruI and XhoI and ligate into the EcoRV and XhoI-digested JS1 vector. Note that NruI and EcoRV are both blunt end enzymes and thus these sites are compatible.

3.4. Lentiviral Vector Production

1. Trypsinize HEK 293T cells on day 1 in the morning and resuspend the cells in DMEM/10%FCS without antibiotics (see Note 9). Pass the cells through a cell strainer to remove clumps of cells. Count the cells and seed 6.0×10^5 cells per transfection in a 6-well plate in 2 ml of culture medium without antibiotics.
2. On day 2, transfect the cells in the afternoon using Lipofectamine. Prepare the DNA–Lipofectamine complexes in OptiMEM for each transfection as follows (see Notes 10 and 11): Dilute 0.95 μg of lentiviral vector construct, 0.6 μg of pSYNGP, 0.33 μg of pVSV-G, and 0.25 μg of pRSV-Rev in 250 μl OptiMEM.
3. When the lentiviral vector encodes an anti-HIV-1 shRNA that target the genomic RNA (vector targeting), add 2.9 μg of competitor or inhibitor plasmids (see Subheading 2.4, item 9), or 100 nM of siRNA against Dicer to the cotransfection (see Notes 12 and 13). This will prevent that vector targeting siRNAs are incorporated into RISC and precludes targeting of the vector RNA genome.
4. Mix the Lipofectamine gently before use and dilute for each well 6 μl Lipofectamine in 250 μl OptiMEM (see Note 10) and incubate for 5 min at room temperature.

5. Combine the DNA and Lipofectamine mixtures and mix gently. Incubate for 20 min at room temperature to allow the formation of DNA–Lipofectamine complexes.
6. Replace the medium on the cells with 500 μl DMEM/10% FCS without antibiotics.
7. Add the DNA–Lipofectamine–OptiMEM mix dropwise to the cells.
8. In the morning of day 3, replace the transfection medium with 2 ml of OptiMEM$^+$ medium (see Note 14).
9. On day 4, harvest the produced lentiviral vector (see Notes 15 and 16). Spin down detached cellular debris by low speed centrifugation (316–493 × *g*) for 10 min and filter the vector-containing supernatant with a 0.45 μm filter. The highest titers are obtained at day 4. Optionally, lentiviral vectors can be harvested at day 5, but the transduction titer is generally about tenfold lower compared to day 4.
10. The cleared viral supernatant can be concentrated with a centrifugal filter (MWCO 100 kD) at 3,500 × *g* (see Note 17), according to the instructions of the manufacturer. The centrifugation time is dependent on the desired end volume.
11. Aliquot the viral supernatants and store the stocks at −80°C.
12. Determine the production of lentiviral vector particles (capsid titer) by CA-p24 ELISA as described previously (58). To determine the transduction titer, we transduce the SupT1 T cell line with a dilution series of the lentiviral vector, which express eGFP (Fig. 2) (see Note 18). Three days posttransduction, cells were analyzed with FACS to determine the transduction efficiency by detecting the percentage of eGFP-positive cells.

4. Notes

1. Other polymerase III promoters could be used to express shRNAs or e-shRNAs. DNA plasmids containing the human U6 (59), or 7SK (60) polymerase III promoters and the human U1 polymerase II promoter (61) are commercially available. pSilencer 2.0-U6 (Ambion Inc., Austin, TX, USA), psiRNA-h7SKhygro (Invivogen), and the pGeneClip-Basic Vector (Promega Corp., Madison, WI, USA) contain respectively the U6, 7SK, and U1 promoter (18).
2. Annealing of oligonucleotides can also be performed in a heat block at 94°C for 5 min, 80°C for 4 min, 75°C for 4 min, and 70°C for 10 min, and then turn off the heat block to slowly

cool the samples to room temperature. In exceptional cases, annealing may be unsuccessful due to self-annealing of the oligonucleotides. To prevent self-annealing, lower the temperature stepwise in a PCR machine (94°C for 5 min, 85°C for 4 min, 82°C for 4 min, 80°C for 4 min, 78°C for 4 min, etc.).

3. It is not necessary to CIP-treat the digested vector because the DNA ends are not compatible. But if the vector will be CIP-treated, one should phosphorylate the annealed oligonucleotides.

4. Addition of 1 M Betaine in the sequence reaction will improve the quality of sequencing templates, such as hairpin structures and GC-rich sequences (62).

5. Construction of the e-shRNA cassettes requires multiple oligonucleotides because of their size limit (usually 80 nt). Longer oligonucleotides can be ordered, but they are significantly more expensive and require a longer production time.

6. Destabilizing G-U base pairs can be introduced in the hairpin stem of the e-shRNA by modification of the passenger strand (by introducing A to G mutations or C to T mutations), thus maintaining the guide strand sequence. We usually introduce three G-U base pairs per shRNA unit when we make e-shRNAs with two or more siRNAs to facilitate cloning and sequencing without affecting the RNAi activity. In the e3-63 molecule (Fig. 4b), we inserted G-U at base pair position 5, 14, and 17 in the base shRNA unit, positions 3, 8, and 16 in the middle unit, and positions 1, 5, and 13 in the top shRNA unit.

7. The pcDNA6.2-GW/EmGFP-miR vector is designed to express engineered pre-miRNAs based on the murine miR-155 sequence. The vector is engineered in such a way that annealed oligonucleotides can be directly cloned into the linearized vector with miR-155 flanking sequences on both sides of the pre-miRNA. In our case, we used other miRNA backbones; therefore, we removed the miR-155 flanking sequences by digestion with BamHI and XhoI. Subsequently, we generate an artificial miRNA in the miRNA backbone of interest and insert the whole artificial pri-miRNA into the BamHI and XhoI site of the vector.

8. We removed the EmGFP encoding sequence from the pcDNA6.2-GW/EmGFP-miR vector because our JS1 lentiviral vector already encodes eGFP.

9. For lentiviral vector production, cells are seeded in medium without antibiotics. Wrap the plates in saran wrap to diminish the chance of contamination.

10. For lentiviral vector production, make a master mix of the packaging constructs. This can also be done for the Lipofectamine–OptiMEM mix.
11. A low lentiviral vector titer may be caused by poor quality of the plasmids, but it may also be affected by the inserted transgene. Make fresh plasmids and repeat the lentiviral vector production and titer measurement. For plasmid DNA isolation, we use the BIOKE NucleoBond Xtra Midi/Maxi kit. DNA concentration of the plasmid is determined by UV spectrophotometry and quantitative analysis on an agarose gel. The ratio of absorbance at 260 and 280 nm is used to assess the purity of DNA. Pure DNA generally has a ratio of between 1.8 and 2.0. Lower ratios are obtained when DNA plasmids are contaminated with proteins. A second measure for the purity of the DNA is the ratio of absorbance at 260 and 230 nm. This ratio is commonly between 2.0 and 2.2 for pure DNA. Lower ratios are caused by contamination with organic chemical compounds.
12. To inhibit or saturate the RNAi pathway, we also used other RNAi pathway competitor/inhibitor constructs, including a luciferase reporter with the corresponding target sequence to provide an RNAi target decoy, a plasmid encoding Adenovirus VA RNA to inhibit Dicer (63), or a plasmid encoding the RNAi suppressor protein VP35 of Ebola virus (64, 65). However, with these constructs we did not observe a positive effect on the vector titer.
13. Titer reduction could also be due to inefficient nuclear transport of the vector RNA genome. We cotransfected a plasmid-expressing CRM1 (chromosome region maintenance 1) (66), which is involved in nuclear transport of spliced and unspliced RNA. CRM1 overexpression did not affect the vector titer, indicating that restricted RNA export is not the cause of a low vector titer.
14. Production in OptiMEM improves vector titer about twofold as compared to standard DMEM with 10% serum. In addition, the absence of serum improves the subsequent filtration and concentration steps.
15. During handling of lentiviral vectors, be careful with handling infectious material and use gloves and filter tips to prevent contamination.
16. When harvesting lentiviral vectors, carefully remove the viral supernatant and avoid pipetting of detached cells.
17. Alternatively, ultracentrifugation can be used to concentrate the lentiviral vector particles.
18. The vector titer should be determined on the eventual target cells because the transduction efficiency may differ among different cell types.

References

1. Waterhouse, P. M., Wang, M. B., Lough, T. (2001) Gene silencing as an adaptive defence against viruses. *Nature* **411**, 834–42.
2. Voinnet, O. (2001) RNA silencing as a plant immune system against viruses. *Trends Genet* **17**, 449–59.
3. Wilkins, C., Dishongh, R., Moore, S. C., Whitt, M. A., Chow, M., Machaca, K. (2005) RNA interference is an antiviral defence mechanism in Caenorhabditis elegans. *Nature* **436**, 1044–7.
4. Wang, X. H., Aliyari, R., Li, W. X. et al. (2006) RNA interference directs innate immunity against viruses in adult Drosophila. *Science* **312**, 452–4.
5. Segers, G. C., Zhang, X., Deng, F., Sun, Q., Nuss, D. L. (2007) Evidence that RNA silencing functions as an antiviral defense mechanism in fungi. *Proc Natl Acad Sci U S A* **104**, 12902–6.
6. Bernstein, E., Caudy, A. A., Hammond, S. M., Hannon, G. J. (2001) Role for a bidentate ribonuclease in the initiation step of RNA interference. *Nature* **409**, 363–6.
7. Yi, R., Qin, Y., Macara, I. G., Cullen, B. R. (2003) Exportin-5 mediates the nuclear export of pre-microRNAs and short hairpin RNAs. *Genes Dev* **17**, 3011–6.
8. Lund, E., Guttinger, S., Calado, A., Dahlberg, J. E., Kutay, U. (2004) Nuclear export of microRNA precursors. *Science* **303**, 95–8.
9. Bartel, D. P. (2004) MicroRNAs: genomics, biogenesis, mechanism, and function. *Cell* **116**, 281–97.
10. Filipowicz, W. (2005) RNAi: the nuts and bolts of the RISC machine. *Cell* **122**, 17–20.
11. Haasnoot, J., Westerhout, E. M., Berkhout, B. (2007) RNA interference against viruses: strike and counterstrike. *Nat Biotechnol* **25**, 1435–43.
12. Haasnoot, P. C. J., Berkhout, B. (2006) RNA interference: Its use as antiviral therapy. Handbook of Experimental Pharmacology. Heidelberg: Springer, 117–50.
13. Kim, D. H., Rossi, J. J. (2007) Strategies for silencing human disease using RNA interference. *Nat Rev Genet* **8**, 173–84.
14. Paddison, P. J., Caudy, A. A., Bernstein, E., Hannon, G. J., Conklin, D. S. (2002) Short hairpin RNAs (shRNAs) induce sequence-specific silencing in mammalian cells. *Genes Dev* **16**, 948–58.
15. Brummelkamp, T. R., Bernards, R., Agami, R. (2002) A system for stable expression of short interfering RNAs in mammalian cells. *Science* **296**, 550–3.
16. Zeng, Y., Yi, R., Cullen, B. R. (2003) MicroRNAs and small interfering RNAs can inhibit mRNA expression by similar mechanisms. *Proc Natl Acad Sci U S A* **100**, 9779–84.
17. Ter Brake, O., Konstantinova, P., Ceylan, M., Berkhout, B. (2006) Silencing of HIV-1 with RNA interference: a multiple shRNA approach. *Mol Ther* **14**, 883–92.
18. Ter Brake, O., 't Hooft, K., Liu, Y. P., Centlivre, M., von Eije, K. J., Berkhout, B. (2008) Lentiviral vector design for multiple shRNA expression and durable HIV-1 inhibition. *Mol Ther* **16**, 557–64.
19. Liu, Y. P., Haasnoot, J., Berkhout, B. (2007) Design of extended short hairpin RNAs for HIV-1 inhibition. *Nucleic Acids Res* **35**, 5683–93.
20. Liu, Y. P., Haasnoot, J., Ter Brake, O., Berkhout, B., Konstantinova, P. (2008) Inhibition of HIV-1 by multiple siRNAs expressed from a single microRNA polycistron. *Nucleic Acids Res* **36**, 2811–24.
21. Liu, Y. P., von Eije, K. J., Schopman, N. C. et al. (2009) Combinatorial RNAi against HIV-1 using extended short hairpin RNAs. *Mol Ther* **17**, 1712–23.
22. Liu, Y. P., Gruber, J., Haasnoot, J., Konstantinova, P., Berkhout, B. (2009) RNAi-mediated inhibition of HIV-1 by targeting partially complementary viral sequences. *Nucleic Acids Res* **37**, 6194–204.
23. Banerjea, A., Li, M. J., Bauer, G. et al. (2003) Inhibition of HIV-1 by lentiviral vector-transduced siRNAs in T lymphocytes differentiated in SCID-hu mice and CD34+ progenitor cell-derived macrophages. *Mol Ther* **8**, 62–71.
24. Boden, D., Pusch, O., Lee, F., Tucker, L., Ramratnam, B. (2004) Efficient gene transfer of HIV-1-specific short hairpin RNA into human lymphocytic cells using recombinant adeno-associated virus vectors. *Mol Ther* **9**, 396–402.
25. Das, A. T., Brummelkamp, T. R., Westerhout, E. M. et al. (2004) Human immunodeficiency virus type 1 escapes from RNA interference-mediated inhibition. *J Virol* **78**, 2601–5.
26. Lee, S. K., Dykxhoorn, D. M., Kumar, P. et al. (2005) Lentiviral delivery of short hairpin RNAs protects CD4 T cells from multiple clades and primary isolates of HIV. *Blood* **106**, 818–26.
27. Boden, D., Pusch, O., Silbermann, R., Lee, F., Tucker, L., Ramratnam, B. (2004) Enhanced gene silencing of HIV-1 specific siRNA using microRNA designed hairpins. *Nucleic Acids Res* **32**, 1154–8.

28. Boden, D., Pusch, O., Lee, F., Tucker, L., Ramratnam, B. (2003) Human immunodeficiency virus type 1 escape from RNA interference. *J Virol* 77, 11531–5.
29. Westerhout, E. M., Ooms, M., Vink, M., Das, A. T., Berkhout, B. (2005) HIV-1 can escape from RNA interference by evolving an alternative structure in its RNA genome. *Nucleic Acids Res* **33**, 796–804.
30. Harper, S. Q., Staber, P. D., He, X. et al. (2005) RNA interference improves motor and neuropathological abnormalities in a Huntington's disease mouse model. *Proc Natl Acad Sci U S A* **102**, 5820–5.
31. Ralph, G. S., Radcliffe, P. A., Day, D. M. et al. (2005) Silencing mutant SOD1 using RNAi protects against neurodegeneration and extends survival in an ALS model. *Nat Med* **11**, 429–33.
32. Gimeno, R., Weijer, K., Voordouw, A. et al. (2004) Monitoring the effect of gene silencing by RNA interference in human CD34+ cells injected into newborn RAG2-/- gammac-/- mice: functional inactivation of p53 in developing T cells. *Blood* **104**, 3886–93.
33. Van den Haute, C., Eggermont, K., Nuttin, B., Debyser, Z., Baekelandt, V. (2003) Lentiviral vector-mediated delivery of short hairpin RNA results in persistent knockdown of gene expression in mouse brain. *Hum Gene Ther* **14**, 1799–807.
34. Greber, U. F., Fassati, A. (2003) Nuclear import of viral DNA genomes. *Traffic* **4**, 136–43.
35. Follenzi, A., Battaglia, M., Lombardo, A., Annoni, A., Roncarolo, M. G., Naldini, L. (2004) Targeting lentiviral vector expression to hepatocytes limits transgene-specific immune response and establishes long-term expression of human antihemophilic factor IX in mice. *Blood* **103**, 3700–9.
36. Kordower, J. H., Emborg, M. E., Bloch, J. et al. (2000) Neurodegeneration prevented by lentiviral vector delivery of GDNF in primate models of Parkinson's disease. *Science* **290**, 767–73.
37. Miyoshi, H., Smith, K. A., Mosier, D. E., Verma, I. M., Torbett, B. E. (1999) Transduction of human CD34+ cells that mediate long-term engraftment of NOD/SCID mice by HIV vectors. *Science* **283**, 682–6.
38. Naldini, L., Blomer, U., Gallay, P. et al. (1996) In vivo gene delivery and stable transduction of nondividing cells by a lentiviral vector. *Science* **272**, 263–7.
39. Nguyen, T. H., Oberholzer, J., Birraux, J., Majno, P., Morel, P., Trono, D. (2002) Highly efficient lentiviral vector-mediated transduction of nondividing, fully reimplantable primary hepatocytes. *Mol Ther* **6**, 199–209.
40. Kafri, T., Blomer, U., Peterson, D. A., Gage, F. H., Verma, I. M. (1997) Sustained expression of genes delivered directly into liver and muscle by lentiviral vectors. *Nat Genet* **17**, 314–7.
41. Laufs, S., Guenechea, G., Gonzalez-Murillo, A. et al. (2006) Lentiviral vector integration sites in human NOD/SCID repopulating cells. *J Gene Med* **8**, 1197–207.
42. Montini, E., Cesana, D., Schmidt, M. et al. (2006) Hematopoietic stem cell gene transfer in a tumor-prone mouse model uncovers low genotoxicity of lentiviral vector integration. *Nat Biotechnol* **24**, 687–96.
43. Montini, E., Cesana, D., Schmidt, M. et al. (2009) The genotoxic potential of retroviral vectors is strongly modulated by vector design and integration site selection in a mouse model of HSC gene therapy. *J Clin Invest* **119**, 964–75.
44. Cartier, N., Hacein-Bey-Abina, S., Bartholomae, C. C. et al. (2009) Hematopoietic stem cell gene therapy with a lentiviral vector in X-linked adrenoleukodystrophy. *Science* **326**, 818–23.
45. Anderson, J., Li, M. J., Palmer, B. et al. (2007) Safety and efficacy of a lentiviral vector containing three anti-HIV genes – CCR5 ribozyme, tat-rev siRNA, and TAR decoy – in SCID-hu mouse-derived T cells. *Mol Ther* **15**, 1182–8.
46. Henry, S. D., van der, W. P., Metselaar, H. J., Tilanus, H. W., Scholte, B. J., van der Laan, L. J. (2006) Simultaneous targeting of HCV replication and viral binding with a single lentiviral vector containing multiple RNA interference expression cassettes. *Mol Ther* **14**, 485–93.
47. Poluri, A., Sutton, R. E. (2007) Titers of HIV-based vectors encoding shRNAs are reduced by a Dicer-dependent mechanism. *Mol Ther* **16**, 378–86.
48. Ter Brake, O., Berkhout, B. (2007) Lentiviral vectors that carry anti-HIV shRNAs: problems and solutions. *J Gene Med* **9**, 743–50.
49. Zhou, D., Zhang, J., Wang, C. et al. (2009) A method for detecting and preventing negative RNA interference in preparation of lentiviral vectors for siRNA delivery. *RNA* **15**, 732–40.
50. Liu, Y. P., Vink, M. A., Westerink, J. T. et al. (2010) Titers of lentiviral vectors encoding shRNAs and miRNAs are reduced by different mechanisms that require distinct repair strategies. *RNA* **16**, 1328–39.

51. Tafer, H., Ameres, S. L., Obernosterer, G. et al. (2008) The impact of target site accessibility on the design of effective siRNAs. *Nat Biotechnol* **26**, 578–83.
52. Obernosterer, G., Tafer, H., Martinez, J. (2008) Target site effects in the RNA interference and microRNA pathways. *Biochem Soc Trans* **36**, 1216–9.
53. Ameres, S. L., Martinez, J., Schroeder, R. (2007) Molecular basis for target RNA recognition and cleavage by human RISC. *Cell* **130**, 101–12.
54. Brown, K. M., Chu, C. Y., Rana, T. M. (2005) Target accessibility dictates the potency of human RISC. *Nat Struct Mol Biol* **12**, 469–70.
55. Seppen, J., Rijnberg, M., Cooreman, M. P., Oude Elferink, R. P. (2002) Lentiviral vectors for efficient transduction of isolated primary quiescent hepatocytes. *J Hepatol* **36**, 459–65.
56. Kotsopoulou, E., Kim, V. N., Kingsman, A. J., Kingsman, S. M., Mitrophanous, K. A. (2000) A Rev-independent human immunodeficiency virus type 1 (HIV-1)-based vector that exploits a codon-optimized HIV-1 gag-pol gene. *J Virol* **74**, 4839–52.
57. Dull, T., Zufferey, R., Kelly, M. et al. (1998) A third-generation lentivirus vector with a conditional packaging system. *J Virol* **72**, 8463–71.
58. Jeeninga, R. E., Hoogenkamp, M., Armand-Ugon, M., de Baar, M., Verhoef, K., Berkhout, B. (2000) Functional differences between the long terminal repeat transcriptional promoters of HIV-1 subtypes A through G. *J Virol* **74**, 3740–51.
59. Yu, J. Y., DeRuiter, S. L., Turner, D. L. (2002) RNA interference by expression of short-interfering RNAs and hairpin RNAs in mammalian cells. *Proc Natl Acad Sci U S A* **99**, 6047–52.
60. Koper-Emde, D., Herrmann, L., Sandrock, B., Benecke, B. J. (2004) RNA interference by small hairpin RNAs synthesised under control of the human 7S K RNA promoter. *Biol Chem* **385**, 791–4.
61. Denti, M. A., Rosa, A., Sthandier, O., De Angelis, F. G., Bozzoni, I. (2004) A new vector, based on the PolII promoter of the U1 snRNA gene, for the expression of siRNAs in mammalian cells. *Mol Ther* **10**, 191–9.
62. Haqqi, T., Zhao, X., Panciu, A., Yadav, S. P. (2002) Sequencing in the presence of betaine: improvement in sequencing of the localized repeat sequence regions. *J Biomol Tech* **13**, 265–71.
63. Andersson, M. G., Haasnoot, P. C. J., Xu, N., Berenjian, S., Berkhout, B., Akusjarvi, G. (2005) Suppression of RNA interference by adenovirus virus-associated RNA. *J Virol* **79**, 9556–65.
64. de Vries, W., Haasnoot, J., van der Velden, J. et al. (2008) Increased virus replication in mammalian cells by blocking intracellular innate defense responses. *Gene Ther* **15**, 545–52.
65. Haasnoot, J., de Vries, W., Geutjes, E. J., Prins, M., de Haan, P., Berkhout, B. (2007) The Ebola virus VP35 protein is a suppressor of RNA silencing. *PLoS Pathog* **3**, e86.
66. Popa, I., Harris, M. E., Donello, J. E., Hope, T. J. (2002) CRM1-dependent function of a cis-acting RNA export element. *Mol Cell Biol* **22**, 2057–67.

Chapter 19

Production of Multicopy shRNA Lentiviral Vectors for Antiviral Therapy

Scot D. Henry, Quiwei Pan, and Luc J.W. van der Laan

Abstract

For effective RNA interference (RNAi)-based therapies against viral infection, particularly highly mutational viruses like HCV and HIV, combinational strategies that target multiple regions within a viral genome are required to prevent resistance. The use of lentiviral vectors for combinatorial RNAi (coRNAi) offers possibilities to deliver multiple short hairpin RNA (shRNA) sequences simultaneously to individual cells while maintaining high expression levels required to suppress viral replication. By applying coRNAi, one can impart either a protective strategy, i.e., treatment prior to infection, or a long-term treatment postinfection without the eventuality of mutational outgrowth due to incomplete selection pressure. In this chapter, we provide a detailed description of the methods available to create coRNAi vectors and discuss some of the current problems and technical limitations.

Key words: RNA interference, Combinatorial, Lentiviral vectors, siRNA, Promoters

1. Introduction

Current antiviral therapy for chronic Hepatitis C virus (HCV) infection is only effective in approximately half of patients and associated with serious adverse side effects. These shortcomings urge the need to develop alternative treatment strategies. RNA interference (RNAi) represents a promising new therapeutic approach which could be used as a single treatment or in combination with interferon-α (1). If RNAi therapies are to be utilized as an effective treatment or for prevention of disease, long-term, stable siRNA expression needs to be achieved. Raw small interfering RNA (siRNA) or plasmid-encoded short hairpin RNA (shRNA) transfections elicit only short-term silencing, whereas integrating self-inactivating lentiviral vectors (2) that encode for shRNA, can induce long-term and continuous gene-silencing (3).

Ronald P. van Rij (ed.), *Antiviral RNAi: Concepts, Methods, and Applications*, Methods in Molecular Biology, vol. 721,
DOI 10.1007/978-1-61779-037-9_19, © Springer Science+Business Media, LLC 2011

Vector-derived shRNA is a single stranded RNA containing complementary sequences separated by a loop sequence that can fold into a predicted double-stranded hairpin. This structure can be recognized by the microRNA (miRNA) processing enzyme Dicer and cleaved into biologically active siRNA.

Many viruses mutate at high rates (3, 4) and can therefore rapidly develop resistance to monotherapies. HCV has been estimated to generate mutants at a rate of $1.4–1.9 \times 10^{-3}$ base substitutions per genome site per year (5, 6) and rapidly generates and/or selects escape mutants with monotherapy of ribavirin (7, 8). Similarly, the treatment with a single siRNA results in selection of resistant quasispecies of HCV that rapidly outgrow the wild type viral strain (9, 10). In order to prevent these events in a therapeutic setting, multicopy-based combinatorial RNAi (coRNAi) would need to be delivered, either sequentially or simultaneously. If delivered sequentially, the risk remains that a single cell may not receive the multiple siRNA pool and consequently its viral population could still develop resistance and outgrow the combination therapy. Simultaneous delivery, in contrast, ensures that each cell receives all forms of the combination of siRNAs.

Targets of siRNA can be viral sequences or host cell factors that the virus uses to replicate. Combinations of both viral and cellular targets (11, 12) are possible with multiple siRNAs delivered by lentiviral vectors. Host cell factors involved in infection are by themselves not prone to mutation and therefore may represent good therapeutic targets for coRNAi to help prevent resistance (13). However, targeting host cell factors may have drawbacks including potential deleterious effects by impeding normal cellular functions. Moreover, viruses can develop resistance even when host cell factors are targeted, as was recently shown for inhibition of host cyclophilins (14).

Functional shRNA have a characteristic stem-loop structure (Fig. 1a). shRNA constructs are designed as two approximately 64 nt

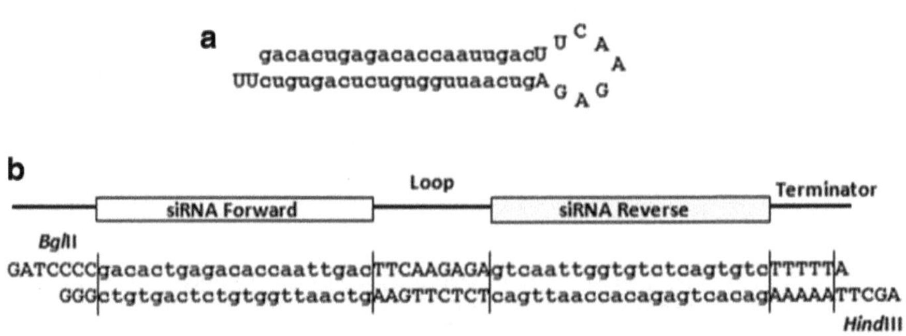

Fig. 1. (**a**) Structure and sequence of a fully folded shRNA. (**b**) Annealed oligonucleotide structure of shRNA sequences. Designed sequences contain both the *Bgl* II and *Hind* III sites for cloning directly into the pSPS72-H1 plasmid. The forward and reverse sequences are linked by a 9 nt loop structure and end in the poly T pol III termination sequence to ensure that RNA of the correct size is formed.

oligonucleotides. These oligonucleotides include two insertion restriction sites, 21 nt forward siRNA sequence, 9–12 nt loop structure, 21 nt reverse siRNA sequence, and poly U terminator. Annealing the forward and reverse DNA oligonucleotides generates the double-stranded shRNA expression construct (Fig. 1b), which can be subsequently cloned into a transfer vector that contains a polymerase III promoter, such as the H1 or U6 promoters.

Several strategies have been developed to express multiple shRNAs through the use of infinitely expandable restriction sites (11, 15). We were able to generate vectors containing up to eight H1-shRNA cassettes (unpublished data and not tested for efficacy). Mcintyre et al. have generated vectors that expressed 11 shRNAs using a unique PCR-based clonal expansion method (15), which have not yet been tested for efficacy (16). However, this later study together with the study of Brake et al. (17) showed evidence that the use of the same promoter to generate multiple shRNA cassettes may result in recombination-mediated repeat deletion (stuttering) of shRNA cassettes. Although the gene-silencing efficacy was not entirely diminished when applied in sufficient shRNA copies, this issue could potentially be met with resistance development if applied to treat viral infection in patients. To circumvent this issue, different promoters could be used to express individual shRNAs (17).

Extended short hairpin RNAs (eshRNA) expressing two siRNA sequences (18) and long hairpin RNAs (lhRNA) expressing more than two siRNA sequences (19) can be used as alternative strategies to simultaneously express multiple siRNAs (Fig. 2a). These long double-stranded RNA structures are relatively easy to produce because they require only one promoter and often only one round of cloning is required, thus avoiding the stuttering issue of multiple promoter repeats (16, 20, 21). When using a contiguous sequence, independent siRNAs are generated by dicer, which preferably cleaves every 21 bp beginning at the loop stem. One serious limitation, however, is that this configuration results in a positional effect in that the first (closest to the hairpin) sequence gets processed most effectively, the second less so, and

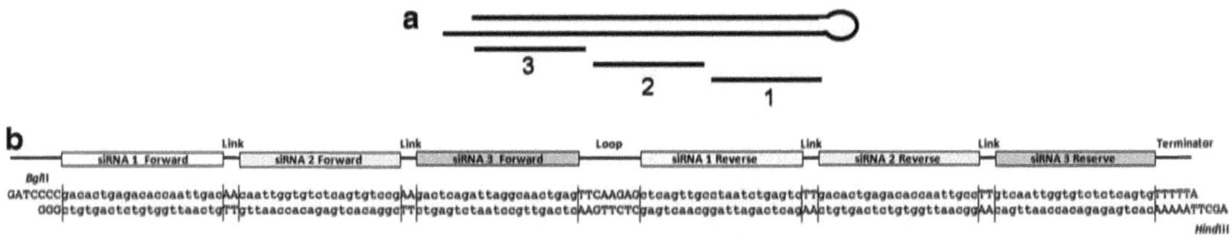

Fig. 2. (**a**) Long Hairpin RNA (lhRNA) schematic detailing three siRNA sequences coding from a contiguous hairpin structure. (**b**) Functional diagram of an lhRNA with three siRNA sequences as an annealed oligo, including *Bgl* II and *Hind* III restriction sites, spacer links, 9 bp loop sequence and, pol III poly T terminator sequence.

the third and subsequent positions becoming almost ineffective (22–24). To alleviate this problem, small linkers can be placed between the desired siRNA sequence in a stacking method which largely removes the positional effect (Fig. 2b), though this is only effective up to a stack of three siRNA sequences (24, 25). Stacked lhRNAs or eshRNAs could then be utilized in combinations under the control of independent promoters to produce many iterations of multiple siRNA expression, a particularly attractive idea for antiviral therapies.

Endogenous miRNA backbones have gained recent focus in terms of application in coRNAi strategies (26, 27). Particular advantages are their increased flexibility in allowing for conditional expression from a polymerase II promoter as well as the expression of multiple siRNA sequences or coexpression of siRNA with protein from a single transcript (18, 19, 28–36). Sequences need to be carefully designed to mimic the natural structural features of the pre-miRNA, such as mismatches, bulges, and thermodynamic stability. Sequences incorporated into miRNA-like structures tend to be less effective than the same sequence delivered under a pol III promoter. Furthermore, as miRNAs are known to be more nonspecific in their targets (37), due to the wobbles, mismatch base pairing, and internal instabilities. Therefore, the incidence of off-target effects using this method may be increased. We have previously developed multicopy shRNA sequence vectors designed to target two independent regions of the HCV genome as well as the host cell receptor CD81. These vectors were as effective as their individual components with the added advantage of single vector delivered coRNAi (11). We describe here one method of producing lentiviral vectors capable of expressing single and multiple shRNAs sequences simultaneously, specifically to treat highly mutational viral targets.

2. Materials

2.1. Plasmids and Oligonucleotide Preparation

1. Transfer Plasmid pSP72 (Promega).
2. pSuper-H1 (Oligoengine).
3. Third generation Lentiviral vectors (Addgene).
    (a) pMDLg/RRE. This plasmid encodes gag, the virion main structural proteins; pol, responsible for the retrovirus-specific enzymes; and RRE, a binding site for the Rev protein which facilitates export of the RNA from the nucleus.
    (b) pRSV-REV. Rev cDNA expressing plasmid in which the joined second and third exons of HIV-1 rev are under the transcriptional control of the RSV U3 promoter.

(c) pMD2.G. This plasmid encodes the envelope G protein from vesicular stomatitis virus (VSV), which is used for its high stability and broad tropism.

(d) pND-CAG-GFP-WPRE. This plasmid contain the woodchuck hepatitis virus posttranscriptional regulatory element (WPRE) and enhanced green fluorescent protein (GFP) cDNA, expression of which is driven by a composite CAG promoter (consisting of the cytomegalovirus immediate early enhancer, chicken beta-actin promoter, and rabbit beta-globin intron) (see Note 1).

4. Annealing buffer: 100 mM NaCl, 50 mM HEPES, pH 7.4.
5. Plasmid Purification kit: SV MiniPrep kit (Promega) and EndoFree Plasmid Maxi Kit; (Qiagen) (see Note 2).
6. Gel Extraction kit (GeneJet Gel Extraction Kit; Fermentas).

2.2. Bacterial Culture

1. Lennox Broth (LB): 10 g/l Tryptone, 5 g/l $NaCl_2$, 5 g/l Yeast Extract.
2. LB Agar (LBA); 10 g/l Tryptone, 5 g/l $NaCl_2$, 5 g/l Yeast Extract, 15 g/l Agar.
3. Ampicillin (^{AMP}); 100 mg/ml, store at −20°C for up to a year. Use at a final concentration of 100 μg/ml in LB or LBA.
4. Bacterial cells; DH-5α (Invitrogen), DH-10β (Invitrogen), STBL3 (Invitrogen), or JM109 (Promega) (see Note 3).
5. Competent Cell Wash Buffer: 100 mM Tris–HCl, pH 8.0, 20 mM $CaCl_2$, 100 mM $MgCl_2$. Store at room temperature, chill to 4°C before use.
6. Competent Cell Freeze Buffer: 15% Glycerol, 10 mM Tris–HCl pH 8.0, 20 mM $CaCl_2$, 100 mM $MgCl_2$. Store at room temperature, chill to 4°C before use.

2.3. Cloning

1. T4 polynucleotide kinase (PNK) and 10× PNK buffer (Fermentas).
2. rATP (Promega).
3. Shrimp Alkaline Phosphatase (SAP) (Fermentas).
4. T4 DNA Ligase (Fermentas).
5. Various restriction enzymes (*Hind* III, *Bgl* II, *Xho* I, *Sal* I, FastDigest, Fermentas).
6. Agarose.
7. Ethidium Bromide (Sigma) or SYBR Green (Invitrogen).

2.4. Viral Plasmid Transfection

1. Complete Dulbecco's Modified Eagle's Medium (cDMEM): DMEM medium (Gibco/BRL) supplemented with 10% fetal bovine serum (FBS), 2 mM L-glutamate, and 100 U Penicillin/100 μg Streptomycin.

2. Trypsin-EDTA: 0.25% trypsin, 1 mM ethylenediamine tetra-acetic acid (EDTA) (Gibco/BRL).
3. HEK-293T (ATCC).
4. Polyethylenimine (PEI, average MW ~25,000 LS, Sigma, see Note 4).
5. HEPES Buffered Saline (HBS): 20 mM HEPES, 150 mM NaCl.
6. Cell fixative: 4% paraformaldehyde (PFA) in phosphate buffered saline (PBS), pH 7.4–7.6.

2.5. Polymerase Chain Reaction

1. PCR reagents (Taq polymerase, dNTPs, $MgCl_2$, PCR buffer).
2. Phenol/chloroform/isoamylalcohol (25:24:1 ratio).
3. H1 Promoter primers: Forward: 5′-TTGCATGTCGCTAT GTGTTCTGGG-3′; Reverse: 5′-GCCCGGTACCCAGCTT TTGT-3′.
4. Ethidium bromide (Sigma) or SYBR Green (Invitrogen).
5. Agarose.
6. Gel Extraction kit (GeneJet Gel Extraction Kit; Fermentas).

3. Methods

3.1. Annealing of Oligonucleotide-Based shRNA Duplexes

Many online programs or methods (see Note 5) are available for prediction of optimal siRNA target sequences. Each method uses different criteria and algorithms to determine optimal siRNA sequences. Generally, siRNA sequence follow the following rules: (a) Low to moderate G/C content (36–52%), $T_m < 60°C$, (b) Low internal stability at the sense 3′ end, (c) At least one A/T duplex between position 15 and 19, and (d) Lack of internal repeats. Some algorithms also incorporate the secondary structures of the target sequence and chance of off-target effects. There is no guarantee that the sequence will have potency in RNAi, unless the manufacturer has quality controlled their product. Each siRNA sequence will need to be tested for efficacy in your experimental system.

In many viral systems, target sequences often need to be selected through exhaustive stepwise screening of multiple siRNAs. A few methods have been developed recently to make this process more efficient (15, 38, 39). For brevity, we provide the protocol for expression of two or more from a single lentiviral vector, assuming that the siRNA sequences are known.

1. Obtain two DNA oligonucleotides for hairpin RNA expression. In our experience oligonucleotides obtained commercially are pure enough without gel purification for efficient ligation if dissolved in sterile, nuclease-free H_2O (see Note 6).

2. Dilute the original stock to a concentration of 10 mg/ml. Prepare from this stock a 3 mg/ml working stock.
3. Add 1 μl of each working stock of oligonucleotide to 48 μl of annealing buffer. Flick to mix and centrifuge briefly at maximum speed in a tabletop centrifuge.
4. Using either a PCR machine or a benchtop microincubator, create a program that will heat to 90°C for 5 min, and then cool to 70°C for 10 min. Cool the annealed oligonucleotides stepwise to 4°C (lowering the temperature gradually over the course of 20–30 min). The annealed oligonucleotides can be used immediately in the following steps or stored at −20°C.

3.2. Phosphorylation of Annealed Oligonucleotides

1. If your oligonucleotides were purchased without a 5′ phosphate modification, perform a forward reaction with T4 polynucleotide kinase (T4 PNK) to phosphorylate the annealed oligonucleotides. Mix:
   (a) 2 μl 10× reaction buffer.
   (b) 1 μl T4 PNK.
   (c) 20 pmol rATP.
   (d) 1–20 μl Annealed oligonucleotides (0.06–0.85 μg).
   (e) Nuclease-free H_2O to a final volume of 20 μl.
2. Incubate at 37°C for at least 30 min.
3. Purify by gel extraction or by a commercial DNA cleanup column (see Note 7).

3.3. Dephosphorylation of Linear Transfer Vector

To ensure that multiple repetitive cloning steps can be achieved, your transfer vector will need a pair of compatible restriction sites that leaves one intact while the other is destroyed after a successful ligation. In our laboratory, we chose the combination of *Xho*I and *Sal*I, as these are common restriction sites available in many plasmids. To enable a shRNA ready promoter plasmid we modified the pSP72 plasmid with an H1 cassette from the pSuper vector. This was done to ensure that the entire H1-shRNA cassette could be removed via the *Xho*I/*Sal*I cut enabling multiple insertions into the *Xho*I site of the lentiviral backbone plasmid with the resulting *Xho*I site that is created after a successful insertion of a new cassette available for a subsequent insertion. Briefly, the pSuper plasmid was digested with *Pst*I/*Hind*III. The fragment including the human H1 promoter and a small portion of the multiple cloning site (MCS) was gel extracted and ligated in pSP72 cut with the same enzymes. The resulting pSP72-H1 plasmid was digested with *Sma*I and *Hpa*I to remove the second *Bgl*II site and 83 bp of the newly inserted MCS, in preparation for later ligations (Fig. 3).

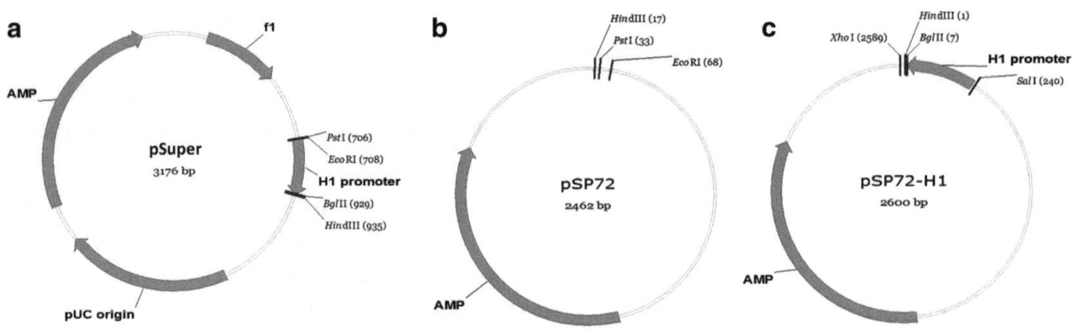

Fig. 3. Cloning strategy to remove the H1 promoter from pSuper (**a**) and ligation into pSP72 (**b**) to form the pSP72-H1 (**c**) plasmid ready to accept annealed oligonucleotide siRNA sequences.

1. Linearize your plasmid of interest with restriction enzymes that match your designed oligonucleotides. In this example, we used *Bgl*II and *HindIII* on the 5′ and 3′ end of the annealed oligonucleotides (see Note 8).
2. Dephosphorylation of the vector may be required to ensure an efficient ligation with minimal background. Mix:
   (a) Up to 10 pmol linear transfer plasmid.
   (b) 5 μl 10× Reaction buffer.
   (c) 1 U SAP.
   (d) Water to a final volume of 20 μl.
3. Incubate for 30 min at 37°C.
4. Purify by phenol/chloroform extraction or by a commercial DNA cleanup column (see Note 7).

3.4. Ligation into Promoter Transfer Vector (pSP72-H1)

1. Assemble the ligation reaction by adding a 1–8 molar excess of vector over insert. In our experience, a molar excess of 5–8 works well with little background. Include a negative control reaction without oligonucleotides. Mix:
   (a) 1–8 μl (up to 15 μl) Annealed phosphorylated oligonucleotides.
   (b) Linear dephosphorylated transfer vector.
   (c) 1 μl 10× T4 DNA ligase buffer.
   (d) 1 μl T4 DNA ligase.
   (e) Nuclease-free H_2O to a final volume of 20 μl.
2. Incubate overnight at 12°C.
3. Prior to transformation, ligation reaction mix may be treated with a self-ligation rebuilt restriction site to ensure that self-ligations do not transform, thereby reducing background colonies. In the case of pSuper and the pSp72-H1 derivative thereof, the *Bgl*II site is destroyed upon successful ligation

to the insert, whereas a self-ligation would retain the site. As most salt concentrations in ligase buffers are compatible with the activity of most restriction enzymes, it is possible to simply add the enzyme directly to the mix. Add 1.0 μl *Bgl*II to ligation mix, and incubate for 30 min at 37°C and heat inactivate for 5 min at 70°C.

3.5. Transformation

1. Thaw competent cells (see Note 9) on ice. Add the entire ligation reaction and mix well with a pipette.
2. Incubate on ice for 30–45 min with occasional swirling (do not vortex).
3. Incubate at 42°C in a waterbath for 90 s. Do not move, swirl, or mix during the heat-shock.
4. Place the sample immediately on ice and cool for at least 90 s.
5. Add 1 ml of antibiotic-free LB broth and incubate for 30–60 min at 37°C with gentle shaking (see Note 10).
6. Spin briefly at maximum speed in a microcentrifuge, discard supernatant with a gentle flick into a waste container (see Note 11). Resuspend the pellet in the remaining LB (approx. 75 μl) and spread evenly on LBA^{AMP} petri dishes.
7. Grow overnight at 37°C.

3.6. PCR Clonal Screen

1. Pick individual colonies from the LBA^{AMP} plate with a side-swipe motion with a clean 10 μl pipette tip or autoclaved toothpick (see Note 12). Place the colony into 20 μl of nuclease-free water in a microcentrifuge tube and swirl the tip within the water to resuspend the colony.
2. Vortex to fully resuspend the colony. Perform PCR using H1 promoter primers at an annealing temperature of 52°C. In our experience, 3 μl of this resuspension will generate a good PCR signal within 35 cycles (see Note 13).
3. Run the PCR products on a 2.0% agarose gel with either ethidium bromide or SYBR green visualizing dyes (see Note 14). The PCR products will provide a 145 bp amplicon without an insert and approximately a 203 bp fragment with the shRNA insert.
4. Inoculate 5 ml of LB^{AMP} with positive clones, incubate overnight at 37°C, and purify the plasmid. Confirm the PCR results with a restriction digest using enzymes that will indicate the orientation of the insert. Additionally, the clones should be confirmed by sequence analyses.

3.7. Cloning of the shRNA Cassette into a Lentiviral Vector

1. Remove the H1-shRNA cassette using *Sal*I/*Xho*I digestion, and purify the fragment from gel.
2. The cassette can be sequentially inserted into the *Xho*I site of the lentiviral vector. We used the pND-CAG/GFP/WPRE

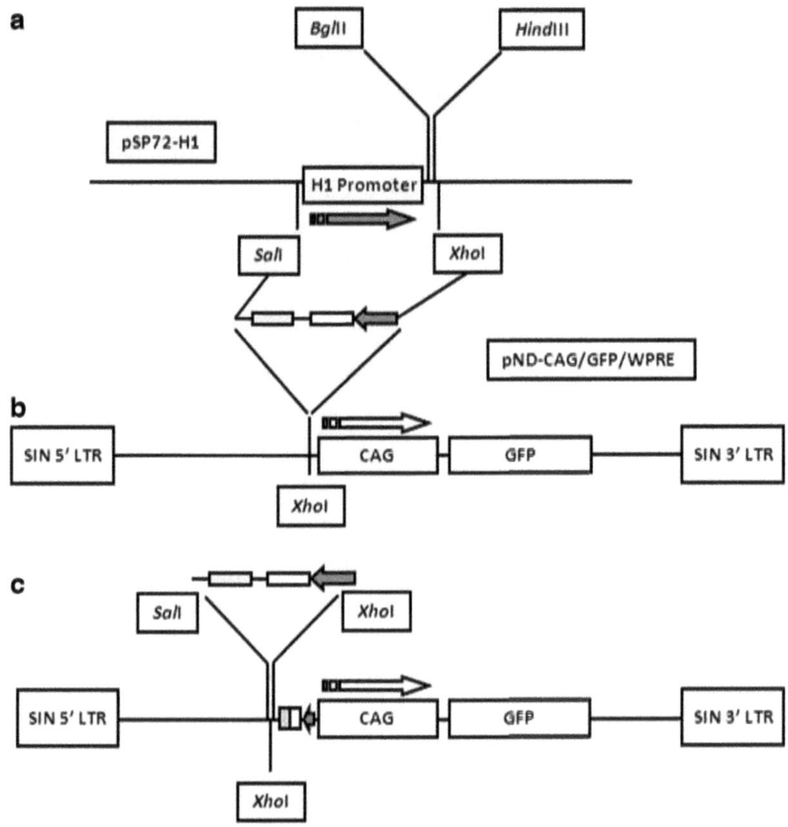

Fig. 4. Cloning strategy for subsequent cloning of multiple shRNA sequences in the lentiviral vector pND/CAG/GFP/WPRE. (**a**) Each H1-shRNA cassette is excised from the pSP72-H1 vector containing the shRNA sequence via *Xho* I/*Sal* I digestion. (**b**) The cassette is ligated to an *Xho* I-digested pND/CAG/GFP/WPRE vector in opposite orientation to the CAG promoter. (**c**) Additional H1-shRNA cassettes can be placed into the reformed *Xho* I site at the downstream end of the shRNA cassette.

and inserted the H1-shRNA cassette in opposite orientation to the CAG reading direction (Fig. 4).

3. Analyze insertion and orientation of the shRNA cassette by restriction analysis with *Nde*I/*Xho*I. After successful insertion of the cassette, one downstream *Xho*I site is available for insertion of multiple cassettes (Fig. 5).
4. By repeating this process, plasmids containing double and triple shRNA cassettes can be created (see Note 15 and Fig. 4).

3.8. PEI/HBS-Mediated Transfection of Lentiviral Plasmids (see Note 16)

1. Grow HEK-293T cells to 80–100% confluence in 10 cm dishes (Table 1).
2. Prepare the transfection mixture for one 10-cm dish. Combine:
   (a) 20 μg of the pND lentiviral vector with the shRNA of interest.
   (b) 10 μg pMDLg/RRE.

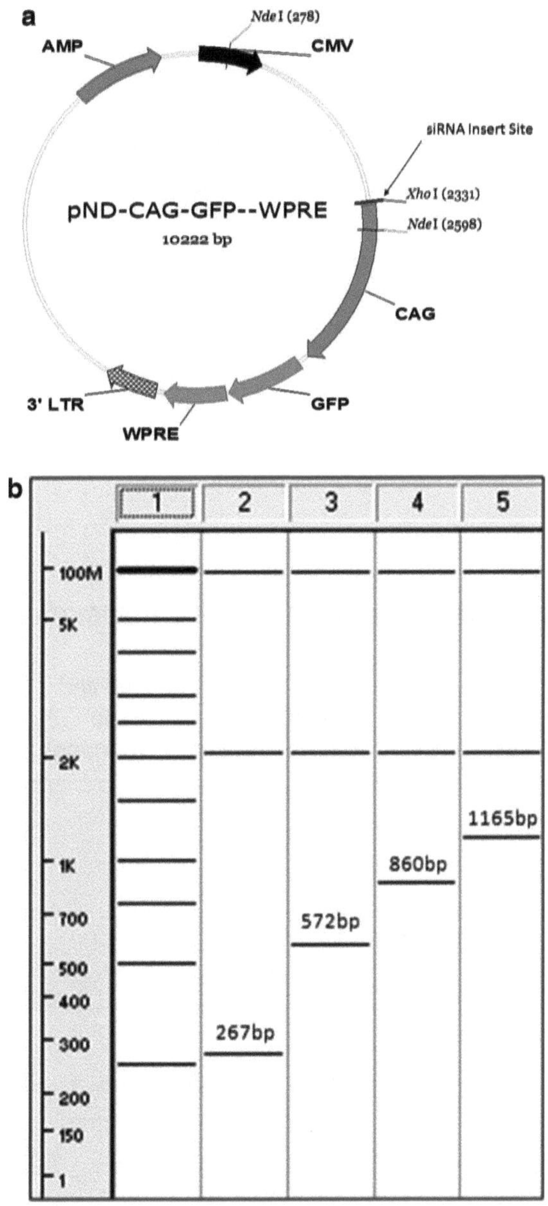

Fig. 5. The *Nde* I/*Xho* I restriction digestion of pND plasmids containing the lentiviral vector with 1, 2, and 3 H1-shRNA cassettes. (**a**) Orientation of *Nde* I and *Xho* I sites used to monitor insertion and orientation of H1-shRNA cassettes. (**b**) Vector NTI virtual gel application detailing *1* BenchTop 1KB DNA ladder (Promega); *2* pND-CAG/GFP/WPRE vector without inserts; *3* pND-CAG/GFP/WPRE vector with a single H1-shRNA cassette; *4* pND-CAG/GFP/WPRE vector with two H1-shRNA cassettes; *5* pND/CAG/GFP/WPRE vector with three H1-shRNA cassettes after digestion with *Nde* I/*Xho* I and analysis on a 2% agarose gel. (**c**) Plasmid digests showing actual pND-AG/GFP/WPRE with various numbers of shRNA cassettes. When the pND-CAG/GFP/WPRE plasmid is cut with *Nde* I/*Xho* I and contains no shRNA cassette the smallest fragment is 276 bp, with one H1-shRNA cassette the smallest fragment is 572 bp, with two cassettes; 860 bp, with three cassettes; 1,165 bp.

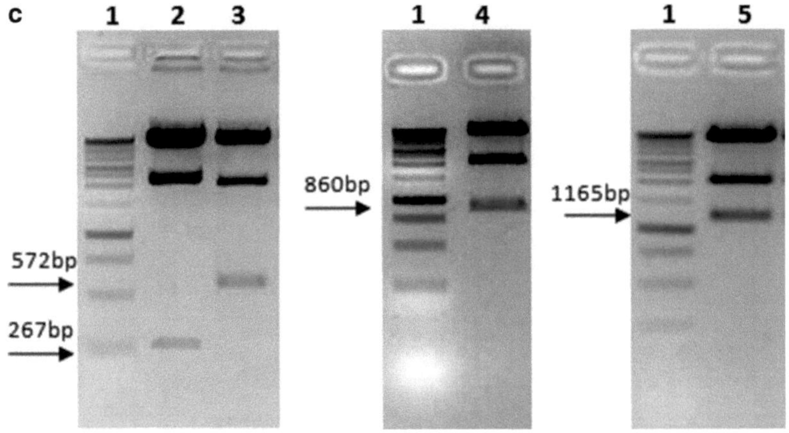

Fig. 5. (continued)

Table 1
Upscaling of third generation lentiviral preparations

	10 cm Dish	225 cm^2 Flask (Nunc)	500 cm^2 Plate (Corning)	Triple flask (Nunc)	Hyper flask (Corning)	Five cell stack (Corning)
Growth area (cm^2)	78	225	550	675	1,720	3,180
Plasmid (μγ)						
pMD2.G	6.0	17.3	38.4	51.6	132.1	244.2
pMDLg/RRE	10.0	28.8	64.0	86.0	220.2	407.0
pRSV-REV	5.0	14.4	32.0	43.0	110.1	203.5
pND-CAG/GPF/WPRE	20.0	57.6	128.0	172.0	440.3	814.1
Total (μg)	41.0	118.11	262.4	352.6	902.7	1668.9
HBS (μl)	250.0	500.0	1000.0	1500.0	3500.0	4000.0
PEI (3 μg/μg DNA)[a]	123.0	354.2	787.2	1057.8	2708.0	5006.6
HBS (μl)	250.0	500.0	1000.0	1500.0	3500.0	4000.0

Chart for the scale-up of lentivirus production detailing total amount of plasmid (in μg) and the total volume of HBS used to create a transfection mix of all four plasmids from the third generation lentiviral system

[a]This is a μg amount for PEI and will need to be adjusted depending on your final concentration prep of PEI e.g., final concentration of PEI 36.4 μg/μl: 123 μg required for a 10 cm prep; 3.4 μl into 246.4 μl HBS

(c) 5 μg pRSV-REV.

(d) 6 μg pMD2.G.

(e) HBS to a final volume of 250 μl.

3. Add 3 μg of PEI per μg DNA to 250 μl of HBS. While slowly vortexing, add the PEI/HBS drop-wise (see Note 17).

The solution will become a bit cloudy/milky as small precipitates begin to form.

4. Incubate the transfection mixture for 15 min at room temperature.
5. Remove the growth medium from the 293T cells and cover the cells with 10 ml of DMEM without FCS as serum may negatively influence the transfection efficiency.
6. Add the freshly prepared transfection mixture drop-wise while gently swirling the dish to allow even mixing (see Note 18).
7. Incubate for 3–6 h at 37°C in a 5% CO_2-buffered incubator, and replace the medium with 10 ml of fresh cDMEM and incubate overnight in a CO_2 incubator.
8. Collect the virus-containing supernatant at 48 and 72 h after transfection by carefully transferring the medium into a collection vessel. Filter the medium immediately through a 0.45 μm cellulose acetate filter (see Note 19) to remove cell debris. The collected medium may be stored for several days at 4°C, or it can be processed directly.
9. Pellet the viral vector particles by ultracentrifugation at 50,000 × *g* for 2 h at 10°C.
10. Aspirate the supernatant and resuspend the pellet in PBS supplemented with 1.0% BSA (or cDMEM) in ~0.1% volume of the final volume of the medium collected. As we harvest 10 ml of supernatant at two timepoints, this corresponds in our protocol to 20 μl for a 10 cm plate.
11. Store in small aliquots to minimize freeze–thaw cycling (see Note 20). Store at −80°C.

3.9. Titration of Lentiviral Vectors

It is important to determine viral titers in the most effective manner to minimize loss of vector and to determine the amount of lentiviral vector required for transduction (see Note 21). The following procedure requires a lentiviral vector that contains a fluorescent reporter gene. In our example, the pND-CAG-GFP-WPRE vector backbone contains a GFP protein expressed from the CAG promoter.

1. Plate 10,000 293T cells in 200 μl of cDMEM medium in a 96-wells plate. Incubate overnight at 37°C or for 6–8 h until all cells have adhered.
2. Transduce the cells with 1, 3, 5, 7, and 10 μl of concentrated viral stock, and incubate for 6 h.
3. Remove the medium, carefully wash the cells with PBS, and replace with fresh medium.
4. After overnight incubation in a CO_2 incubator, detach the cells by Trypsin-EDTA treatment.

5. Pellet the cells in a FACS tube for 5 min at $750 \times g$ and wash at least twice in warm PBS supplemented with 0.05% BSA.
6. Resuspend the cells in 4% PFA, incubate for 10–15 min at room temperature, and wash twice with PBS (see Note 22).
7. Resuspend the cells in 200–500 μl of PBS/0.05% BSA and perform flowcytometric analysis to quantify the GFP-positive cells. The viral titer is calculated as the number of transduced cells divided by the volume of the stock. For example, if 1 μl transduces 10% of the cells in 10,000 cells, the titer is 1,000 transducing units (TU)/μl, or 1×10^6 TU/ml (see Note 23).

4. Notes

1. The third generation packaging system requires a TAT independent lentiviral vector with a chimeric 5′ LTR in which the HIV promoter was replaced with a CMV or RSV promoter. Many different lentiviral vector systems are available, both commercially and from academic laboratories. We chose a third generation packaging system for its safety, efficacy, and ease of use.
2. Plasmid can be purified by phenol/chloroform extraction. Purification kits, however, produce much cleaner and more concentrated preparations. It is import to use an endotoxin-free plasmid purification kit for plasmids that will be used for transfection as even small amounts of endotoxin will reduce transfection efficiencies.
3. These bacterial strains have different properties: DH-5α (Invitrogen): a versatile strain that can be used in many everyday cloning applications. Mutations in DH5α increase insert stability and improve the quality of plasmid DNA preparations. DH-10β (Invitrogen): high transformation efficiency makes them ideal for generating cDNA or genomic libraries. STBL3 (Invitrogen): reduced frequency of unwanted homologous recombination and well-suited for cloning of unstable DNA. JM109 (Promega): high transformation efficiency, perfect for subcloning, T-vector cloning, and production of single stranded phagemids. While all these strains can be used, we recommend STBL3 as these cells display a reduced frequency of recombination and are therefore excellent for cloning repetitive sequences as used in coRNAi applications.
4. Prepare a 100 mg/ml PEI solution in water and dialyze for 4–5 days (using a dialysis membrane with a ~14 kDa cut-off) to remove toxic, low molecular weight polymers. During

dialysis, the volume will increase as water osmotically enters the dialysis tube. Keep track of the change in volume and recalculate the concentration of PEI in your final solution. Use this concentration to calculate the required amounts of PEI in your experiments. Typically the final concentration ranges from 25 to 60 mg/ml.

5. Online siRNA prediction tools are available at: http://web.mit.edu/sirna/, www.ambion.com/techlib/misc/siRNA_finder.html, www.rnainterference.org/HumanSequences.html, http://sirecords.biolead.org/, http://sivirus.rnai.jp/, http://optirna.unl.edu/

   All online sources have their own merits, and the choice for a specific website depends on personal preference and experience. Some of these websites may already provide validated siRNA sequences against your targets.

6. Unless otherwise stated, all solutions should be prepared with water that has a resistance of 18.2 MΩ-cm and total organic content of less than five parts per billion.

7. Commercial kits are available for gel extraction or PCR cleanup, both will result in slight losses of oligonucleotide (10–20%), but the downstream applications are fairly sensitive to salt concentrations. Fermentas PNK buffers are compatible with Fermentas' ligase after addition of 5 mM rATP.

8. The restriction sites will differ, dependent on the promoter vector into which you are cloning an shRNA cassette.

9. Heat-shock or electro-competent cells can be purchased directly from a commercial supplier, but can easily be prepared in an academic laboratory. In our experience, heat-shock procedures result in fewer dead cells and easier routine transformations. The following protocol is for producing heat-shock competent cells: Obtain a small quantity of plasmid-free bacterial cells of your preference (see Note 3). Inoculate 5 ml of LB medium without antibiotics and incubate overnight at 37°C in a shaking incubator at approximately 100–120 rpm. The following morning, dilute the overnight culture into 250 ml of LB without antibiotics and grow, while shaking at 37°C, to midlog phase or an OD620 nm of 0.3–0.4. This will take approximately 2–3 h. Centrifuge the entire culture volume at 3,000 × *g* for 5 min at 4°C and decant the supernatant. Resuspend the pellet in 25 ml of cold competent cell wash buffer. Incubate on ice for at least 30 min. Centrifuge the entire volume at 3,000 × *g* for 5 min at 4°C and decant the supernatant. Resuspend the pellet in 5 ml of Competent Cell Freeze Buffer. Aliquot in volumes of five times the volume of your routine ligation reactions (e.g., when performing ligations in a final volume of 20 μl, aliquots of 100 μl competent

cells would be appropriate). Competent cells can be stored at −80°C for up to a year.

10. Recovery from heat-shock allows for replication of the plasmid and transcription and translation of the antibiotic resistance marker, prior to the addition of antibiotic selection.
11. A vacuum device may be used, but there is a risk of losing part of the pellet and retaining too little LB to resuspend the pellet.
12. In our experience a clean nuclease-free pipette tip allows for easier resuspension of the colony than a toothpick.
13. As most of the transfer vectors do not allow for β-galactasidase-based blue/white colony screening, the colonies will need to be analyzed by a PCR-based clonal screen or via more laborious restriction analyses. The PCR method can screen 96 clones (or 384 for higher throughput) in little over 2.5 h. Colony PCR cannot determine the orientation of the insert, but with unidirectional ligations, such as described in this chapter, this is not a major problem.
14. We have not been successful in adapting this procedure in a qPCR-format with SYBR Green reagent. The genomic and plasmid DNA interferes with the SYBR Green signal. Melt curves will therefore provide little good information about the presence of an insert in the plasmid.
15. This procedure is not very efficient and in our experience many rounds of cloning were needed to obtain a single positive clone. Be persistent; do not omit essential steps such as dephosphorylation of the vector. In our experience, dephosporylation may result in a fivefold higher ratio of positive to negative clones.
16. Many different transfection procedures and reagents are available, some of which may be more effective in specific cell types. The choice of transfection reagent depends on personal preference and experimental verification of the effectiveness of the method in your specific cell line. We chose a combination of PEI and HBS for transfection of lentiviral vectors (see Note 4). In a four plasmid transfection system, it is important to use an effective carrier to transfer all the plasmids into the producer cells. The PEI/HBS method is cheap and, if the PEI is dialyzed properly, the procedure provides relatively high transfection efficiencies and a relatively low toxicity (293T cells, for example, tolerating over 5 h of serum-free PEI transfection before cell death begins to occur). Calcium-phosphate precipitation works well with single plasmid transfections, but the chance of delivering all four plasmids into the same cell is lower than with micelle-like reagents, such as

PEI, Lipofectamin (Invitrogen), TurboFect (Fermentas), or with the nanoparticle reagent XFect (Clonetech).

17. Always add the PEI/HBS to the DNA/HBS solution and NOT the other way round! The optimal DNA:PEI ratio is 1 μg DNA: 2–4 μg PEI. For new batches of PEI, the PEI:DNA weight ratio should be titrated.
18. It is possible to add the DNA–PEI–HBS solution all at once without any loss of transfection efficiency, when using, for example, Triple Flask (Nunc), HyperFlask, or CellSTACK Culture Chambers (Corning). Be sure to gently mix the solutions before returning the flasks to a horizontal position.
19. For cleaner virus preparations use a 0.2 μm filter to remove more cellular debris and some of the concentrated FCS from the centrifugation, which may affect the transduction efficiency and cytotoxicity. If your cells are sensitive to any perturbation in culture, we recommend using a 0.2-μm filter to ensure the highest purity of the viral preparation. A 0.2 μm filter will, however, reduce the total yield.
20. Viral preparations can be thawed and refrozen three times without significant loss in viral titers. The half-life of the infectious particles is approximately 18 months when frozen at −80°C. This timeline can be prolonged by storing stocks in liquid nitrogen.
21. Several methods are available for titration (40), such as qPCR, FACS, or western blot. An understanding of the level of infectivity of your viral preparations is critical for consistency in your experiments. In our experience, titration by flowcytometric analyses on either producer cells or target cells is rapid and relatively easy.
22. This step results in complete cell fixation and complete loss of infectivity.
23. We observed variability in the percentage of transduced cell when calculating the required volume of lentivirus as described in Subheading 3.9, step 5. We have used a different methodology to determine the volume of lentiviral stock required to obtain a given transduction percentage. Plot the percentage of GFP-positive cells against the volume of lentiviral vector and generate a linear relationship equation ($y = mx + b$, where m = slope and b = Y-intercept). This will provide an estimated titer (transducing units (TU)/μl) that can be extrapolated to the volume required to achieve a given percentage of transduction. This value may differ from the traditional TU/ml, but works well in our experience, especially for experiments in which transduction efficiencies nearing 100% were crucial.

Acknowledgments

The authors would like to thank Dr. Bob Scholte, Prof. Hugo Tilanus Prof. Herold Metselaar, and Prof. Harry Janssen for general support. This study has been supported financially by the Erasmus MC Translational Research Fund and the Liver Research Foundation (SLO) Rotterdam.

References

1. Pan, Q., Henry, S. D., Metselaar, H. J., Scholte, B., Kwekkeboom, J., Tilanus, H. W., Janssen, H. L., and van der Laan, L. J. (2009) Combined antiviral activity of interferon-alpha and RNA interference directed against hepatitis C without affecting vector delivery and gene silencing. *J Mol Med* **87**, 713–22.
2. Dull, T., Zufferey, R., Kelly, M., Mandel, R. J., Nguyen, M., Trono, D., and Naldini, L. (1998) A third-generation lentivirus vector with a conditional packaging system. *J Virol* **72**, 8463–71.
3. Boden, D., Pusch, O., Lee, F., Tucker, L., and Ramratnam, B. (2003) Human immunodeficiency virus type 1 escape from RNA interference. *J Virol* **77**, 11531–5.
4. Gitlin, L., Karelsky, S., and Andino, R. (2002) Short interfering RNA confers intracellular antiviral immunity in human cells. *Nature* **418**, 430–4.
5. Ogata, N., Alter, H. J., Miller, R. H., and Purcell, R. H. (1991) Nucleotide sequence and mutation rate of the H strain of hepatitis C virus. *Proc Natl Acad Sci U S A* **88**, 3392–6.
6. Okamoto, H., Kojima, M., Okada, S., Yoshizawa, H., Iizuka, H., Tanaka, T., Muchmore, E. E., Peterson, D. A., Ito, Y., and Mishiro, S. (1992) Genetic drift of hepatitis C virus during an 8.2-year infection in a chimpanzee: variability and stability. *Virology* **190**, 894–9.
7. Young, K. C., Lindsay, K. L., Lee, K. J., Liu, W. C., He, J. W., Milstein, S. L., and Lai, M. M. (2003) Identification of a ribavirin-resistant NS5B mutation of hepatitis C virus during ribavirin monotherapy. *Hepatology* **38**, 869–78.
8. Kanda, T., Yokosuka, O., Imazeki, F., Tanaka, M., Shino, Y., Shimada, H., Tomonaga, T., Nomura, F., Nagao, K., Ochiai, T., and Saisho, H. (2004) Inhibition of subgenomic hepatitis C virus RNA in Huh-7 cells: ribavirin induces mutagenesis in HCV RNA. *J Viral Hepat* **11**, 479–87.
9. Konishi, M., Wu, C. H., Kaito, M., Hayashi, K., Watanabe, S., Adachi, Y., and Wu, G. Y. (2006) siRNA-resistance in treated HCV replicon cells is correlated with the development of specific HCV mutations. *J Viral Hepat* **13**, 756–61.
10. Wilson, J. A., and Richardson, C. D. (2005) Hepatitis C virus replicons escape RNA interference induced by a short interfering RNA directed against the NS5b coding region. *J Virol* **79**, 7050–8.
11. Henry, S. D., van der Wegen, P., Metselaar, H. J., Tilanus, H. W., Scholte, B. J., and van der Laan, L. J. (2006) Simultaneous targeting of HCV replication and viral binding with a single lentiviral vector containing multiple RNA interference expression cassettes. *Mol Ther* **14**, 485–93.
12. Korf, M., Jarczak, D., Beger, C., Manns, M. P., and Kruger, M. (2005) Inhibition of hepatitis C virus translation and subgenomic replication by siRNAs directed against highly conserved HCV sequence and cellular HCV cofactors. *J Hepatol* **43**, 225–34.
13. Novina, C. D., Murray, M. F., Dykxhoorn, D. M., Beresford, P. J., Riess, J., Lee, S. K., Collman, R. G., Lieberman, J., Shankar, P., and Sharp, P. A. (2002) siRNA-directed inhibition of HIV-1 infection. *Nat Med* **8**, 681–6.
14. Ciesek, S., Steinmann, E., Wedemeyer, H., Manns, M. P., Neyts, J., Tautz, N., Madan, V., Bartenschlager, R., von Hahn, T., and Pietschmann, T. (2009) Cyclosporine A inhibits hepatitis C virus nonstructural protein 2 through cyclophilin A. *Hepatology* **50**, 1638–45.
15. McIntyre, G. J., Groneman, J. L., Tran, A., and Applegate, T. L. (2008) An infinitely expandable cloning strategy plus repeat-proof PCR for working with multiple shRNA. *PLoS ONE* **3**, e3827.

16. McIntyre, G. J., Yu, Y. H., Tran, A., Jaramillo, A. B., Arndt, A. J., Millington, M. L., Boyd, M. P., Elliott, F. A., Shen, S. W., Murray, J. M., and Applegate, T. L. (2009) Cassette deletion in multiple shRNA lentiviral vectors for HIV-1 and its impact on treatment success. *Virol J* **6**, 184.
17. Brake, O. T., Hooft, K., Liu, Y. P., Centlivre, M., Jasmijn von Eije, K., and Berkhout, B. (2008) Lentiviral vector design for multiple shRNA expression and durable HIV-1 inhibition. *Mol Ther* **16**, 557–64.
18. Gou, D., Weng, T., Wang, Y., Wang, Z., Zhang, H., Gao, L., Chen, Z., Wang, P., and Liu, L. (2007) A novel approach for the construction of multiple shRNA expression vectors. *J Gene Med* **9**, 751–63.
19. Weinberg, M. S., Ely, A., Barichievy, S., Crowther, C., Mufamadi, S., Carmona, S., and Arbuthnot, P. (2007) Specific inhibition of HBV replication in vitro and in vivo with expressed long hairpin RNA. *Mol Ther* **15**, 534–41.
20. Sano, M., Li, H., Nakanishi, M., and Rossi, J. J. (2008) Expression of long anti-HIV-1 hairpin RNAs for the generation of multiple siRNAs: advantages and limitations. *Mol Ther* **16**, 170–7.
21. Ter Brake, O., and Berkhout, B. (2007) Lentiviral vectors that carry anti-HIV shRNAs: problems and solutions. *J Gene Med* **9**, 743–50.
22. Konstantinova, P., de Vries, W., Haasnoot, J., Ter Brake, O., de Haan, P., and Berkhout, B. (2006) Inhibition of human immunodeficiency virus type 1 by RNA interference using long-hairpin RNA. *Gene Ther* **13**, 1403–13.
23. Saayman, S., Barichievy, S., Capovilla, A., Morris, K. V., Arbuthnot, P., and Weinberg, M. S. (2008) The efficacy of generating three independent anti-HIV-1 siRNAs from a single U6 RNA Pol III-expressed long hairpin RNA. *PLoS ONE* **3**, e2602.
24. Liu, Y. P., von Eije, K. J., Schopman, N. C., Westerink, J. T., ter Brake, O., Haasnoot, J., and Berkhout, B. (2009) Combinatorial RNAi against HIV-1 using extended short hairpin RNAs. *Mol Ther* **17**, 1712–23.
25. Barichievy, S., Saayman, S., von Eije, K. J., Morris, K. V., Arbuthnot, P., and Weinberg, M. S. (2007) The inhibitory efficacy of RNA POL III-expressed long hairpin RNAs targeted to untranslated regions of the HIV-1 5′ long terminal repeat. *Oligonucleotides* **17**, 419–31.
26. Pan, Q., Tilanus, H. W., Janssen, H. L., and van der Laan, L. J. W. (2009) Prospects of RNAi and microRNA-based therapies for hepatitis C. *Expert Opin Biol Ther* **9**, 713–24.
27. Pan, Q., Henry, S. D., Scholte, B. J., Tilanus, H. W., Janssen, H. L., and van der Laan, L. J. W. (2007) New therapeutic opportunities for Hepatitis C based on small RNA. *World J Gastroenterol* **13**, 4431–6.
28. Aagaard, L. A., Zhang, J., von Eije, K. J., Li, H., Saetrom, P., Amarzguioui, M., and Rossi, J. J. (2008) Engineering and optimization of the miR-106b cluster for ectopic expression of multiplexed anti-HIV RNAs. *Gene Ther* **15**, 1536–49.
29. Boden, D., Pusch, O., Silbermann, R., Lee, F., Tucker, L., and Ramratnam, B. (2004) Enhanced gene silencing of HIV-1 specific siRNA using microRNA designed hairpins. *Nucleic Acids Res* **32**, 1154–8.
30. Cullen, B. R. (2006) Viruses and microRNAs. *Nat Genet* **38 Suppl 1**, S25–30.
31. Shan, Z. X., Lin, Q. X., Yang, M., Deng, C. Y., Kuang, S. J., Zhou, Z. L., Xiao, D. Z., Liu, X. Y., Lin, S. G., and Yu, X. Y. (2009) A quick and efficient approach for gene silencing by using triple putative microRNA-based short hairpin RNAs. *Mol Cell Biochem* **323**, 81–9.
32. McManus, M. T., Petersen, C. P., Haines, B. B., Chen, J., and Sharp, P. A. (2002) Gene silencing using micro-RNA designed hairpins. *RNA* **8**, 842–50.
33. Shan, Z., Lin, Q., Deng, C., Li, X., Huang, W., Tan, H., Fu, Y., Yang, M., and Yu, X. Y. (2009) An efficient method to enhance gene silencing by using precursor microRNA designed small hairpin RNAs. *Mol Biol Rep* **36**, 1483–9.
34. Shin, K. J., Wall, E. A., Zavzavadjian, J. R., Santat, L. A., Liu, J., Hwang, J. I., Rebres, R., Roach, T., Seaman, W., Simon, M. I., and Fraser, I. D. (2006) A single lentiviral vector platform for microRNA-based conditional RNA interference and coordinated transgene expression. *Proc Natl Acad Sci U S A* **103**, 13759–64.
35. Sun, D., Melegari, M., Sridhar, S., Rogler, C. E., and Zhu, L. (2006) Multi-miRNA hairpin method that improves gene knockdown efficiency and provides linked multi-gene knockdown. *Biotechniques* **41**, 59–63.
36. Zhu, X., Santat, L. A., Chang, M. S., Liu, J., Zavzavadjian, J. R., Wall, E. A., Kivork, C., Simon, M. I., and Fraser, I. D. (2007) A versatile approach to multiple gene RNA interference using microRNA-based short hairpin RNAs. *BMC Mol Biol* **8**, 98.
37. Lim, L. P., Lau, N. C., Garrett-Engele, P., Grimson, A., Schelter, J. M., Castle, J., Bartel,

D. P., Linsley, P. S., and Johnson, J. M. (2005) Microarray analysis shows that some microRNAs downregulate large numbers of target mRNAs. *Nature* **433**, 769–73.

38. Dinh, A., and Mo, Y. Y. (2005) Alternative approach to generate shRNA from cDNA. *Biotechniques* **38**, 629–32.
39. Wu, H., Dinh, A., and Mo, Y. Y. (2007) Generation of shRNAs from randomized oligonucleotides. *Biol Proced Online* **9**, 9–17.
40. Lizee, G., Aerts, J. L., Gonzales, M. I., Chinnasamy, N., Morgan, R. A., and Topalian, S. L. (2003) Real-time quantitative reverse transcriptase-polymerase chain reaction as a method for determining lentiviral vector titers and measuring transgene expression. *Hum Gene Ther* **14**, 497–507.

Chapter 20

Intranasal Delivery of Antiviral siRNA

Sailen Barik

Abstract

Intranasal administration of synthetic siRNA is an effective modality of RNAi delivery for the prevention and therapy of respiratory diseases, including pulmonary infections. Vehicles used for nasal siRNA delivery include established as well as novel reagents, many of which have been recently optimized. In general, they all promote significant uptake of siRNA into the lower respiratory tract, including the lung. When properly designed and optimized, these siRNAs offer significant protection against respiratory viruses such as influenza virus, parainfluenza virus and respiratory syncytial virus (RSV). Nasally administered siRNA remains within the lung and does not access systemic blood flow, as judged by its absence in other major organs such as liver, heart, kidney, and skeletal muscle. Adverse immune reaction is generally not encountered, especially when immunogenic and/or off-target siRNA sequences and toxic vehicles are avoided. In fact, siRNA against RSV has entered Phase II clinical trials in human with promising results. Here, we provide a standardized procedure for using the nose as a specific route for siRNA delivery into the lung of laboratory animals. It should be clear that this simple and efficient system has enormous potential for therapeutics.

Key words: Intranasal, Antiviral, Influenza, Parainfluenza, RNAi, RSV, siRNA

1. Introduction

RNA interference (RNAi) is a normal physiological mechanism for RNA-guided regulation of gene expression, common in all higher eukaryotes (1). In this pathway, double-stranded RNA (dsRNA) silences the expression of genes with sequences that are complementary to the antisense strand of the dsRNA (2). In one therapeutic approach to harness the power of RNAi, a class of synthetic dsRNAs, known as siRNA (short or small interfering RNA), is used (3, 4). The siRNA is 21–23 nucleotides long and may be designed to contain 2-nt long 3′ extensions. The antisense strand (also called "guide strand") of the siRNA engages

Ronald P. van Rij (ed.), *Antiviral RNAi: Concepts, Methods, and Applications*, Methods in Molecular Biology, vol. 721, DOI 10.1007/978-1-61779-037-9_20,

into a multiprotein RNAi-induced silencing complex (RISC). A catalytic component of RISC, Argonaute-2, then specifically cleaves the target RNA strand within RISC, thus destroying the target (2). Clearly, RNAi offers high specificity and efficacy in turning off the expression of target genes.

The first successful antiviral application of siRNA in mammals was demonstrated against RSV (4) in cell culture, and the efficacy was later translated into the mouse model of RSV infection (5). RSV is a nonsegmented single-stranded RNA virus of the *Paramyxoviridae* family. In addition to RSV, a number of serious and life-threatening human and domestic animal viruses (e.g., mumps, measles, parainfluenza, borna, rabies, ebola, Hendra and Nipah) belong to this family (6). Of note, the influenza (Flu) virus, belonging to the *Orthomyxoviridae* family, is also an RNA virus but contains a segmented genome. Thus, an intranasal RNAi approach can also be effective against all respiratory viral diseases (7, 8), including Flu (9) and SARS (10).

The nasal route represents an improved siRNA delivery regimen compared to systemic approaches such as intravenous, intraperitoneal, or hydrodynamic tail vein injections, since it is convenient, noninvasive, and therapeutically effective. It may also prove useful in treating debilitating noninfectious respiratory diseases such as chronic obstructive pulmonary disease (COPD), cystic fibrosis, asthma, and perhaps some forms of lung cancer, when combined with existing anticancer drugs.

2. Materials

2.1. siRNA

1. siRNA protected with 2′-ACE (5′-silyl-2′-acetoxy ethyl orthoester) chemistry (see Notes 1–4).
2. DEPC-treated water.
3. RNase-free aerosol resistant pipette tips.
4. RNase-free microfuge tubes.
5. 9.5 M Ammonium acetate, made in DEPC-treated water.
6. 95 and 100% Ethanol.
7. 2′-Deprotection buffer: 100 mM acetic acid adjusted to pH 3.8 using TEMED (Dharmacon).
8. siRNA buffer: 20 mM KCl, 6 mM HEPES-KOH, pH 7.5, 0.2 mM $MgCl_2$, made in DEPC-treated water.

2.2. siRNA Transfection Reagents

1. TransIT-TKO siRNA transfection reagent (Mirus Bio Corporation).
2. Opti-MEM I Reduced-Serum Medium (Invitrogen).
3. RNase-free microfuge tubes (Ambion).

2.3. Animals and Anesthesia Reagents

1. BALB/c mice, 6–8 weeks old, averaging 18 g in weight (Charles River Laboratories).
2. 5 mg/mL Sodium pentobarbital (Nembutal).
3. 25-Gauge single-use needles (VWR).
4. 1 cc (mL) Single-use syringes with BD Luer-Lok tip (VWR).
5. RNase-free gel-loading microcapillary tips.

3. Methods

3.1. siRNA 2′-Deprotection, Annealing, and Desalting

1. Add 200 μL of 2′-deprotection buffer to tube containing 0.1 μmol 2′-ACE protected, single-stranded RNA (see Note 3).
2. Combine the two complementary strands of RNA, vortex for 10 s and centrifuge for 30 s at 10,000 × *g*.
3. Heat the mixture at 60°C for 45 min.
4. Remove from heat and centrifuge briefly, 5–10 s.
5. Allow solution to cool to room temperature over 30 min.
6. Add 40 μL of 9.5 M ammonium acetate and 1.5 mL of 100% ethanol to the 400 μL of siRNA duplex solution, vortex.
7. Place the tubes at −20°C for more than 16 h or at −70°C for 2 h.
8. Centrifuge at 14,000 × *g* for 30 min at 4°C.
9. Carefully remove the supernatant away from the pellet.
10. Rinse the pellet with 200 μL of ice-cold 95% ethanol.
11. Dry under vacuum using Speed-Vac.
12. The dry pellet can be stored at −20°C until use or resuspended in the siRNA buffer or another appropriately buffered solution.

3.2. siRNA-Vehicle Complex Formulation

Prepare the siRNA-vehicle complex immediately before nasal administration. Determine the optimal siRNA amount by titering from 3 to 15 nmol per mouse.

1. In a sterile, RNase-free plastic tube, add 35 μL of Opti-MEM reduced-serum medium.
2. Add 5 μL of the TransIT-TKO transfection reagent into the tube containing Opti-MEM medium.
3. Mix thoroughly by vortexing for 10 s.
4. Incubate at room temperature for 10 min.

5. Add the desired amount of siRNA in 1 μL of siRNA buffer to the diluted TransIT-TKO reagent.
6. Carefully mix by gentle pipetting.
7. Incubate at room temperature for 20 min.

3.3. Animal Anesthesia

Prior to nasal administration of siRNA, mice (see Note 5) must be anesthetized to minimize any pain or discomfort. Nembutal is administered by intraperitoneal (IP) injection at a recommended drug dosage of 50 mg/kg.

1. Gently lift the mouse by the tail and place it on a cage lid.
2. Grip the loose skin of the neck to immobilize the head of the animal.
3. With the head immobilized, extend the tail to draw the skin tight over the abdomen by gripping the tail with your little finger.
4. The animal should be held in a head-down position.
5. Disinfect injection site with 95% ethanol.
6. Insert the 25-gauge hypodermic needle into the lower right or left quadrant of the abdomen, and inject the recommended amount of Nembutal in a volume of 0.2–0.25 mL (depending on the weight of the mouse).
7. Place animal back into the cage and wait until anesthesia takes effect.
8. The animal is ready for siRNA administration when no voluntary movement is observed.

3.4. siRNA Administration

1. Place the anesthetized mouse on a laboratory towel with face up.
2. With head immobilized, insert microcapillary tip containing the siRNA transfection reagent complexes into nostril (see Note 6).
3. Instill the solution slowly over a 2–3-min period, allowing the mouse to breathe the liquid in (see Notes 7 and 8).
4. Place animal back into the cage and monitor for at least 45 min to avoid depression of cardiac and/or respiratory functions.
5. Test for the desired RNAi effect at appropriate intervals. For antiviral studies, instill virus through the nostril as well. For human RSV, which does not infect mice well, use 10^7–10^8 virus particles per animal, and measure standard lung titer assay and clinical symptoms, such as body weight and respiration rate.

4. Notes

1. It is assumed that the reader has expertise in designing siRNA against a given target. We generally follow the design suite offered by the Dharmacon (Thermo Scientific) and the Whitehead Institute (MIT, Cambridge, MA) web pages. We also try to avoid the relatively strong immunogenic sequence: 5′GUCCUUCAA3′ (11).
2. Essentially all RNA viral genes are essential for robust virus growth, and hence the siRNA designed against any viral mRNA will inhibit virus replication. Nonetheless, we generally target components of the viral RNA synthesis machinery; for *Paramyxoviridae*, they are: L (Large protein; RNA-dependent RNA polymerase), P (Phosphoprotein; a transcription factor for L), and N (Nucleocapsid protein; essential for the template function of the viral genomic RNA).
3. For reasons of animal ethics and cost-saving, we always test a new siRNA sequence in cell culture for antiviral potency by measuring the progeny viral titer. Use a range of concentrations from 2 to 200 nM; a potent siRNA should exhibit 90% inhibition at low nM concentrations (e.g., at a concentration below 20 nM). With this criteria and using the design recommendations in Note 1, we have experienced a success rate of 80% (8 out of 10 siRNAs) over the years.
4. Steps in Subheading 3.1 are not needed if fully deprotected, annealed, ready-to-use siRNA is purchased commercially, which are more expensive. To save cost, we usually purchase 2′-ACE protected RNA and deprotect it. The 2′-ACE modification increases stability of the RNA during shipment and storage.
5. We have described a simple method of nasal delivery of siRNA in the mouse model, but it can be successfully scaled up or down to use in other laboratory animals.
6. To diminish any possible toxic effect of delivery reagents, siRNA can be introduced "naked," i.e., without transfection reagent, which exhibits about 70–80% activity of the reagent-complexed siRNA. It is best not to use polyethyleneimine (PEI). Although PEI is often used to form complexes with nucleic acids, mice do not tolerate PEI through the nose, although delivery through other avenues, such as intravenous, are well tolerated. In our attempts, essentially all mice died within minutes of inhalation of PEI-siRNA complex. New generation of transfection reagents are constantly being reported (12, 13) that may provide better delivery.

7. Administration of excessive liquid will "drown" the mouse and cause death. Try to keep the total volume under 45 μL in routine application although up to 100 μL may be tolerated.
8. The intranasal procedure also works in an aerosolized application. We have used small homemade enclosures in which the anesthetized mouse is placed and the siRNA complex (made as in Subheading 3.2) is sprayed by a hand-held nebulizer (the common type used as an inhaler by asthmatics). In this method, however, a larger amount of siRNA is needed because most of the mist gets wasted and only a small fraction is inhaled by the animal. To optimize cost vs. benefit, it is recommended that various amounts of mist and duration of exposure is tested and optimized for a given enclosure volume. If this method is to be routinely used, a commercial motorized nebulizer can be easily optimized. It is recommended to check with the local pediatric department for the exact model, vendor, and usage. Modify the system by removing the plastic face cup (or mask) at the delivery end and insert the tube directly into the mouse enclosure. Reagent cost can be reduced by designing a smaller face cup to snugly fit the mouse nose area.

References

1. Fire, A., Xu, S., Montgomery, M. K., Kostas, S. A., Driver, S. E., and Mello, C. C. (1998) Potent and specific genetic interference by double-stranded RNA in *Caenorhabditis elegans. Nature* **391**, 806–811.
2. Sashital, D. G. and Doudna, J. A. (2010) Structural insights into RNA interference. *Curr. Opin. Struct. Biol.* **20**, 90–97.
3. Elbashir, S. M., Martinez, J., Patkaniowska, A., Lendeckel, W., and Tuschl, T. (2001) Functional anatomy of siRNAs for mediating efficient RNAi in *Drosophila melanogaster* embryo lysate. *EMBO J.* **20**, 6877–6888.
4. Bitko, V. and Barik, S. (2001) Phenotypic silencing of cytoplasmic genes using sequence-specific double-stranded short interfering RNA and its application in the reverse genetics of wild type negative-strand RNA viruses. *BMC Microbiol.* **1**, 34.
5. Bitko, V., Musiyenko, A., Shulyayeva, O., and Barik, S. (2004) Inhibition of respiratory viruses by nasally administered siRNA. *Nat. Med.* **11**, 50–55.
6. Mackie, P. L. (2003) The classification of viruses infecting the respiratory tract. *Paediatr. Respir. Rev.* **4**, 84–90.
7. Barik, S. (2004) Control of nonsegmented negative-strand RNA virus replication by siRNA. *Virus Res.* **102**, 27–35.
8. Barik, S. (2005) Silence of the transcripts: RNA interference in medicine. *J. Mol. Med.* **83**, 764–773.
9. Ge, Q., McManus, M. T., Nguyen, T., et al. (2003) RNA interference of influenza virus production by directly targeting mRNA for degradation and indirectly inhibiting all viral RNA transcription. *Proc. Natl. Acad. Sci. USA* **100**, 2718–2723.
10. Li, B. J., Tang, Q., Cheng, D., et al. (2005) Using siRNA in prophylactic and therapeutic regimens against SARS coronavirus in Rhesus macaque. *Nat. Med.* **11**, 944–951.
11. Hornung, V., Guenthner-Biller, M., Bourquin, C., et al (2005) Sequence-specific potent induction of IFN-alpha by short interfering RNA in plasmacytoid dendritic cells through TLR7. *Nat. Med.* **11**, 263–270.
12. Akinc, A., Zumbuehl, A., Goldberg, M., et al. (2008) A combinatorial library of lipid-like materials for delivery of RNAi therapeutics. *Nat. Biotechnol.* **26**, 561–569.
13. Semple, S. C., Akinc, A., Chen, J., et al. (2010) Rational design of cationic lipids for siRNA delivery. *Nat. Biotechnol.* **28**, 172–176.

Chapter 21

Antibody-Mediated Delivery of siRNAs for Anti-HIV Therapy

Sang-Soo Kim, Sandesh Subramanya, Dan Peer, Motomu Shimaoka, and Premlata Shankar

Abstract

RNA interference (RNAi) is a potent and specific gene silencing mechanism that utilizes small double-stranded RNA intermediates (small interfering RNAs or siRNAs) to target homologous mRNA sequences for degradation. The therapeutic potential of RNAi for HIV infection has been demonstrated in many studies. However, successful clinical application of RNAi is contingent on developing practical strategies to deliver siRNA to the desired target cells and tissues. Recently, there has been significant progress towards developing reagents that selectively deliver exogenous siRNA to immune cells that are targeted by HIV or involved in viral pathogenesis, such as T cells, macrophages, and dendritic cells. Here, we describe details of two antibody-based strategies for *systemic* delivery of siRNA either specifically to T cells via the CD7 receptor or to multiple immune cell types via LFA-1, present on all leukocytes.

Key words: RNA interference, Small interfering RNA, Targeted delivery, Cationic peptide, Liposome, scFvCD7, Integrin, LFA-1 (lymphocyte function-associated antigen-1), I-tsNPs (integrin-targeted stabilized nanoparticles), HIV

1. Introduction

Despite the success of highly active antiretroviral therapy (HAART) in ameliorating HIV infection, the significant practical problems associated with life-long treatment, including drug-induced toxicity, patient noncompliance and development of drug resistance, point to a clear need for alternative therapeutic approaches for controlling the infection (1) RNA interference (RNAi) is a potent posttranscriptional gene silencing mechanism with great potential for therapeutic intervention in HIV infection. The effector moieties in the process are small interfering RNA (siRNA) molecules generated by Dicer processing of long

Ronald P. van Rij (ed.), *Antiviral RNAi: Concepts, Methods, and Applications*, Methods in Molecular Biology, vol. 721, DOI 10.1007/978-1-61779-037-9_21, © Springer Science+Business Media, LLC 2011

double-stranded RNA. These 21–23 bp long siRNAs are loaded onto a multiprotein complex called RNA-induced silencing complex (RISC). During loading, the sense "passenger" strand is cleaved by an endonuclease, Argonaute 2 (Ago2) after which the antisense strand guides RISC to the target mRNA for sequence-specific degradation by the same endonuclease (2). RNAi can be induced artificially by the introduction of chemically synthesized siRNAs or by intracellular generation of siRNA from vector-driven expression of precursor short hairpin RNAs (shRNAs) (3). Silencing effects of siRNA are short-lived (few days) due to dilution by cell division, whereas shRNA-mediated gene silencing lasts longer because the siRNA is continuously generated inside the cell. Nevertheless, siRNAs are very attractive for therapy because of the ease of design and synthesis and versatility that allows simultaneous use of multiple siRNAs or change of sequences to accommodate virus mutations. More importantly, siRNA treatment can be stopped immediately in case of unanticipated toxicity. However, major limitations for the therapeutic use of siRNAs include poor stability, large size, and strong negative charge that prevents easy cellular uptake. Thus, there is a critical need to develop reagents for in vivo delivery of siRNA to appropriate cells and tissues. Targeted delivery has the major advantage of focusing and increasing the therapeutic index of candidate siRNAs at the affected local region while minimizing the potential off-target effects in nontarget cells and tissues (4). Liposomes, polymers, peptides, and aptamers have been recently used for systemic delivery of siRNA to immune cell types, such as T cells, macrophages, and dendritic cells, that are important in HIV pathogenesis (5).

Targeting cellular ligands using antibodies or peptides is a particularly attractive strategy for siRNA delivery as the complex is actively internalized into target cells where the siRNA induces the desired biological effect (6–11). Monomeric antibody (encompassing two heavy and two light chains) (8, 11) or antibody fragments like single chain antibody variable fragment (scFv; a fusion of the variable regions of the heavy and light chain linked together with a short linker) (7) or the antigen-binding Fab fragment (6) have been used for delivery of antiviral and host gene-directed siRNAs as potential therapy for HIV-1 infection. The siRNAs have been either encapsulated in a liposomal formulation that has been chemically coupled to the antibody (8, 11) or piggy-backed on a fusion protein that harbors a nucleic acid-binding moiety like the cationic protein protamine (6) or nona-d-arginine (9dR) (7).

Two antibody-based approaches (illustrated in Fig. 1), one for targeted siRNA delivery to T cells using a scFv to CD7 fused to nona-d-arginine (9dR) residues (scFvCD7-9dR) (7) and another for broader targeting of multiple immune cell types using lymphocyte function-associated antigen-1 (LFA-1) integrin-targeted and stabilized nanoparticles (LFA-1 I-tsNPs) (11) are described below.

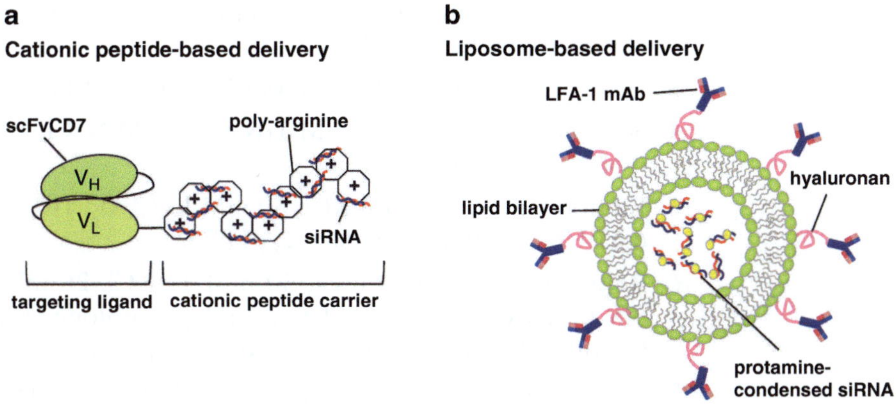

Fig. 1. Strategies for antibody-mediated in vivo siRNA delivery for anti-HIV therapy. These delivery vehicles comprise a targeting and a cargo moiety. (**a**) In cationic peptide-based delivery system, the siRNA is either covalently conjugated with or noncovalently bound to a positively charged peptide such as polyarginine or protamine (cationic peptide siRNA carrier) to which the targeting peptide or antibody moieties are conjugated (targeting ligand). (**b**) In a liposome-based delivery vehicle, the siRNA is condensed with cationic peptide such as protamine and encapsulated within the aqueous core of a unilamellar lipid bilayer. The outer surface of the lipid is conjugated with stabilizing molecules, such as hyaluronan and a targeting agent, such as an antibody to a specific cell-surface antigen. Liposome-based delivery vehicle has a superior capacity to entrap siRNA molecules compared to cationic peptide-based delivery vehicle.

2. Materials

2.1. CD7 Cloning and Production of CD7 Single Chain Variable Fragment (scFvCD7) in E. coli

1. pET 26b(+) vector (Addgene, MA).
2. pAK400/scFvCD7-GFP plasmid (12) (single chain variable fragment of CD7 antibody cloned in pAK400 vector) (Gift from George Fey, University of Erlangen, Germany).
3. Qiagen Gel extraction kit (Qiagen, CA).
4. BL21 Star (DE3) *E. coli* competent cells (Invitrogen, CA).
5. Kanamycin (Sigma, MO). Prepare a 50 mg/mL stock solution in MilliQ water. Store 1 mL aliquots at –20°C.
6. LB agar (Invitrogen). 32.0 g/L of MilliQ water. Autoclave at 121°C for 15 min. Add kanamycin to a final concentration of 50 μg/mL right before pouring plates, when the agar reaches a temperature of 40°C.
7. Luria Bertani (LB) medium: Bacto-Tryptone (10 g), Bacto-yeast extract (5 g), NaCl (10 g). Make volume to 1 L with MilliQ water. Adjust pH to 7.0 and autoclave.
8. 1 mM IPTG (Invitrogen) stock solution in RNase-free water distilled water.

2.2. CD7 Purification and Refolding of scFvCD7-Cys

1. Lysis buffer: 0.3 mg/mL lysozyme, 50 mM Tris, 1 mM EDTA, 100 mM NaCl, 0.5 mM DTT, and 1 mM phenylmethylsulfonyl fluoride, pH 8.0 (Sigma).

2. Bio Scale Mini Profinity immobilized metal affinity chromatography (Bio-Rad).
3. Wash buffer: 20 mM Tris–Cl, 0.5 M NaCl, pH 8.0.
4. Solubilization buffer: 20 mM Tris–Cl, 0.5 M NaCl, 10% glycerol, 8 M urea, pH 8.0.
5. Elution buffer: 300 mM KCl, 50 mM KH_2PO_4, 10% glycerol, 6 M Urea, 250 mM imidazole, pH 7.9.
6. Refolding buffer: 1 M urea, 50 mM Tris–Cl, 20 mM NaCl, 0.8 mM KCl, 1 mM EDTA, 2 mM glutathione (reduced), 0.4 mM glutathione disulfide (oxidized) pH 8.1.
7. Slize-A-Lyzer Dialysis cassettes (10,000 MWCO; Pierce, IL).
8. Centricon centrifugal filter units (Millipore, MA).
9. BCA Protein assay kit (Pierce).
10. Jurkat T cells.
11. RPMI 1640 medium (Gibco, CA).
12. Dulbecco's phosphate-buffered saline (DPBS) 1× sterile, without calcium and magnesium (Gibco, CA).
13. Anti-human CD7-PE, CD3-FITC, and CD4-PECy5 antibodies (BD PharMingen).
14. FACS buffer: 2% heat-inactivated FBS in 1× PBS. Store buffer at 4°C.
15. Fixing buffer: 2% paraformaldehyde in 1× PBS.
16. Flow cytometer: BD FACS Canto II (Becton, Dickinson & Company).

2.3. Conjugation of scFvCD7-Cys to 9-D-Arginine

1. Cys(Npys)-(D-Arg)9 peptide (Mol. Wt 1.68 kD; Anaspec, CA). Dissolve to a final concentration of 1 mg/mL in 0.1 M phosphate buffer (pH 5.5). Store as 1 mL aliquots at −20°C.
2. Slize-A-Lyzer Dialysis cassettes (10,000 MWCO; Pierce, IL).
3. Thiol and sulfide quantization assay kit (Molecular Probes).

2.4. Formation of scFvCD7-9dR/siRNA Complex

1. RPMI 1640 medium (Gibco, CA) (see Subheading 2.2, item 11).
2. siRNAs (Dharmacon Inc., CO) (see Note 1).

2.5. Preparation of Lymphocyte Function-Associated Antigen-1 (LFA-1) Integrin-Targeted and Stabilized Nanoparticles (LFA-1 I-tsNPs)

1. Phosphatidylcholine (PC) (Avanti Polar Lipids, Inc., AL).
2. 1,2-dipalmitoyl-sn-glycero-3-phosphoethanolamine (DPPE) (Avanti Polar Lipids, Inc., AL).
3. Cholesterol (Chol) (Avanti Polar Lipids, Inc., AL).
4. Rotary evaporator (Buchi Corporation, Switzerland).
5. HEPES-buffered saline (HBS): 10 mM HEPES, 151 mM NaCl, 4.7 mM KCl, 2 mM $CaCl_2$, 1.2 mM $MgCl_2$, 7.8 mM glucose, pH 7.4.

6. Thermobarrel Lipex extruder (Lipex biomembranes Inc., Vancouver, BC, Canada).
7. Nucleopore membranes with 0.1–1 μm pore size.
8. Hyaluronan (HA, 751 kD, intrinsic viscosity, 16 dL/g, Genzyme Corp, Cambridge, MA): Stored as a powder at −20°C freezer in a desiccator.
9. 1 M Sodium acetate, pH 4.5: Dilute to 0.1 M with double distilled H_2O.
10. 1 M Borate, pH 9.0: Diluted to 0.1 M with H_2O.
11. 1-(3-dimethylaminopropyl)-3-ethylcarbodimide hydrochloride (EDAC, Sigma-Aldrich). Prepare fresh each time.
12. *N*-hydroxysuccinimide (NHS, Fluka, MI). Prepare fresh each time.
13. TS1/22 monoclonal antibody against human integrin LFA-1. This antibody can be purified from a hybridoma cell line (ATCC).
14. 1 M ethanolamine hydrochloride, pH 8.5 (Sigma-Aldrich, MI).
15. Sepharose CL-4B beads (Amersham Biosciences).
16. Alpha 1-2 LD plus lyophilizer (Christ, Osterode, Germany).

2.6. siRNA Entrapment in Nanoparticles

1. Full-length human recombinant protamine (Abnova, Taipei City, Taiwan). Prepare a 50 mg/mL stock solution in RNase-free deionized water.
2. RNase-free distilled water.
3. siRNAs (Dharmacon Inc., CO) (see Note 1).

2.7. In Vitro Transfection of siRNA with scFvCD7-9R or LFA-1 I-tsNP

1. Jurkat T cells (see Subheading 2.2, item 10).
2. RPMI 1640 medium (Gibco, CA).
3. Fetal Bovine Serum (FBS): Heat inactivate at 56°C for 30 min. Let cool at room temperature for 30 min and add to a final concentration of 10% to RPMI 1640 medium.
4. Penicillin/streptomycin/glutamine supplement (Gibco, CA): Add 5 mL of the 100× stock to 500 mL RPMI 1640 medium to obtain a 1× final concentration.
5. RNeasy mini kit (Qiagen, CA).

2.8. In Vivo Administration of siRNA

1. D-Glucose solution (prepare a 20% stock D-glucose solution in MilliQ water and filter sterilize). Store as 1 mL aliquots in −20°C (schematic in Fig. 2).
2. Human cell-engrafted humanized mice (see Note 2).
3. Microtainer tubes with EDTA (BD Biosciences, CA).
4. Ficoll-Paque Premium (1.077 g/mL) (GE Healthcare, NJ).

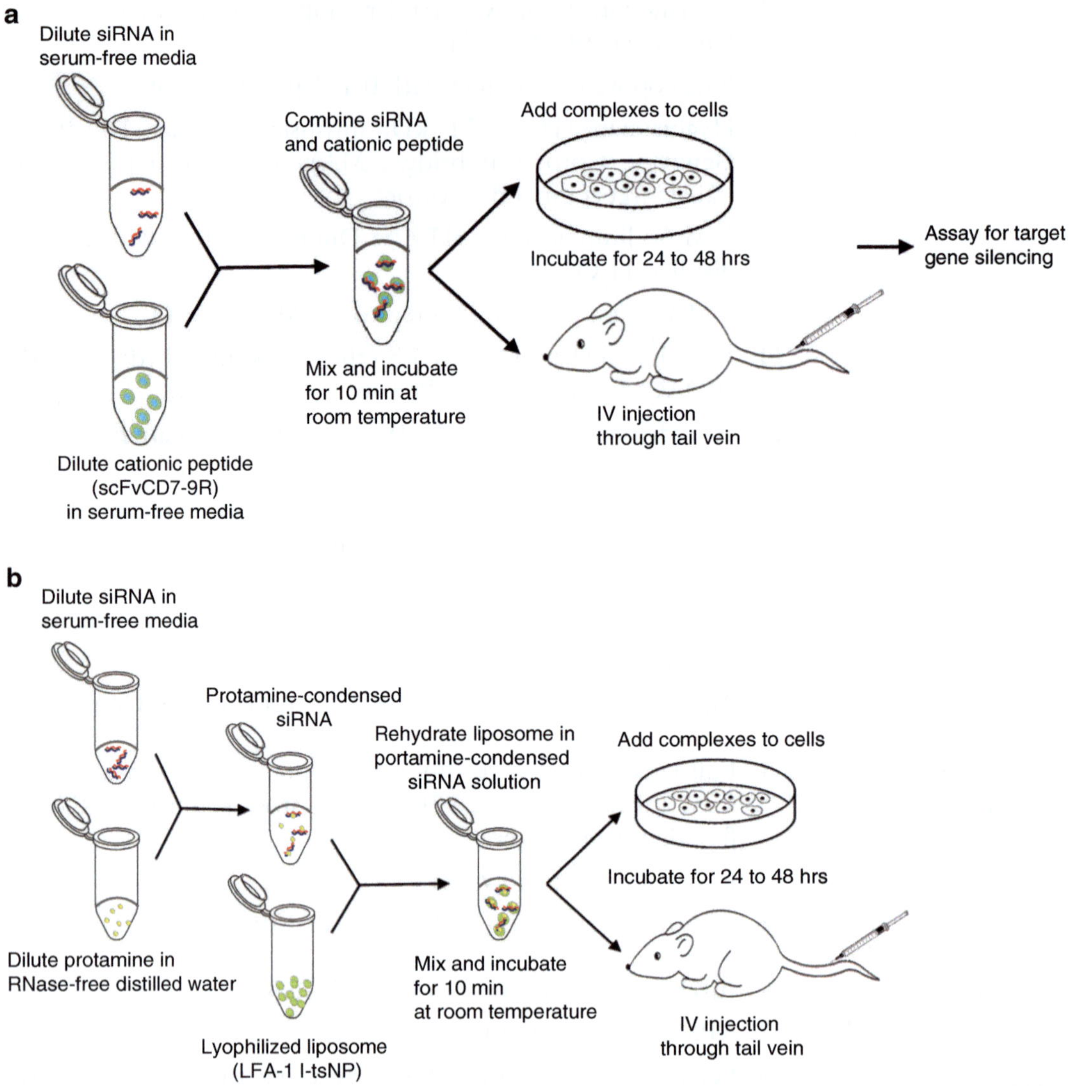

Fig. 2. Schematic representation of generation of siRNA complexes with cationic peptide- or liposome-based delivery systems. (**a**) In a cationic peptide-based delivery system, the siRNA is diluted in serum-free media and mixed with the delivery reagent to form a complex. (**b**) In the liposome-based delivery vehicle, the siRNA is condensed with the cationic peptide, protamine. siRNA is encapsulated within the liposome by rehydrating lyophilized liposome in protamine-condensed siRNA solution.

5. FACS buffer (see Subheading 2.2, item 14).
6. Fixing buffer (see Subheading 2.2, item 15).
7. Immunomagentic beads: CD3, CD19, CD14, CD11b, and CD11c (StemCell Technologies, Vancouver, BC).
8. AMPLICOR HIV-1 MONITOR Test, v1.5 (Roche Diagnostics Corporation, IN).

3. Methods

3.1. Cloning and Production of CD7 Single Chain Variable Fragment (scFvCD7) in E. coli

1. Amplify the coding region of CD7 scFv from the pAK400/scFvCD7-GFP plasmid (12) by polymerase chain reaction (PCR) using a reverse primer to introduce a cysteine residue in frame at the C-terminus (see Note 3). Run the PCR product on a 1% agarose gel, purify the amplified fragment using a gel extraction kit, digest with the appropriate restriction enzymes, and ligate into pET26b (+) vector to obtain pET26b/scFvCD7-Cys.
2. Transform the plasmid (pET26b/scFvCD7-Cys) into BL21 Star (DE3) *E. coli* competent cells and plate on LB agar plates with 50 μg/mL kanamycin. Incubate overnight at 37°C incubator to obtain transformants.
3. Generate a starter culture by inoculating an isolated colony in 10 mL of LB medium. Incubate at 37°C for 4 h in a shaker at 225 rpm.
4. Inoculate 5 mL of starter culture in 1 L LB medium and incubate at 37°C with shaking until $OD_{600}=1.6$.
5. Induce recombinant protein expression with 1 mM IPTG at 30°C for 14–16 h.

3.2. Purification and Refolding of scFvCD7-Cys

1. Pellet bacterial cells by centrifuging at 4,500 × *g* for 10 min.
2. Lyse cells with lysis buffer for 20 min at 4°C.
3. Sonicate the lysate five times for 1 min each (with 2 min rest on ice between pulses).
4. Separate the insoluble fraction by centrifuging at 15,000 × *g* for 15 min at 4°C.
5. Solubilize the inclusion bodies in insoluble fraction with solubilization buffer at 4°C.
6. Centrifuge at 75,000 × *g* for 20 min at 4°C and use solubilized fraction for purification.
7. Purify the recombinant protein by FPLC with Bio Scale Mini Profinity immobilized metal affinity chromatography and elute into 1 mL fractions with elution buffer. Fractions 2–6 typically contain the desired protein. Absorbance at 280 nm is used to detect presence of protein in the eluted fractions. Pool eluted fractions.
8. Dialyze against 1 L refolding buffer using Slide-A-Lyzer dialysis cassette for 18 h at 4°C.
9. Dialyze the refolded protein against 3 L of 1× PBS (pH 7.4) at 4°C for 24 h with four buffer changes for complete equilibration.

10. Immediately concentrate the sample on Centricon centrifugal filter units and measure protein concentration using a BCA Protein assay kit. Recombinant proteins are typically stored in aliquots at –80°C. Repeated (two or more) freeze–thaw cycles should be avoided.
11. Verify cell-specific binding by flow cytometry. Preincubate 5×10^5 Jurkat T cells with purified scFvCD7-Cys (20 μg/mL) in RPMI medium at 37°C for 4 h. Centrifuge cells at 2,300 × *g* in a microcentrifuge for 5 min at room temperature. Wash cells once with 1 mL of 1× PBS and stain with anti-human CD7-PE, CD3-FITC, and CD4-PECy5 antibodies using dilution of 1:100 in FACS buffer for 30 min on ice. Wash cells again with 1 mL of 1× PBS and resuspend in 100 μL of FACS fixing buffer. Analyze samples (in fixing buffer) on a flow cytometer. If there is specific binding, expression of CD3 and CD4 should remain unchanged, whereas CD7 expression should be completely blocked (Fig. 3a).

3.3. Conjugation of scFvCD7-Cys to 9-D-Arginine

1. Add 4 mL of 0.1 M phosphate buffer (pH 5.5) to 1 mL of scFvCD7-Cys (1 mg/mL stock).
2. Add Cys(Npys)-(D-Arg)9 peptide (1 mg/mL) dropwise to the scFvCD7-Cys solution at a molar ratio of 10 to 1 and gently stir with a magnetic stirrer for 4 h at room temperature. A molar ratio of 10:1 corresponds to 1.68 mg of scFvCD7-Cys for each milligram of Cys(Npys)-(D-Arg)9 peptide.
3. Remove unconjugated Cys(Npys)-(D-Arg)9 peptide by dialysis overnight in 2 L of 1× PBS, pH 7.4 at 4°C with four buffer changes using Slide-A-Lyzer dialysis cassette.
4. Measure conjugation efficiency using a thiol and sulfide quantization assay (see Note 4).

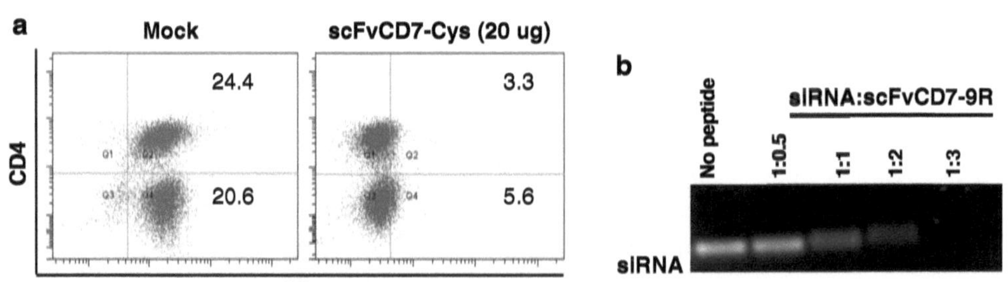

Fig. 3. (**a**) Binding of scFvCD7-Cys to CD7. Jurkat cells (1×10^5 cells) were incubated with scFvCD7-Cys (20 μg) or mock incubated for 2 h and stained with antibodies to CD4 and CD7. Flow cytometric analysis demonstrates CD7 expression on mock-treated cells which is markedly reduced on cells treated with scFVCD7-Cys. Levels of CD4 remain unchanged. (**b**) siRNA binding to scFvCD7-9dR. siRNA (100 pmol) were incubated with scFvCD7-9R at indicated molar ratios for 20 min and analyzed on 2% agarose gels. The position of the unbound siRNA is indicated. Decrease of fluorescence intensity at higher molar ratios is due to quenching of the fluorescence with binding of siRNA to the carrier, indicating retention of the complexed siRNA in the well.

3.4. Formation of scFvCD7-9dR/siRNA Complex

1. Dilute siRNA (200 pmol) in 50 μL serum-free RPMI medium and mix gently. Dilute scFvCD7-9dR in 50 μL serum-free RPMI medium at various molar ratios (for example, a molar ratio of 1:10 corresponds to 1.3 μg of siRNA to 30 μg of scFvCD7-9R). Mix gently and incubate for 5 min at room temperature (see Notes 5 and 6).
2. Add siRNA dropwise to the scFvCD7-9dR solution. Mix gently and incubate for 15–20 min to allow complex formation at room temperature.
3. To determine optimal binding ratio, analyze complexes of siRNA and scFvCD7-9dR at various molar ratios by 2% agarose gel electrophoresis with uncomplexed siRNA as negative control (Fig. 3b). With increasing amounts of scFvCD7-9R more siRNA-protein complexes form as a result of which the complexed siRNA runs slower than the uncomplexed siRNA. The molar ratio at which siRNA is optimally retarded in the gel or undetectable (due to quenching of the nucleic acid-binding dye) would be used for in vitro and in vivo experiments.

3.5. Preparation of LFA-1 I-tsNPs

1. Prepare multilamellar vesicles (MLV) composed of soybean PC (phosphatidylcholine), DPPE (1,2-dipalmitoyl-sn-glycero-3-phosphoethanolamine), and Chol (cholesterol) at a molar ratio of 3:1:1 using the lipid-film method (see Notes 7 and 8). Weigh the appropriate amounts of lipids to a final concentration of 40 mg/mL in a round-bottom flask.
2. Dissolve the lipids in the round-bottom flask with 20 mL of 96% ethanol (to obtain a 40 mg/mL final lipids concentrations) by stirring for 30 min at 65°C. Evaporate the ethanol using a rotary evaporator until complete dryness and look for the appearance of a thin film layer on the round-bottom flask. Upon completion of the dryness process, blow nitrogen into the round-bottle flask for 10 min (see Note 9).
3. Hydrate the lipid film with 20 mL of HBS (or PBS), pH 7.4 to generate MLV at 40 mg/mL concentration. Thoroughly vortex until a thin milky liposome suspension is formed and shake in a shaker incubator for 2 h at 37°C to generate MLV.
4. Extrude the resulting MLV into small unilamellar vesicles (SUV) using a Thermobarrel Lipex extruder with circulating water bath at 65°C under nitrogen pressures of 300–550 psi. Carry out the extrusion in a stepwise manner using progressively decreasing pore-sized membranes (from 1, 0.8, 0.6, 0.4, 0.2, to 0.1 μm), with 10 cycles per pore-size (see Note 10).
5. Surface-modify SUV with high molecular weight hyaluronan (HA) by dissolving 20 mg HA in 10 mL of 0.1 M sodium acetate buffer pH 4.5. Then, stir at 37°C for 30 min to fully dissolve the HA. Add 400 mg of EDAC, and stir for 2 h at 37°C.

6. Centrifuge the extruded SUV for 1 h in an ultracentrifuge at 640,000 × *g* (4°C) and resuspend pellet with 10 mL of 0.1 M borate buffer, pH 9.0.
7. Combine the activated HA with the SUV suspension in a 1:1 volume ratio and incubate overnight at 37°C, with gentle stirring. Separate the resulting HA-coated SUV from unbound HA by ultracentrifugation (640,000 × *g*, 4°C, for 1 h for each wash). Resuspend the pelleted HA-SUV in 20 mL of HBS (or PBS), pH 7.4. Repeat three cycles of this washing step. Resuspend with 20 mL of HBS (or PBS), pH 7.4.
8. mAb TS1/22 is coated to HA-coated SUV using an amine-coupling method. First, activate the HA-coated SUV by mixing 50 μL of the HA-SUV suspension with 200 μL of 400 mmol/L EDAC and 200 μL of 100 mmol/L NHS (in this order) and incubate for 20 min at room temperature with gentle stirring using a stir bar.
9. Mix the EDAC-NHS-activated HA-coated SUV with 50 μL of 10 mg/mL TS1/22 mAb in HBS (or PBS), pH 7.4 and incubate for 150 min at room temperature with gentle stirring. At the end of the incubation, add 20 μL of 1 M ethanolamine HCl (pH 8.5) to block the reactive residues and stir for additional 10 min at room temperature prior to purification.
10. Purify resulting I-tsNP and IgG-sNP by size exclusion column chromatography packed with sepharose CL-4B beads equilibrated with HBS (or PBS), pH 7.4 to remove unattached mAbs.
11. Lyophilize the purified I-tsNPs using lyophilizer and store vials (~200 μg lipid/vial of 200 μL volume) at −80°C until use.

3.6. siRNA Entrapment in Nanoparticles

1. Dissolve full-length human recombinant protamine in DEPC-treated water to a concentration of 5 mg/mL.
2. Add 16 μL of siRNA (100 pmol/μL) to 176 μL serum-free RPMI media and incubate for 5 min at room temperature.
3. Add 8 μL of protamine solution to siRNA at a molar ratio of 1:5 (siRNA:protamine) and incubate for 20 min at room temperature to form protamine-condensed siRNA complexes.
4. Rehydrate one vial of lyophilized I-tsNPs by adding 200 μL protamine-condensed siRNA solution.
5. Mix gently and incubate for 10 min at room temperature. This siRNA entrapment procedure should be performed immediately before use.

3.7. In Vitro Transfection of siRNA with scFvCD7-9R or LFA-1 I-tsNP

1. Seed peripheral blood mononuclear cells (PBMCs) or Jurkat T cells at 2×10^5 cells/well in 300 μL of serum-free RPMI medium two hours before transfection (see Notes 11 and 12) (schematic in Fig. 2).
2. For transfection with scFvCD7-9R, add 50 μL/well containing 200 pmol of siCD4 or any other test siRNA complexed to scFvCD7-9R at the predetermined optimal molar ratio at 37°C in a 5% CO_2 incubator for 4 h.
3. For TS1/22 I-tsNP entrapped siRNA, add 50 μL/well and culture at 37°C in a 5% CO_2 incubator for 4 h.
4. Include appropriate controls, such as cells with no treatment; cells with siRNAs alone; cells with negative control siRNA (e.g., silencer firefly Luciferase siRNA).
5. Centrifuge cells in a microcentrifuge at $2{,}300 \times g$ for 5 min, remove supernatants, resuspend the pellet in fresh RPMI medium containing 10% FBS, and further incubate at 37°C under 5% CO_2 for 48–72 h.
6. Harvest cells for flow cytometry to determine CD4 gene silencing at the protein level or silencing of other genes as appropriate. Use harvested cells also to extract total RNA by RNeasy mini kit for quantitative real-time PCR to assess relevant gene knockdown (see Note 13).

3.8. In Vivo Administration of siRNA

1. Mix siRNA (50 μg) with scFvCD7-9R at optimal molar ratio (see Subheading 3.4, step 3) in a total volume of 200 μL in d-glucose solution (5% final concentration) (schematic in Fig. 2).
2. Intravenously inject scFvCD7-9dR/siRNA into the lateral tail vein of mouse using 1-cc syringe with a 27-G needle (see Notes 14 and 15).
3. For in vivo nanoparticle administration, inject 200 μL of I-tsNPs/siRNA complex into the lateral tail vein of mouse using 1-cc syringe with 27-G needle attached (see Notes 14 and 15).
4. For HIV challenge experiments in mice, inject 100 μL of HIV_{BaL} (~30,000 $TCID_{50}$) intraperitoneally (see Note 16).
5. Bleed mice retro-orbitally at weekly intervals and collect blood into EDTA-treated tubes (see Note 17).
6. Separate plasma from peripheral blood mononuclear cells (PBMCs) by Ficoll-Paque isodensity gradient centrifugation. Use plasma to quantitate viral loads with the Amplicor HIV-1 Monitor Test.
7. Use PBMCs for monitoring CD4 T cell levels by staining with CD4 antibody in a volume of 100 μL for 30 min at 4°C. Wash cells in 1 mL of FACS buffer and pellet by centrifugation

in a microcentrifuge at 2,300 × *g* for 5 min at room temperature. Resuspend the pellet in 100 μL of fixing buffer and analyze by flow cytometry.

8. Terminate experiments by euthanizing animals (usually after 4–6 weeks) and harvest spleen, liver, and peripheral blood. Single cell suspensions are obtained by crushing the tissues and passing the cell suspension through a 40-μm filter. Cells are then layered on Ficoll-Paque and centrifuged at 800 × *g* for 30 min. Mononuclear cells obtained from the interface are washed twice with 1× PBS. Select specific human immune subsets by immunomagnetic separation with cell-specific magnetic beads: CD3 for T cells, CD19 for B cells, CD14 for monocytes, CD11b for macrophages, and CD11c for dendritic cells. Use purified cells (>93%) for total RNA extraction for gene knockdown studies.

4. Notes

1. Antisense strand sequence of siRNA to host genes are siCD4: 5′-GAUCAAGAGACUCCUCAGUdTdT-3′; and siCCR5: 5′-GUCAGUAUCAAUUCUGGAAdTdT-3′. Antisense strand sequences of siRNA to viral genes are siVif: 5′-GUU CAGA AGUACACAUCCCdTdT-3′; siTat: 5′-UAUGGCAGGAAG AAGCGGAdTdT-3′ Antisense strand sequence of siRNA to an irrelevant *luciferase* gene is siLuc: 5′-UCGAAGUAC UCAGCGUAAGdTdT-3′. The use of 3′ terminal dTdT overhang is preferred as it makes the siRNA duplex more resistant to exonuclease activity (13).
2. Humanized mice can be generated in the laboratory using NOD.cg-*PrkdcscidIL2rgtm/Wjl/Sz* (NOD/SCIDIL2r$\gamma^{-/-}$) mice (the Jackson Laboratory, Bar Harbor, ME) or similar immunodeficient mice strains. Three different types of human cell-engrafted mice are currently available for studying HIV infection. (1) Hu-PBL mice (immunodeficient mice transplanted with human peripheral blood leukocytes) can be generated by intraperitoneally injecting 10^7 PBMCs (per mouse) freshly isolated from HIV-seronegative donors into 4–6-week-old mice after sublethal irradiation (2–3 Gy) (14), (2) Hu-HSC mice (Immunodeficient mice transplanted with human hematopoietic stem cells) can be generated by intravenously injecting T cell-depleted $CD34^+$ cord blood cells (3×10^4 cells/mouse) into 1–2-day-old neonatal mice after irradiation (1 Gy) (15). (3) BLT mice (Bone Marrow Liver Thymic mice) can be generated by implanting human fetal thymus and liver tissues after conditioning the mice with

sublethal (2–3 Gy) whole-body irradiation (16) followed by intravenous injection of hematopoeitic stem cells ($CD34^+$) ($2–5 \times 10^5$ per mouse) isolated from fetal liver of the same donor.

3. Cys residue at C-terminal end of scFvCD7 allows conjugation to a nona-d-arginine (9dR) peptide through coupling with sulfhydryl groups. Maleimide chemistry can also be used to couple a free sulfhydryl group of a cysteine residue.
4. The Thiol and Sulfide Quantitation Kit provides an ultrasensitive colorimetric assay for quantitating protein thiols. In this assay, thiols or inorganic sulfides reduce a disulfide-inhibited derivative of papain to stoichiometrically release the active enzyme. The activity of the enzyme is then measured using the chromogenic papain substrate *N*-benzoyl-l-arginine, *p*-nitroanilide (L-BAPNA). The thiol (or sulfide) concentration of experimental samples is determined against a standard curve generated by plotting absorbance values of l-cysteine standards at 410 nm.
5. All in vitro and in vivo procedures must use RNase-free techniques and supplies (e.g., tubes, tips, and reagents).
6. Do not vortex the complex. Mix gently with tapping.
7. The lipid-film method is a traditional method of lipid formulation that involves dissolving lipids in a volatile solvent, evaporating the solvent to obtain a thin film, and hydrating the dried thin film with an aqueous vehicle to obtain the crude liposome formulation.
8. Multilamellar vesicles (MLVs) are usually formed by mechanical dispersion of dried lipid in an aqueous buffer. MLV is composed of multiple lipid bilayers. Small unilamellar vesicles (SUVs) comprise of an internal aqueous space enclosed by a single lipid bilayer. SUVs are much smaller structure and are more homogenous in size than MLVs.
9. MLV preparation: the water bath of the rotary evaporator is set at 65°C and the chiller is set at least at −10°C. These settings are important for efficient evaporation. Following evaporation of ethanol, pass nitrogen or argon gas for 20 min to completely remove traces of ethanol and prevent oxidation of lipids.
10. The MLV suspension before the extrusion step can be stored at 4°C. Make sure that the MLV is warmed to 37°C immediately before extrusion.
11. All media and solutions for cell culture should to be warmed to 37°C in a water bath.
12. It is very important to keep the cell density from being too low or too high during transfection. Optimal cell density

should be determined experimentally. Cells should be homogeneously distributed in each well.

13. siRNAs exert their effects at the mRNA level. Therefore, to validate gene knockdown, measuring target mRNA levels using quantitative real-time PCR method is preferred. Knockdown can also be verified by measuring corresponding protein levels by enzymatic or cell-based assay.
14. Mix siRNA with delivery reagent just before injection. Once siRNA-liposome or siRNA-cationic peptide complexes are mixed, do not freeze or store the complexed solution.
15. The maximum injection volume is 10% of body weight.
16. HIV infected humanized mice should be handled with proper personal protective equipment at animal biosafety level 3 (ABSL3), in accordance with governmental and institutional biosafety guidelines.
17. Do not use heparinized tubes for quantitation of viral load using AMPLICOR HIV-1 MONITOR Test, v1.5 (Roche Diagnostics Corporation, IN).

References

1. Rossi, J. J., June, C. H., and Kohn, D. B. (2007) Genetic therapies against HIV, *Nat Biotechnol* *25*, 1444–1454.
2. Hammond, S. M., Caudy, A. A., and Hannon, G. J. (2001) Post-transcriptional gene silencing by double-stranded RNA, *Nat Rev Genet* *2*, 110–119.
3. Dykxhoorn, D. M., and Lieberman, J. (2006) Running interference: prospects and obstacles to using small interfering RNAs as small molecule drugs, *Annu Rev Biomed Eng* *8*, 377–402.
4. Kim, D. H., and Rossi, J. J. (2007) Strategies for silencing human disease using RNA interference, *Nat Rev Genet* *8*, 173–184.
5. Kim, S. S., Garg, H., Joshi, A., and Manjunath, N. (2009) Strategies for targeted nonviral delivery of siRNAs in vivo, *Trends Mol Med* *15*, 491–500.
6. Song, E., Zhu, P., Lee, S. K., Chowdhury, D., Kussman, S., Dykxhoorn, D. M., Feng, Y., Palliser, D., Weiner, D. B., Shankar, P., Marasco, W. A., and Lieberman, J. (2005) Antibody mediated in vivo delivery of small interfering RNAs via cell-surface receptors, *Nat Biotechnol* *23*, 709–717.
7. Kumar, P., Ban, H. S., Kim, S. S., Wu, H., Pearson, T., Greiner, D. L., Laouar, A., Yao, J., Haridas, V., Habiro, K., Yang, Y. G., Jeong, J. H., Lee, K. Y., Kim, Y. H., Kim, S. W., Peipp, M., Fey, G. H., Manjunath, N., Shultz, L. D., Lee, S. K., and Shankar, P. (2008) T cell-specific siRNA delivery suppresses HIV-1 infection in humanized mice, *Cell* ***134***, 577–586.
8. Peer, D., Zhu, P., Carman, C. V., Lieberman, J., and Shimaoka, M. (2007) Selective gene silencing in activated leukocytes by targeting siRNAs to the integrin lymphocyte function-associated antigen-1, *Proc Natl Acad Sci U S A* ***104***, 4095–4100.
9. Peer, D., Park, E. J., Morishita, Y., Carman, C. V., and Shimaoka, M. (2008) Systemic leukocyte-directed siRNA delivery revealing cyclin D1 as an anti-inflammatory target, *Science* *319*, 627–630.
10. Subramanya, S., Kim, S. S., Abraham, S., Yao, J., Kumar, M., Kumar, P., Haridas, V., Lee, S. K., Shultz, L. D., Greiner, D., N, M., and Shankar, P. (2010) Targeted delivery of small interfering RNA to human dendritic cells to suppress dengue virus infection and associated proinflammatory cytokine production, *J Virol* ***84***, 2490–2501.
11. Kim, S. S., Peer, D., Kumar, P., Subramanya, S., Wu, H., Asthana, D., Habiro, K., Yang, Y. G., Manjunath, N., Shimaoka, M., and Shankar, P. (2010) RNAi-mediated CCR5 silencing by LFA-1-targeted nanoparticles prevents HIV infection in BLT mice, *Mol Ther* ***18***, 370–376.

12. Peipp, M., Kupers, H., Saul, D., Schlierf, B., Greil, J., Zunino, S. J., Gramatzki, M., and Fey, G. H. (2002) A recombinant CD7-specific single-chain immunotoxin is a potent inducer of apoptosis in acute leukemic T cells, *Cancer Res* ***62***, 2848–2855.
13. Elbashir, S. M., Harborth, J., Lendeckel, W., Yalcin, A., Weber, K., and Tuschl, T. (2001) Duplexes of 21-nucleotide RNAs mediate RNA interference in cultured mammalian cells, *Nature* ***411***, 494–498.
14. Nakata, H., Maeda, K., Miyakawa, T., Shibayama, S., Matsuo, M., Takaoka, Y., Ito, M., Koyanagi, Y., and Mitsuya, H. (2005) Potent anti-R5 human immunodeficiency virus type 1 effects of a CCR5 antagonist, AK602/ONO4128/GW873140, in a novel human peripheral blood mononuclear cell nonobese diabetic-SCID, interleukin-2 receptor gamma-chain-knocked-out AIDS mouse model, *J Virol* ***79***, 2087–2096.
15. Ishikawa, F., Yasukawa, M., Lyons, B., Yoshida, S., Miyamoto, T., Yoshimoto, G., Watanabe, T., Akashi, K., Shultz, L. D., and Harada, M. (2005) Development of functional human blood and immune systems in NOD/SCID/IL2 receptor {gamma} chain(null) mice, *Blood* ***106***, 1565–1573.
16. Lan, P., Tonomura, N., Shimizu, A., Wang, S., and Yang, Y. G. (2006) Reconstitution of a functional human immune system in immunodeficient mice through combined human fetal thymus/liver and CD34+ cell transplantation, *Blood* ***108***, 487–492.

Chapter 22

Aptamer-Targeted RNAi for HIV-1 Therapy

Jiehua Zhou and John J. Rossi

Abstract

The highly specific mechanism of RNA (RNAi) that inhibits the expression of disease genes is increasingly being harnessed to develop a new class of therapeutics for a wide variety of human maladies. The successful use of small interfering RNAs (siRNAs) for therapeutic purposes requires safe and efficient delivery to specific cells and tissues. Herein, we demonstrate novel cell type-specific dual inhibitory function anti-gp120 aptamer-siRNA delivery systems for HIV-1 therapy, in which both the aptamer and the siRNA portions have potent anti-HIV activities. The envelope glycoprotein is expressed on the surface of HIV-1 infected cells, allowing binding and internalization of the aptamer-siRNA chimeric molecules. The Dicer substrate siRNA delivered by the aptamers is functionally processed by Dicer, resulting in specific inhibition of HIV-1 replication and infectivity in cultured CEM T-cells and primary blood mononuclear cells. Our results provide a set of novel aptamer-targeted RNAi therapeutics to combat HIV and further validate the use of anti-gp120 aptamers for delivery of Dicer substrate siRNAs.

Key words: RNA interference, Anti-gp120 aptamer, Cell type-specific delivery system, Aptamer-targeted RNAi

1. Introduction

RNA interference (RNAi) is a powerful endogenous process triggered by small interfering RNA (siRNA) duplexes 21–25 nt in length, which result in sequence-specific posttranscriptional gene silencing (1, 2). RNAi is rapidly becoming one of the methods of choice for gene function studies, and is also being exploited for therapeutic applications (3–5). Since the first in vivo evidence of RNAi-based therapeutic efficacy in a mouse disease model shown in 2003 (6), several promising siRNAs drugs against human diseases, such as aged-related macular degeneration (AMD) and respiratory syncytial virus (RSV) have already been evaluated in clinical trials with encouraging safety profiles and efficacy (5). Also along this line, novel RNAi-based antiviral therapeutics present

Ronald P. van Rij (ed.), *Antiviral RNAi: Concepts, Methods, and Applications*, Methods in Molecular Biology, vol. 721,
DOI 10.1007/978-1-61779-037-9_22, © Springer Science+Business Media, LLC 2011

considerable promise for treatment of human immunodeficiency virus (HIV) infection as shown in many previous studies (7–9). These may reduce drug resistance and toxicity risks associated with long-term HAART treatment (10). However, effective delivery of siRNA therapeutics into virus susceptible and/or infected cells has been challenging. In this regard, a targeted delivery of siRNA to the relevant target cells or tissues, hence promoting cellular uptake and reducing off-target effects or other unwanted side effects, is highly desirable for RNAi-based antiviral therapeutic safety and efficacy.

HIV-1 infection is initiated by the interactions between the external envelope glycoprotein gp120 of HIV and the human cell surface receptor CD4, subsequently leading to fusion of the viral membrane with the target cell membrane (11–13). We have previously demonstrated that the gp120 expressed on the surface of HIV infected cells can be used as a target for aptamer-mediated delivery of anti-HIV siRNAs and several new 2′-fluorine-substituted aptamers that bind to the HIV-$1_{\text{Ba-L}}$ gp120 protein with low nanomolar affinity have been isolated from a RNA library by the SELEX procedure (systematic evolution of ligands by exponential enrichment) (14, 15). In the present chapter, we describe a series of new dual inhibitory function anti-gp120 aptamer-siRNA chimeras in which both the aptamer and the siRNA portions have potent anti-HIV activities. The flow cytometric analysis and confocal microscopy confirmed that aptamers and aptamer-siRNA chimeras specifically bind to and are internalized into cells expressing HIV gp160. The aptamer-targeted RNAi efficacy on HIV-1 suppression was further validated by HIV p24 ELISA and quantitative real-time PCR. The Dicer substrate siRNA delivered by the anti-gp120 aptamers is functionally processed by Dicer, resulting in specific inhibition of HIV-1 replication and infectivity in cultured CEM T-cells and primary blood mononuclear cells. Importantly, we have introduced a "sticky" sequence onto a chemically synthesized aptamer which facilitates attachment of the Dicer substrate siRNAs for potential multiplexing of different siRNAs with the same aptamer species. Therefore, this aptamer-targeted RNAi strategy provides a new paradigm for delivery of anti-HIV siRNAs by allowing selective delivery to HIV infected cells and dual function inhibition of HIV replication and spread, showing considerable promise for systemic anti-HIV therapy.

2. Materials

2.1 Generation of Aptamer and Chimera RNAs by In Vitro Transcription

1. The single-stranded DNA template, relevant primer sets for PCR, siRNAs. and antisense strand RNAs (Integrated DNA Technologies, Coralville, IA). Make stock solution with water and store in aliquots at −20°C (see Note 1).

2. Taq PCR polymerase and buffer (New England BioLabs). Store in aliquots at –20°C.
3. dNTP (Roche). Make 10 mM dNTP mixture and store in aliquots at –20°C.
4. QIAquick Gel purification Kit (QIAGEN).
5. DuraScribe T7 transcription Kit (EPICENTRE Biotechnologies). The kit includes the 2′-fluorine-2′-deoxyribonucleoside-5′-triphosphates, 2′-fluorine-CTP (2′-F-dCTP), and 2′-fluorine-UTP (2′-F-dUTP). Store at –20°C.
6. Bio-Spin 30 Columns (Bio-Rad). Store at 4°C.
7. Acid phenol:chloroform (5:1) solution (pH 4.5, Ambion). Phenol and chloroform are human health hazards. Take appropriate measures to prevent exposure.
8. Chloroform:isopropanol (24:1) solution (Sigma).
9. 20 mg/mL Glycogen (Roche). Store in aliquots at –20°C.

2.2. Generation of Aptamer-Stick-siRNA Chimeras

1. Aptamer-stick, stick-sense, and stick-antisense RNA are chemically synthesized and purified with HPLC by the Synthetic and Biopolymer Chemistry Core in the City of Hope (see Note 2). Store in aliquots at –20°C.
2. HEPES buffer: 100 mM Hepes, pH 7.4. Use NaOH adjust pH value. Store at room temperature.
3. Refold buffer (5× HBS): 50 mM Hepes, pH 7.4, 750 mM NaCl, 5 mM $MgCl_2$, 5 mM $CaCl_2$, 13.5 mM KCl. Store in aliquots at –20°C.
4. 1× HBS buffer: 10 mM Hepes, pH 7.4, 150 mM NaCl, 1 mM $MgCl_2$, 1 mM $CaCl_2$, 2.7 mM KCl. Store in aliquots at –20°C.

2.3. Gel Shift Assays and Determination of Dissociation Constants

1. Forty percent acrylamide/bis-acrylamide solution (AccuGel 19:1, National Diagnostics). This is a neurotoxin when unpolymerized. Take appropriate measures to prevent exposure.
2. *N,N,N,N′*-Tetramethyl-ethylenediamine (TEMED, Sigma). Store at 4°C.
3. Ammonium persulfate (APS) (Sigma): prepare 10% solution in water and immediately freeze in single use (150 μL) aliquots at –20°C.
4. 10× TBE (National Diagnostics).
5. Running buffer (1× TBE): dilute 10× TBE with water. Store at room temperature.
6. Calf intestinal phosphatase (CIP), T4 polynucleotide kinase, and their buffers (New England BioLabs). Store at –20°C.
7. Gamma-^{32}P-ATP (MP Biomedical). This is a radioactive isotope. Take appropriate measures to prevent exposure. Store at –20°C.

8. HIV-1_{Ba-L} gp120 protein (AIDS Research and Reference Reagent Program). Store in aliquots at –80°C.
9. Native loading buffer (4×): 10 mM Tris–HCl, pH 7.5; 1 mM EDTA, 0.1% Bromophenol Blue, 0.1% Xylene Cyanol FF, 0.1% Orange G, 40% Glycerol. Store in aliquots at –20°C.
10. Binding buffer: 10 mM Hepes, pH 7.4, 150 mM NaCl, 1 mM $MgCl_2$, 1 mM $CaCl_2$, 2.7 mM KCl, 0.01% BSA, 10 mM DTT (see Note 3). Store in aliquots at –20°C.

2.4. Cell-Surface Binding Studies by Flow Cytometry

1. CHO-WT gp160 and CHO-EE control cells (AIDS Research and Reference Reagent Program). Culture in GMEM-S medium (glutamine-deficient minimal essential medium with 400 μM methionine sulfoximine (MSX)) (Gibco, Invitrogen) in a humidified 5% CO_2 incubator at 37°C (see Note 4).
2. Silencer siRNA Labeling Kit (Ambion). Store at –20°C.
3. Trypsin/EDTA 1× solution (Irvine Scientific, CA). Store in aliquots at 4°C.
4. Washing buffer: Dulbecco's phosphate buffered saline (DPBS, Cellgro, Mediatech). Store in aliquots at 4°C.
5. Binding buffer: 10 mM Hepes, pH 7.4, 150 mM NaCl, 1 mM $MgCl_2$, 1 mM $CaCl_2$, 2.7 mM KCl, 0.01% BSA. Store in aliquots at –20°C.

2.5. Live-Cell Confocal Microscopy

1. 35 mm Plates for live-cell confocal microscopy (Glass Bottom Dish, MatTek, MA).
2. Hoechst 33342 (Molecular Probes, Invitrogen, CA): prepare 0.15 mg/mL solution in water and store in aliquots at 4°C.
3. Zeiss LSM 510 Meta Inverted 2 photon confocal microscope system.

2.6. HIV-1 Challenge and p24 Antigen Assay

1. CCR5-CEM cells (ATCC). Culture in RPMI-1640 (Cellgro, Mediatech Inc.) supplemented with 10% fetal bovine serum (FBS, HyClone), L-glutamine and 1× Pen-Strep (Gibco, Invitrogen), in a humidified 5% CO_2 incubator at 37°C.
2. Peripheral blood mononuclear cells (PBMCs) from healthy volunteers. Isolate PBMCs from whole blood by centrifugation through Ficoll-Hypaque solution (Histopaque-1077, Sigma) and deplete from CD8 T-cells (cytotoxic/suppressor cells) using Dynabeads CD8 (Invitrogen, CA). Culture PBMCs in T-cell active medium (BioE, St. Paul, MN) in a humidified 5% CO_2 incubator at 37°C.
3. HIV-1 IIIB, NL4-3, JR-FL, and Bal virus (AIDS Research and Reference Reagent Program). Store in aliquots at –80°C (see Note 5).
4. HIV-1 p24 Antigen ELISA kit (Beckman Coulter, CA). Store at 4°C.

2.7. Quantitative RT-PCR Assay

1. The relevant PCR primer (Integrated DNA Technologies, Coralville, IA). Prepare 5 mM stock solution in water and store in aliquots at –20°C.
2. STAT-60 reagent (TEL-TEST, Friendswood, TX). This reagent containing phenol and guanidine is a human health hazard. Take appropriate measures to prevent exposure. Store in aliquots at 4°C.
3. DNA-free kit (Ambion, CA). Store at –20°C.
4. RNase inhibitor (Promega). Store at –20°C.
5. Moloney murine leukemia virus reverse transcriptase (MML-RT) and random primers (Invitrogen) Store at –20°C.
6. 2× iQ SyberGreen MasterMix (BIO-RAD) for PCR. Store at –20°C.

3. Methods

Several 2′-Fluoro substituted aptamers that specifically bind to the HIV-$1_{Ba\text{-}L}$ gp120 protein with nanomolar affinity have been successfully isolated from an 81 nt RNA library via a nitrocellulose membrane-based SELEX. From two of them, two different types of dual inhibitory function anti-gp120 aptamer-siRNA conjugates were constructed and evaluated for HIV-1 inhibition. One of these is a covalent aptamer-siRNA chimera (Subheading 3.1) and the other is an aptamer with a GC rich bridge that facilitates the interchange of different RNAs with the same aptamer (Subheading 3.2). First, the binding affinities of these conjugates for gp120 was assessed by using a gel shift assay (Subheading 3.3) and flow cytometry (Subheading 3.4), indicating that the chimeras maintain approximately the same binding affinites as the aptamers alone. The time-course images collected via confocal microscopy showed that Cy3-labeled aptamer-siRNA conjugates can be successfully internalized into the cytoplasm of cells expressing gp160 (Subheading 3.5). Furthermore, HIV-1 challenge assays were performed to verify the anti-HIV activity of the chimeras in inhibiting HIV-1 replication in cell cultures (Subheading 3.6). Meanwhile, to confirm that the siRNA component was functioning along with the aptamer, following internalization of the chimeras in infected cells, the relative levels of inhibition of target gene expression were evaluated by quantitative RT-PCR expression assays (Subheading 3.7).

3.1. Generation of Aptamer and Chimera RNAs by In Vitro Transcription

1. Directly generate double-stranded DNA template through PCR (see Note 6) using 2 μM each of 5′- and 3′-primers, along with 2 mM $MgCl_2$ and 200 μM of each dNTP. Recover the resulting PCR products using a QIAquick Gel purification Kit.

2. Transcribe chimera sense strand from its PCR generated DNA templates using the DuraScription Kit. In the transcription reaction mixture, replace CTP and UTP with 2′-F-CTP and 2′-F-UTP to produce RNA that is resistant to RNase A degradation. Set up a 20 μL transcription reaction, by mixing:
   (a) 1 μg of DNA template.
   (b) 2 μL 10× buffer.
   (c) 2 μL 50 mM ATP.
   (d) 2 μL 50 mM GTP.
   (e) 2 μL 50 mM 2′-Fl-dCTP.
   (f) 2 μL 50 mM 2′-Fl-dUTP.
   (g) 2 μL 100 mM DTT.
   (h) 2 μL T7 RNA polymerase.

   Incubate at 37°C for 6 h (see Note 7), and purify with Bio-Spin 30 Columns following phenol extraction and ethanol precipitation.
3. In order to avoid interferon response, further treat transcribed RNA with CIP to remove the initiating 5′-triphosphate. Set up a total volume of 60 μL, by mixing:
   (a) 3 μg of transcripts.
   (b) 6 μL Buffer 3.
   (c) 0.25 μL CIP.
   (d) Water to a final volume of 60 μL.

   Incubate at 37°C for 60 min. After phenol/chloroform exaction and ethanol precipitation, resuspend RNA pellet into water.
4. To prepare the chimeras, combine the chimeras harboring only the sense strand RNA with the appropriate antisense RNA in refolding buffer (see Note 8), heat mixture at 95°C for 3 min and then slowly cool to 37°C. Continue incubation at 37°C for 10 min. (A schematic anti-HIV gp120 aptamer-siRNA chimera is shown in Fig. 1a.)

3.2. Generation of Aptamer-Stick-siRNA Chimeras

1. Refold the aptamer-stick RNA in refolding buffer as described above (see Subheading 3.1, step 5).
2. Anneal the Sense-stick or Antisense-stick strand with its complementary partner using the same molar amounts as the corresponding partner strand to form the stick-siRNAs, heat mixture at 95°C for 3 min and then slowly cool to 37°C.
3. Incubate the same amount of the refolded aptamer-stick with the stick-siRNA at 37°C for 10 min in 1× HBS buffer to form the aptamer-stick-siRNAs. (A schematic anti-HIV gp120 aptamer-stick-siRNA conjugate is shown in Fig. 1b.)

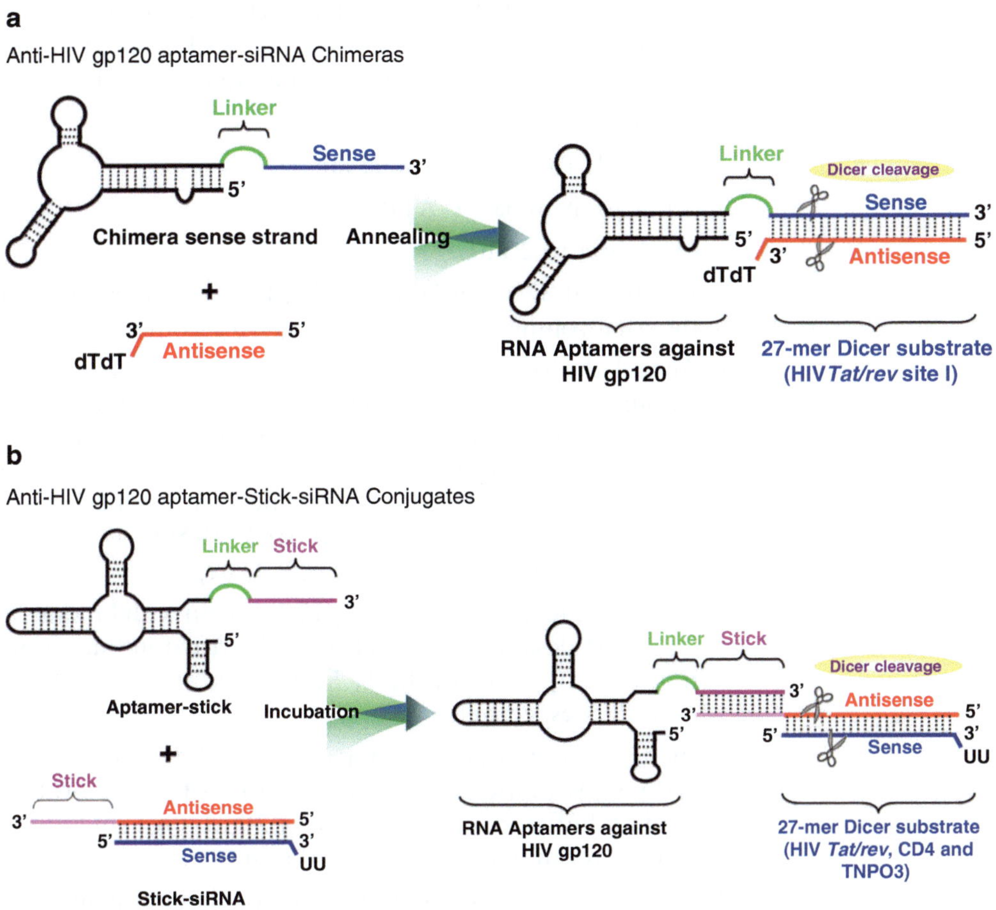

Fig. 1. Aptamer-targeted RNA interference. (**a**) Schematic anti-HIV gp120 aptamer-siRNA chimeras. The 2′-Fluoro modified chimera sense single strand was transcribed, followed by annealing of the complementary siRNA antisense strand to complete the chimeric molecule. The aptamer region is responsible for binding to gp120; the 27-mer Dicer substrate RNA duplex targets a common exon of HIV-1 *tat/rev*. A linker (UU) is introduced between the aptamer and siRNA. After Dicer processing, the resultant 21-mer siRNAs are incorporated into RISC to enter the RNAi pathway. (**b**) Schematic anti-HIV gp120 aptamer-stick-siRNA conjugates. The anti-gp120 aptamer and the 27mer siRNA targeting HIV-1 *tat/rev*, CD4, or TNPO3 are shown. Either the antisense or sense strand of 27-mer Dicer substrate RNA duplex is linked to the aptamer portion via the "stick" sequence, which consists of 16 nt appended to the aptamer 3′-end allowing complementary interaction of one of the two siRNA strands with the aptamer. After a simple incubation, they form stable base-pairing. A linker of seven three-carbons (chemical bond $CH_2CH_2CH_2$ between the aptamer RNA and the stick portion is used to avoid steric interaction of the stick with the aptamer.

3.3. Gel Shift Assays and Determination of Dissociation Constants

1. Prepare a 25 mL of 5% nondenaturing polyacrylamide gel by mixing 2.5 mL of 10× TBE buffer, 3.125 mL of 40% acrylamide/bis solution, 19.375 mL of water, 150 μL of 10% APS solution, and 30 μL of TEMED. The gel should polymerize in about 30 min. Carefully remove the comb and use a 30-mL syringe fitted with a needle to wash the wells with running buffer (1× TBE).
2. Complete the assembly of the gel unit and connect to a power supply. Preelectrophorese the gel for 1 h at 180 V at 4°C.

3. 5′-End-P^{32}-labeled labeling: Heat 10 pmol of CIP treated chimera sense strand RNA or aptamer-stick RNA at 95°C for 5 min and then chill on the ice. Subsequently, add:
   (a) 2 μL PNK buffer.
   (b) 1 μL T4 polynucleotide kinase.
   (c) 1 μL gamma-^{32}P-ATP.
   (d) Water to 20 μL.

   Incubate at 37°C for 30 min, add 20 μL of water , and purify the total volume (40 μL) using a G-50 column.
4. Anneal the corresponding antisense strand with equimolar amounts of 5′-end-labeled chimera sense strand in refolding buffer to form the aptamer-siRNA chimeras as described above (see Subheading 3.1, step 5). Similarly, incubate the refolded 5′-end-labeled aptamer-stick with annealed stick-siRNA to prepare the aptamer-stick-siRNA conjugates as described above (see Subheading 3.2, step 3).
5. Serially dilute the HIV-1 $_{\text{Ba-L}}$ gp120 protein with binding buffer to the desired concentrations. The reaction final concentrations of gp120 are 0, 1, 5, 10, 20, 40, 80, 160, 320, 640 nM, respectively.
6. Incubate a constant amount of 5′-^{32}P-end-labeled RNA (10 nM) with increasing concentrations of gp120 protein in binding buffer in a total volume of 20 μL on a rotating platform at room temperature for 30 min.
7. After incubation, add 5 μL of native loading buffer into 20 μL of binding reaction and run in a 5% nondenaturing polyacrylamide gel at 180 V at 4°C for 2 h, until the second dye (Bromophenol Blue) runs at middle of gel.
8. Wrap gel with sealwrap and directly expose the gel to a Phosphor image screen and quantify the radioactive signal using a Typhoon scanner.
9. Calculate the dissociation constants using nonlinear curve regression with a Graph Pad Prism. An example result is shown in Fig. 2.

3.4. Cell-Surface Binding Studies by Flow Cytometry

1. Generate fluorescent aptamer and chimeras using the Silencer siRNA Labeling Kit (see Note 9). Add the following reagents in order:
   (a) 22.5 μL Nuclease-free water.
   (b) 5 μL 10× Labeling buffer.
   (c) 15 μL RNA (5 μg).
   (d) 7.5 μL Labeling dye.

   Incubate at 37°C for 1 h.

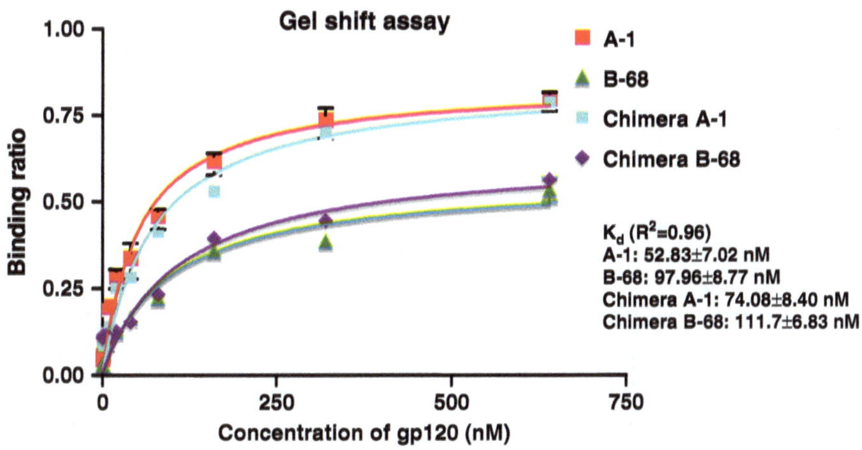

Fig. 2. Binding activity detection and binding curve from a gel shift assay. The 5′-end P32 labeled aptamers (A-1 and B-68) or chimeras (Ch A-1 and Ch B-68) were incubated with the increasing amounts of gp120 protein. The binding reaction mixtures were analyzed by a gel mobility shift assay. Both aptamers and aptamer-siRNA chimera specifically bind the $HIV_{Ba\text{-}L}$ gp120 protein. The aptamer-siRNA chimeric RNAs and the corresponding parental aptamers have comparable K_d values. Data represent the average of three replicates.

2. After incubation, add 5.0 μL (0.1 vol) 5 M NaCl and 125 μL (2.5 vol) cold 100% EtOH, and mix thoroughly. Incubate at −20°C for 60 min. And then centrifuge at top speed in a microcentrifuge (12,000×*g*) at 4°C for 20 min. Remove supernatant and wash pellet with 175 μL of 70% EtOH. Air dry pellet in the dark. And then suspend labeled RNA in 15 μL of nuclease-free water.
3. Measure the absorbance of the labeled RNA at 260 nm and at the absorbance maximum for the fluorescent dye (Cy3 at 550 nm).
4. Calculate the base:dye ratio and RNA concentration according to the calculator provided by http://www.ambion.com/techlib/append/base_dye.html (see Note 10).
5. Combine an equal amount of Cy3-labeled chimeras sense stand and antisense strand and refold the mixture in refolding buffer as described above (see Subheading 3.1, step 4). For the aptamer-stick-siRNA, incubate the refolded Cy3-labeled aptamer-stick with the siRNA-stick to form the aptamer-stick-siRNA conjugated as described above (see Subheading 3.2, step 3).
6. Wash the CHO-WT gp160 or CHO-EE control cells with prewarmed washing buffer, trypsinize, and detach cells from the plates (see Note 11).
7. After washing cells twice with 500 μL binding buffer, resuspend cell pellets in binding buffer and incubate at 37°C

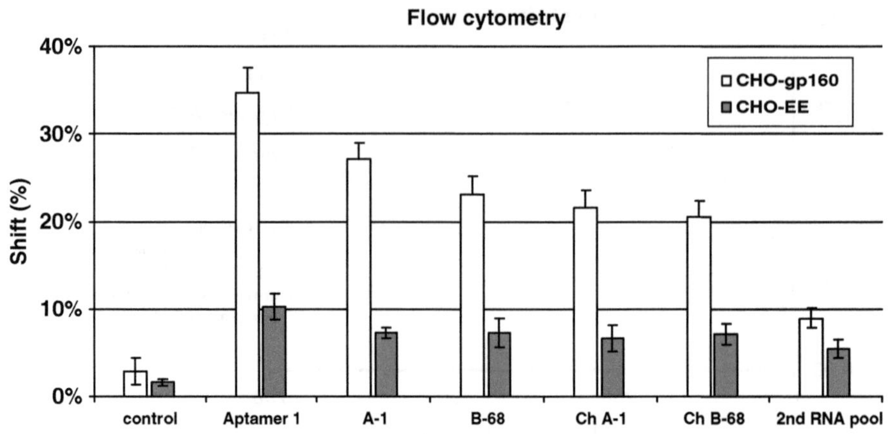

Fig. 3. Cell-type specific binding studies. Cy3-labeled experimental RNAs were tested for binding to CHO-WT gp160 cells and CHO-EE control cells. Cell surface binding of Cy3-labeled RNAs was assessed by flow cytometry. The aptamer and chimeras showed cell-type specific binding affinity. The second RNA pool and irrelevant RNA were used as negative controls. Data represent the average of two replicates.

for 30 min and then pellet cells and resuspend in 50 μL of prewarmed binding buffer containing 400 nM Cy3-labeled experimental RNAs (see Note 12).

8. After incubation at 37°C for 40 min, wash cells three times with 500 μL of prewarmed binding buffer, and resuspend cells in 350 μL of binding buffer prewarmed to 37°C and immediately analyze by flow cytometry. An example result is shown in Fig. 3.

3.5. Internalization and Intracellular Localization Studies by Live-Cell Real-Time Confocal Microscopy

1. One day before the assay, grow CHO-WT gp160 and CHO-EE control cells in 35-mm plate with seeding at 0.3×10^6 in 2 mL of GMEM-S medium to allow about 70% confluence in 24 h.
2. On the day of the experiments, wash cells with 1 mL of prewarmed PBS and then incubate with 1 mL of prewarmed complete growth medium for 30 min at 37°C.
3. Prepare Cy3-labeled aptamer-siRNA chimeras as described in Subheading 3.4, step 5.
4. Incubate the refolded aptamer-stick with the siRNA-stick containing 5′-Cy3-labeled sense strand to form the aptamer-stick-siRNA conjugated as described in Subheading 3.2, step 3.
5. For aptamer-siRNA chimera: Stain the cells with 0.15 mg/mL Hoechst 33342 (nuclear dye for live cells) for 15 min at 37°C. Immediately, wash out dye with 1.0 mL fresh medium twice and replace 2 mL prewarmed fresh medium. Subsequently, add the Cy3-labeled aptamer-siRNA chimera at a 100 nM final concentration into the media and incubate for live-cell confocal microscopy in a 5% CO_2 microscopy incubator at 37°C.

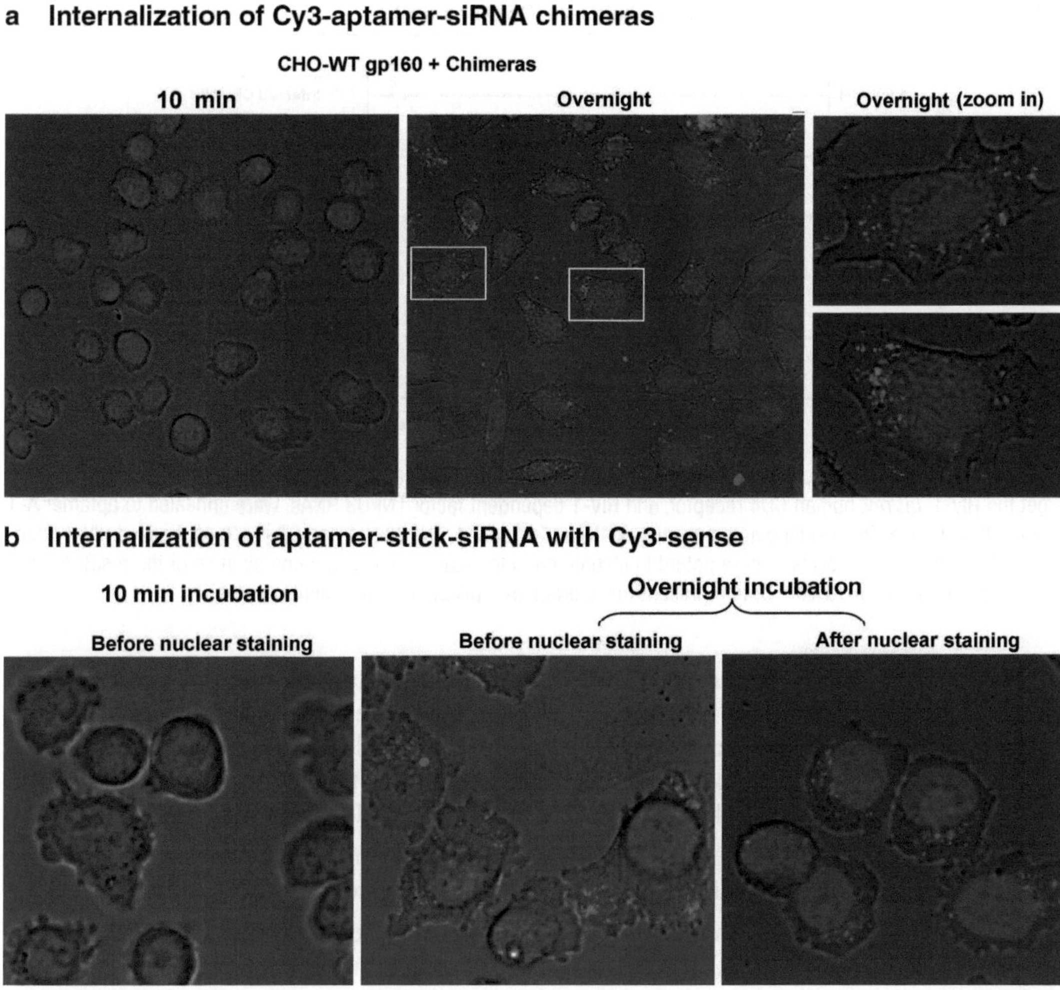

Fig. 4. Internalization and intracellular localization analyses. (**a**) CHO-gp160 cells were grown in 35 mm plates. Before incubation with 100 nM of Cy3-labeled chimeras, cells were stained with Hoechst 33342 (nuclear dye for live cells) and then analyzed using real-time confocal microscopy. (**b**) CHO-gp160 cells were grown in 35 mm plates and incubated in culture medium with a 100 nM concentration of aptamer-stick-siRNA containing a 5′-Cy3-labeled sense strand for real-time live-cell confocal microscopy analysis. After overnight incubation, cells were stained with Hoechst 33342 and then analyzed by confocal microscopy.

Collect the images every 15 min on a confocal microscope under water immersion at 40× magnification. An example result is shown in Fig. 4 (see Note 13 for an alternative approach).

3.6. HIV-1 Challenge and p24 Antigen Assay

1. Infect CCR5-CEM cells or human PBMC with HIV viruses (IIIB, NL4-3 or Bal) at an MOI of 0.001 or 0.005. At 24 h postinfection, gently wash the cells three times with PBS to remove free virus, and incubate in a 5% CO_2 incubator at 37°C for another 4 days.

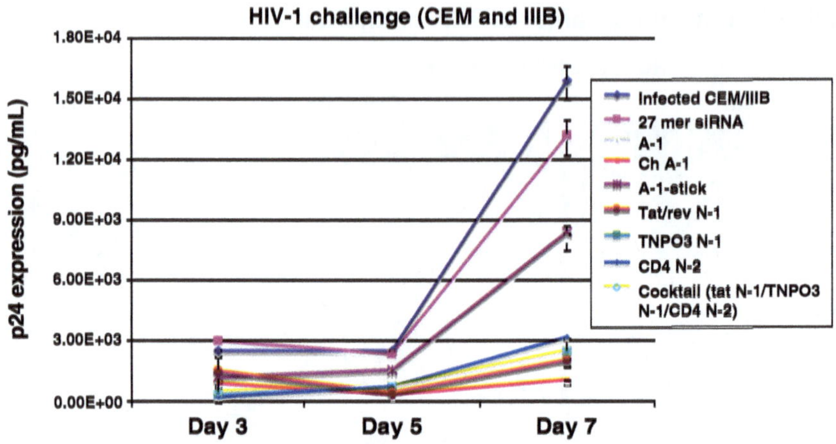

Fig. 5. Inhibition on HIV-1 infection mediated by aptamer-mediated siRNA delivery system. Three different siRNAs which target the HIV-1 *Tat/rev*, human CD4 receptor, and HIV-1 dependent factor TNPO3 RNAs, were annealed to aptamer A-1 via the "stick" bridge. The resulting aptamer-stick-siRNAs (*Tat/rev* N-1, TNPO3 N-1 and CD4 N-2) neutralized HIV-1 infection in CEM cell culture, providing more potent inhibition than the aptamer alone. The combination of three siRNAs also suppressed the HIV-1 replication. Data represent the average of triplicate measurements of p24.

2. Prior to RNA treatments, gently wash the infected cells with PBS three times to remove free virus. Mix 2×10^4 infected cells and 3×10^4 uninfected cells and incubate with refolded experimental RNAs at 400 nM final concentration in 96-well plates at 37°C (100 μL per well, triplex assay).
3. Collect 10 μL of culture supernatants per well at different times (3, 5, 7 and 9 days) and store at −20°C until p24 assay.
4. Perform the p24 antigen analyses using a HIV-1 p24 Antigen ELISA kit (see Note 14). An example result is shown in Fig. 5.

3.7. Analysis of siRNA Function by Quantitative RT-PCR Assay

1. Infect CCR5-CEM cells or human PBMC with HIV viruses (IIIB, NL4-3 or Bal) and incubate with the experimental RNAs (400 nM) as described above (see Subheadings 3.6, steps 1 and 2)
2. After 7 days of treatment, pellet cells and isolate total RNAs with STAT-60, and then treat the total RNAs with DNase I to remove genomic DNA; Mix the following reagents:
   (a) 8 μL Nuclease-free water.
   (b) 1.5 μL 10× DNase buffer.
   (c) 4 μL RNA (2 μg).
   (d) 0.5 μL RNain inhibitor.
   (e) 1.0 μL RNase-free DNase I.

   Incubate at 37°C for 1 h, heat at 80°C for 10 min to inactivate DNase I, and immediately chill the reaction on the ice.
3. Add 2 μL of 50 ng/μL random primer and 1 μL of 10 mM dNTP into the reaction mixture, and heat at 65°C for 5 min.

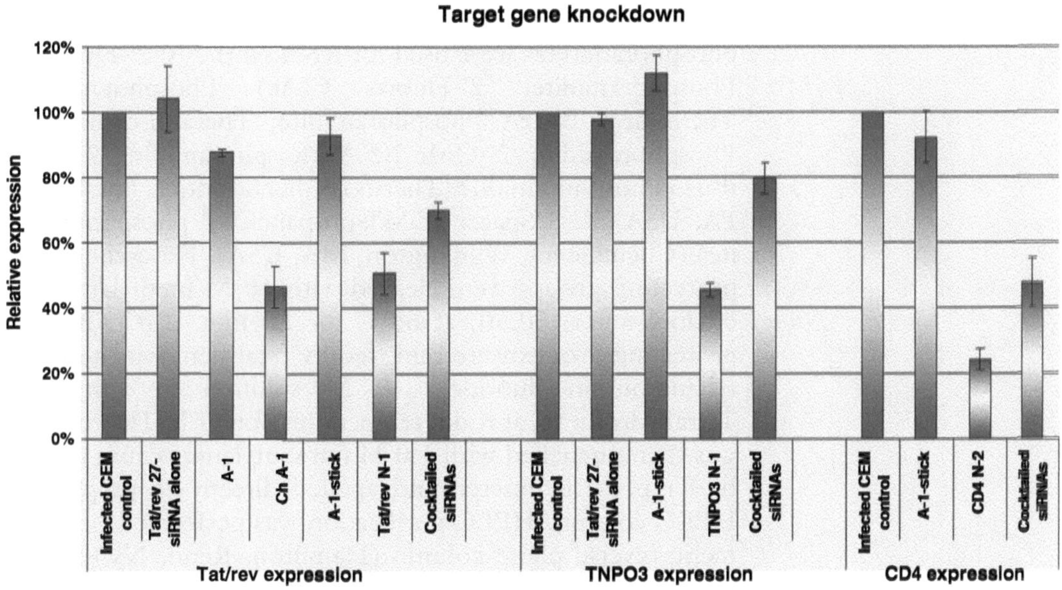

Fig. 6. siRNAs delivered by aptamers downregulate target gene expression in the HIV infected CEM cells (HIV-1 *tat/rev*, TNPO3 and CD4). Data represent the average of three replicates.

Immediately, chill the reaction on the ice. Subsequently, add the following reagents:

(a) 5 μL 5×First strand buffer.

(b) 2.5 μL 0.1 M DTT.

(c) 0.5 μL RNasin inhibitor.

(d) 1.0 μL MMLV-RT.

Incubate the reaction (total volume of 27 μL) at 25°C for 10 min and at 37°C for 1 h. Heat the mixture at 70°C for 15 min to inactivate reverse transcriptase and chill on ice. The cDNA is ready for qRT-PCR analysis (see Note 15).

4. Analyze expression of the target genes by quantitative RT-PCR using 2× iQ SyberGreen Mastermix and specific primer sets at a final concentration of 400 nM (triplex assay). Use the GAPDH expression as an internal control for normalization of the qPCR data. An example result is shown in Fig. 6.

4. Notes

1. Unless stated otherwise, all chemicals were purchased from Sigma-Aldrich and all solutions should be prepared in water that has a resistivity of 18.2 MΩ-cm and autoclaved. This standard is referred to as "water" in this text.

2. RNA Synthesis was performed on a GE Healthcare OligoPilot 10 plus Synthesizer on a 10 μmol scale, in a DMT OFF mode,

using standard phosphoramidite protocols. The following phosphoramidites were used for RNA synthesis: 2′-Fluoro U Phosphoramidite; 2′-Fluoro C(Ac) Phosphoramidite; TheraPure® Bz rA Phosphoramidite; TheraPure® iBu rG Phosphoramidite; 2′-OMe Bz A Phosphoramidite; 2′-OMe iBu G Phosphoramidite (Thermo Fisher Scientific, Pittsburgh, PA, USA), C-3 Spacer (DMT-propane-diol phosphoramidite) (ChemGenes, Wilmington, MA, USA). Exocyclic amine protecting groups were cleaved with 10 M methylamine in ethanol-water, 50:50, at 55°C for 20 min. 2′-*O*-TBDMSi protecting groups were removed by treatment with tetrabutylammonium fluoride, 1.0 M solution in anhydrous Tetrahydrofuran, at room temperature for 17 h. The reaction was then quenched with 1.0 M tetrabutylammonium acetate buffer pH 7.0, filtered and applied directly to preparative HPLC column. HPLC purification was performed on polymeric reverse phase column (Hamilton, Reno, NV) in tetrabutylammonium acetate buffers pH 7.0 (ref (1)). Collected fractions were analyzed on 12% acrylamide denaturing 8 M urea gel. Fractions containing the product were pooled, concentrated to dryness under vacuum, precipitated with ethanol in the presence of sodium chloride. Final purification was performed by electrophoresis in a 10% polyacrylamide/8 M urea. Before loading samples were heat denatured and electrophoresis was performed at an elevated temperature (55°C) to minimize secondary structure. Band of the product was located by UV shadowing, cut out, minced, eluted twice in 0.1 M ammonium acetate, then concentrated and desalted by HPLC chromatography (TEAA buffers, PRP column) and ethanol precipitated according to standard procedures.

3. Since DTT is not stable for long-term storage, prepare fresh binding buffer by adding DTT before use.
4. GMEM-S Propagation Medium for CHO-Env Cells: To make 1 L: 704 mL sterile water; 100 mL of 10× MEM without L-glutamine (Gibco); 100 mL of FBS (can use supplemented bovine calf serum); 36 mL of 7.5% sodium bicarbonate (Gibco); 20 mL of 50× nucleosides (Sigma – see below); 10 mL of 100× glutamate and asparagine solution (Sigma – see below); 10 mL of 100 mM sodium pyruvate (Gibco); 10 mL of penicillin-streptomycin 5,000 U/mL (Gibco); 10 mL of nonessential amino acids (Gibco). Sigma reagents for 50× nucleoside mix: 35 mg of adenosine; 35 mg of guanosine; 35 mg of cytidine; 35 mg of uridine; 12 mg of thymidine; Prepare in 100 mL ddH$_2$O and filter-sterilize through a 2 μm filter. Store frozen. Sigma reagents for 100× glutamate and asparagine solution: 600 mg L-glutamic acid; 600 mg L-asparagine; prepare in 100 mL ddH$_2$O, filter-sterilize,

and store frozen. For complete selection medium, supplement with 400 μM MSX (Sigma). Prepare MSX stock at a concentration of 18 mg/mL in medium, filter-sterilize, and store at –20°C.

5. Propagate the virus stock solution and titer on PBMCs. HIV-1 virus is infectious and poses a hazard to human health. Handle with proper personal protective equipment in a BSL-2/3 laboratory, according to institutional and governmental biosafety guidelines.
6. 100 μL PCR reaction contains 100 pmol of oligo 1, 100 pmol of oligo 2, 20 nmol of dNTP, 1×Taq buffer, 0.6 μL of Taq polymerase. Conduct 10 PCR cycle as following: 94°C, 0 min; 94°C, 1 min, 58.5°C, 1 min, 72°C, 1 min; 72°C, 7 min; 4°C, keep!. PCR primers sequences: A-1 oligo 1: 5′-TAA TAC GAC TCA CTA TAG GGA GGA CGA TGC GGA ATT GAG GGA CCA CGC GCT GCT TGT-3′; A-1 oligo 2: 5′-TCG GGC GAG TCG TCT GCC ATC ACG ACA AAC TGC TTA TCA CAA CAA GCA GCG CGT GGT-3′; B-68 oligo 1: 5′-TAA TAC GAC TCA CTA TAG GGA GGA CGA TGC GGA CAT AGT AAT GAC ACG GAG GAT GGA-3′; B-68 oligo 2: 5′-TCG GGC GAG TCG TCT GAC CGT CAA GAG ATG GCT GTT TTT TCT CCA TCC TCC GTG TCA-3′; Ch A-1 oligo 1: 5′-TAA TAC GAC TCA CTA TAG GGA GGA CGA TGC GGA ATT GAG GGA CCA CGC GCT GCT TGT TGT GAT AAG CAG TTT-3′; Ch A-1 oligo 2: 5′-TGA TGA GCT CTT CGT CGC TGT CTC CGC AAT CGG GCG AGT CGT CTG CCA TCA CGA CAA ACT GCT TAT CAC AA-3′. Ch B-68 oligo 1: 5′-TAA TAC GAC TCA CTA TAG GGA GGA CGA TGC GGA CAT AGT AAT GAC ACG GAG GAT GGA GAA AAA ACA GCC ATC-3′; Ch B-68 oligo 2: 5′-TGA TGA GCT CTT CGT CGC TGT CTC CGC AAT CGG GCG AGT CGT CTG ACC GTC AAG AGA TGG CTG TTT TTT CT-3′.
7. Mix all the reagents at room temperature and finally add T7 RNA polymerase. Increase the amount of DNA template (1–4 μg) and reaction time (overnight incubation) will improve yield of RNA.
8. Perform RNA refolding step in 1× HBS buffer. For example: mix 10 μL of 10 μM chimera sense strand, 10 μL of 10 μM antisense strand and 5 μL refolding buffer (5× HBS) in a final volume of 25 μL.
9. Fluorescent dye reagent and Cy3-labeled RNA is sensitive to light and should be limited the exposure to light for entire procedure.
10. Detected by spectrophotometry, the base:dye ratio can be calculated, which should be lower than 200. The lower the

base:dye ratio, the more dye molecules are present on the RNA, indicating an effective labeling. Dye molecules (Cy3, Cy5, or fluorescein) covalently attach to the guanine base of the RNA. The dye molecule can also attach to the adenosine base of RNA, however, this reaction is very rare.

11. Since gp120 protein is easily shed from cell surface, cells are gently and quickly (1–2 min) trypsinized with a small amount of trypsin-EDTA solution.
12. Washing step: spin cells at 2× 100×*g* for 5 min at 25°C and limit the exposure to light for entire procedure.
13. An alternative approach for confocal microscopy: Directly add the Cy3-labeled aptamer-stick-siRNA conjugate at a 100 nM final concentration into the media and incubate for live-cell confocal microscopy in a 5% CO_2 microscopy incubator at 37°C. After 16 h of incubation and imaging, stain the cells with 0.15 mg/mL Hoechst 33342 and collect images again as described in Subheading 3.5, step 5). An example result for aptamer-stick-siRNA is shown in Fig. 4b.
14. Allow all reagents to reach room temperature before use. Label test tubes to be used for the preparation of standards and specimens. If the entire 96 well plate will not be used, remove surplus strips from the plate frame. Place surplus strips and desiccant into the Resealable plastic bag (provided by the manufacturer), seal and store at 4°C.
15. In this case, reverse transcribe 2 μg of total RNA into cDNA (27 μL). Add 173 μL of water and use 10 μL of diluted cDNA in a qRT-PCR reaction of 25 μL. For qRT-PCR, add 12.5 μL of 2× iQ SyberGreen Mastermix, 2 μL of 5 mM primer mix, and 0.5 μL of water.

Acknowledgments

This work was supported by grants from the National Institutes of Health AI29329 and HL07470 awarded to J.J.R. The HIV-$1_{Ba\text{-}L}$ gp120 protein, HIV-1 Bal virus, and CHO-EE and CHO-WT gp160 cell lines (16, 17) were obtained through the NIH AIDS Research and Reference Reagent Program, Division of AIDS, NIAID, NIH.

References

1. Fire, A., Xu, S., Montgomery, M. K., Kostas, S. A., Driver, S. E., and Mello, C. C. (1998) Potent and specific genetic interference by double-stranded RNA in *Caenorhabditis elegans. Nature 391*, 806–11.
2. Zamore, P. D., Tuschl, T., Sharp, P. A., and Bartel, D. P. (2000) RNAi: double-stranded RNA directs the ATP-dependent cleavage of mRNA at 21 to 23 nucleotide intervals. *Cell 101*, 25–33.

3. Castanotto, D., and Rossi, J. J. (2009) The promises and pitfalls of RNA-interference-based therapeutics. *Nature* ***457***, 426–33.
4. Kim, D. H., and Rossi, J. J. (2007) Strategies for silencing human disease using RNA interference. *Nat Rev Genet* ***8***, 173–84.
5. de Fougerolles, A., Vornlocher, H. P., Maraganore, J., and Lieberman, J. (2007) Interfering with disease: a progress report on siRNA-based therapeutics. *Nat Rev Drug Discov* ***6***, 443–53.
6. Song, E., Lee, S. K., Wang, J., Ince, N., Ouyang, N., Min, J., Chen, J., Shankar, P., and Lieberman, J. (2003) RNA interference targeting Fas protects mice from fulminant hepatitis. *Nat Med* ***9***, 347–51.
7. Scherer, L., Rossi, J. J., and Weinberg, M. S. (2007) Progress and prospects: RNA-based therapies for treatment of HIV infection. *Gene Ther* ***14***, 1057–64.
8. Martinez, M. A. (2009) Progress in the therapeutic applications of siRNAs against HIV-1. *Methods Mol Biol* ***487***, 343–68.
9. Tsygankov, A. Y. (2009) Current developments in anti-HIV/AIDS gene therapy. *Curr Opin Investig Drugs* ***10***, 137–49.
10. Podlekareva, D., Mocroft, A., Dragsted, U. B., Ledergerber, B., Beniowski, M., Lazzarin, A., Weber, J., Clumeck, N., Vetter, N., Phillips, A., and Lundgren, J. D. (2006) Factors associated with the development of opportunistic infections in HIV-1-infected adults with high CD4+ cell counts: a EuroSIDA study. *J Infect Dis* ***194***, 633–41.
11. Kilby, J. M., and Eron, J. J. (2003) Novel therapies based on mechanisms of HIV-1 cell entry. *N Engl J Med* ***348***, 2228–38.
12. Kwong, P. D., Wyatt, R., Robinson, J., Sweet, R. W., Sodroski, J., and Hendrickson, W. A. (1998) Structure of an HIV gp120 envelope glycoprotein in complex with the CD4 receptor and a neutralizing human antibody. *Nature* ***393***, 648–59.
13. Wyatt, R., and Sodroski, J. (1998) The HIV-1 envelope glycoproteins: fusogens, antigens, and immunogens. *Science* ***280***, 1884–8.
14. Zhou, J., Li, H., Li, S., Zaia, J., and Rossi, J. J. (2008) Novel dual inhibitory function aptamer-siRNA delivery system for HIV-1 therapy. *Mol Ther* ***16***, 1481–9.
15. Zhou, J., Swiderski, P., Li, H., Zhang, J., Neff, C. P., Akkina, R., and Rossi, J. J. (2009) Selection, characterization and application of new RNA HIV gp 120 aptamers for facile delivery of Dicer substrate siRNAs into HIV infected cells. *Nucleic Acids Res* ***37***, 3094–109.
16. Weiss, C. D., and White, J. M. (1993) Characterization of stable Chinese hamster ovary cells expressing wild-type, secreted, and glycosylphosphatidylinositol-anchored human immunodeficiency virus type 1 envelope glycoprotein. *J Virol* ***67***, 7060–6.
17. Vodicka, M. A., Goh, W. C., Wu, L. I., Rogel, M. E., Bartz, S. R., Schweickart, V. L., Raport, C. J., and Emerman, M. (1997) Indicator cell lines for detection of primary strains of human and simian immunodeficiency viruses. *Virology* ***233***, 193–8.

Part V

RNAi Screens to Study Virus–Host Interactions

Chapter 23

RNAi Screening for Host Factors Involved in Viral Infection Using *Drosophila* Cells

Sara Cherry

Abstract

Since viral pathogens represent a significant threat to human health, a better understanding of the cellular factors that impact infection would facilitate the development of therapeutics. The recent advent of RNA interference (RNAi) technology coupled with the ease and efficiency of RNAi in *Drosophila* cell culture has led to the widespread use of this experimental system for high-throughput RNAi screening of host factors required for viral infection [Cherry et al., *Genes Dev* 19:445–452, 2005; Hao et al., *Nature* 45:890–893, 2008; Sessions et al., *Nature* 458:1047–1050, 2009]. Here, we describe the use of this system for the identification of host factors that impact viral infection.

Key words: *Drosophila*, High-throughput screening, RNAi, dsRNA, Cell-based assay, Virus, Innate immunity, Host factors

1. Introduction

The unique nature of viral pathogens, in that they are obligate intracellular parasites with limited coding capacity, requires that they coopt host proteins and processes to complete their life cycle. The identification of these cellular factors has been hampered by the lack of robust, genome-wide loss-of-function screening strategies. The development of simple RNAi methodology coupled with the complete genome sequence led to the optimization of genome-wide cell-based screens in *Drosophila* cells (1). Importantly, a wide variety of viruses can infect *Drosophila* cells including a number of mammalian viruses of medical and agricultural importance such as vesicular stomatitis virus, Influenza virus, Dengue virus, and West Nile virus to name a few (2–5). Moreover, previous screens have revealed that host factors can be identified that control various steps in the virus life cycle including entry,

Ronald P. van Rij (ed.), *Antiviral RNAi: Concepts, Methods, and Applications*, Methods in Molecular Biology, vol. 721, DOI 10.1007/978-1-61779-037-9_23, © Springer Science+Business Media, LLC 2011

translation, and RNA replication (6, 7). And lastly, many of the factors that are limiting for viral replication in *Drosophila* cell culture are also limiting in human cells infected with these viruses (2, 3, 6, 7). Therefore, the identification of these limiting factors presents novel targets for antiviral therapeutics.

Here, we present a generalized protocol for a high-throughput RNAi screen to identify cellular factors involved in viral infection using Vesicular Stomatitis virus as an example.

2. Materials

2.1. dsRNA Synthesis

1. Primers (with fused T7-polymerase promoter sequences).
2. Genomic DNA or cDNA from *Drosophila*.
3. Taq polymerase and buffer.
4. dNTPs.
5. MEGAscript T7 kit (Ambion).
6. RNeasy Purification kit (Qiagen).
7. Aluminum seals (Corning).

2.2. RNAi and Virus Infection in Cell Culture

1. *Drosophila* cells (S2 or others available at https://dgrc.cgb.indiana.edu/) (see Note 1).
2. Schneider's *Drosophila* Medium with L-Glutamine (Invitrogen).
3. Heat inactivated fetal bovine serum (FBS, SIGMA).
4. 100 U/mL penicillin, 100 μg/mL streptomycin.
5. Complete Schneider's medium: Schneider's *Drosophila* Medium supplemented with 10% heat inactivated FBS, penicillin, and streptomycin.
6. Multiwell tissue culture plates. This protocol can be performed in different formats (Table 1). Perform genome-wide RNAi screens in 384 well plates, black with clear bottom (Corning 3712).

Table 1
Parameters for RNAi

Plate type	Cells/well	dsRNA/well	SFM/well	Complete media/well
6	2.5×10^6	5 μg	1 mL	2 mL
24	2×10^6	2 μg	200 μL	400 μL
96	50×10^3	750 ng	50 μL	100 μL
384	15×10^3	250 ng	10 μL	20 μL

7. VSV-GFP virus. Prepare sufficient amounts of high titer VSV-GFP (e.g., 1×10^8 pfu/mL) for a complete screen to avoid lot-to-lot variability. We use a GFP expressing virus when available to avoid the need for immunostaining.

2.3. Staining

1. 24 Pin vacuum manifold (Drummond).
2. Fix solution: 4% formaldehyde in PBS. Prepare by diluting a 37% stock solution.
3. Nuclear staining solution: 5 μg/mL Hoechst 33342 in PBS/0.1% Triton. Prepare fresh from a frozen 5 mg/mL Hoechst 33342 stock solution.
4. Wash buffer: 0.1% Triton in PBS.
5. Aluminum seals (Corning).

2.4. Imaging and Image Analysis

1. Automated microscope (ImageXpress Micro, Molecular Devices) with a 20× objective.
2. MetaXpress Software with the Cell Scoring module.

2.5. Genome-Wide Screen

1. 384 well plates, black with clear bottom (Corning 3712).
2. Well Mate (Matrix) microplate dispenser for all liquid handling steps in 96 or 384 well plates.
3. Genome-wide dsRNA library (Ambion) (see Note 2).

3. Methods

3.1. dsRNA Synthesis

1. Design gene-specific primers containing the T7 promoter sequence at their 5′ ends using Snapdragon (http://www.flyrnai.org/cgi-bin/RNAi_find_primers.pl). Typically, we use primers against an exon present in all isoforms that is 200–500 bp in length that has no overlaps of 19 bp with any other gene in *Drosophila*.
2. Perform a standard 50 μL PCR reaction with 250 ng of genomic DNA (or 50 ng of plasmid).

   PCR program (see Note 3):

   (a) 94°C for 3 min.

   (b) 94°C for 30 s.

   (c) 60°C for 45 s.

   (d) 72°C for 1 min.

   (e) Repeat steps (b)–(d), 8 times.

   (f) 94°C for 30 s.

   (g) 60°C (decrease the annealing temperature by1°C in each cycle) for 45 s.

(h) 72°C for 1 min.
(i) Repeat steps (f)–(h), 5 times.
(j) 94°C for 30 s.
(k) 55°C for 45 s.
(l) 72°C for 1 min.
(m) Repeat steps (j)–(l), 20 times.
(n) 72°C for 10 min.
(o) 4°C Hold.

3. Run 5 μL of PCR reaction on a 1% agarose gel to verify the presence of a single band.
4. Use 1 μL of the first PCR to perform a second round of PCR using the same conditions if a robust band is not detected in the first round of PCR. Verify that there is a single band on an agarose gel.
5. Use 8 μL of the second round PCR in a 20 μL in vitro transcription reaction using the MEGAscript T7 kit, according to the manufacturer's instructions.
6. Incubate at 37°C for 16 h.
7. Add 79 μL water and 1 μL DNase included in the transcription kit and incubate at 37°C for 1 h.
8. Purify dsRNA using RNAeasy column following the manufacturer's protocol, elute the RNA in 50 μL of water.
9. Run 1 μL on a 1% agarose gel to confirm a single band and determine the concentration using a spectrophotometer. Expect a total yield of 40–80 μg per reaction.
10. No annealing is required.
11. Seal plates and store at −80°C.

3.2. RNAi in Tissue Culture Plates

1. Grow *Drosophila* cells at 28°C in complete Schneider's media. Cells should be in log phase for experiments (see Note 4).
2. Pellet cells at 500 × *g* and resuspend in serum-free Schneider's media (see Table 1).
3. Add cells to wells prealiquoted with dsRNA (see Table 1).
4. Spin cells onto plates at 500 × *g* for 1 min.
5. Incubate in incubator for 45 min.
6. Add complete Schneider's media (see Table 1).
7. Incubate for 3 days in humidified chamber (see Note 5).

3.3. Infection

1. Infect cells by adding 10 μL of virus suspension in serum-free media at an MOI of 0.33. Spin virus onto plates at 500 × *g* for 1 min (see Note 6).

2. Incubate for 20 h in a humidified chamber in an incubator. A single round of VSV-GFP infection in S2 cells takes approximately ~20 h (see Note 6).

3.4. Staining

1. Spin down plates at 500 × *g* for 1 min before each step.
2. Gently remove media using vacuum manifold for each step (see Note 7).
3. Fix cells in 15 μL of fix solution for 10 min at room temperature.
4. Add 30 μL PBS/0.1% triton per well to wash.
5. Remove PBS, add 30 μL of nuclear staining solution, and incubate for 10 min at room temperature.
6. Wash wells with 30 μL of PBS/0.1% triton and leave wells in PBS for storage.
7. Seal the plates with aluminum seals and store at 4°C until analysis.

3.5. Imaging and Image analysis

1. Capture at least three images per well at a magnification of 20× in both the Hoechst channel and the virus channel (GFP, Fig. 1).
2. Use the cell scoring module in MetaXpress software to segment the image and calculate the total number of cells per well (Hoechst$^+$) and the number of infected cells (GFP$^+$). The percentage of infected cells is a good measure of infectivity (see Notes 8 and 9).
3. Software can also be used to determine the intensity of GFP per cell, which is a surrogate for the level of replication per cell.

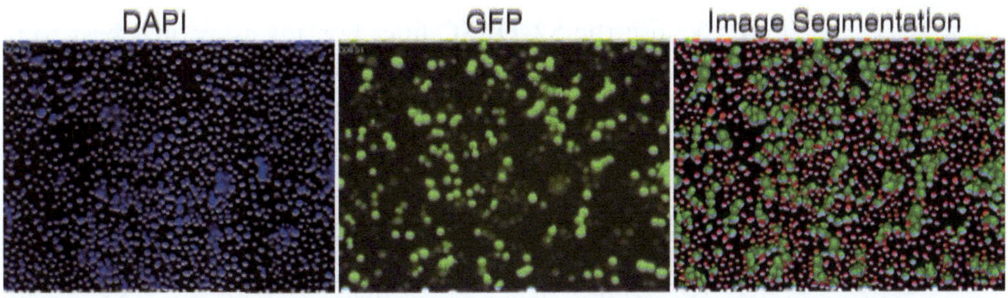

Fig. 1. Image analysis and segmentation. *Drosophila* cells are seeded in 384 well plates such that there is a single infected monolayer at the endpoint of the assay. 17,000 S2 cells were seeded and 3 days later infected with VSV-GFP (MOI = 0.33) for 20 h. The cells were fixed, processed for microscopy and counterstained with Hoescht 33342. Cells were imaged at 20× using ImageXpressMicro software in the DAPI and GFP channels shown. The image was segmented using the Cell Scoring Module and the segmented image is shown. Nuclei are shown in *red* and the infected cells in *green*. The infected nuclei are shown as the *brighter green* within the *lighter green* infected cytoplasm. The segmented image is used to derive metrics including the number of cells, the number of infected cells, and the intensity of infection.

4. Analysis of cell number can be used to remove cytotoxic candidates. We calculate the mean and standard deviation of the cell counts and exclude wells with a robust Z score <–2 which is typically ~15% of change.

3.6. Screening

Once the assay has been optimized according to Notes 9–12, the screening can begin (see Notes 13 and 14).

1. One plate from the screen is used for a pilot screen. Seed some of the empty wells with control dsRNAs (see Notes 10 and 11). Perform the pilot screen duplicate.
2. Statistical tests are performed to determine if the positive candidates can be identified that impact viral infection but not cell viability (see Note 15).
3. Next, the genome-wide screen is performed in duplicate. Screen sixteen 384 well plates at a time over a few weeks.
4. Wells that have a robust Z score for percent infection >2 or <–2 in duplicate, and do not affect cell viability (Z score >–2) are considered as potential candidates from the primary screen.
5. Positive candidates identified in the screen are validated using independent reagents. Another set of dsRNAs are generated against the genes of interest from another region of the mRNA.
6. Those genes for which two independent dsRNAs have the same phenotype can be considered for further study.

4. Notes

1. We use the S2-DRSC line. For any assay, it is best to test a few different lines to determine which is most amenable to your biological read-out.
2. Multiple sources of genome-wide dsRNA libraries are available (DRSC, Open Biosystems, etc).
3. We use this PCR program to increase the specificity of the reaction.
4. *Drosophila* cells are semiadherent and are passaged without trypinization every 4 days by pipeting the cells off the flasks and diluting the cells 1:10.
5. Since the volume of 384 well is small, evaporation can have large effects on the biology leading to artifactual changes in the read-outs of the edge wells. To overcome this, the plates should be maintained under high humidity. We use Tupperware

containers lined with water soaked paper towels in a humidified incubator.

6. The amount of virus used to achieve a given percent infection is empirically determined. The target percent infection and time of infection is also optimized using a positive control to maximize the fold changes.
7. Since the *Drosophila* cells are semiadherent, the fix step should be done very gently. In addition, fix can be directly added to the wells without removing supernatant.
8. The metrics used to identify hits rely on the quantification of the images. Therefore, the image analysis needs to be robust. The segmentation parameters that call a cell GFP positive should be set using both negative (uninfected wells) and positive controls (infected wells).
9. The number of cells per well needs to be optimized for the assay length. Serial dilutions of cells across the plate are seeded and fixed at the end time of the assay. Cells are processed and imaged. A single monolayer with no cell piling is optimal for image segmentation.
10. Controls are used to validate that RNAi is functional. Cells are treated with dsRNA against a nontargeting control (e.g., luciferase) and a cellular gene required for cell viability (e.g., dIAP). More than 99% of the cells should be dead at the end point of the assay upon dIAP knockdown.
11. The level of infection needs to be optimized for the assay. Both the MOI and the time of infection should be queried. Two different dsRNAs are used for this. First, a dsRNA against the viral antigen probed is used (e.g., GFP) which should lead to a decrease in viral infection as measured by GFP^+ cells. Second, a cellular gene required for infection is used (e.g., Rab5) to ensure that viral infection is sensitive to loss of cellular factors. The MOI and time-point most sensitive to GFP and Rab5 dsRNA treatments is the one chosen for screening.
12. Once the assay has a coefficient of variation (CV) of <15%, and a Z factor >0.25 for percent infection a pilot screen can be initiated. CV = Standard deviation/average of an entire plate. Z factor = (1−(3SD(sample)+3SD(control))/absolute value of (average(sample)−average(control))).
13. Standardization of the reagents used for screening is essential. We test lots of Schneider's media, fetal calf serum, virus innocula, and staining reagents. Then we aliquot and use the same batch for the entire screen.
14. Plate-to-plate and day-to-day variability are minimized through the use of automated liquid handling (WellMate, Matrix) for all liquid additions to multiwell plates.

15. There are a variety of different statistical tests to identify positive candidates during screening. We favor robust statistics whereby we use the plate median and interquartile range to identify outliers or "positive candidates." The percentage infection is log transformed. The median is calculated. Then calculate the interquartile range for the log transformed data. Interquartile Range = (QUARTILE(numbers,3)–QUARTILE (numbers,1)) in Excel. Next, calculate the robust *Z* score = (log transformed %–median)/(interquartile range*0.74).

Acknowledgments

This work was supported by NIH grants R01AI074951 and U54AI057168.

References

1. Cherry, S. (2008) Genomic RNAi screening in Drosophila S2 cells: what have we learned about host-pathogen interactions? *Curr Opin Microbiol* **11**, 262–270.
2. Hao, L., Sakurai, A., Watanabe, T., Sorensen, E., Nidom, C. A., Newton, M. A., Ahlquist, P., and Kawaoka, Y. (2008) Drosophila RNAi screen identifies host genes important for influenza virus replication. *Nature* **454,** 890–893.
3. Sessions, O. M., Barrows, N. J., Souza-Neto, J. A., Robinson, T. J., Hershey, C. L., Rodgers M. A., Ramirez, J. L., Dimopoulos, G., Yang, P. L., Pearson, J. L., and Garcia-Blanco, M. A. (2009) Discovery of insect and human dengue virus host factors. *Nature* **458,** 1047–1050.
4. Shelly, S., Lukinova, N., Bambina, S., Berman, A., and Cherry, S. (2009) Autophagy is an essential component of Drosophila immunity against vesicular stomatitis virus. *Immunity* **30,** 588–598.
5. Chotkowski, H. L., Ciota, A. T., Jia, Y., Puig-Basagoiti, F., Kramer, L. D., Shi, P. Y., and Glaser, R. L. (2008) West Nile virus infection of *Drosophila melanogaster* induces a protective RNAi response. *Virology* **377,** 197–206.
6. Cherry, S., Doukas, T., Armknecht, S., Whelan, S., Wang, H., Sarnow, P., and Perrimon, N. (2005) Genome-wide RNAi screen reveals a specific sensitivity of IRES-containing RNA viruses to host translation inhibition. *Genes Dev* **19,** 445–452.
7. Cherry, S., Kunte, A., Wang, H., Coyne, C., Rawson, R. B., and Perrimon, N. (2006) COPI activity coupled with fatty acid biosynthesis is required for viral replication. *PLoS Pathog* **2,** e102.

Chapter 24

Genome-Wide RNAi Screen for Viral Replication in Mammalian Cell Culture

Bhupesh K. Prusty, Alexander Karlas, Thomas F. Meyer, and Thomas Rudel

Abstract

Influenza infections are considered a global threat to public health and cause seasonal epidemics and recurring pandemics. High mutation rates facilitate the generation of viral escape mutants rendering vaccines and drugs directed against virus-encoded targets ineffective. One alternative approach that could prevent viral escape is the targeting of host cell determinants that are temporarily dispensable for the host but crucial for virus replication. Here, we report a genome-wide RNAi screening approach in mammalian cell culture system that led us to the identification of several host cell genes influencing influenza A virus replication. Interestingly, the majority of the identified host gene products are indispensable for viral replication of a broad range of influenza viruses ranging from the highly pathogenic avian H5N1 strain to the current pandemic swine-origin H1N1 strain. Our results provide a new approach to explore virus–host interactions and to identify promising antiviral targets.

Key words: Influenza A virus, H1N1, H5N1, RNAi, siRNA

1. Introduction

The ability of viruses, particularly influenza viruses, to mutate rapidly is a bottleneck for the efficient design of vaccines against virus-encoded targets. Several host cell factors influence the fate of influenza virus–host interaction and thereby decide the fate of the virus and its survival inside the host cell (1). Comprehensive knowledge of such critical host cell determinants could provide valuable insight into the molecular mechanisms of viral replication and facilitate the development of novel drugs that target host cell factors and are thus less prone to select for resistant viral mutants. To identify such host cell factors important for influenza virus infection of human cells, we developed a genome-wide

Ronald P. van Rij (ed.), *Antiviral RNAi: Concepts, Methods, and Applications*, Methods in Molecular Biology, vol. 721, DOI 10.1007/978-1-61779-037-9_24,

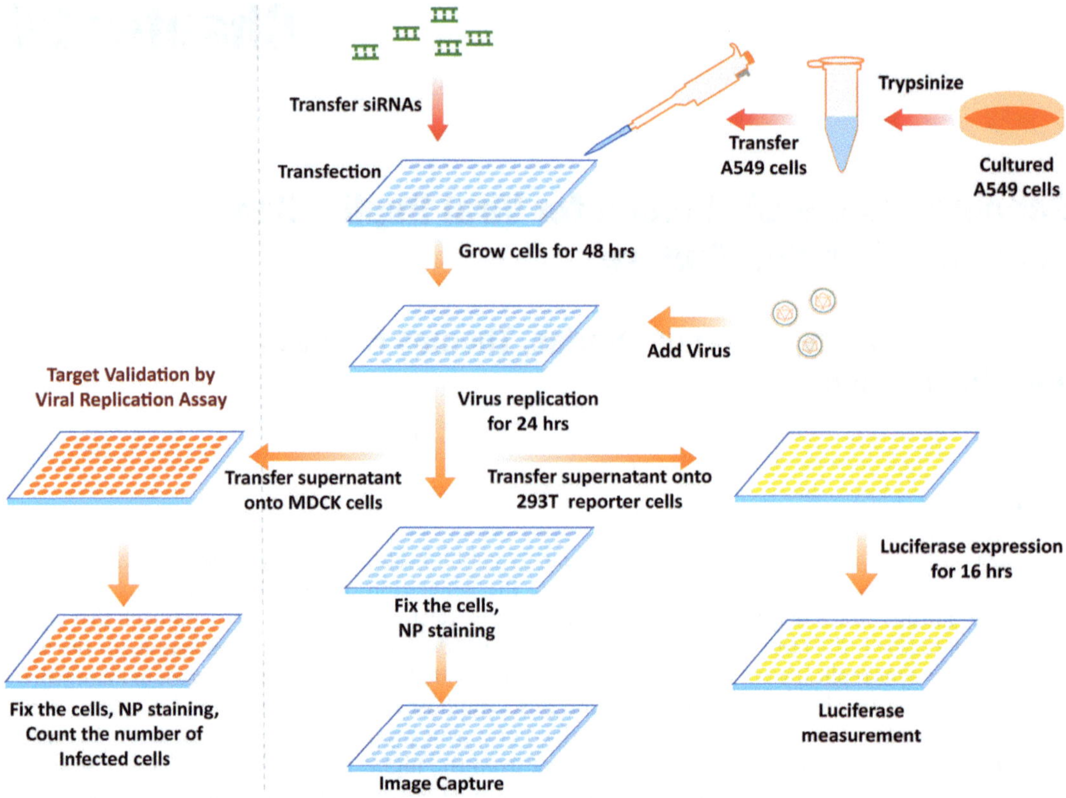

Fig. 1. Outline of the genome-wide RNAi screening procedures for viral replication in mammalian cell culture. Briefly, A549 human lung epithelial cells are transfected with siRNAs 48 h prior to infection with influenza A virus and are stained for viral NP at 24 h postinfection. Virus containing supernatant from infected A549 cells is transferred onto luciferase reporter cells to quantify virus replication. For target validation and titration of viral replication, viral supernatants from infected A549 cells are serially diluted and added to fresh MDCK cells followed by fixing and staining of viral nuclear protein.

RNAi screen using a two-step approach (Fig. 1). Recently, we used such a genome-wide RNA interference (RNAi) screen, which resulted in the identification of 287 human host cell genes influencing influenza A virus replication (2). Important among these factors are the SON DNA binding protein (SON), which is important for normal trafficking of influenza virions to late endosomes early in infection and CDC-like kinase 1 (CLK1), which reduces influenza virus replication by more than two fold as a result of impaired splicing of the viral M2 messenger RNA (2).

In this chapter, we provide a detailed protocol of this genome-wide RNAi screening for the dissection of virus–host interactions and the identification of drug targets for several types of influenza viruses. For this, we use three parameters: the percentage of infected cells as well as the total number of infected cells, both determined by immunofluorescence microscopy, and luciferase expression analysis. Immunofluorescence staining of viral nuclear protein (NP) is used to calculate for number of virus-infected cells in a total host cell population. Influenza A virus-inducible reporter genes are used to quantify the viral replication activity in influenza virus-infected cells. During hit validation the total number of virus particles released into the supernatant was determined by performing a viral replication assay. We use Allstars neutral siRNAs as a control in our experiments. We validate our siRNA-mediated approach by using a known inhibitory control siRNA against the viral NP.

First, A549 human lung epithelial cells are transfected with siRNAs 48 h prior to infection with influenza A virus (A/WSN/33) and are stained with a virus-specific antibody at 24 h postinfection (h p.i.) to monitor the infection rate of the initially infected cells (see Subheadings 3.1–3.3 and 3.5) Second, virus supernatants are transferred onto luciferase reporter cells at 24 h p.i. to quantify virus replication (see Subheading 3.4 and Fig. 1). To verify assay reliability, an siRNA directed against influenza nuclear protein (NP) mRNA, with a known capability to inhibit influenza replication is tested (1). Knockdown of NP effectively blocks viral replication, which can be assessed by immunofluorescence staining (Fig. 2a) and the luciferase reporter assay (Fig. 2b). This NP control siRNA is included in 10 out of 384 wells on each screening plates, the nontargeting neutral siRNA control "Allstars" is included in 20 wells to allow proper normalization.

2. Materials

2.1. Cell Culture and Virus

1. The human lung epithelial cell line, A549 (ATCC number, CCL-185), the human embryonic kidney cell line, 293T (ATCC number, CRL-11268), and the Madin Darby Canine Kidney cell line, MDCK (ATCC number, CCL-34) (ATCC-LGC Promochem, Germany). Primary normal human bronchial/tracheal epithelial cells, NHBE (CC-2541, Lonza, Cologne, Germany) (see Note 1).
2. Dulbecco's Modified Eagle's Medium (DMEM). Supplement with 4 mM L-glutamine, 100 U/ml penicillin/streptomycin and 10% fetal calf serum (henceforth named as complete media). For A549 cell culture, use an additional supplement of 4 mM sodium pyruvate.
3. Clonetics bronchial epithelial growth media (BEGM) BulletKit (CC-3170) (Lonza, Cologne, Germany).
4. Growth supplements for primary epithelial cells (included in the BulletKit): Bovine pituitary extract (BPE), Hydrocortisone, human epidermal growth factor (hEGF), epinephrine, transferrin, insulin, retinoic acid, triidothyronine, GA-1000.

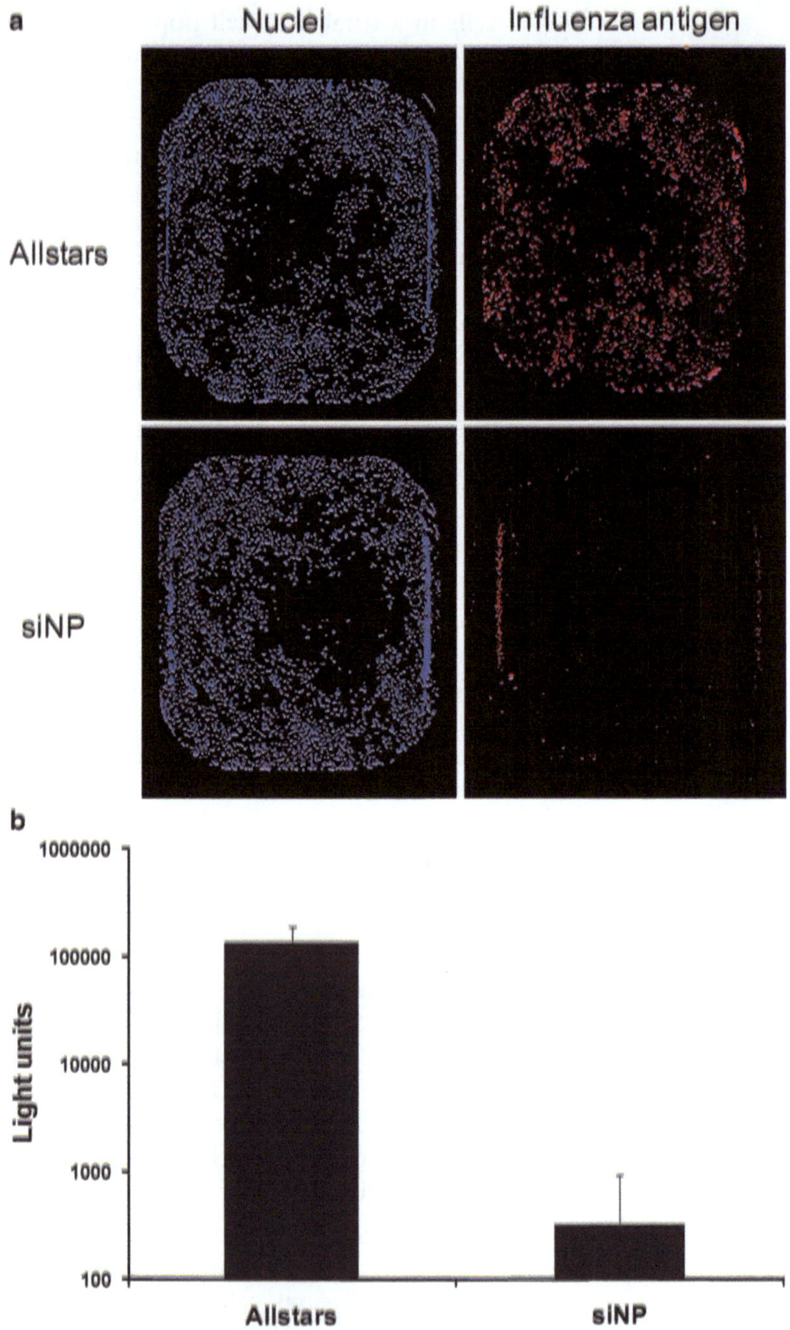

Fig. 2. Controls used during the primary screening process. (**a**) Depicted are representative images of the nontargeting (Allstars) and inhibitory (siNP) control samples, stained with an anti-NP antibody and analyzed by Acumen eX3 cytometer. (**b**) Graph depicts light units exerted by the corresponding supernatants transferred onto 293T reporter cells.

All the supplements are added at 0.5/500 ml growth medium except for BPE, which is added at 2/500 ml of medium.

5. Trypsin dissolved in DMEM medium at a concentration of 1 μg/ml. Trypsin solution should be stored at 4°C.

6. Opti-MEM solution (Invitrogen).
7. Roswell Park Memorial Institute-1640 (RPMI-1640) is supplemented with 25 mM HEPES.

2.2. siRNA Screening

1. HiPerFect transfection agent (Qiagen) (see Note 2).
2. Acumen eX3 cytometer (TTP LabTech, Melbourn, UK).
3. Biomek FXP laboratory Automation Workstation (Beckman Coulter), located in a biosafety level 2 (BSL2) cabinet.
4. The druggable and the genome-wide siRNA library (Human Druggable Genome siRNA Set V2.0, and Human Whole Genome siRNA set V 1.0; Qiagen, Hilden, Germany).
5. Control siRNAs. An siRNA targeting influenza nuclear protein, siNP (5′-AAGGAUCUUAUUUCUUCGGAG-3′), and the nontargeting neutral siRNA control Allstars (Qiagen) (see Note 3).

2.3. Virus Infection

1. Phosphate buffered saline (PBS): PBS can be made as a 10× stock solution. To prepare 1 L of 10× PBS, dissolve 80 g of NaCl, 2 g of KCl, 14.4 g of Na_2HPO_4, and 2.4 g of KH_2PO_4 in 800 mL of H_2O. Adjust to pH 7.4 with HCl, and then add H_2O to 1 L. Dispense the solution into aliquots and sterilize by autoclaving. Store PBS at room temperature (RT).
2. Infection buffer: PBS supplemented with 0.2% (w/v) bovine serum albumin (BSA). Store buffer at 4°C.
3. Postinfection culture medium: DMEM supplemented with 0.2% BSA, 4 mM L-glutamine, and 100 U/ml penicillin/streptomycin.

2.4. Luciferase Reporter Assay

1. FluA firefly luciferase plasmid, pRep4-fluA (3).
2. GFP-expressing plasmid, pLVTHM (Addgene, Cambridge, MA, USA).
3. Bright-Glo firefly luciferase substrate (Promega, Madison) (see Notes 4–6).
4. Envision multilabel plate reader (PerkinElmer, see Note 7).
5. Transfection medium: DMEM supplemented with 10% FCS and 4 mM L-glutamine (see Note 8).

2.5. Indirect Immunofluorescence Labeling

1. Fixing solution: 3.7% (w/v) para-formaldehyde in PBS (see Note 9).
2. Cell permeabilization solution: 0.3% (v/v) triton X-100, 10% FCS in PBS. Store at 4°C.
3. Primary antibody against the viral nucleoprotein, NP (clone AA5H, AbD Serotec). Antibodies can be stored at 4°C.
4. Cy3 conjugated secondary goat antibody directed against mouse IgG (Dianova, Hamburg, Germany). Dissolve lyophilized

antibody in 1.1 ml glycerine and 1.1 ml H_2O. Store in aliquots at –20°C.

5. Antibody dilution buffer: 10% (v/v) FCS, 0.1% (v/v) Tween20 in PBS. Store buffer at 4°C.
6. Cellular DNA staining dye: 0.1% Hoechst dye (Sigma-Aldrich) in water. Store at 4°C.
7. Olympus Automated microscope (Olympus, Soft Imaging Solutions, München, Germany).
8. DAPI and Cy3 filter sets (AHF-Analysetechnik, Tübingen, Germany).
9. ScanR analysis software (Olympus, Soft Imaging Solutions, München, Germany).

3. Methods

3.1. Cell Culture, Virus Propagation, and Titration

1. Grow 293T cells and MDCK cells in DMEM complete media.
2. Grow the A549 cells in DMEM complete medium, with an additional supplement of 4 mM sodium pyruvate, at 37°C and 5% CO_2 (see Notes 10 and 11).
3. Grow the primary NHBE cells in Clonetics BEGM media with appropriate supplements.
4. Grow the influenza virus strains A/WSN/33 (H1N1) and A/PR8/34 (H1N1) in allantoic cavities of 11-days-old embryonated chicken eggs (4).
5. Titrate the viral stocks by standard plaque assay on MDCK cells using an agar overlay medium (5).

3.2. siRNA Screening

1. Dissolve all the siRNAs to a final concentration of 200 nM with the siRNA dissolving solution provided by the manufacturer (see Note 12).
2. Array 4 μl/well of each of the siRNAs into 384-well plates. We use Biomek FXP laboratory Automation Workstation (Beckman Coulter) for all multiwell pipetting in order to avoid any errors.
3. Screen four siRNAs per gene for the druggable genome (6), and two siRNAs per gene for the whole genome library, three times independently. In our system, we screened a combination of the druggable and the genome-wide siRNA library targeting approximately 17,000 annotated genes and 6,000 predicted genes.
4. Add 8 μl of DMEM medium containing 0.35 μl HiPerFect to each of the well. DMEM medium should be at RT.

5. Shake the plates on a tabletop shaker for 10 min at room temperature (RT).
6. After incubation at RT for 10 min, add a cell suspension of 28 μl medium containing 500 A549 cells to make a final siRNA concentration of 20 nM.
7. Mix the cells by gentle swirling and incubate at 37°C and 5% CO_2 for 48 h before virus infection.

3.3. Virus Infection

1. Remove the medium and wash the cells once with 40 μl of PBS per well.
2. Infect the cells with influenza virus at an MOI of 0.12 in 15 μl of infection buffer for 60 min at RT (see Note 13).
3. Wash the cells with 40 μl of infection buffer and incubate for 23 h at 37°C in 40 μl of postinfection culture media (for A549 cells) or BEGM with supplements (for NHBE cells). All these steps should be done under BSL2 flow cabinet.
4. At 24 h p.i., transfer 12.5 μl of supernatant onto freshly seeded 293T reporter cells (1×10^4 cells per well in volume of 25 μl), incubate for 16 h at 37°C and 5% CO_2, and then measure luciferase activities (see Subheading 3.4). Analyze the remaining A549 cells by immunofluorescence (see Subheading 3.5).

3.4. Luciferase Reporter Assay

Influenza A virus-inducible reporter genes have been successfully utilized by Lutz and coworkers for the detection of influenza A virus replication and quantification of virus titers (7). In these vectors, the RNA polymerase I promoter/terminator cassette was used to express RNA transcripts encoding green fluorescence protein or firefly luciferase flanked by the untranslated regions of the influenza A/WSN/33 NP segment (7). Influenza A virus-specific reporter gene activity can be detected in cells after reconstitution of the influenza A virus polymerase complex from cDNA or after virus infection. We use 293T reporter cells containing pRep4-FluA plasmid (containing the firefly reporter) for the quantification of influenza A virus replication.

1. Transfect 293T cells in batches with the pREP4-FluA plasmid. For the transfection, suspend 1×10^7 293T cells in 10 ml of transfection medium in a large T150 cell culture flask.
2. Prepare transfection mixes in two steps. First add 30 μl of Lipofectamine 2000 to 1 ml of Opti-MEM solution in one eppendorf tube. Subsequently, in a second tube, add 16 μg of the pREP4-FluA luc plasmid DNA to an additional 1 ml Opti-MEM solution.
3. After 5 min incubation at RT, mix both the solutions with gentle pipetting, vortex briefly, and incubate at RT for 20 min. Finally, add the transfection mix dropwise to the cell suspension.

4. To determine transfection success, cotransfect cells with 1.6 μg of a GFP-expressing plasmid (see Subheading 2.4).
5. After 24 h, wash the cells once with prewarmed PBS.
6. Add 1 ml of trypsin to the cells and keep in a 37°C incubator for 3–5 min.
7. Stop the trypsin activity by adding complete DMEM medium.
8. Count the cells with a Neubauer cell counting chamber.
9. Seed the detached 293T into 384-well plates at a concentration of 1×10^4 cells per well in a volume of 25 μl of postinfection culture medium per well.
10. Directly after cell seeding, add 12.5 μl of virus-containing supernatant to the cells.
11. At 16 h p.i., equilibrate the cultured cells by incubating for 15 min at RT.
12. Thirty minutes prior to measuring luminescence, transfer the contents of one bottle of Bright-Glo Buffer to one bottle of Bright-Glo Substrate and mix it by inversion until the substrate is thoroughly dissolved (see Note 7).
13. Subsequently, add 10 μl of Bright-Glo firefly luciferase substrate to the cell culture media in each well, swirl gently for a minimum of 2 min and measure luciferase activity using a multilabel plate reader.

3.5. Indirect Immunofluorescence Labeling

1. At 24 h p.i., remove and discard the virus-containing media from the A549 cells (see Subheading 3.3). Wash once with 100 μl of DMEM without serum.
2. Remove the media at low aspiration speed followed by two more washes with 100 μl of PBS.
3. Fix the cells with 40 μl of 3.7% formaldehyde for 20 min at RT.
4. Discard the fixing solution and wash the cells thoroughly for three times with PBS.
5. Permeabilize and block the cells for 10 min in 40 μl of cell permeabilization solution at RT followed by two washes in PBS at RT.
6. Discard the PBS solution and replace it with 15 μl of primary antibody containing solution.
7. Incubate the samples sequentially with a primary antibody against the viral nucleoprotein (NP) diluted 1:10,000 in antibody dilution buffer for 1 h at RT.
8. Remove the primary antibody containing solution and wash the cells three times with PBS for 5 min each.
9. Remove the PBS and incubate with 15 μl of the secondary Cy3 conjugated antibody directed against mouse IgG (1:100)

in antibody dilution buffer along with 0.1% Hoechst dye. Hoechst dye is used to stain cellular DNA. Incubate the samples in the secondary antibody containing solution for 1 h at RT in the dark.

10. Wash the samples 3–5 times for 5 min each with 50 μl volume of PBS. Finally, cells should be stored in 50 μl of PBS at 4°C until analysis.
11. Determine the numbers of influenza infected cells and the total number of host cells using either an Olympus automated microscope or the Acumen eX3 microplate cytometer. Take the images with DAPI and Cy3 filter sets.
12. Use ScanR analysis software to automatically identify and quantify influenza NP and cell nuclei.
13. Determine the total number of cells (DAPI signal) and the number of infected cells (Cy3) and calculate the percentage of infected cells.
14. Exclude toxic siRNAs, which reduce the total cell numbers (<750 cells/well) upon transfection (72 h post transfection).
15. Perform data analysis as described in Note 14.

3.6. siRNA Validation Experiment

The aim of the validation experiments is to confirm the dataset that was obtained during the primary screen. Therefore, we performed it only for a subset of genes. The reason for doing it was to identify at what stage of replication the individual factor is involved in. This is relevant only for the follow-up experiments and not for the screening procedure itself. Validation should be performed by using four independent siRNAs, if possible. In order to identify host cell factors that are essential for a variety of influenza strains different influenza A viruses can be used for infections, e.g., the A/WSN/33 or the swine-origin A/Hamburg/04/2009 virus. During the validation three independent experiments should be performed to obtain significant data.

1. For siRNA validation experiments, transfect A549 cells with the desired siRNAs. For each well of a 96-well plate, mix 20 μl of a 100 nM siRNA dilution in DMEM without supplements into 1 μl of HiPerFect + 9 μl DMEM medium and incubate for 10 min at RT.
2. Stop complex formation by adding of 25 μl of DMEM complete medium.
3. Seed 3000 A549 cells in 50 μl of DMEM complete medium into each well and incubate at 37°C and 5% CO_2 for 6 h.
4. Exchange the medium with fresh DMEM complete medium and incubate the cells for an additional 48 h using the same growth conditions.

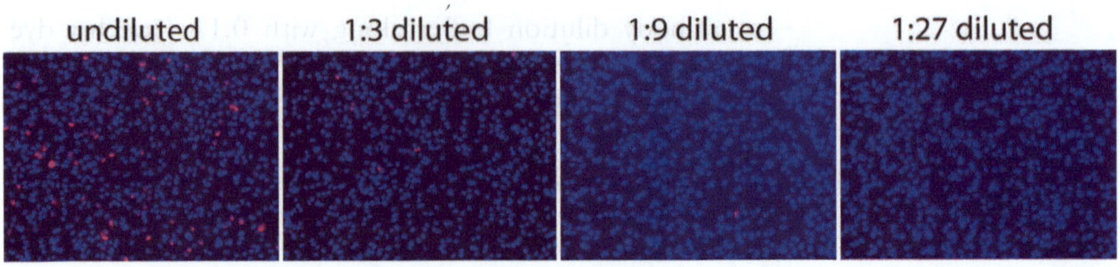

Fig. 3. Quantification of infectious viral particles by analyzing serial dilutions of influenza virus containing supernatants. A549 cells were transfected with the neutral control siRNAs (Allstars) and 48 h later infected with influenza A/WSN/33 (MOI 0.001). Supernatants were removed 48 h p.i. and transferred either undiluted or diluted (as indicated above) to MDCK cells. Viral antigen was detected by immunofluorescence 6 h p.i. using an NP specific antibody.

5. Infect the cells with the desired virus at MOI 0.001 for 48 h.
6. At 48 h p.i., analyze the supernatants by the viral replication assay (see Subheading 3.7).

3.7. Viral Replication Assay

Viral replication assay can be used to obtain a quantitative validation of the identified genes from the siRNA screen (see Note 15). To this end, the number of viruses released from siRNA transfected and influenza virus-infected A549 cells is determined by immunofluorescence staining.

1. Seed 5,000 or 12,000 MDCK cells in 384- or 96-well plates, respectively, in DMEM complete media.
2. One day later, wash the cells twice with PBS.
3. Infect the cells with a dilution series of cell culture supernatants (dilution: 1:1, 1:3, 1:9 and 1:27) and incubate at RT for 1 h.
4. Add infection buffer (40 or 100 μl/well in 384- or 96-well plates, respectively) and incubate the plates at 37°C and 5% CO_2 for 6 h, followed by fixation with 3.7% para-formaldehyde, antibody staining, and automated image processing as described in Subheading 3.5.
5. Count the total numbers of infected MDCK cells. Calculate the infection rate as a percentage of the Allstars transfected control (Fig. 3).

4. Notes

1. Additional information about the primary NHBE cells can be found at http://www.lonza.com.
2. Transfection reagents are highly sensitive to rapid and repeated temperature changes. Therefore, it should be stored at 4°C and while working it should be kept on ice.

3. Control siRNAs should be included in all screening plates.
4. The validity of luciferase reporter assays should be checked prior to the main experiment and the concentration of reporter plasmids required for good reporter activity should be titrated by repeated experiments.
5. Since luciferase activity is temperature-dependent, the temperature of the Bright-Glo™ reagent should be kept constant while quantifying luminescence. This is achieved most easily by using reagent that is equilibrated to RT. Equilibration of the reagent prior to use is unnecessary when the buffer is stored at RT. If the reagent is stored frozen after reconstitution, it must be thawed at temperatures below 25°C to ensure performance. Mix well after thawing. The most convenient and effective method to thaw or temperature equilibrate cold reagent is to place it in a water bath at RT.
6. Other firefly luciferase substrates can also be used for the reporter activity measurement.
7. Many different models of luminometers are available for measuring reporter activity. Luciferase reading measurement should be adjusted according to the machine used.
8. Antibiotics should not be used in transfection media as they can be toxic to the cells during the process of transfection and can cause extensive cell death.
9. Prepare fresh 3.7% para-formaldehyde solution in PBS before each experiment. The solution has to be carefully heated under a fume hood to ensure that the para-formaldehyde powder dissolves properly. Cool to RT for use.
10. It is very important that the cells should not be overgrown before seeding into the siRNA containing media. Confluency of 80–90% is recommended for siRNA transfection. In addition, cells should be checked at regular intervals for mycoplasma contamination.
11. A549 cells should be washed once with warm (37°C PBS) and then trypsinized with prewarmed 1× trypsin solution. Cells should not be kept with trypsin for more than 3–5 min. Add fresh DMEM with FCS to the cells and mix properly with gentle pipetting. Dilute the cells to obtain a final concentration of 500 cells/28 μl.
12. Diluted siRNA stocks should be stored at –80°C and multiple freeze–thaw cycle should always be avoided. Aliquot stocks into small amounts to maintain the siRNA quality for an extended period of time.
13. Perform all infection experiments with H1N1 viruses (including the pandemic swine-origin influenza A virus) at biosafety levels 2 (BSL 2) conditions, whereas the highly pathogenic H5N1 viruses require BSL 3 conditions.

14. This type of sequential genome-wide RNAi screen identifies several different human genes, which are designated as primary hits. To maximize the robustness of the hit selection, raw screening data from all three parameters should be subjected to a series of stringent statistical analysis. First, nonexpressed genes should be excluded by determining whether genes are constitutively or inducibly expressed via microarray gene expression profile of noninfected and infected host cells. Second, toxic siRNAs should be excluded by quantifying the loss of viable cells (after siRNA transfections) with a microscopical assay. siRNAs can be defined as toxic if the total cell number dropped below 750 cells per well at 48 h after transfection. Third, nontoxic siRNAs targeting expressed genes should be analyzed using R-package cellHTS (8), followed by feeding the obtained *Z*-scores (indicating how many standard deviations an observation is above or below the mean) into redundant siRNA activity (RSA) analysis (7). By utilizing the RSA algorithm, which identifies multiple active siRNAs, the collective behavior of all wells targeting a gene is taken into consideration in order to circumvent potential off-target effects. All genes phenotypically affected by at least two different siRNAs (even if only two siRNAs are available for a given gene, as is the case for the whole genome library) should be selected for further analysis. Finally, Genedata Screener should be used to identify genes with a robust *Z*-score below −2. Validation of primary hits can then be done using several different molecular techniques including the viral replication assay specified in Subheadings 3.6 and 3.7.
15. All experimental procedures (Subheadings 3.1–3.7) can be repeated in primary normal human bronchial epithelial cells (NHBE) in order to mimic in vivo conditions.

Acknowledgements

The authors would like to thank N. Machuy, Y. Shin, K.-P. Pleissner, A. Mäurer, and T. Wolff for significant scientific input in setting up the screen, M. Drabkina, G. Heins, D. Khalil, and D. Manntz for technical support. We also thank A. Pekosz for providing the influenza luciferase reporter construct and S. Becker for the pandemic H1N1 influenza strain. This work was supported through the EU FP6 project RIGHT (LSHB-CT-2004-005276), ERA-Net Pathogenomics (grant no. 0313938A) and a grant from FCI to T.F.M., and the RiNA network Berlin to T.R. and T.F.M.

References

1. Ge, Q., McManus, M. T., Nguyen, T., Shen, C. H., Sharp, P. A., Eisen, H. N., and Chen, J. (2003) RNA interference of influenza virus production by directly targeting mRNA for degradation and indirectly inhibiting all viral RNA transcription. *Proc. Natl. Acad. Sci. USA.* **100**, 2718–2723.
2. Karlas, A., Machuy, N., Shin, Y., Pleissner, K., Artarini, A. et al. (2009) Genome-wide RNAi screen identifies human host factors crucial for influenza virus replication. *Nature* **463**, 818–824.
3. Lutz, A., Dyall, J., Olivo, P. D., and Pekosz, A. (2005) Virus-inducible reporter genes as a tool for detecting and quantifying influenza a virus replication. *J. Virol. Methods* **126**, 13–20.
4. Woolcock, P. R. (2008) Avian influenza virus isolation and propagation in chicken eggs. *Methods Mol. Biol.* **436**, 35–46.
5. Opitz, B., Rejaibi, A., Dauber, B., Eckhard, J., Vinzing, M., Schmeck, B., Hippenstiel, S., Suttorp, N., and Wolff, T. (2007) IFNbeta induction by influenza A virus is mediated by RIG-1 which is regulated by the viral NS1 protein. *Cell Microbiol.* **9**, 930–938.
6. Hopkins, A.L., and Groom, C. R. (2002) The druggable genome. *Nat. Rev. Drug Discov.* **1**, 727–730.
7. Konig, R., Chiang, C. Y., Tu, B. P., Yan, S.F., DeJesus, P. D., Romero, A., Bergauer, T., Orth, A., Krueger, U., Zhou, Y., and Chanda, S. K. (2007) A probability-based approach for the analysis of large-scale RNAi screen. *Nat. Methods* **4**, 847–849.
8. Boutros, M., Bras, L., and Huber W. (2006) Analysis of cell-based RNAi screens. *Genome Biol.* 7, R66.1–R66.11.

Chapter 25

RNAi Screening in Mammalian Cells to Identify Novel Host Cell Molecules Involved in the Regulation of Viral Infections

Carolyn B. Coyne and Sara Cherry

Abstract

It is clear that viral entry, replication, and spread is a complex process involving a dialog between the virus and the targeted host cell. Viruses have evolved highly specific strategies to hijack cellular factors to promote their internalization, initiate their replication, and facilitate their eventual spread. However, the identification of many of these host cell molecules has been hindered by the requirement for robust genome-scale loss-of-function assays that are capable of targeting a wide variety of host factors. The more recent use of genome-scale or genome-wide RNA interference (RNAi) screens have extended our knowledge of the complex interplay between a virus and host and have implicated a wide variety of cellular factors required for infection of a number of viruses. Here, we describe an approach to target mammalian host cell factors involved in regulating viral infections by the use of a genome-scale RNAi library screen.

Key words: High-throughput screening, RNAi, Host factors

1. Introduction

Most viruses have evolved strategies to coopt host cell factors to promote a variety of events throughout their infectious life cycles. These events can be associated with modification(s) of the host cell that facilitate endocytic uptake, initiate viral replication, and/or manipulate apoptotic and innate immune signaling. The identification of these host cell factors has remained elusive largely in part due to the lack of robust high-throughput screening assays that target a wide variety of host genes simultaneously. The advent of genome-scale RNAi screening approaches has provided a substantial step forward in our ability to identify host cell factors involved in regulating viral infections in mammalian cells. Several

Ronald P. van Rij (ed.), *Antiviral RNAi: Concepts, Methods, and Applications*, Methods in Molecular Biology, vol. 721, DOI 10.1007/978-1-61779-037-9_25,

genome-wide RNAi screens have implicated a wide variety of cellular factors required for infection of a number of viruses (1–7). Although the functions of many of these signaling molecules remains to be defined, the results clearly implicate intracellular cell signaling as a key strategy used by viruses during infection.

Here, we present a generalized protocol for a high-throughput RNAi screen to identify mammalian cellular factors involved in viral infection using Coxsackievirus B3 and human osteosarcoma (U2OS) cells as examples.

2. Materials

2.1. siRNA and Transfections

1. 384-Well plates, black with clear bottom (Corning 3712).
2. 12 pin vacuum manifold (Drummond).
3. Automated liquid handling device (Well-mate, Thermo Fisher).
4. Genome-scale library of duplex siRNAs (Druggable Genome, Ambion) (see Note 1).
5. siRNA duplex optimized to reduce viral receptor expression by >80% (Ambion) (see Note 2).
6. Opti-MEM reduced serum medium (Invitrogen).
7. Hiperfect Transfection reagent (Qiagen).

2.2. Cell Culture

1. U2OS cells (ATCC) are a human osteosarcoma cell line that serves as an ideal cell type for high-throughput screening strategies as they are highly adherent and are highly transfectable by most commercially available transfection reagents (see Note 3).
2. Dulbecco's Modified Eagle Medium High Glucose, containing 4,500 mg/L D-glucose, and sodium pyruvate (DME-H, Invitrogen). Supplement with 10% fetal bovine serum (FBS, Invitrogen), 100 μg/mL penicillin, and 100 μg/mL streptomycin (Invitrogen).
3. Trypsin/EDTA: 0.25% trypsin, 1 mM ethylenediamine tetraacetic acid (EDTA).

2.3. Infections

1. Coxsackievirus B3 is a member of the picornavirus family of small positive sense RNA viruses. This virus is commercially available (ATCC) and can be expanded and titred in HeLa cells as described previously (8).

2.4. Immunofluorescent Staining

1. Ice cold methanol/acetone (3:1 ratio).
2. Phosphate buffered saline (PBS), 137 mM NaCl, 2.7 mM KCl, 4.3 mM Na_2HPO_4, 1.47 mM.

3. KH_2PO_4, adjust to a final pH of 7.4.
4. 0.1% Triton-X 100 in PBS.
5. Primary antibody recognizing the enterovirus viral coat protein VP1 (anti-VP1) monoclonal antibody (NCL-ENTERO, Novocastra Laboratories) (see Note 4).
6. Secondary antibody: Alexa Fluor-488 Goat anti-mouse IgG (Invitrogen).
7. 4′,6-Diamidino-2-phenylindole (DAPI), a blue-fluorescent nucleic acid stain to identify nuclei. Dissolve 10 mg in 2 mL of deionized water for a final concentration of 5 mg/mL. This stock can be stored at <20°C for long-term storage. To counterstain nuclei, dilute the 5 mg/mL stock to make a 300 nM working solution in PBS.
8. Adhesive sticker (Fisher Scientific).

2.5. Microscope and Software for Automated Image Acquisition and Analysis

1. ImageXpress Micro (Molecular Devices) with a 10× objective.
2. Automated image analysis software (MetaXpress, Leica).

3. Methods

3.1. Complexing siRNAs and Transfection Reagent

1. Using an automated liquid handling station (which is beneficial to avoid well to well variability), array a pool of siRNAs (three siRNAs per target gene) in 384-well plates (25 nM final concentration) (see Note 5).
2. Create a mix of Hiperfect transfection reagent with Opti-MEM (0.5 μL Hiperfect and 9.5 μL Opti-MEM per well). To limit errors in pipetting and to account for void volume in Well-Mate tubing, mix sufficient reagent for 400 wells/plate (thus, 200 μL Hiperfect in 3.8 mL of Opti-MEM per 384-well plate). This solution is stable for ~1 h at room temperature.
3. Add 10 μL of the Hiperfect/Opti-MEM mix to each well.
4. Spin plates at 800 × g for 1 min (see Note 6).
5. Incubate for 10 min at room temperature to form siRNA:Hiperfect complexes.

3.2. Preparation and Transfection of Cells

1. Grow U2OS cells at 37°C and 5% CO_2 in complete DMEM-H medium. Cells should be grown to approximately 80% confluence.

2. Wash cells briefly with PBS and dislodge with trypsin/EDTA.
3. Pellet cells (800 × *g*, 3 min) and resuspend in medium at a concentration of 80,000 cells/mL.
4. Add 25 μL of the cell suspension (2,000 cells/well) to wells containing precomplexed siRNAs.
5. Spin plates (800 × *g*, 1 min).
6. Incubate at 37°C and 5% CO_2 for 48 h (see Note 7).

3.3. Infections

1. Prepare virus stocks (Coxsackievirus B3-RD strain, ~1×10^{11} pfu/mL) (see Note 8).
2. Aspirate medium from the wells using a vacuum manifold.
3. Infect cells by adding 25 μL virus suspension in DMEM-H at an MOI of 0.5–1 (see Note 8) which leads to ~20–30% infection level.
4. Spin plates at 800 × *g* for 1 min.
5. Incubate for 8 h at 37°C and 5% CO_2 (see Note 7).

3.4. Immuno-fluorescent Staining

1. Aspirate virus-containing medium from wells using a vacuum manifold.
2. Wash wells briefly with 30 μL PBS to remove residual medium.
3. Add 30 μL ice cold 3:1 methanol:acetone to each well to fix the cells.
4. Spin plates at 800 × *g* for 1 min.
5. Incubate for 5 min at room temperature.
6. Aspirate methanol:acetone using vacuum manifold and add 50 μL PBS per well. At this point, the plates can be stored at 4°C for up to 1 week or stained immediately.
7. Aspirate PBS using a vacuum manifold and add 25 μL of primary antibody diluted 1:500 in PBS.
8. Spin plates at 800 × *g* for 1 min.
9. Incubate at room temperature for 1 h.
10. Aspirate and wash wells with 50 μL of PBS. Repeat this wash step three times (for a total of four washes).
11. Add 20 μL of secondary antibody diluted 1:2,000 in PBS containing 300 nM DAPI.
12. Spin plates at 800 × *g* for 1 min.
13. Aspirate and wash wells with 50 μL of PBS. Repeat this wash step three times.

14. After the final wash, add 30 μL of PBS for storage and cover the plate with adhesive sticker. The fluorescent signal is stable for several weeks if the plates are stored at 4°C.

3.5. Image Acquisition and Analysis

1. Capture at least three images per well in both the DAPI channel and the virus channel (488 nm).
2. Use automated image analysis software to calculate the number of cells ($Dapi^+$) and the number of infected cells ($VP1^+$). The ratio of these values ($VP1^+/DAPI^+$) is the level of infection.
3. Calculate robust *Z*-scores for each well (see Note 9).
4. "Hits" can be identified as those wells that exhibited a change in infection (either decrease or increase) by ≥2 standard deviations from calculated Z-scores (see Note 9).
5. Analysis of cell number can be used to remove siRNAs with cytotoxic effects.

Toxic siRNAs are excluded based upon decreased cell viability as measured by a robust *Z*-score ≤−2. In general, this corresponds to a decrease in cell number of more that 30% (Fig. 1).

3.6. Limiting Well-to-Well and Plate-to-Plate Variability

1. Plate-to-plate and day-to-day variability are minimized via the use of automated liquid handling.
2. Cell number should be optimized prior to RNAi screening in order to eliminate any cell piling that may occur throughout the course of the screen (there should be a single monolayer of cells to ensure proper image acquisition).
3. The level of infection should also be optimized (MOI and time of infection) in order to achieve 30–50% infection. This will allow for the identification of hits that have both an inhibitory and stimulatory effect on virus infection levels.

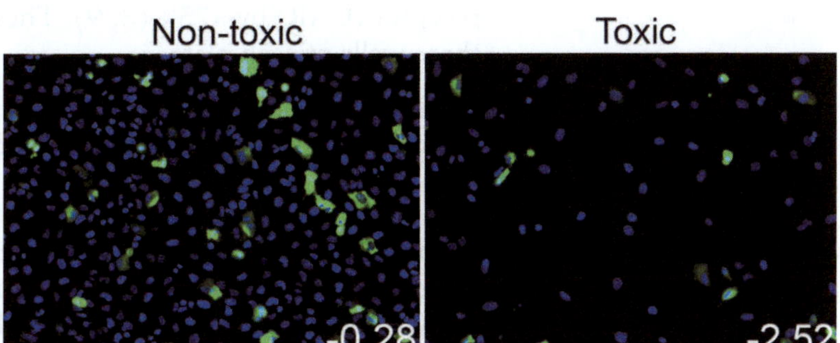

Fig. 1. Identification of toxic siRNAs. Representative fluorescence images of cytotoxic siRNAs. DAPI-stained nuclei of nontoxic (*left*) and toxic (*right*) siRNAs. Corresponding *Z*-scores are shown in white. *Blue*, DAPI-stained nuclei and green, VP1.

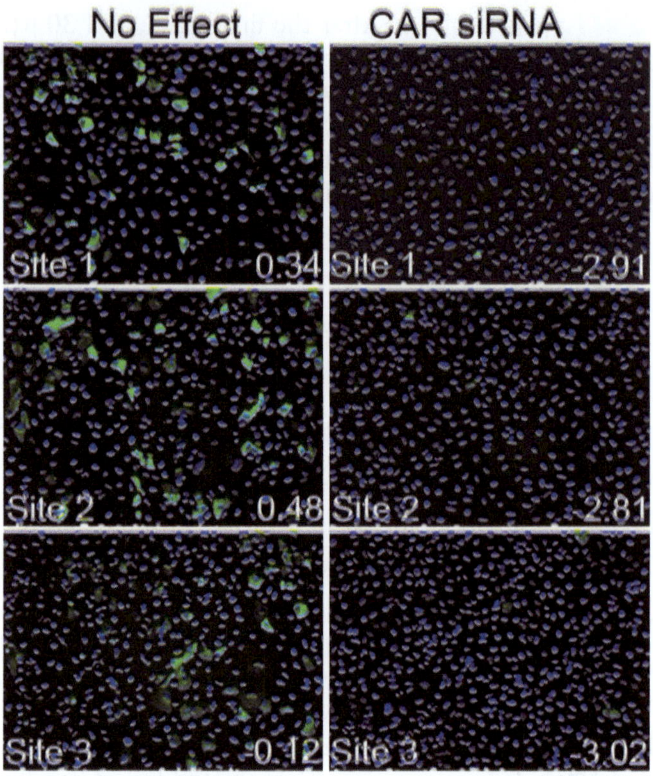

Fig. 2. Receptor siRNAs serve as effective internal controls. Shown are three independent immunofluorescence images from a single well of CVB-infected U2OS cells transfected with an siRNA which had no effect (*left*) or transfected with positive control CAR siRNA (*right*). Corresponding *Z*-scores (*bottom right*) are shown in *white text*. *Blue*, DAPI-stained nuclei, and *green*, VP1.

4. To control for variability between plates/wells (and to control for siRNA transfection efficiency between plates/wells), it is ideal to include preoptimized siRNAs that reduce expression of viral receptors, if possible. We previously used siRNAs that reduce the expression of the coxsackievirus and adenovirus receptor (CAR) by >75% (8, 9). These siRNAs are spotted in three wells of each plate.

These wells should be identified as hits in the screen and should exhibit minimal variability in *Z*-scores between plates (Fig. 2).

3.7. Validation and Follow-Up

1. Positive candidates identified in the initial screen (primary screen) should be validated by performing a secondary screen analysis using independent siRNAs designed against another region of the mRNA (this will minimize effects induced due to off-target events). A gene is included in follow-up studies when it has been identified as a hit in both primary and secondary screens.

2. Additional follow-up experiments to confirm positive candidates include pharmacological inhibitors and/or dominant-negative mutants. Multiple methods to validate a given hit will ensure that a hit is genuine.
3. Once a positive candidate is validated by the above-described means, the function of that gene in regulating virus infection can begin to be dissected.

4. Notes

1. The Ambion Druggable Genome RNAi library is a commercially available library containing siRNAs targeting a varying amount of human genes (V3 targets ~7,000 genes whereas V4 targets ~9,000 genes), with four individual siRNAs per target. Many of these siRNAs have been tested for efficacy. However, some siRNAs are not specific to given genes owing to redundancy and sequence homology (generally within members of related gene families). The library targets a wide variety of molecules involved in metabolism, intracellular signaling (including serine/threonine and tyrosine kinases), small GTP-binding proteins and their effector molecules, phosphatases, proteases, a variety of ion channels, caspase and caspase-related molecules, and molecular motor-related targets. The library is shipped as 0.25 nmol/siRNA lyophilized powder in 96-well plates (267 total plates for V3 and 356 total plates for V4). SiRNAs are reconstituted in nuclease-free water at the desired final concentration (an automated website from Ambion for this calculation can be found at: http://www.ambion.com/techlib/append/oligo_dilution.html). For example, to achieve a 1 μM stock of siRNA, each well would be reconstituted with 250 μL of nuclease-free water. Plates are then stored at –80°C.
2. CAR siRNA sequence: sense, 5′-GGUGGAUCAAGUGAUUAUU-3′ and antisense, 5′-AAUAAUCACUUGAUCCACC-3′.
3. It is imperative to begin any RNAi screen with cells that are healthy and have been maintained properly. We test our cells for mycoplasma contamination biweekly by PCR to ensure that they do not exhibit any growth or infection anomalies.
4. As the screening strategy relies on immunofluorescent means to detect viral replication, it is important to optimize staining conditions (including optimal fixation conditions and dilution of primary and secondary antibodies) prior to performing any screen.

5. Individual siRNAs (reconstituted as described in Note 1) should be pooled together using an arrayer robot to avoid errors (such as those associated with manual pipetting) that might alter siRNA concentrations (for example, TekBench from TekCel). The final volume of each pooled set of siRNAs should be 25 nM (thus, 6.25 nM final concentration for each individual siRNA).
6. To ensure that there is no solution remaining on the walls of the well, it is helpful to perform a "tap spin" for a brief period at a low *g*-force to ensure that all media/solution covers the bottom of the well (this is particularly critical when a low volume is added to each well). These "tap spins" appear at a variety of steps throughout the screening protocol. This is most easily accomplished in a swinging bucket rotor cell culture centrifuge with the appropriate multiwell plate adaptors (such as the Eppendorf 5804).
7. To avoid loss of medium due to evaporation throughout the experiment, it is ideal to maintain plates under high humidity. We accomplish this by using large Tupperware containers lined with water soaked paper towels.
8. CVB3 can be expanded, purified, and titered as described (8). It is best to prepare a "master mix" of media: virus at MOI = 1 to avoid pipetting areas that could contribute to plate-to-plate variability in infection levels. For example, as each well is infected with 25 μL of media containing virus (MOI = 1), each plate will require 9.6 mL of virus-containing media (assuming 384-well plates). If the RNAi screen totals 50 384-well plates, a total of 480 mL of virus-containing media would be required.
9. A robust *Z*-score is a statistical measure that quantifies the distance (in standard deviations) that a data point is from the median (and median absolute deviation) of a data set. In RNAi high-throughput screening, the calculation of *Z*-scores does not rely on control siRNAs, but instead assumes that the majority of siRNAs will have no effect (on virus infection, in this case) and can thus serve as controls. Thus, to calculate *Z*-scores, first perform a log transformation of all infection levels from each site of each well of a plate [log (VP1/DAPI)] (using Excel or another equivalent statisical software) and transform. Then, calculate the median and interquartile ranges (which represents the lowest 25% and the highest 25% of infection levels) of these data. Then, calculate *Z*-scores based on these data. A *Z*-score equal to 0 has the exact same value as the mean of the plate (although in a large data set, this is a rare value), whereas a *Z*-score equal to 1 is exactly one standard-deviation above the mean (and −1 is therefore one

standard-deviation below the mean). Malo et al. (10) provide an excellent review describing the use of robust statistics in high-throughput screening and the benefits and pitfalls of these techniques.

Acknowledgments

This work was supported by grants from the NIH [R01AI081759 (CBC) and R01AI074951, U54AI057168 (SC)].

References

1. Pelkmans, L., Fava, E., Grabner, H., Hannus, M., Habermann, B. et al. (2005) Genome-wide analysis of human kinases in clathrin- and caveolae/raft-mediated endocytosis. *Nature* **436**, 78–86
2. Brass, A. L., Dykxhoorn, D. M., Benita, Y., Yan, N., Engelman, A. et al. (2008) Identification of host proteins required for HIV infection through a functional genomic screen. *Science* **319**, 921–926
3. Krishnan, M. N., Ng, A., Sukumaran, B., Gilfoy, F. D., Uchil, P. D. et al. (2008) RNA interference screen for human genes associated with West Nile virus infection. *Nature* **455**, 242–245
4. Hao, L., Sakurai, A., Watanabe, T., Sorensen, E., Nidom, C. A. et al. (2008) Drosophila RNAi screen identifies host genes important for influenza virus replication. *Nature* **454**, 890–893
5. Tai, A. W., Benita, Y., Peng, L. F., Kim, S. S., Sakamoto, N. et al. (2009) A functional genomic screen identifies cellular cofactors of hepatitis C virus replication. *Cell Host Microbe* **5**, 298–307
6. Sessions, O. M., Barrows, N. J., Souza-Neto, J. A., Robinson, T. J., Hershey, C. L. et al. (2009) Discovery of insect and human dengue virus host factors. *Nature* **458**, 1047–1050
7. Cherry, S., Kunte, A., Wang, H., Coyne, C., Rawson, R. B. et al. (2006) COPI activity coupled with fatty acid biosynthesis is required for viral replication. *PLoS Pathog* **2**, e102
8. Coyne, C. B., Bergelson, J. M. (2006) Virus-induced Abl and Fyn kinase signals permit coxsackievirus entry through epithelial tight junctions. *Cell* **124**, 119–131
9. Coyne, C. B., Kim, K. S., Bergelson, J. M. (2007) Poliovirus entry into human brain microvascular cells requires receptor-induced activation of SHP-2. *EMBO J* **26**, 4016–4028
10. Malo, N., Hanley, J. A., Cerquozzi, S., Pelletier, J., Nadon, R. (2006) Statistical practice in high-throughput screening data analysis. *Nat Biotechnol* **24**, 167–175

ERRATUM

Production of Multicopy shRNA Lentiviral Vectors for Antiviral Therapy

Scot D. Henry, Qiuwei Pan, and Luc J.W. van der Laan

Ronald P. van Rij (ed.), *Antiviral RNAi: Concepts, Methods, and Applications*, Methods in Molecular Biology, vol. 721, DOI 10.1007/978-1-61779-037-9, © Springer Science+Business Media, LLC 2011

DOI 10.1007/978-1-61779-037-9_26

The name of the second contributing author of Chapter 19 "Quiwei Pan" is misspelled. The proper spelling of his name is "Qiuwei Pan". The "i" comes before the "u" in his first name.

The online version of the original chapter can be found at http://dx.doi.org/10.1007/978-1-61779-037-9_19

INDEX

A

1A............ 5, 12–14, 201, 202, 204, 211, 216, 233, 239–241
Adenovirus 25, 34, 36, 183–185, 187–189, 191, 308, 402
Aedes aegypti 7, 8, 254
Agarose formaldehyde gel 157–158, 164, 165
Ago–1 6
Ago–2 .... 4–7, 9, 10, 12, 13, 201, 202, 216
293-Ago2 cell line 184, 187–188
Agrobacterium tumefaciens 247, 251
A549 human lung epithelial cells 384–385
AIDS 30, 52, 68, 70, 73, 358, 370
American nodavirus 8, 10
Anesthesia 335, 336
Anopheles gambiae 7
Antisense oligonucleotides 36, 78
Antiviral defense 3–8, 16–18, 183, 201, 202, 232, 254
Antiviral immunity 8, 216
Antiviral RNAi 3, 5, 7, 8, 11, 14, 17, 18, 28, 107, 201, 216, 224, 288
Antiviral therapy 313–330
Apoptosis 35, 44, 48, 54, 55, 58, 60, 61
Aptamer 72, 91, 355–370
Arbovirus 14
Argonaute 3–5, 24, 26, 78, 184, 201, 215, 232, 240, 241, 334, 340
Asco virus 154
ATP regenerating system 242
Automated image analysis 399, 401
Automated liquid handling 381, 398, 399, 401
Automated microscope 377, 388, 391

B

B2 12
Band shift assay 250
Base modifications 83, 87
Beet western yellows virus 12
Binary vectors 251
Birnaviridae 7, 8, 10

C

C33A cells 187
Cap labeling 235, 239
Cationic peptide 341, 344, 352
CCR5 .... 68, 69, 71–73, 358, 365, 366
$CD4^+$ T cells 30
$CD34^+$ T cells 294–296, 350, 351
Cell free S2 cell extract 232, 233, 241
Cell viability assay 269, 272, 284–286
CEM T-cells 356
Cytomegalovirus 32, 48, 50, 173, 174, 181, 317
CHO cells 358, 363–365, 368, 370
Cholesterol conjugation 90
Chromosome region maintenance 1(CRM1) 308
Cis-acting replication element (CRE) 269
Coefficient of variation 381
Combinatorial RNAi 294, 295, 298, 301, 314
Confocal microscopy 356, 358, 359, 364–365, 370
Coxsackievirus 28, 269, 288, 398, 400, 402
C3PO 5, 6, 232
Creatine phosphate 186, 193, 197, 218, 222, 225–227, 235, 242
Creatine phosphokinase 186, 235
Cricket paralysis virus 5, 7, 12, 14, 201, 233
Cross-linking immune precipitation (CLIP) assays 47
Cucumber mosaic virus 12
Culex pipiens quinquefasciatus 8,10
CXCR4 .... 69, 71
Cymbidium ringspot virus 11, 12
Cytoplasmic extracts 189, 191

D

Danger signal 3, 6
Deep sequencing 18, 24, 27, 28, 51, 121
Denaturing polyacrylamide gel electrophoresis 239, 241
Dengue virus 7, 27, 375
DGCR8 .... 26, 44, 294
Dicer 3–5, 11, 17, 26, 28, 29, 33, 34, 44, 45, 78, 91, 143, 184, 191, 201, 202, 204, 211, 215–228, 232, 293, 294, 300, 305, 308, 314, 315, 339, 355, 356, 361
Dicer–2 .... 4, 5, 201, 202, 204, 215, 216, 227

Ronald P. van Rij (ed.), *Antiviral RNAi: Concepts, Methods, and Applications*, Methods in Molecular Biology, vol. 721, DOI 10.1007/978-1-61779-037-9, © Springer Science+Business Media, LLC 2011

Dicer assay............215–228
Dicer-substrate siRNAs............78
Dicistroviridae............7, 8, 13
Dissociation constant............247, 251
Drosha............26, 44, 45, 51, 57, 143, 216, 294, 297, 298, 300
Drosophila
A virus............10
birnavirus............8, 10
C virus............5, 7, 8, 12, 14, 201, 216, 233
melanogaster............4, 26, 124, 126, 129, 216, 231–243
S2 cell extract............215–228, 236
S2 cells............189, 216, 218–219, 221–224, 231, 233, 254
totivirus............8, 10
X virus............7, 8
Druggable genome............387, 388, 398, 403
dsRNA
feeding............202, 204, 205, 208–212
uptake............7, 253–262
Dual-luciferase reporter assay............206, 269, 271, 280–281, 284

E

Ebola virus............25, 33, 35, 308
Echovirus............282, 288
E. coli............175, 224–225, 251, 252, 279, 281, 290, 341, 345
Electrophoretic mobility shift assay (EMSA)............245, 249
Encephalo-myocarditis virus............28
Enterovirus............269, 399
Epstein Barr virus............47, 48, 147
Ethylene-bridged nucleic acid............81
Exportin 5............26, 44, 294
Extended shRNA............294, 295, 299, 303–304

F

FASTA............144–145
FASTQ............133
Firefly luciferase............202–212, 234, 237, 280, 349, 387, 389, 390, 393
FLAG-epitope............184
Flaviviridae............7, 8, 10
Flock House virus (FHV)............5, 7, 9, 10, 13, 15, 17, 33, 201, 202, 224
Flow cytometry............346, 349, 350, 358, 359, 362–364
2′-Fluoro substitution............81

G

G418............184, 187, 188, 194, 195
Gel mobility shift assay............245–252, 363
Gel retardation assay............245
Gel shift assay............357–359, 361–363
Gene therapy............293–308
Genome-wide dsRNA library............377
Genome-wide RNAi screen............376, 383–394, 398
Glycine-tryptophane (GW) motifs............13
gp120............68, 72, 356, 358–363, 370
Green fluorescent protein (GFP)............33, 203–209, 211, 269–271, 275, 278–284, 290, 317, 321–323, 325, 326, 329, 341, 345, 377, 379, 381, 387, 390
GST............219, 224–2257
Guanylyltransferase............235

H

HA-epitope............184
HeLa cell line............31, 187, 281, 285, 287, 398
Hematopoietic stem cells............72, 294, 295, 350
Hemolymph............14, 253
Hen-1............6
Hepatitis C virus (HCV)............25, 27–31, 33, 35, 36, 267, 313, 314, 316
Herpes simplex virus............28, 48, 50
Herpesviruses............47–51, 54, 55, 147, 151, 173
Highly active antiretroviral therapy (HAART)............68, 69, 72, 339, 356
High-throughput RNAi............376, 398
High-throughput screening............397, 398, 404, 405
High-throughput sequencing............47, 123
HIV-1............25, 27–35, 68–73, 269, 293–308, 316, 340, 344, 349, 352, 355–370
Holo-RISC............232, 233, 241
H1 promoter............298, 304, 318–321
Human adenovirus type 5............184
Human embryonic kidney (HEK) 293 T cells............299
Humanized mice............71, 343, 350, 352
Human osteosarcoma (U2OS) cells............398
Human T lymphotropic virus type I............33
Hybridization............27, 151, 157, 160–163, 165, 168, 169, 174–176, 178–181, 276

I

Illumina sequencing............108, 123, 124, 127
IMD............16, 17
Immune evasion............52, 53, 153
Immunofluorescence microscopy............384
Immunoprecipitation............59, 179, 185, 190, 195, 196
Influenza virus............25, 32, 33, 35, 268, 375, 383–385, 388, 389, 392
Insect............7–9, 11–14, 27, 33, 108–109, 201, 202, 232
Insect virus............33
Interferon............27, 78, 86, 313, 360
Intranasal delivery............90, 333–338
In vitro transcription............170, 194, 196, 203–207, 210, 211, 217, 220–221, 226, 234, 237–239, 251, 258, 269, 356, 357, 359–360, 378

J

Jak-STAT pathway ... 17
Jurkat T cells ... 342, 343, 346, 349

K

Kaposi's sarcoma-associated herpesvirus ... 48

L

Latency ... 29, 30, 35, 49–50, 52–55, 57–59
Latent infection ... 163
Lentiviral vectors ... 36, 72–73, 293–308, 313–330
lhRNA. *See* Long hairpin RNA
Liposome ... 68, 90, 340, 341, 344, 347, 351, 352
LNA. *See* Locked nucleic acid
Locked nucleic acid ... 36, 81–83, 85, 87, 89, 92, 269, 287
Long hairpin RNA (lhRNA) ... 315
Long terminal repeat (LTR) ... 34, 69, 326
Loquacious ... 4, 5, 216
Luciferase ... 34, 47, 54, 55. 58, 59, 194, 202, 204, 209, 211, 234, 235, 278, 301, 308, 350, 381, 384, 385, 387, 389–390, 393, 394
Luminometer ... 206, 209, 212, 271, 284, 393
Luteoviridae ... 12
Lymphocytic choriomeningitis virus ... 28
Lytic infection ... 55, 154, 163

M

Madin Darby Canine Kidney (MDCK) cell line ... 384, 385, 388, 392
Marek's disease virus (MDV) ... 33, 48–52, 62
Massive parallel sequencing ... 8
Melanoma differentiation-associated gene 5 (Mda5) ... 86, 204, 212
Merkel cell polyomavirus ... 48, 51
2′-methoxyethyl substitution ... 81
m^7GpppG-cap ... 185, 191, 194
Mice ... 28, 32, 58, 71, 78, 85, 87, 90–92, 178, 335–337, 343, 349, 350, 352
Microarray analyses ... 144, 154
Microinjection ... 255, 259
Microprocessor ... 143, 147, 216
microRNA ... 6, 16, 24–26, 29–32, 34–36, 43–46, 162, 163, 179, 314
microRNA target prediction ... 31, 46
Microscopy ... 364, 379
miR* ... 6
miRNA star ... 6
mivaRNAs ... 184, 194
Mosquito ... 8, 14, 15, 108, 254
Mouse cytomegalovirus ... 48, 173, 174, 181
Murine gamma-herpesvirus type 68 ... 48
Murine polyomavirus ... 48, 51

N

Nano-D-arginine ... 340, 351
Native acrylamide gel ... 245, 248, 250
Native gel electrophoresis ... 250
ncRNA ... 165
Neomycin ... 187
Nicotiana benthamiana ... 33, 34, 246, 247, 249
Nodaviridae ... 7, 8, 10
Non-coding RNA ... 168
Non-denaturing polyacrylamide gel ... 361, 362
Noravirus ... 8, 10
Normal human bronchial/tracheal epithelial cells . 385, 394
Northern blot analysis ... 153–170, 184
Novoalign ... 124–126, 128, 129, 133–135, 137, 139
NS3 ... 34
Nuclease protection assay ... 173–181

O

Off-target effect ... 79, 81, 88–89, 94, 256, 269, 273, 287, 316, 318, 340, 356, 394
2′O-Methyl substitution ... 81
O'nyong-nyong virus (ONNV) ... 7, 14
Orthomyxoviridae ... 334

P

P19 ... 11, 12, 34
P38 ... 12, 13
p122 ... 246, 247, 249–251
PACT ... 26, 294
PAN ... 165
p24 antigen assay ... 358, 365–366
Parainfluenza virus (PIV) ... 334
Paramyxoviridae ... 334, 337
Pasha ... 44, 216
PAZ. *See* Piwi/Argonaute/Zwile
PAZ domain ... 4, 17, 232
P bodies ... 26, 30, 46
PCR amplification ... 112, 118–119, 194, 237, 278
Peripheral blood mononuclear cells ... 31, 68, 71, 86, 349, 358
Perl script ... 124
Pharmacokinetics ... 89–90, 92
Phosphodiester backbone modifications ... 80–81
Phosphorothioates ... 269
Phosphothioate modifications ... 80
Pig-B Reagent ... 155, 158, 166
piRNA ... 15, 16
Piwi/Argonaute/Zwile (PAZ) ... 4, 12, 17, 232
Piwi-interacting RNAs ... 6, 15–16, 18, 24
Plant virus ... 12–13, 33, 34
Plaque assay ... 388

Plaque reduction assay....................269, 272, 284, 286–288
Polerovirus....................12, 246
Poliovirus....................27
Polyacrylamide gel electrophoresis....................219, 223, 237, 239, 241
Polyhistidine tag....................211
Polymerase III promoter....................295, 306, 315
Polyomavirus (PyV)....................48, 51, 53
Potato virus X....................33
pre-miRNAs....................25, 34, 78, 144, 147–148, 150, 151, 216, 294, 307
Primate foamy virus type 1....................28
Primer extension assays....................154
pri-miRNA....................25, 44, 45, 143, 303, 304, 307
Protamine....................71, 340, 341, 343, 344, 348
Proteinase K....................186, 193, 235, 240
Pseudo uracil....................83
454 pyrosequencing....................108

Q

Quantitative RT-PCR....................359, 366–367

R

Radio-labeled GTP capped mRNA....................236–238
Radiolabeled probe....................154, 163, 169
Radiolabeled size marker....................109–110, 114, 121
R2D2....................4, 6, 7, 44, 216, 232
Real-time PCR (RT-PCR)....................154, 174, 349, 352, 356, 359
Renilla luciferase reporter....................202, 204, 209
Reovirus....................246
Respiratory syncytial virus (RSV)....................68, 92, 268, 355
Retinoic acid inducible gene (RIG-I)....................35, 86, 88
Reverse transcription....................112, 118, 121, 154, 254–255, 257, 270, 278, 280
Rhadinovirus....................48, 50, 52
Rhesus lymphocryptovirus....................48, 50
Rhesus monkey rhadinovirus....................48
Ribose modification....................82–83
Rice hoja blanca virus (RHBV)....................34
RISC. *See* RNA-induced silencing complex
RISC assembly....................13, 233, 239
RISC loading complex....................4, 6, 7, 26, 57, 108
RNA
  isolation....................155, 158–159, 166, 168, 254, 256–257
  silencing suppressors....................24
  structure....................145, 315
RNA-dependent RNA polymerase....................7, 9, 216, 282, 337
RNAfold....................149, 276
RNAi
  library....................403
  reporter....................202–204, 206, 212
  screening....................375–382, 384, 397–405
  suppression....................11, 13, 14, 202
RNA-induced silencing complex....................3–6, 11, 12, 17, 18, 24, 27, 34, 44–47, 59, 78, 79, 82–84, 88, 89, 179, 183–198, 201, 202, 204, 215, 216, 231–233, 235, 240–242, 268, 287, 293, 294, 297, 298, 300, 305, 334, 340, 361
RNase
  contamination....................158, 242
  protection assays....................154
RNaseIII....................4, 215
RNaseZap....................108, 155–160, 164, 254
RSV. *See* Respiratory syncytial virus
RT-PCR. *See* Real-time PCR

S

SARS coronavirus (SARS-CoV)....................32, 267
S2 cell extract....................215–228
S2 cell line....................9–10, 202, 210
Schneider 2 cell line....................202, 218
SDS-PAGE....................196, 228
Seed sequence....................45, 52, 54, 55, 57, 58, 60, 89
SELEX....................356, 359
Sequencing by ligation....................108
Sequencing-by-synthesis....................108
S15 extracts....................185, 187, 190
Short hairpin RNA (shRNAs)....................27, 73, 78, 294, 296–298, 300–301, 306, 313, 315, 316, 340
Sigma3....................246
Simian agent 12....................48
Simian polyomaviruses....................183
Simian virus 40....................48
Sindbis virus....................7, 8
Single chain antibody variable fragment....................340
siRISC....................231–232
7SK....................306
Slicer activity....................6, 12, 13, 231–243
Slicer assay....................235, 239–241
Small interfering RNAs (siRNA)
  delivery....................68, 70–73, 87, 90, 91, 334, 340, 341, 366
  immunogenicity....................80, 86–88
  library....................88, 387, 388
Small RNA isolation....................179
Small RNA library....................109, 123
Small RNA profile....................8–11
S1 Nuclease protection assay....................173–181
Solexa....................108, 111, 117, 129, 130
SOLiD....................108
Spermidine....................196, 226, 234, 255, 261
S2R+ cell line....................203–206, 208, 210
Stripping blot....................158
SupT1 T cell line....................299, 306
SV40....................48, 51–53, 56
Systemic silencing....................7

T

TAR binding protein (TRBP)............26, 44, 294
TAR RNA decoy............68, 73
Tas............28, 29, 33, 34
Tat............25, 31, 33, 34, 69, 71, 73, 361, 366, 367
293T cell line............299
Tetraviridae............8, 10
2-thio uracil............83
Titration............180, 325–326, 329, 384, 388
Tobacco mosaic tobamovirus (TMV)............246, 247
Togaviridae............7
Toll............16, 17, 87
Toll-like receptors (TLR)............87
Tombusviridae............11, 13
Total RNA isolation............155, 158–159, 166, 254, 256–257
Totiviridae............8, 10
T7 promoter............194, 205–207, 211, 217, 220, 234, 236, 237, 256, 377
Transactivation response (TAR) element............34
Transduction............71, 296, 306, 308, 325, 329
Transfection............31, 54, 90, 91, 165, 187, 195, 202, 204, 205, 208–212, 231, 241, 270–272, 278, 280–282, 285, 286, 288, 290, 305–306, 313, 317–318, 322–326, 328–329, 334–337, 343, 349, 351, 387, 389–394, 398–400, 402
Trizol............108, 113, 166, 254, 256–257
Trypan blue............189, 233, 236
Tumorigenesis............33, 60, 153
Turnip crinkle virus (TCV)............12, 13

U

UNIX............125–127
3′-Untranslated regions............261
U6 promoter............73
UV shadowing............238, 368

V

vaccinia virus............28, 33, 35
Vago............17
V5 epitope............211
Vesicular stomatitis virus (VSV)............27–29, 36, 296, 300, 317, 375–377, 379
Vir-1............17
Viral escape............68, 69, 73, 268–269, 294
Viral microRNA............153–170
Viral miRNA............32, 43–61, 143–151, 153, 154, 174–176, 178, 181
Viral siRNA............3, 6, 123–141, 224
Viral small RNA............107–121, 129, 184
Viral suppressors of RNAi (VSR)............7, 9, 12–15, 17, 201–212, 216
Virus-associated RNA (VA RNA)............184, 191, 194, 308
Virus-derived small interfering RNA (v-siRNAs)............3, 6–10, 13, 15, 17, 18, 26–28, 107–108, 124, 201
Virus-induced gene silencing (VIGS)............183
Visitor............123–142
VMir............144–150
VP35............25, 33, 35, 308
VSV. *See* Vesicular stomatitis virus

W

Western blotting............188, 196, 272, 283
West Nile virus (WNV)............8, 10, 27, 375

Z

Z score............380, 382, 394, 401–402, 404

MIX
Papier aus verantwortungsvollen Quellen
Paper from responsible sources
FSC® C105338

If you have any concerns about our products,
you can contact us on
ProductSafety@springernature.com

In case Publisher is established outside the EU,
the EU authorized representative is:
Springer Nature Customer Service Center GmbH
Europaplatz 3, 69115 Heidelberg, Germany

Printed by Libri Plureos GmbH
in Hamburg, Germany